(continued on back)

AQUATIC CHEMICAL KINETICS

AQUATIC CHEMICAL KINETICS

Reaction Rates of Processes in Natural Waters

Edited by

WERNER STUMM

Swiss Federal Institute of Technology (ETH)
Zürich, Switzerland

A Wiley-Interscience Publication
John Wiley & Sons, Inc.
New York / Chichester / Brisbane / Toronto / Singapore

Library of Congress Cataloging-in-Publication Data

Aquatic chemical kinetics: reaction rates of processes in natural
    waters/edited by Werner Stumm.
        p.   cm.—— (Environmental science and technology)
    "A Wiley-Interscience publication."
    Includes bibliographical references.
    ISBN 0-471-51029-7
    1. Water chemistry.   2. Chemical reaction, Rate of.   I. Stumm,
Werner, 1924–   .II. Series.
GB855.A64   1990
551.48—dc20   89-70486
                                                                CIP

Printed in the United States of America

10 9 8 7 6 5 4 3 2 1

CONTRIBUTORS

PATRICK L. BREZONIK, Department of Civil and Mineral Engineering, University of Minnesota, Minneapolis, Minnesota

BOŽENA ĆOSOVIĆ, Rudjer Boškovic Institute, Center for Marine Research, Zagreb, Croatia, Yugoslavia

GERALD V. GIBBS, Department of Geological Sciences, Virginia Polytechnic Institute and State University, Blacksburg, Virginia

PHILIP M. GSCHWEND, Ralph M. Parsons Laboratory for Water Resources and Hydrodynamics, Massachusetts Institute of Technology, Cambridge, Massachusetts

JANET G. HERING, Institute for Water Resources and Water Pollution Control (EAWAG), Dübendorf, Switzerland; Swiss Federal Institute of Technology (ETH), Zürich, Switzerland

MICHAEL R. HOFFMANN, Department of Environmental Engineering Science, California Institute of Technology, Pasadena, California

JÜRG HOIGNÉ, Institute for Water Resources and Water Pollution Control (EAWAG), Dübendorf, Switzerland; Swiss Federal Institute of Technology (ETH), Zürich, Switzerland

ANTONIO C. LASAGA, Kline Geology Laboratory, Yale University, New Haven, Connecticut

ABRAHAM LERMAN, Department of Geological Sciences, Northwestern University, Evanston, Illinois

GEORGE W. LUTHER, III, College of Marine Studies, University of Delaware, Lewes, Delaware

FRANÇOIS M. M. MOREL, Ralph M. Parsons Laboratory for Water Resources and Hydrodynamics, Massachusetts Institute of Technology, Cambridge, Massachusetts

JAMES J. MORGAN, Department of Environmental Engineering Science, California Institute of Technology, Pasadena, California

CHARLES R. O'MELIA, Department of Geography and Environmental Engineering, The Johns Hopkins University, Baltimore, Maryland

NEIL M. PRICE, Ralph M. Parsons Laboratory for Water Resources and Hydrodynamics, Massachusetts Institute of Technology, Cambridge, Massachusetts

JERALD L. SCHNOOR, Department of Civil and Environmental Engineering, The University of Iowa, Iowa City, Iowa

JACQUES SCHOTT, Laboratoire de Géochimie, Université Paul-Sabatier, Toulouse, France

RENÉ P. SCHWARZENBACH, Swiss Federal Institute for Water Resources and Water Pollution Control (EAWAG), Dübendorf, Switzerland; Swiss Federal Institute of Technology (ETH), Zürich, Switzerland

ALAN T. STONE, Department of Geography and Environmental Engineering, The Johns Hopkins University, Baltimore, Maryland

WERNER STUMM, Institute for Water Resources and Water Pollution Control (EAWAG), Dübendorf, Switzerland; Swiss Federal Institute of Technology (ETH), Zürich, Switzerland

BARBARA SULZBERGER, Institute for Water Resources and Water Pollution Control (EAWAG), Dübendorf, Switzerland; Swiss Federal Institute of Technology (ETH), Zürich, Switzerland

BERNHARD WEHRLI, Lake Research Laboratory, Institute for Water Resources and Water Pollution Control (EAWAG), Kastanienbaum, Switzerland; Swiss Federal Institute of Technology (ETH), Zürich, Switzerland

ERICH WIELAND, Institute for Water Resources and Water Pollution Control (EAWAG), Dübendorf, Switzerland; Swiss Federal Institute of Technology (ETH), Zürich, Switzerland

ROLAND WOLLAST, Laboratoire d'Océanographie, Université Libre de Bruxelles, Brussels, Belgium

SERIES PREFACE
Environmental Science and Technology

The Environmental Science and Technology Series of Monographs, Textbooks, and Advances is devoted to the study of the quality of the environment and to the technology of its conservation. Environmental science therefore relates to the chemical, physical, and biological changes in the environment through contamination or modification, to the physical nature and biological behavior of air, water, soil, food, and waste as they are affected by man's agricultural, industrial, and social activities, and to the application of science and technology to the control and improvement of environmental quality.

The deterioration of environmental quality, which began when man first collected into villages and utilized fire, has existed as a serious problem under the ever-increasing impacts of exponentially increasing population and of industrializing society. Environmental contamination of air, water, soil, and food has become a threat to the continued existence of many plant and animal communities of the ecosystem and may ultimately threaten the very survival of the human race.

It seems clear that if we are to preserve for future generations some semblance of the biological order of the world of the past and hope to improve on the deteriorating standards of urban public health, environmental science and technology must quickly come to play a dominant role in designing our social and industrial structure for tomorrow. Scientifically rigorous criteria of environmental quality must be developed. Based in part on these criteria, realistic standards must be established and our technological progress must be tailored to meet them. It is obvious that civilization will continue to require increasing amounts of fuel, transportation, industrial chemicals, fertilizers, pesticides, and countless other products; and that it will continue to produce waste products of all descriptions. What is urgently needed is a total systems approach to modern civilization through which the pooled talents of scientists and engineers, in cooperation with social scientists and the medical profession, can be focused on the development of order and equilibrium in the presently disparate segments of the human environment. Most of the skills and tools that are needed are already in existence. We surely have a right to hope a technology that has created such manifold environment problems is also capable of solving them. It is our hope

vii

that this Series in Environmental Sciences and Technology will not only serve to make this challenge more explicit to the established professionals, but that it also will help to stimulate the student toward the career opportunities in this vital area.

ROBERT L. METCALF
WERNER STUMM

PREFACE

The objectives of this book are (1) to treat features of chemical kinetics in aqueous solutions and in the context of aquatic systems (oceans, fresh water, atmospheric water, and soil), (2) to strengthen our understanding of reaction mechanisms and of specific reaction rates in natural waters and in water technology, and (3) to stimulate innovative research in aquatic chemical kinetics.

The authors—physical and inorganic chemists, surface and colloid chemists, geochemists, oceanographers, aquatic chemists, chemical engineers, and environmental engineers—have attempted to write their chapters in such a way as to provide a teaching book and to assist the readers (students; geochemists; physical chemists; air, water, and soil scientists; and environmental engineers) in understanding general principles; emphasis is on explanation and intellectual stimulation rather than on extensive documentation. The information given should also be helpful in guiding research in aquatic chemistry and in applying kinetics to the exploration of naturally occurring processes and in developing new engineering practices.

In this volume we progress from simple concepts and laboratory studies to applications in natural water, soil, and geochemical systems. We start by introducing kinetics as a discipline and giving a set of basic principles emphasizing the elementary reaction as a basic unit of chemical processes. Then we discuss the environmental factors that are of importance in controlling the rate of chemical transformations and illustrate from a mechanistic point of view the kinetics of chemical catalysis in the areas of cloud chemistry, groundwater chemistry, and water treatment processes. We show how to use linear free-energy relationships—to bridge the gap between kinetics and equilibria—especially for reactions of homologous series of compounds in order to procure kinetic information on reactions that have not been determined in the laboratory. Such information is especially useful in the chemical transformation of chemical pollutants and in redox processes. We address the question of whether in some instances the rates of biogeochemical reactions may be influenced, or even controlled, by the rates of metal coordination reactions.

An appreciation of the role of solid-water interfaces and surface-controlled reactions is a prerequisite for understanding many important processes in natural systems, and especially the contributions of physicochemical and biological reactions. Thus, special attention is paid to the kinetics of surface reactions. The discussion spans the range from ab initio quantum mechanical calculations and frontier-molecular-orbital theories to extracellular enzymatic reactions and includes the adsorption of organic solutes and redox processes occurring at these surfaces. It is shown that the geochemical cycling of electrons is not only mediated by microorganisms but is of importance at particle–water interfaces, especially at the sediment–water interface due to strong redox gradients and in surface waters due to heterogeneous photoredox processes. This volume also reflects the great progress achieved in recent years in the study of kinetics of the dissolution of oxide and carbonate minerals and the weathering of minerals.

Finally, we demonstrate in discussions on weathering rates in the field, on the kinetics of colloid chemical processes, and on the role of surficial transport processes in geochemical and biogeochemical processes that spatial and temporal heterogeneities and chemical versus transport time scales need to be assessed in order to treat the dynamics of real systems.

Most of the authors met in March 1989 in Switzerland for a workshop. Background papers formed the basis for the discussions. However, this book is not the "proceedings of a conference", instead, it is the offspring of the workshop and its stimulating discourses.

I am most grateful to many colleagues who have reviewed individual chapters and have given useful advice. Credit for the creation of this volume is, of course, primarily due to its authors.

WERNER STUMM

Zürich, Switzerland
January 1990

CONTENTS

AQUATIC CHEMICAL KINETICS

1

KINETICS OF CHEMICAL TRANSFORMATIONS IN THE ENVIRONMENT

Alan T. Stone

Department of Geography and Environmental Engineering,
The Johns Hopkins University, Baltimore, Maryland

and

James J. Morgan

Department of Environmental Engineering Science, California Institute of
Technology, Pasadena, California

1. INTRODUCTION

Environmental chemists are most often concerned with the response of an environmental system to change. This change may be natural (such as the diurnal cycle of solar irradiation) or caused by human intervention (such as the dispersion of a pesticide). Since change is such a major concern, it should not be surprising that chemical kinetics is an integral component of models of natural systems. Intrinsically "kinetic" questions concerning the nature and behavior of natural systems include:

> When will the maximum concentration of a pollutant appear in a system, and how high will it be?
>
> When will the minimum concentration of an important nutrient occur, and how low will it be?
>
> What is the residence time of a particular element or species?
>
> Will a given compound be accumulated or exported from an open system? How is the ability of various physical processes to transport a compound dependent on its chemical form?

In this chapter, we will discuss (1) the basic "unit" of chemical kinetics, the elementary reaction; (2) collections of elementary reactions that represent entire

1

chemical processes; (3) dynamic models that represent complete natural and engineered systems; (4) unique characteristics of surface chemical reactions; and (5) kinds of kinetic information and how they can be used to answer questions such as the ones listed above.

Chemical kinetics can be examined on several levels of sophistication (Gardiner, 1969; Denbigh and Turner, 1984). The first level is qualitative and based solely on prior practical experience; if a certain set of chemical conditions exists, a particular outcome is observed. Experiments can be performed that systematically catalog factors influencing rates of chemical reactions. The next level attempts to capture the chemical dynamics of a system in a quantitative description; a set of equations is developed using experimentally derived rate constants allowing reaction rates to be predicted over a range of chemical conditions. Whether or not extrapolations accurately predict chemical reactions under unexplored chemical conditions depends on how well the set of kinetic equations and rate constants captures the true dynamics of the system. On the most fundamental level, chemical kinetics is a molecular description of chemical reactions. A series of encounters between chemical species is hypothesized, and the level of agreement between the proposed mechanism and experimental findings is critically examined. In special circumstances, the molecular description of chemical reactions allows generalizations to be made concerning the reaction behavior of an entire class of compounds. These generalizations provide the basis for structure–reactivity relationships, which yield quantitative predictions concerning rates of unexplored chemical reactions.

A mechanism is a set of postulated molecular events that results in the observed conversion of reactants to products (Gardiner, 1969). Proposed mechanisms are important statements about the dynamics of a chemical process. As we shall see, mechanisms imply certain relationships between physical and chemical properties of a system (species concentrations, temperature, ionic strength, etc.) and rates of chemical transformations. As long as these relationships are consistent with experimental evidence, proposed mechanisms are considered useful. The provisional nature of all chemical mechanisms is important to recognize; as new experimental evidence is acquired, proposed mechanisms are tested with greater scrutiny. At some point, all mechanisms may have to be discarded in favor of new proposed mechanisms that agree more favorably with experimental evidence.

2.   THE BASIC "UNIT" OF CHEMICAL PROCESSES: THE ELEMENTARY REACTION

2.1.   Reaction Mechanisms

Assume, for the moment, that a proposed mechanism has been provided. What does this mechanism tell us about the course and rate of a chemical process, and about the influence of various physical and chemical factors?

We begin by considering an important environmental reaction, the base-catalyzed hydrolysis of carboxylic acid esters (Tinsley, 1979):

Reaction Mechanism

$$\underset{\substack{\| \\ \text{O}}}{\text{RC}}\!\!-\!\!\text{OR}' \;+\; \text{OH}^- \;\xrightarrow{\;k_1\;}\; \underset{\substack{| \\ \text{OH}}}{\overset{\substack{\text{O}^- \\ |}}{\text{RC}}}\!\!-\!\!\text{OR}' \tag{1}$$

$$\underset{\substack{| \\ \text{OH}}}{\overset{\substack{\text{O}^- \\ |}}{\text{RC}}}\!\!-\!\!\text{OR}' \;\xrightarrow{\;k_{-1}\;}\; \underset{\substack{\| \\ \text{O}}}{\text{RC}}\!\!-\!\!\text{OR}' \;+\; \text{OH}^- \tag{2}$$

$$\underset{\substack{| \\ \text{OH}}}{\overset{\substack{\text{O}^- \\ |}}{\text{RC}}}\!\!-\!\!\text{OR}' \;\xrightarrow{\;k_2\;}\; \underset{\substack{\| \\ \text{O}}}{\text{RC}}\!\!-\!\!\text{OH} \;+\; \text{R}'\text{O}^- \tag{3}$$

$$\underset{\substack{\| \\ \text{O}}}{\text{RC}}\!\!-\!\!\text{OH} \;+\; \text{R}'\text{O}^- \;\xrightarrow{\;k_{-2}\;}\; \underset{\substack{| \\ \text{OH}}}{\overset{\substack{\text{O}^- \\ |}}{\text{R}\!-\!\text{C}}}\!\!-\!\!\text{OR}' \tag{4}$$

$$\underset{\substack{\| \\ \text{O}}}{\text{RC}}\!\!-\!\!\text{OH} \;+\; \text{R}'\text{O}^- \;\xrightarrow{\;k_3\;}\; \underset{\substack{\| \\ \text{O}}}{\text{RC}}\!\!-\!\!\text{O}^- \;+\; \text{R}'\text{OH} \tag{5}$$

$$\underset{\substack{\| \\ \text{O}}}{\text{RCO}^-} \;+\; \text{R}'\text{OH} \;\xrightarrow{\;k_{-3}\;}\; \underset{\substack{\| \\ \text{O}}}{\text{RCOH}} \;+\; \text{R}'\text{O}^- \tag{6}$$

Overall Reaction

$$\underset{\substack{\| \\ \text{O}}}{\text{RC}}\!\!-\!\!\text{OR}' \;+\; \text{OH}^- \;\longrightarrow\; \underset{\substack{\| \\ \text{O}}}{\text{RCO}^-} \;+\; \text{R}'\text{OH} \tag{7}$$

Each step or molecular event in a reaction mechanism is called an "elementary reaction". Each elementary reaction listed above is balanced for both mass and charge. The rate at which an elementary reaction takes place is proportional to the concentration of each species participating in the molecular event: increasing participant concentrations yields a proportional increase in encounter frequency. This observation, called the principle of mass action (Gardiner, 1969) is the basis for quantitative treatment of reaction kinetics. Reaction 7, which represents the overall reaction stoichiometry, is also balanced for mass and charge. Reaction 7 is a composite of several molecular events; it cannot be used to make fundamental statements concerning reaction mechanism and rate.

Rates of each reaction step can be calculated by use of the principle of mass action and values of pertinent rate constants. The rate of the hydroxide ion addition to the ester is given by

$$r_1 = k_1 [RCOOR'] [OH^-] \tag{8}$$

Changes in concentrations for particular chemical species are found by accounting for both production and consumption. Net loss of ester, for example, is given by

$$-\frac{d[RCOOR']}{dt} = r_1 - r_{-1}$$
$$= k_1 [RCOOR'][OH^-] - k_{-1}[RC(O^-)(OH)(OR')] \tag{9}$$

2.2.   Concentration versus Time

We are often concerned with changes in concentration as a function of time, requiring that rate equations be integrated. This requires that boundary conditions be taken into account, such as the concentrations of species at the onset of reaction ($t=0$). Equation 9 cannot be integrated without some additional work, since changes in $[RCOOR']$ and $[RC(O^-)(OH)(OR')]$ are interconnected with changes in concentrations of other reaction species. We will leave the discussion of processes involving two or more elementary reactions for a later section.

Most elementary reactions involve either one or two reactants. Elementary reactions involving three species are infrequent, because the likelihood of simultaneous three-body encounter is small. In closed, well-mixed chemical systems, the integration of rate equations is straightforward. Results of integration for some important rate laws are listed in Table 1, which gives the concentration of reactant A as a function of time. First-order reactions are particularly simple; the rate constant k has units of s^{-1}, and its reciprocal value $(1/k)$ provides a measure of a characteristic time for reaction. It is common to speak in terms of the half-life $(t_{1/2})$ for reaction, the time required for 50% of the reactant to be consumed. When

$$[A] = \tfrac{1}{2}[A]_0, \qquad \tfrac{1}{2} = \exp(-kt_{1/2}) \tag{10}$$

$$t_{1/2} = \frac{0.693}{k} \tag{11}$$

For first-order reactions in closed vessels, the half-life is independent of the initial reactant concentration. Defining characteristic times for second- and third-order reactions is somewhat complicated in that concentration units appear in the reaction rate constant k. Integrated expressions are available in a number of

TABLE 1. Analytical Solutions to Differential Equations Describing Elementary Reactions

Boundary conditions: at $t=0$, $[A]=[A]_0$
$[B]=[B]_0$

First-order reaction k (units s^{-1})

$$A \xrightarrow{k} P, \qquad \frac{d[A]}{dt} = -k[A], \qquad [A]=[A]_0\, e^{-kt}$$

Second-Order Reactions k (units $M^{-1}\, s^{-1}$)

$$A+A \xrightarrow{k} P, \qquad \frac{d[A]}{dt} = -_2k[A]^2, \qquad [A] = \frac{[A]_0}{1+_2k[A]_0 t}$$

$$A+B \xrightarrow{k} P, \qquad \frac{d[A]}{dt} = -k[A][B],$$

$$[A] = \frac{[A]_0([A]_0 - [B]_0)}{[A]_0 - [B]_0\, e^{-k([A]_0 - [B]_0)t}}$$

Third-Order Reactions k (units $M^{-2}\, s^{-1}$)

$$A+A+A \xrightarrow{k} P, \qquad \frac{d[A]}{dt} = -_3k[A]^3, \qquad [A] = \frac{[A]_0}{\sqrt{1+6kt[A]_0^2 kt}}$$

standard references (e.g., Capellos and Bielski, 1980; Laidler, 1987; Moore and Pearson, 1981).

If the chemical kinetics of interest can be defined by a single elementary reaction, but the reaction occurs in an open system, the differential equation for reaction must be elaborated accordingly. In a later section, we will discuss how both chemical reaction and mass transport can be accounted for in calculating changes in species concentrations as a function of time.

2.3. Theory of Elementary Reactions, ACT

Elementary reactions are distinguished from one another by the chemical characteristics of the participating reactants and their modes of interaction with one another. Fundamental distinctions are made between unimolecular reactions (e.g., A→products), bimolecular reactions (e.g., A+B→products), and those occurring in homogeneous solution and those occurring at an interface (heterogeneous reaction). All elementary reactions are, however, representations of single molecular events, and therefore rate constants should respond in similar, predictable ways to changes in the physical characteristics of the system that affect molecular motion: temperature, pressure, and ionic strength. Activated-complex theory (ACT), also referred to as *transition-state theory* (TST), was developed to explore these relationships.

The ACT begins by postulating an activated complex for each elementary reaction, the high-energy ground-state species formed from the encounter of reactant molecules. An elementary bimolecular reaction

$$A + B \xrightarrow{k} \text{products} \tag{12}$$

can be viewed as the formation of the activated complex (AB), and its eventual decay to form products:

$$A + B \rightleftharpoons (AB)^{\neq} \qquad K^{\neq} \tag{13}$$

$$(AB)^{\neq} \longrightarrow \text{products} \tag{14}$$

where A, B, and $(AB)^{\neq}$ are in local equilibrium with one another, and $K^{\neq}$ is a kind of equilibrium constant. Decay of the activated complex to form products is simply related to the vibrational frequency of the species imparted by thermal energy (Gardiner, 1969):

$$R = v[(AB)^{\neq}] = \frac{k_B T}{h} [(AB)^{\neq}] \tag{15}$$

where k_B is Boltzmann's constant and h is Planck's constant. This is a useful formulation, since intrinsically chemical aspects of the reaction are contained in the value of $K^{\neq}$. The TST was developed originally by Eyring and others on the basis of statistical mechanics [see, e.g., Lasaga (1983) or Moore and Pearson (1981)].

The rate of product formation (e.g., in moles per liter per second) is related to the concentration (in moles per liter) of the activated complex $(AB)^{\neq}$. Because $K^{\neq}$ is a thermodynamic quantity, it is related to the activity of the species involved in reaction 13.

$$K^{\neq} = \frac{\{(AB)^{\neq}\}}{\{A\}\{B\}} = \frac{[(AB)^{\neq}]\gamma_{\neq}}{[A][B]\gamma_A \cdot \gamma_B} \tag{16}$$

where γ_A, γ_B, and $\gamma_{\neq}$ are activity coefficients for species A, B, and $(AB)^{\neq}$, respectively. Equations 15 and 16 can now be combined in order to find the rate of product formation as a function of the concentrations of reactants A and B:

$$R = \frac{k_B T}{h} \frac{\gamma_A \gamma_B}{\gamma_{\neq}} K^{\neq} [A][B] = k[A][B] \tag{17}$$

with the rate constant for the reaction given by

$$k = \frac{k_B T}{h} \frac{\gamma_A \gamma_B}{\gamma_{\neq}} K^{\neq} \tag{18}$$

(in liters per mole per second). Equations 17 and 18 indicate the connection between the ACT rate for an elementary bimolecular reaction and the second-order rate constant k.

Ionic Strength.   The effect of ionic strength on rates of elementary reactions readily follows. Using Eq. 18, we can let k_0 be the value of the second-order rate constant in the reference state, such as an infinitely dilute solution (where all the activity coefficients are unity); k is the rate constant at any specified ionic strength:

$$k = k_0 \frac{\gamma_A \cdot \gamma_B}{\gamma_{\neq}} \tag{19}$$

From the ionic strength, values of γ_A, γ_B, and $\gamma_{\neq}$ can be calculated using the Davies equation (Stumm and Morgan, 1981, p. 135). (The charge of the activated complex is known; it is simply the sum of the charge of the two reactants.) Activity coefficients for anions and cations typically decrease as the ionic strength is increased. According to Eq. 19, increasing the ionic strength (1) lowers the reaction rate between a cation and anion, (2) raises the reaction rate between like-charged species, and (3) has little effect on reaction rate when one or both of the reactants is uncharged.

Temperature.   The effect of temperature on rate constants for elementary reactions will now be examined. To assist in the interpretation of experimental information, Arrhenius (1889) postulated the following relationship:

$$k = A e^{-E_a/RT} \tag{20}$$

$A = A(T)$ and E_a are referred to as the *Arrhenius parameters*. The logarithmic form of Eq. 20

$$\ln k = \ln A - \frac{E_a}{RT} \tag{21}$$

suggests plotting logarithms of experimental rate constants versus reciprocal absolute temperatures $(1/T)$ to estimate the preexponential factors A and activation energies E_a. We can relate the Arrhenius parameters to ACT by postulating a Gibbs free energy of activation, $\Delta G^{\circ \neq}$, related to $K^{\neq}$ in the following manner:

$$\Delta G^{\circ \neq} = \Delta H^{\circ \neq} - T \Delta S^{\circ \neq} = -RT \ln K^{\neq} \tag{22}$$

Equation 18 can now be rewritten in terms of $\Delta G^{\circ \neq}$, $\Delta H^{\circ \neq}$, and $\Delta S^{\circ \neq}$:

$$k = \frac{k_B T}{h} \frac{\gamma_A \gamma_B}{\gamma_{\neq}} e^{-\Delta G^{\circ \neq}/RT} = \frac{k_B T}{h} \frac{\gamma_A \gamma_B}{\gamma_{\neq}} e^{\Delta S^{\circ \neq}/R} e^{-\Delta H^{\circ \neq}/RT} \tag{23}$$

For an elementary reaction, comparison of the Arrhenius equation (Eq. 20) with the corresponding ACT equation (Eq. 23) (and with $\gamma_A = \gamma_B = \gamma_{\neq} = 1.0$) yields the following values for the Arrhenius parameters:

$$A = \frac{k_B T}{h} \, e^{\Delta S^{\circ \neq}/R}, \quad E_a = \Delta H^{\circ \neq} + RT \tag{24}$$

Thus, the Arrhenius equation, predicting a linear relationship between $\ln k$ and $1/T$, is confirmed by the ACT treatment.

3. SIMPLE COLLECTIONS OF ELEMENTARY REACTIONS

Mechanisms for most chemical processes involve two or more elementary reactions. Our goal is to determine concentrations of reactants, intermediates, and products as a function of time. In order to do this, we must know the rate constants for all pertinent elementary reactions. The principle of mass action is used to write differential equations expressing rates of change for each chemical involved in the process. These differential equations are then integrated with the help of stoichiometric relationships and an appropriate set of boundary conditions (initial concentrations, for example). For simple cases, analytical solutions are readily obtained. Complex sets of elementary reactions may require numerical solutions.

3.1. Reactions in Series

Two first-order elementary reactions in series are

$$A \xrightarrow{k_1} B \xrightarrow{k_2} C \tag{25}$$

From the principle of mass action, rates of the first and second steps are given by

$$r_1 = k_1 [A] = -d[A]/dt \tag{26}$$
$$r_2 = k_2 [B] = d[C]/dt \tag{27}$$

How do the concentrations of the three species change as a function of time? Only one process acts on A; it is consumed by the first elementary reaction. Equation 26 can be integrated directly, giving typical first-order decay in A:

$$[A] = [A]_0 \, e^{-k_1 t} \tag{28}$$

Two processes act on B; it is produced by the first elementary reaction, but consumed by the second:

$$d[B]/dt = k_1[A] - k_2[B] \tag{29}$$

Combining Eqs. 28 and 29 yields a differential equation that can be readily integrated:

$$d[B]/dt = k_1[A]_0 e^{-k_1 t} - k_2[B] \tag{30}$$

$$[B] = \frac{k_1[A]_0}{k_2 - k_1}(e^{-k_1 t} - e^{-k_2 t}) \tag{31}$$

As expected, the dynamic behavior of [B] depends on the relative magnitudes of k_1 and k_2. When $k_1 \gg k_2$, the maximum value of [B] will be high; when $k_1 \ll k_2$, the maximum value of [B] will be low.

Only one process acts on C; it is produced by the second elementary reaction. The concentration of C as a function of time is found by inserting Eq. 31 into Eq. 27, or by taking advantage of the mass-balance equation:

$$[A]_0 + [B]_0 + [C]_0 = [A] + [B] + [C] \tag{32}$$

$$[C] = [C]_0 + ([A]_0 - [A]) + ([B]_0 - [B]) \tag{33}$$

Thus, the concentrations of all reactants, intermediates, and products have been determined as a function of time.

Three reactions in series will now be considered:

$$A \xrightarrow{k_1} B \xrightarrow{k_2} C \xrightarrow{k_3} D \tag{34}$$

$$r_1 = k_1[A] = -d[A]/dt \tag{35}$$

$$r_2 = k_2[B] \tag{36}$$

$$r_3 = k_3[C] = d[D]/dt \tag{37}$$

$$d[B]/dt = r_1 - r_2 = k_1[A] - k_2[B] \tag{38}$$

$$d[C]/dt = r_2 - r_3 = k_2[B] - k_3[C] \tag{39}$$

The equations are considerably more complex than in the preceding case, but an analytical solution can still be found [Szabo (1969) and Capellos and Bielski (1980) provide useful compilations of analytical solutions]. The mass balance equation, and its derivative with respect to time, are useful in solving these equations.

$$[A]_0 + [B]_0 + [C]_0 + [D]_0 = [A] + [B] + [C] + [D] \tag{40}$$

$$0 = d[A]/dt + d[B]/dt + d[C]/dt + d[D]/dt \tag{41}$$

Species constants and rates of the three contributing elementary reactions are shown in Figures 1a for the case when $k_1 = k_2 = k_3 = 0.1$ day^{-1} and $[B]_0 = [C]_0 = [D]_0 = 0$. As the reaction progresses, the predominant species shifts from A to B to C, eventually forming D. Reaction rates, proportional to reactant concentrations, are continually changing as the reaction progresses. Rate r_1 decreases exponentially, as the original pool of A is consumed; r_2 and r_3 first grow, as intermediates B and C are produced, but eventually diminish as significant amounts of D are formed.

For comparison, similar calculations are shown in Figures 1b but using a different set of rate constants ($k_1 = 0.02$ day^{-1}, $k_2 = k_3 = 0.10$ day^{-1}). The characteristic time for the first reaction step is now five times longer than characteristic times for the second and third steps:

$$t_1 = 1/k_1 = 50.0 \text{ days} \tag{42}$$

$$t_2 = 1/k_2 = t_3 = 1/k_3 = 10.0 \text{ days} \tag{43}$$

As a consequence, the rates of the second and third elementary reactions are

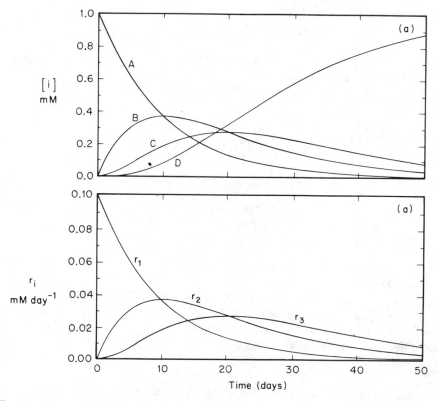

Figure 1a. Consecutive irreversible reactions. Rate constants for the three elementary reactions are the same ($k_1 = k_2 = k_3 = 0.1$ day^{-1}) and $[B]_0 = [C]_0 = [D]_0 = 0$. Reaction rates ($r_1, r_2, r_3$), proportional to substrate concentration, are continually changing and seldom the same.

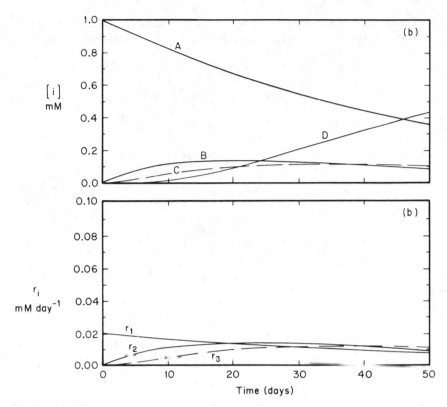

Figure 1b. Consecutive irreversible reactions. The rate constant for the first reaction in séries in small relative to the other three ($k_1 = 0.02$ day^{-1}, $k_2 = k_3 = 0.1$ day^{-1}). The "bottleneck" caused by the rate-limiting step restrains reaction rates for subsequent steps in the reaction.

constrained by the supply of intermediate B coming from the first reaction step. Under these conditions, the first step can be termed the rate-controlling step, exerting the strongest influence on the rate of final product formation.

A generality can be made about all reactions in series. The rate of final product formation is influenced by the rate constants of all prior reactions steps. Overall rates of product formation are most influenced by the step with the longest characteristic time. This step constrains the rates of subsequent steps, despite their larger rate constants.

3.2. Reactions in Parallel

In the mechanism given below, A might be viewed as a pollutant that degrades by two competitive bimolecular reactions, forming two possible reaction products:

$$A + B \xrightarrow{k_1} P_1 \tag{44}$$

$$A + C \xrightarrow{k_2} P_2 \tag{45}$$

Rates of the two competing reactions are proportional to the concentration of the common substrate A and the concentrations of the two competing reactants B and C:

$$r_1 = k_1[A][B] = d[P_1]/dt \tag{46}$$

$$r_2 = k_2[A][C] = d[P_2]/dt \tag{47}$$

$$d[A]/dt = -r_1 - r_2 = -(k_1[B] + k_2[C])[A] \tag{48}$$

A first example of parallel reactions is presented in Figure 2; reactants B and C are present in equal concentrations at the start of reaction, but C is five times more reactive. Concentrations and rates for the two contributing reactions have

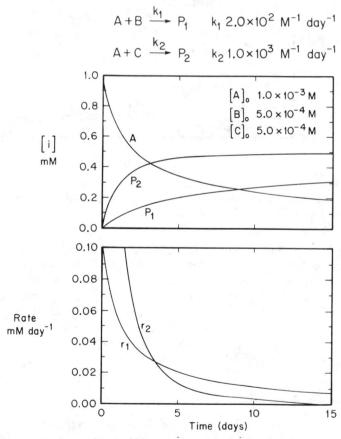

Figure 2. Two second-order parallel irreversible reactions. Rates of the two elementary reactions in parallel are dependent on the concentration of the common substrate ([A]) and the concentrations of the competing reactants ([B] and [C]). For the conditions given, C reacts more quickly than B, but is quickly depleted by reaction. Once this has taken place, reaction of A with B becomes the dominant reaction.

been calculated numerically. At the onset, reaction of A with C predominates, and production of P_2 exceeds the production of P_1. As the reaction progresses, however, C is depleted more quickly than B, causing the production of P_2 to decline relative to the production of P_1. Reaction of A with B grows in importance as the reaction progresses, eventually becoming the predominant pathway. For this particular case, $[A]_0$ is high enough for complete conversion of B to P_1 and of C to P_2 to occur as the reaction reaches completion.

If the supply of A is limiting $[A]_0 < ([B]_0 + [C]_0)$, the final product distribution is determined by the relative magnitude of rate constants k_1 and k_2. If the competitive reactants B and C are initially present at equimolar concentrations, C will consume a disproportionate share of limiting substrate A. It should be remarked that the rates and time scales in this example depend on *two* reactant concentrations, so that it is not possible to identify characteristic times from rate constants alone.

Pseudo-First-Order Treatment. A convenient simplification is possible when the concentrations of reactants B and C are much greater than the substrate A. As the reaction progresses, changes in [B] and [C] are small relative to changes in [A], and the former can be considered effectively constant. Each elementary reaction can then be considered "pseudo-first-order" with respect to A. Let

$$k_1' = k_1[B]_0 \quad \text{and} \quad k_2' = k_2[C]_0 \tag{49}$$

$$r_1 = k_1'[A] = d[P_1]/dt \tag{50}$$

$$r_2 = k_2'[A] = d[P_2]/dt \tag{51}$$

Under these conditions, r_1 and r_2 decrease in proportion to one another as the reaction progresses, and products P_1 and P_2 are produced at a constant ratio to one another. An example illustrating this situation is presented in Figure 3.

Equation 48 is easily integrated under pseudo-first-order conditions, since $(k_1[B] + k_2[C])$ can be considered constant. Then [A], $[P_1]$, and $[P_2]$ as a function of time are

$$[A] = [A]_0 e^{-(k_1[B]+k_2[C])t} = [A]_0 e^{-k_{app}t} \tag{52}$$

$$[P_1] = [P_1]_0 + \frac{k_1[A]_0[B]}{(k_1[B]+k_2[C])}(1 - e^{-(k_1[B]+k_2[C])t}) \tag{53}$$

$$[P_2] = [P_2]_0 + \frac{k_2[A]_0[C]}{(k_1[B]+k_2[C])}(1 - e^{-(k_1[B]+k_2[C])t}) \tag{54}$$

Under pseudo-first-order conditions, reactant A exhibits simple first-order decay. The rate of decay and half-life are determined by the magnitude of k_{app}, the apparent first-order rate constant for loss of A. Since $k_1[B]$ and $k_2[C]$ are in units of reciprocal time, the following half-lives can be defined:

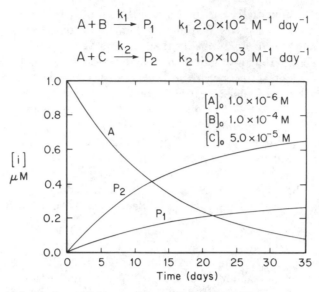

$$A + B \xrightarrow{k_1} P_1 \quad k_1 \ 2.0 \times 10^2 \ M^{-1} \ day^{-1}$$

$$A + C \xrightarrow{k_2} P_2 \quad k_2 \ 1.0 \times 10^3 \ M^{-1} \ day^{-1}$$

Figure 3. Pseudo-first-order parallel irreversible reactions. $[B]_0$ and $[C]_0$ are large relative to $[A]_0$, and can be considered constant. As a consequence, reactions are pseudo-first-order with respect to A, and products P_1 and P_2 are generated at a constant ratio to one another.

Reaction 1:

$$t_{1/2}(1) = 0.693/k_1 [B]_0 \tag{55}$$

Reaction 2:

$$t_{1/2}(2) = 0.693/k_2 [C]_0 \tag{56}$$

Overall:

$$t_{1/2}(A) = 0.693/k_{app} = 0.693/(k_1 [B]_0 + k_2 [C]_0) \tag{57}$$

$$\frac{1}{t_{1/2}(A)} = \frac{1}{t_{1/2}(1)} + \frac{1}{t_{1/2}(2)} \tag{58}$$

According to Eq. 58, when $t_{1/2}(1) \gg t_{1/2}(2)$, the overall half-life for loss of A is approximately equal to $t_{1/2}(2)$. In other words, the competitive reaction with the shortest half-life has the strongest influence on the half-life of the common substrate. It is often possible to apply pseudo-first-order kinetic methods to characterize time scales of complex reactions.

Product yields under pseudo-first-order conditions can also be calculated using Eqs. 53 and 54:

Fractional yield of P_1:

$$\frac{[P_1]}{[P_1] + [P_2]} = \frac{k_1 [B]_0}{k_1 [B]_0 + k_2 [C]_0} \tag{59}$$

Fractional yield of P_2:

$$\frac{[P_2]}{[P_1]+[P_2]}=\frac{k_2[C]_0}{k_1[B]_0+k_2[C]_0} \tag{60}$$

When $[B]_0$ is increased at constant $[C]_0$, the product distribution shifts in favor of P_1.

These findings can be generalized to all parallel reactions. The competitive reaction with the highest rate constant and shortest characteristic time has the strongest effect on the decay rate of the common substrate. When the concentration of the common substrate is limiting, the competitive reaction with the highest rate constant also generates the greatest amount of reaction product.

3.3. Reversible Reactions

Many chemical reactions important in the water environment are reversible. When the reaction of interest is far from equilibrium, the concentration of the product is small, and the rate of the back reaction is low relative to the rate of the forward reaction. Thus, far from equilibrium the back reaction can be ignored, and the reaction can be modeled as an irreversible process. As equilibrium is approached, however, the rate of the back reaction becomes significant, and can no longer be ignored.

Consider the following simple reversible reaction:

$$A \underset{k_2}{\overset{k_1}{\rightleftharpoons}} B \tag{61}$$

for which the rates of change can be described by

$$-d[A]/dt=d[B]/dt=k_1[A]-k_2[B]=r_1-r_2 \tag{62}$$

If we assume that $[B]_0=0$ at the onset of reaction, Eq. 62 can be integrated to obtain $[B]$ as a function of time

$$[B]=\frac{k_1[A]_0}{k_1+k_2}(1-e^{-(k_1+k_2)t}) \tag{63}$$

The equilibrium concentration of B can be calculated from Eq. 63, by letting t approach infinity:

$$[B]_{eq}=\frac{k_1[A]_0}{k_1+k_2} \tag{64}$$

The characteristic time for approach to equilibrium is given by the reciprocal of the *sum* of the rate constants: $1/(k_1+k_2)$. This result has important consequences. In Figure 4, two cases are considered where the sum of the rate

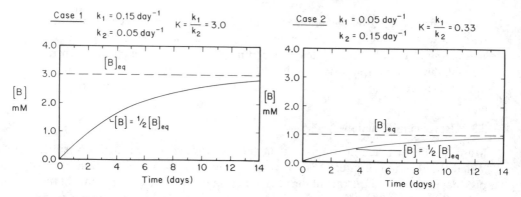

Figure 4. Single reversible reaction. In going from case 1 to case 2, the ratio of the two rate constants (k_1/k_2) is varied while keeping their sum $(k_1 + k_2)$ the same. As a consequence, the characteristic time for attaining equilibrium $[(k_1 + k_2)^{-1}]$ is unchanged, but the position of the final equilibrium is different.

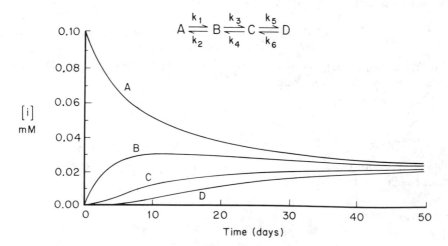

Figure 5. Consecutive reversible reactions. The individual rate constants, $k_1 k_2 \cdots k_6$, are all set equal to 0.1 day^{-1}. The initial concentration of A is 0.10 mM. Concentrations of all species approach their final equilibrium values. In this case, the equilibrium concentration ratios are unity.

constants is the same but relative values are different. In case 1, the final equilibrium position favors product B over reactant A. In case 2, equilibrium lies more in favor of reactant A. Because the sum of rate constants is the same, progress toward the final equilibrium position occurs at the same rate in both cases. Thus, the ratio of rate constants $(K = k_1/k_2)$ determines the final equilibrium position, while the sum of rate constants $(k_1 + k_2)$ determines how quickly that position is approached.

Three reversible reactions in series are presented in Figure 5. In contrast to the case of consecutive irreversible reactions presented earlier (Fig. 1), all four species coexist at the final equilibrium position. The product concentrations [B], [C], and [D] grow as the reaction progresses without overshooting their final equilibrium position.

3.4. Combined Chemical Kinetics and Mass Transport: A CSTR Model Lake

In order to explore the dynamics of open systems, it is necessary to formulate a material balance equation that accounts for all processes that work to elevate or depress the concentrations of chemical species of interest. Physical and chemical processes have been included in Eq. 65 (biological processes have not been considered) (Imboden and Schwarzenbach, 1985):

Change of moles within volume element	=	moles entering volume element	−	moles leaving volume element	±	moles produced or consumed by chemical reaction

$$(65)$$

Overall changes in concentration depend on the relative magnitude of production and consumption terms. For closed systems, the first two terms on the right-hand side are equal to zero, and the material balance equation exhibits the simple forms used in earlier sections.

To illustrate the combined effects of chemical reaction and mass transport, a simple lake model treated as a continuous stirred tank reactor (CSTR) will now be examined. The inflow and outflow rates are constant and equal to one another:

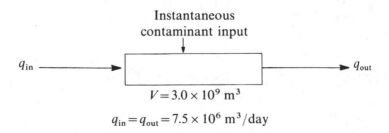

Instantaneous
contaminant input

q_{in} ⟶ q_{out}

$V = 3.0 \times 10^9 \text{ m}^3$

$q_{in} = q_{out} = 7.5 \times 10^6 \text{ m}^3/\text{day}$

In the first example, an instantaneous input of two pollutants is made to the model lake. Pollutant A decays to chemical B according to first order kinetics, while pollutant C is a conservative tracer (no reaction):

$$A \xrightarrow{k} B \qquad (k = 2.5 \times 10^{-3} \text{ days}^{-1}) \tag{66}$$

$$C \qquad \text{(conservative)} \tag{67}$$

Three material balance equations can be written:

$$V d[A]/dt = 0 - q[A] - Vk[A] \tag{68}$$

$$V d[B]/dt = 0 - q[B] + Vk[A] \tag{69}$$

$$V d[C]/dt = 0 - q[C] \tag{70}$$

Integration yields:

$$[A] = [A]_0 \, e^{-(q/V + k)t} \tag{71}$$

$$[B] = [A]_0 \, (e^{-(q/V)t} - e^{-(q/V + k)t})$$

$$[B] = [A]_0 e^{-(q/V)t}(1 - e^{-kt}) \tag{72}$$

$$[C] = [C]_0 e^{-(q/V)t} \tag{73}$$

A comparison of the CSTR model lake with a corresponding closed system is presented in Figure 6. The concentration of conservative pollutant C is constant in the closed system, but decreases exponentially in the open system, because of outflow from the lake. Note that q/V in the exponential term of Eq. 73 has units of reciprocal time. Thus, V/q represents a characteristic time for physical removal, and has a value of 400 days in this example.

The reactive pollutant A will now be considered. In the closed system, the sum ($[A] + [B]$) remains constant, since no loss of material takes place. The concentration of A, however, decreases exponentially in the closed system, because of chemical reaction; the characteristic time for loss of A is given by $1/k$ and has a value of 400 days. In the CSTR model lake, both $[A]$ and the sum ($[A] + [B]$) decrease exponentially, relating to the following characteristic times:

Quantity	Characteristic Time	
$[A]$	$\dfrac{1}{q/V + k} = 200$ days	(74)
($[A] + [B]$)	$\dfrac{1}{q/V} = 400$ days	(75)

Physical removal and consumption by chemical reaction are processes in parallel, causing a decrease of $[A]$ in the CSTR model lake as a function of time.

This kind of calculation can be extended to other reactive pollutants. When the rate constant k for chemical reaction is large ($k \gg q/V$), chemical reaction is

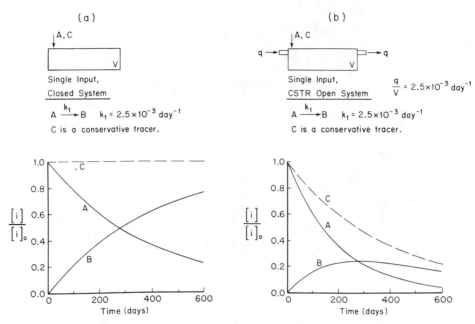

Figure 6. Combined effect of chemical reaction and mass transport. Single instantaneous inputs of A and C are made to: (a) a well-mixed closed system and (b) a continuously stirred tank reactor (CSTR). Pollutant C is conservative tracer; its concentration is constant in the closed system, and decreases in the CSTR because of the outward flux of water. Chemical A is transformed into product B according to first-order kinetics. The sum ([A]+[B]) is constant in the closed system, but decreases in the CSTR. Loss of A from the CSTR arises from both chemical reaction and mass transport.

fast relative to physical removal and becomes the dominant removal mechanism. For exceedingly slow chemical reactions ($k \ll q/V$), physical removal is dominant.

As illustrated in Figure 6, the concentration of reaction product B in the CSTR model lake increases initially, reaches a maximum value, and then decreases. In a natural water, a reaction product may be more toxic to biota than its precursor. Thus, the height of the concentration maximum for a product, such as B, is important. Fast chemical reaction relative to physical removal ($k \gg q/V$) is favorable for temporary buildup of B in the lake.

The foregoing examples have examined the dynamics of a CSTR model lake following a single addition of pollutant. We will now examine results from a steady continuous input of pollutants A and C, reactive and conservative, respectively. The pollutants are added to the inlet water, commencing at $t=0$. As shown in Figure 7a, concentrations of reactive pollutant A, reaction product B, and conservative pollutant C increase in the CSTR lake, until steady-state concentrations are reached. At steady state, inflow and chemical production of a species is matched by outflow and chemical consumption. In Figure 7b, the

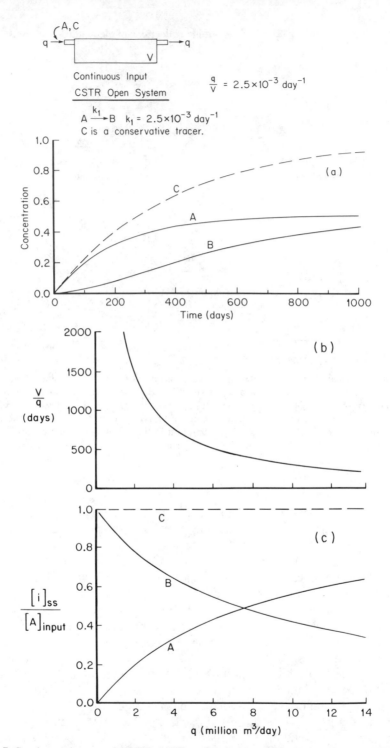

Figure 7 Continuous input model CSTR. (*a*) When chemical species (A, C) are added to a CSTR on a continuous basis, their concentrations and those of products eventually reach steady-state values. (*b*) If the steady flow rate is increased, the fluid residence time of the CSTR, V/q, decreases. (*c*) The residence time of reactant A is thereby decreased and the yield of product B is lowered.

inflow rate to the CSTR model lake is increased, while keeping the concentrations of pollutants A and C in the inflow constant. As the inflow is increased, the residence time in the lake decreases. As a consequence (Fig. 7c), the steady-state concentration of A increases while that of B decreases; less time is available in the lake for chemical reaction to occur.

4.   COMPLEX ENVIRONMENTAL KINETICS

The examples presented in the previous section are exceedingly simple. Vastly more detailed models have been formulated and used to investigate the dynamics of real systems. It is important to justify the level of complexity required in a model. As part of the modeling effort, issues such as the following must be addressed:

To what extent have important physical, chemical, and biological processes been identified and quantified?
Is the available information about a particular natural system sufficient to support a detailed spatial and temporal model?
What assumptions or simplifications can be made to streamline the modeling activity without introducing unnecessary error?

Several excellent reviews discuss the development and implementation of kinetic models of lakes (Imboden and Lerman, 1978; Fischer et al., 1979; Schwarzenbach and Imboden, 1984; Imboden and Schwarzenbach, 1985) and sediments (Berner, 1980). More recently, models have been developed for examining chemical reaction and mass transport in soils (Furrer et al., 1989) and aquifers (Liu and Narashiman, 1989; Jennings, 1987). Our goal here is to highlight central issues relating to the role of chemical kinetics in environmental processes.

4.1.   Characteristic Time Scales

Imboden and Schwarzenbach (1985) have illustrated how the mass-balance equation is a means of accounting for chemical and biological reactions that produce or consume a chemical within a test volume, and for transport processes that import or export the chemical across the boundaries. Each process acting on a chemical can be characterized by an environmental first-order rate constant, expressed in units of time^{-1}. Transport mechanisms include water renewal by rivers, horizontal and vertical turbulent diffusion, advection by lake particles, and settling of particles (Imboden and Schwarzenbach, 1985). Chemical reaction rates and reaction half-lives for a wide variety of reactions have been summarized by Hoffmann (1981), Pankow and Morgan (1981), Morgan and Stone (1985), and Santschi (1988).

Comparison of characteristic times for chemical and biological processes (T_{rxn}) with those of transport processes (T_{phys}) is critical to explaining the dynamic behavior of chemical species in the environment. Fast chemical and biological processes have short characteristic times ($T_{rxn} \ll T_{phys}$) and will proceed to products or to equilibrium in natural systems. Transformation processes with long characteristic times ($T_{rxn} \gg T_{phys}$) are appropriately characterized as slow: chemicals are removed from the test volume before reaction takes place to any significant extent. Reactions for which T_{rxn} and T_{phys} are within an order of magnitude of one another (with T_{rxn} in the range of 10^3 to 10^9 s) require a kinetic description to account for element and chemical species distributions. For species involved in coupled chemical reactions with both short ($T_{rxn} \ll T_{phys}$) and longer ($T_{rxn} \gg T_{phys}$) time scales, quasi-equilibrium descriptions of fast reactions may be combined with kinetic descriptions of slow reactions to yield "constrained equilibrium" or "pseudoequilibrium" models (Morel, 1983; Keck, 1978).

4.2. Import and Export of Chemicals

We are often concerned with the dispersion of pollutants and other chemicals in the environment. Advection and mass flux are indiscriminate transport processes. In the water column of a lake, for example, these processes transport dissolved and particle-bound chemicals equally across the boundaries of the test volume. Settling of particles, in contrast, causes a downward flux of particle-bound chemicals while leaving dissolved chemicals in place. Similarly, surfactants or gases that join rising air bubbles are carried to the surface. These *discriminate* transport processes are very important in a variety of environmental situations:

i. *Discriminate transport processes can deplete chemicals from one environmental compartment, causing them to accumulate in another.* Heavy metals, hydrophobic organic compounds, and other pollutants that have an appreciable affinity for settling particles can be removed from the water column and transported into sediment. Although the water column is cleansed by this process, the sediments become a repository for pollutants.

   When sorption–desorption processes are fast relative to particle settling, the effectiveness of the downward transport is related to the particle concentration, downward particle flux, and the particle–water partition coefficient for the pollutant in question. Rates of sorption–desorption may be limiting in some instances; characteristic times for sorption–desorption then become important, along with the contact time available for interaction with particles.

ii. *Discriminate transport processes can spatially separate potential reactants.* In the photic zone of lakes, photosynthesis generates a strong oxidant (O_2) and a strong reductant (natural organic matter). Natural organic matter is largely particle-bound, and a small proportion of the total production is transported to the sediments. (The remaining portion is oxidized in complex reactions with O_2 in the water column.) The oxidant O_2, in contrast, is not particle-bound; its downward flux is brought about by advection, turbulent mass flux, and molecular diffusion, but not by downward particle settling. The larger flux of reductant into bottom waters and sediments provides an impetus for the development of anoxic conditions.

iii. *Discriminate transport processes can remove reaction products from an environmental compartment, promoting forward reaction.* Products of chemical and biological reactions seldom have the same volatility, solubility, and sorptive characteristics of the reactants. Groundwater inflow may bring dissolved Mn^{2+} and Fe^{2+} into the water column of a lake. Reaction with O_2 generates insoluble hydrous oxide particles, which then settle downward into sediments. Despite a potentially high inward flux of Mn^{2+} and Fe^{2+}, the total concentration of iron and manganese in the water column may be low, a result of an efficient removal process.

In soils, sediments, and aquifers, the transport situation is the reverse of what has been discussed above; solids-bound chemicals (apart from colloidal fractions) are immobile, while dissolved chemicals are mobile. As a consequence, sorption and precipitation retard or prevent dispersion, while desorption and dissolution encourage dispersion. Again, it is important to consider time scales for chemical and biological processes (T_{rxn}) relative to time scales for aqueous-phase transport (T_{phys}). When $T_{rxn} \ll T_{phys}$, quasi-equilibrium can be used to describe chemical reactions in an environmental compartment. In this case, partition coefficients and solubility product constants are used to calculate the proportion of each chemical bound to solids at successive steps in time. When T_{rxn} and T_{phys} are of the same magnitude, rates of sorption, desorption, precipitation, and dissolution must be accounted for, along with rates of flow.

5.   KINETICS AT INTERFACES

Chemical reactions at interfaces are of great importance in many areas of aquatic chemistry. Adsorption and partitioning phenomena are responsible for distributing chemicals among available phases. In the water column, settling particles or rising air bubbles move relative to the aqueous medium. Through such differential transport, adsorption and partitioning are able to influence the dispersion of chemical species. Some chemical reactions must necessarily occur at interfaces because participating reactants do not share a common phase. For example, manganese dioxide $[MnO_2(s)]$ is sparingly soluble in neutral aqueous solution, but able to oxidize many highly soluble species such as oxalate, pyruvate, and ascorbate (Stone and Morgan, 1984b; Stone, 1987) through interfacial reaction. Some chemical reactions may occur slowly in the aqueous phase, but experience acceleration within the chemical microenvironment of an interface. Many biological processes, for example, take advantage of the unique qualities of interfacial reactions; reactive biological molecules are often embedded within a membrane in order to improve the efficiency and yield of biochemical reactions.

The following physical and chemical steps are involved in interfacial reactions:

1. Movement of reactant molecules into the interfacial region by convection, diffusion, or electrical migration
2. Diffusion of reactant molecules within the interfacial region
3. Surface chemical reaction: ligand replacement, electron and group transfer reactions, addition or elimination reactions
4. Outward movement of product molecules from the interfacial region to bulk solution

Overall rates of reaction may depend on rates of one or more of these steps. Many of the important and unique qualities of interfacial reactions arise from strong interconnections between surface chemical reactions and mass transport.

5.1. Transport Terms: Movement in One, Two, and Three Dimensions

Transport is an integral component of all reaction systems. In well-mixed homogeneous solutions, the concentrations of all reactants and products are the same throughout the system, and there is no net movement of chemicals in space. The role of mass transport becomes evident only when chemical reactions are extremely fast. Diffusion determines the encounter frequency of reacting molecules and sets an upward limit on overall rates of reaction. (For example, for a diffusion-controlled bimolecular reaction in water the reaction rate constant is on the order of 10^{10} to $10^{11} \, M^{-1} \, s^{-1}$.) Mass transport plays a pronounced role in surface chemical reactions, since net movement of reactants (from solution to the surface) and products (from the surface to solution) often takes place.

The flux of chemicals to and from surfaces depends on the magnitude of forces causing molecular movement and on the dimensionality of the system. Molecules in solution are transported by the mean motion of water, the advection process (Fischer et al., 1979). Molecules also move relative to the water by diffusion, in response to concentration gradients. For ions, electrostatic forces that contribute to movement are also experienced in regions of changing electrical potential. These three forces are incorporated in the following equation for flux (in one dimension) of a migrating chemical species (Newman, 1973, p. 301):

$$\text{Overall flux} = (\text{advection}) + (\text{Fickian diffusion})$$

$$+ (\text{electrical migration}) \tag{76}$$

$$F_i(x) = v[i]_x - D_i \frac{d[i]_x}{dx} - z_i u_i \mathscr{F} [i]_x \frac{d\psi_x}{dx} \tag{77}$$

where $F_i(x)$ = flux of species i (mol cm^{-2} s^{-1})
v = fluid velocity (cm s^{-1})
$[i]_x$ = species concentration (mol cm^{-3})
D_i = diffusion coefficient of species i (cm^2 s^{-1})
z_i = charge number of species i (equiv. mol^{-1})
u_i = electrical mobility of species i (cm^2 mol J^{-1} s^{-1})
$\mathscr{F}$ = Faraday's constant (96,487 C equiv^{-1})
ψ_x = electrical potential (V)

The subscript x denotes the value of a parameter at a particular distance from the surface.

In the ensuing discussion of surface chemical reactions, the contribution of advection to transport will not be explicitly included. Advection is readily treated on a physical basis without consideration of individual molecular characteristics. In many laboratory situations, bulk solution can be considered "well-mixed" and molecular transport visualized as a migration across a stagnant layer near surfaces where advective transport is negligible.

Surface chemical reactions provide the impetus for molecular movement across interfaces. The acquisition of charge through adsorption of protons, hydroxide ions, and other ions creates electrical potential gradients which bring about electrical migration. Depletion of reactants and accumulation of products by surface chemical reactions create concentration gradients that bring about diffusion. Mass transport rates near an interface are not fixed, however, since concentration gradients and electrical potential gradients can change as a surface chemical reaction progresses.

Consider, for the moment, cations near a negative surface that are not specifically adsorbed or chemically transformed. The cations migrate in response to electrostatic forces, which favor inward movement and accumulation at the surface, and in response to a growing concentration gradient, which favors outward movement. Eventually, an equilibrium distribution is reached where inward electrical migration and outward diffusion exactly cancel one another. At this point, Eq. 77 can be solved by setting the net flux (and the advective flux) of the cation M^{n+} equal to zero, providing the following relationship:

$$[M^{n+}]_x = [M^{n+}]_{\text{Bulk}} e^{-z_i \mathscr{F} \psi_x / RT} \tag{78}$$

In Eq. 78, z_i is the species charge and ψ_x is the electrical potential experienced at a distance x from the surface. Thus, the equilibrium concentration of cations near a positive surface is lower than in bulk solution. Near a negative surface, the equilibrium cation concentration is higher than in bulk solution. The reverse is true for anions.

When adsorption and chemical reaction are taking place, additional terms must be added to Eq. 77 to account for these processes. Figure 8 illustrates

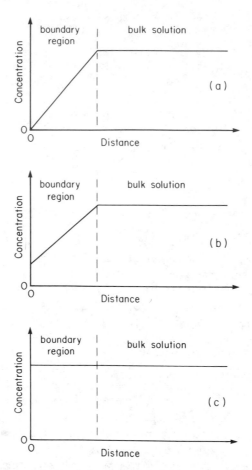

Figure 8. A species present in bulk solution is consumed by a chemical reaction on the available surface. (*a*) Surface chemical reaction is fast relative to molecular transport, and the species is depleted at the surface; the reaction rate is diffusion-controlled. (*b*) Mixed chemical and transport rate control; the concentration diminishes near the surface, but is not fully depleted. (*c*) Transport is fast relative to surface chemical reaction; no depletion is observed; the reaction rate is chemically controlled.

concentration profiles for a chemical reactant consumed by surface chemical reaction. When surface chemical reaction is fast relative to mass transport, the species is depleted at the oxide surface (Fig. 8*a*). When mass transport is fast relative to surface chemical reaction, no depletion is observed (Fig. 8*c*). Situations also exist where characteristic times for chemical reaction and for mass transport are close to one another; then the concentration of reactant drops near the surface, but it is not fully depleted (Fig. 8*b*).

Molecules or ions from homogeneous solution approaching a surface are diffusing in three-dimensional space. Once on a surface, lateral or two-dimensional diffusion may take place, which can be characterized by a two-dimensional diffusion coefficient. Similarly, molecules migrating within pores essentially travel in one-dimensional space. This one- and two-dimensional diffusion is important for many kinds of surface chemical reactions (Hardt, 1979). Consider, for example, two reactants A and B that can react either in the free or in the adsorbed state. In order to calculate the impact of adsorption on reaction rate, we must know the values of the diffusion coefficient in the free and adsorbed state; when diffusion coefficients are of comparable values, the mean time required for encounter is substantially shortened on adsorption (Adam and Delbruck, 1968). This increases overall reaction rate, provided that other factors that influence rate (molecular conformation, local medium characteristics) have not been significantly altered by adsorption. In a different situation, reactant A is present in a fluid streaming past a surface to which the second reactant B is immobilized. Adsorption removes one dimension from the direction of diffusion of A. This, in turn, improves the "catch" of reactant A by B immobilized on the surface (Adam and Delbruck, 1968). In terms of the Arrhenius expression, $k = A \exp(-E_a/RT)$, we may view the "dimensionality" or geometric features of a chemical reaction as reflected in the preexponential terms of the Arrhenius expression, whereas E_a reflects the intrinsic molecular energetics, for instance, of forming the activated complex. The A term has been predicted from ACT (see above) and from collision theory. For details, see Wehrli (Chapter 11, this volume) and Astumian and Schelly (1984).

Reaction of solutes with mineral surface sites is analogous to the immobilized reactant model discussed by Adam and Delbruck (1968). A difference is that many sites line the surface, and lateral movement may bring the adsorbed solute in contact with sites exhibiting higher reaction free energies.

5.2. Adsorption and Surface Chemical Speciation

Whenever two phases come in contact with one another, an interfacial region forms within which physical and chemical characteristics of each phase are disturbed relative to interior (bulk) regions of each phase. At the air–water interface, for example, the directional orientation of water molecules is more pronounced than in bulk solution, in order to compensate for the lack of hydrogen-bonding partners on the gas-phase side of the interface. As a consequence, the dielectric constant and other solvent characteristics that influence chemical reactions are perturbed to some degree. Solute molecules added to air–water or solvent–water systems may reside predominately in one phase or the other, or may concentrate in the interfacial region. Whether or not solute molecules are surface-active depends on the relative energies of possible solute–solute, solute–solvent, and solvent–solvent interactions (Tanford, 1980). The conformation of molecules at interfaces is of great importance: reactive functional groups may be aligned towards or away from each bulk phase, affecting ease of encounter with other reactants. Gas–liquid and liquid–liquid interfaces are very dynamic in nature, making the analysis of interfacial reactions complex.

The mineral–water interface exhibits several important and unique characteristics that are crucial to understanding reactivity. Within the interior of a crystalline phase, atoms are arranged in an orderly fashion, determined by bonding angles and bond strengths. In an amorphous or disordered phase, greater variations in atomic arrangements are observed. The bulk structure of a mineral is broken at the mineral–water interface, representing a discontinuity. The chemistry and three-dimensional structure of surface groups exposed to overlying solution often differ in important respects from those of the solid interior.

Mineral surfaces can be represented using a coordination chemistry approach. Each site consists of a metal center surrounded by inner coordination-sphere ligands, which may be ligands incorporated within the solid, adsorbed water molecules, or adsorbed solutes. The spacing and arrangement of sites may resemble the geometry of the interior mineral phase, or may be distorted because of hydration and rearrangement. Undoubtedly, a range of surface site energies and site conformations exists at any mineral–water interface. Edge or kink sites, for example, may react more quickly than sites that are part of an uninterrupted plane. Different crystal faces of the same mineral have been observed to have distinct adsorption properties and reactivities (Cornell et al., 1974). Surface site heterogeneities are also evident from the following observations: (1) decreasing adsorption affinity as adsorbate surface density is increased, (2) decreased reaction rate per mole of adsorbed reactant as the surface density is increased, and (3) inhibition of adsorption density or reaction rate by nonreacting adsorbate out of proportion to its own surface density.

Spectroscopic methods are often used to determine the conformation and structure of metal complexes. The mineral–water interface, however, provides a formidable challenge to the application of these methods. Very typically, the spectroscopic signal coming from surface sites is swamped by background signals coming from the bulk solid and from aqueous solution. Careful experimental design and the use of techniques exhibiting high selectivity toward the surface are required. Successful applications include ENDOR and EPR spectroscopy of adsorbed metal ions (Motschi and Rudin, 1984; Motschi, 1987) and EXAFS spectroscopy of adsorbed selenium oxyanions (Hayes et al., 1987).

In order to explore the surface charge on mineral surfaces and electrostatic effects, it is necessary to assign a stoichiometry to surface sites. A uniform surface is often postulated, where differences among sites are assumed to arise from differences in stoichiometry. Although this is a simplistic approach, it must suffice until more detailed information is available. In the absence of other specifically adsorbable ions, surface charge arises from the adsorption of protons and hydroxide ions. Surface charge and proton level can then be examined using a three-site model consisting of protonated ($>MeOH_2^+$), neutral ($>MeOH$), and deprotonated ($>MeO^-$) sites (Westall, 1987). A mass-balance equation can be written, where S_T is the total number of available sites (in moles per liter or moles per square meter):

$$S_T = [>MeOH_2^+] + [>MeOH] + [>MeO^-] \qquad (79)$$

These starting assumptions can be extended further, by postulating a set of mass law equations to describe the response of the mineral surface to variation in pH:

$$>MeOH_2^+ = >MeOH + H^+ \text{(surface)} \qquad K_{a1}^s \text{(intr.)} \qquad (80)$$

$$>MeOH = >MeO^- + H^+ \text{(surface)} \qquad K_{a2}^s \text{(intr.)} \qquad (81)$$

where H^+ (surface) refers to a proton located at the surface, within the interfacial region (Davis et al., 1978). Equations 80 and 81 must obviously be approximations for surfaces exhibiting a distribution of site energies; since affinities of various sites for protons and hydroxide ions vary, K_{a1}^s and K_{a2}^s are weighted averages of the microscopic acidity constants exhibited by the collection of surface sites.

Similarly, stoichiometries can be assigned to specifically adsorbed metals and ligands, reflecting the response of the collection of surface sites to changes in pH and adsorbable solute concentrations, for instance:

$$>Me-OH + Pb(H_2O)_6^{2+} = >Me-O-Pb(H_2O)_5^+ + H_3O^+ \qquad K_1^s \qquad (82)$$

$$>Me-OH + Pb(H_2O)_6^{2+} = >MeOH, Pb(H_2O)_6^{2+} \qquad K_2^s \qquad (83)$$

$$>Me-OH_2^+ + SO_4^{2-} = >Me-OSO_3^- + H_2O \qquad K_3^s \qquad (84)$$

$$>Me-OH_2^+ + SO_4^{2-} + H^+ = >Me-OSO_4H + H_2O \qquad K_4^s \qquad (85)$$

Equations 82 and 83 distinguish between inner-sphere and outer-sphere complexes of a surface bound Pb(II). Actual bond lengths, conformations, and site energies for coordination complexes on a real surface may vary from one site to another. The postulated reactions do, however, provide a means of estimating how adsorption density responds in a semiquantitative way to changing medium conditions. See, for example, Schindler and Stumm (1987) for a comprehensive discussion of acid–base and complexation equilibria at oxide–water interfaces.

5.3. Surface Chemical Reaction

Kinetics of surface chemical reactions are governed by the elementary reactions that constitute the reaction mechanism. In accordance with the principle of mass action, rates are proportional to surface concentrations (or surface densities) of participating species: surface sites, specifically adsorbed solutes, and non-specifically adsorbed solutes. The site stoichiometry model presented above is often used to express rate equations for surface chemical reactions.

To illustrate this point, consider the following surface chemical reaction mechanism:

Transport of reactant:

$$A_{Bulk}^- \longrightarrow A_{surface}^- \tag{86}$$

Reactant adsorption:

$$>MeOH_2^+ + A_{surface}^- \underset{k_{-1}}{\overset{k_1}{\rightleftharpoons}} >Me-A + H_2O \tag{87}$$

Surface reaction:

$$>Me-A \underset{k_{-2}}{\overset{k_2}{\rightleftharpoons}} >Me-B \quad \text{slow} \tag{88}$$

Product desorption:

$$>Me-B + H_2O \underset{k_{-3}}{\overset{k_3}{\rightleftharpoons}} >MeOH_2^+ + B_{surface}^- \tag{89}$$

Transport of product:

$$B_{surface}^- \longrightarrow B_{bulk}^- \tag{90}$$

Overall reaction:

$$A_{bulk}^- \longrightarrow B_{bulk}^- \tag{91}$$

Reaction 88 represents the conversion of A^- to B^-. The formation of B^- is directly proportional to $[>Me-A]$, the adsorbed reactant. Estimates of reaction rate based solely on $[A_{bulk}^-]$ are unreliable, since mass transfer, adsorption reactions, and the availability of free sites all influence the surface density of the reacting species $>Me-A$.

Reaction rates are influenced by the stoichiometry of surface species. The relative pH determines the protonation level of unoccupied surface sites ($>MeOH_2^+$, $>MeOH$, and $>MeO^-$) and adsorbed species (such as $>Me-A$, $>Me-AH^+$, $>Me-B$, and $>Me-BH^+$). Two additional reactions can be added to the mechanism presented earlier:

$$>Me-A + H^+ \underset{k_{-a}}{\overset{k_a}{\rightleftharpoons}} >Me-AH^+ \tag{92}$$

$$>Me-AH^+ \underset{k_{-2}^*}{\overset{k_2^*}{\rightleftharpoons}} >Me-BH^+ \tag{93}$$

If k_2^* is greater than k_2, protonation increases the surface chemical reaction rate;

the reaction is acid-catalyzed. Other aspects of surface speciation may also influence reaction rates; coadsorption of major ions (Ca^{2+}, SO_4^{2-}, etc.), natural organic matter, or reaction products may be important in certain instances. The influence of coadsorption can be understood in terms of the rate constants and the relative concentrations (mass-action principle) of all species competing for available surface sites.

In addition to the ideas considered to this point a number of additional concepts and approaches are useful in working with surface chemical reactions. Two of these concern pseudoequilibrium and mass and charge balance.

Assignment of Rate-Limiting Steps and Pseudoequilibrium Steps. Once the rate-determining steps for reactions in series have been identified, reaction steps that take place before or after them can be treated as pseudoequilibrium reactions.

Mass Balance and Charge Balance. In homogeneous phase reactions, reactants, intermediates, and products are uniformly distributed in space. For surface chemical reactions, in contrast, one phase may acquire chemical species or charge at the expense of another. Consider, for example, the adsorption of anion HL^-, which imparts a negative charge to mineral surfaces:

$$>Me–OH + HL^- \text{(surface)} \longrightarrow >Me–L + H_2O \tag{94}$$

The development of negative surface charge is ultimately self-limiting. As the surface charge becomes more negative, electrostatic repulsion lowers the migration rate of HL^- anion toward the surface. Ultimately, if the negative potential of the surface exceeds 100 mV, it becomes exceedingly difficult for the surface to acquire additional negative charge. Charge accumulation limits the rates and final extent of anion adsorption.

Paying close attention to mass and charge balance is particularly useful in studying precipitation and dissolution reactions. Consider the following stoichiometric relations for the dissolution of a metal oxide (Valverde and Wagner, 1976):

$$Me(OH)_3(s) \longrightarrow Me^{3+} \text{(surface)} + 3OH^- \text{(surface)} \tag{95}$$

$$Me^{3+} \text{(surface)} \longrightarrow Me^{3+} \text{(solution)} \tag{96}$$

$$OH^- \text{(surface)} \longrightarrow OH^- \text{(solution)} \tag{97}$$

Overall reaction:

$$Me(OH)_3(s) = Me^{3+} + 3OH^- \tag{98}$$

Is it possible for reactions 95 and 96 to be fast while reaction 97 is slow? This could not take place for very long, because loss of metal ion from the surface would leave behind an accumulation of uncoordinated hydroxide ion and negative charge. In order for appreciable dissolution to take place, either (1) the outward flux of OH^- equals the outward flux of Me^{3+} (OH^- and Me^{3+} may be coordinated to one another) or (2) the outward flux of Me^{3+} is balanced by an inward flux of H^+, which balances the charge and converts OH^- into H_2O.

6. SOURCES OF KINETIC AND MECHANISTIC INFORMATION: INSIGHTS BASED ON ANALOGY

One major objective of aquatic chemistry is to predict rates and pathways of transformation for toxic pollutants added to the environment. To accomplish this, it is desirable to have a complete set of pertinent kinetic information for the pollutant of interest. Rate constants and reaction mechanisms calculated from laboratory rate studies performed over a wide range of physical and chemical conditions provide a strong foundation for predicting rates and reaction pathways in the environment. However, acquiring this information can be quite expensive, and the need for kinetic information relevant to new pollutants or to new circumstances may outstrip our ability to acquire it. As a consequence, we seek other ways of predicting rates and pathways of transformations. Descriptive chemical information may be useful for forecasting in a qualitative manner the course and final outcome of chemical pollution events. In some cases analogies with other more completely studied chemical reactions are available, which can be used as a basis for exploring connections between compound structure, chemical properties, and reactivity in particular environmental situations.

6.1. Empirical Observations

Simple empirical observations can be extremely useful in certain instances. Ship wrecks hundreds of years old have been found in anoxic sediments. In many cases, hulls, oars, and other wooden parts show only minimal degradation. The conclusion can be drawn that degradative pathways of wood under anoxic conditions are extremely slow, and that wood is essentially "inert" under these conditions. In a similar manner, identification of organic pollutants in groundwater systems decades after the last contamination episode provides important evidence concerning the environmental persistence of these compounds. Such observations provide upper limits concerning rate constants for degradation. There are many substances introduced to the environment that do not persist long enough for study of their reactivity in natural surroundings. Their transient presence allows us to place some lower limits on their rates of transformation.

6.2. Thermodynamic Information

When relative thermodynamic information is available, it can be used to determine whether a postulated transformation is feasible. Consider, for example, two possible stoichiometric reactions of manganese with hydrogen peroxide:

$$H_2O_2 + Mn^{2+} \rightleftharpoons MnO_2(s) + 2H^+ \tag{99}$$

$$H_2O_2 + MnO_2(s) + 2H^+ \rightleftharpoons O_2(g) + Mn^{2+} + 2H_2O \tag{100}$$

	$\Delta G°$ (kJ mol^{-1})	ΔG (kJ mol^{-1})
Reaction 1	-103.0	-114.4
Reaction 2	-103.2	-27.2

The Gibbs free-energy change $\Delta G°$ is calculated for standard conditions ($[i] = 1.0\ M$) while ΔG is calculated for more environmentally significant conditions ($[Mn^{2+}] = 1.0 \times 10^{-6}\ M$, $[H_2O_2] = 1.0 \times 10^{-6}\ M$, pH 7.0, 0.2 atm O_2, presence of $MnO_2(s)$]. In both cases, the Gibbs free energy is negative for the chosen conditions, indicating favorable forward reaction. Based on thermodynamic information, oxidation and reduction of manganese can occur simultaneously, bringing about the disproportionation of hydrogen peroxide into molecular oxygen and water.

6.3.   Tentative Mechanisms

Once it is known that a given overall reaction is thermodynamically favorable, the next step is to consider possible reaction mechanisms. Assuming for the moment that reactions of manganese with hydrogen peroxide have not been explored, it is useful to consider analogous reactions of other transition metals. In the classic Fenton reaction, ferrous ion (Fe^{2+}) reduces hydrogen peroxide, generating powerful hydroxyl radical ($OH^•$) oxidant [see Walling (1975)]:

$$H_2O_2 + Fe^{2+} \xrightarrow{k_1} Fe^{3+} + OH^- + OH^• \tag{101}$$

$$OH^• + Fe^{2+} \xrightarrow{k_2} Fe^{3+} + OH^- \tag{102}$$

Manganous ion (Mn^{2+}) should react with hydrogen peroxide in a similar manner, since it shares important characteristics with ferrous ion: (1) manganous ion can be readily oxidized by one (and by two) equivalents to higher valent forms, and (2) manganous ion can form an encounter complex with hydrogen peroxide in a manner analogous to ferrous ion. Reaction of ferric ion (Fe^{3+}) with hydrogen peroxide has also been extensively studied (Walling, 1975):

$$Fe^{3+} + H_2O_2 \longrightarrow H^+ + FeOOH^{2+} \quad (fast) \tag{103}$$

$$FeOOH^{2+} \longrightarrow Fe^{2+} + HO_2^• \tag{104}$$

$$HO_2^{\bullet} \longrightarrow H^+ + O_2^{\bullet} \tag{105}$$

$$O_2^{\bullet} + Fe^{3+} \longrightarrow O_2 + Fe^{2+} \tag{106}$$

As a starting point, we can postulate mechanisms for reactions of manganese with H_2O_2 and other reactive oxygen species that are analogous to reactions 101–106 determined for iron. Implicit in this approach is an appreciation of similarities and differences between the environmental chemistries of iron and manganese.

Once one or more postulated mechanisms have been formulated, we can begin identifying factors that may influence reaction rates. These factors would then be examined as part of an experimental program, and would also be a part of a complete reaction model. Important factors usually include concentrations of participating species, ionic strength, pH, temperature, and character and intensity of radiation (in the case of photochemical reactions).

An important challenge for environmental chemists is to extend reaction kinetics and mechanisms determined within a narrow range of chemical conditions examined in prior research to environmentally significant conditions. The mechanisms presented in reactions 101–106 have an important limitation; they are based on experiments performed in strongly acidic solution. Under the more nearly neutral pH conditions typical of most natural waters, Fe(III) will exist predominately as hydroxy complexes [$FeOH^{2+}$, $Fe(OH)_2^+$, $Fe(OH)_4^-$, etc.] and, most importantly, insoluble oxide–hydroxide phases [$FeOOH(s)$, $Fe_2O_3(s)$, $Fe(OH)_3(s)$, etc.]. Similarly, hydroxy complexes [$MnOH^{2+}$, $Mn(OH)_2^+$, $Mn(OH)_4^-$, etc.] and solid phases [$MnOOH(s)$ and $MnO_2(s)$, etc.] will predominate in the chemistry of Mn(III) and Mn(IV) within the pH range of natural waters. Aquatic chemists concerned with the reaction of H_2O_2 with manganese (and iron) must, therefore, deal with heterogeneous systems, and account for the effects of adsorption phenomena and surface speciation on reaction rates. Many useful analogies exist between reactions of dissolved metal ion complexes and reactions of mineral surfaces that can help guide the study of heterogeneous systems.

6.4. General Trends in Ligand Replacement Reactions of Metal Ion Complexes

Rates of ligand replacement reactions of metal ion complexes are relevant in a number of situations. Assigning a "lability" to a particular metal based on the rate of a representative ligand replacement reaction is an attractive prospect, but one that has important limitations. In this regard, rates of water-exchange reactions have been examined:

$$M(H_2O)_6^{n+} + H_2^*O \rightleftharpoons M(H_2^*O)(H_2O)_5^{n+} + H_2O \tag{107}$$

Water exchange rates can be measured by relaxation methods or by incorporation of isotope-labeled water (Cotton and Wilkinson, 1980). Metals have been categorized on the basis of their rate constants for water exchange:

Class I $(10^9-10^8 \text{ s}^{-1})$
Li^+, Na^+, K^+, Rb^+, Cs^+
Ca^{2+}, Sr^{2+}, Ba^{2+}, Cr^{2+}, Cu^{2+}

Class II $(10^8-10^4 \text{ s}^{-1})$
Mg^{2+}, Mn^{2+}, Fe^{2+}, Co^{2+}, Ni^{2+}

Class III (10^4-1 s^{-1})
Be^{2+}
Al^{3+}, Fe^{3+}, V^{2+}

Class IV $(1-10^{-8} \text{ s}^{-1})$
Co^{3+}
Cr^{3+}, Rh^{3+}, Ir^{3+}, Pt^{2+}

Water exchange of most of the metals listed is consistent with a dissociative (D or I_d) mechanism (Burgess, 1978); the rate-determining step involves cleavage of a metal ion–water bond. As a consequence, rates of water exchange generally decrease as the strength of the metal ion–water bond increases. Bond strength can be roughly approximated using the charge to radius ratio of the metal ion, or more accurately by considering the electronic structure of the aquo complex.

Do rates of other ligand replacement reactions of metals decrease in the order class I > class II > class III > class IV? Although this relationship is observed in many instances, several exceptions exist, and for important chemical reasons. The classification of metals presented above is pertinent to a very restricted set of conditions: (1) all ligands coordinated to the metal ion are water molecules, and (2) the entering and leaving ligands in each reaction are water molecules. When these conditions are not met, we are essentially comparing "apples and oranges"; different coordinated ligands, leaving ligands, and entering ligands may respond very differently to changes in the nature of the central metal ion, resulting in changes in ligand replacement rate. This point is illustrated in Table 2, which

TABLE 2. Rate Constants for Replacement of Coordinated Water by Halide Ions: Hexaquo and Monohydroxo Ions of Cr(III) and Fe(III)

Reactants	k Water Replacement Rate Constant $(M^{-1} \text{s}^{-1})$
$Cr^{III}(H_2O)_6^{3+} + Cl^-$	2.9×10^{-8}
$Cr^{III}(H_2O)_6^{3+} + Br^-$	9.0×10^{-9}
$Cr^{III}(OH)(H_2O)_5^{2+} + Cl^-$	2.8×10^{-5}
$Cr^{III}(OH)(H_2O)_5^{2+} + Br^-$	1.7×10^{-5}
$Fe^{III}(H_2O)_6^{3+} + Cl^-$	9.4×10^0
$Fe^{III}(H_2O)_6^{3+} + Br^-$	3.4×10^0
$Fe^{III}(OH)(H_2O)_5^{2+} + Cl^-$	1.1×10^4
$Fe^{III}(OH)(H_2O)_5^{2+} + Br^-$	3×10^3

collects rate constants for exchange of halide ions with bound water. Although rates of ligand replacement are typically eight orders of magnitude higher for Fe(III) than for Cr(III), (1) the first hydrolysis product of each metal $(M(OH)(H_2O)_5^{n+})$ reacts substantially more rapidly than corresponding hexaquo complexes, and (2) ligand replacement by Cl^- is faster than ligand replacement by Br^-. Thus, small changes in speciation can have a dramatic impact on rates of ligand replacement. Estimates of rates for reactions taking place through dissimilar mechanisms can be off by many orders of magnitude.

6.5.   Organic Functional Groups and Reactivity Generalizations

Functional groups are portions of molecules subject to chemical reaction. In coordination complexes, the central metal ion is a key participant in most reactions that may take place, including ligand replacement, oxidation–reduction, and photochemical reactions. With many organic compounds, the situation is somewhat different; certain functional groups or portions of molecules may be active participants in a chemical reaction, while others may be on the sidelines and have only a minor effect on the reaction. Changes in neighboring molecular structure may prevent a reaction from taking place or cause it to take a different course. A change in structure may alter reaction rates or positions of equilibrium (March, 1985).

To what extent are functional groups within molecules reactive "units" that act independently of other portions of the molecule? This is a particularly pertinent question today, as chemists attempt to develop computer algorithms for predicting physical properties, chemical properties, and reactivity based on molecular structure. Chemists apply intuition based on years of experience when deciding which neighboring group interactions can be ignored in a given situation and which must be accounted for.

Many insecticides and herbicides are in a sense "designer chemicals," which can be engineered to degrade within a specified time period. The substituent groups chosen determine which degradation pathways will be most favorable and the magnitude of rate constants. Chlorpyrifos, ronnel, and several other insecticides share a common phosphorothionate ester group, which is subject to hydrolysis:

The nature of the substituent groups (X, Y, and Z) determines hydrolysis rates of the three ester linkages (1, 2, and 3). Substituents effects arise from both electronic (inductive, field, and resonance) and steric factors (March, 1985). Interaction with the more electronegative sulfur induces a positive charge on the phosphorus atom, rendering it susceptible to nucleophilic attack (Tinsley, 1979). Electron-withdrawing substituents at the X and Y positions substantially improve the capacity of the phenolate anion to act as a leaving group, facilitating hydrolysis of ester linkage 1. Chloro or nitro substituents at the X and Y positions, for example, dramatically promote hydrolysis relative to unsubstituted forms. The aromatic nature of this leaving group favors the expression of electronic effects. Reaction is less sensitive to substituent Z, since aliphatic groups do not convey electronic effects as well. Steric interactions may arise when substituent Y is a bulky group; this substituent is close enough to the phosphorus atom that it may block nucleophilic attack. In this way, substituent groups X, Y, and Z can be selected to bring about the desired time scale of hydrolysis.

Order-of-magnitude generalizations concerning reactivity are often useful in developing models and making predictions about transformation and fate. Some reactions can be classified as fast relative to other pertinent physical and chemical processes, and can be treated as being instantaneous. Many protonation–deprotonation reactions, for example, occur at time scales signifi-cantly below one second, and can therefore be treated as pseudoequilibrium reaction steps.

Rate constants of fast reactions may have to be considered in kinetic models when the product distribution is important. Free radicals, for example, may be consumed by several competitive reactions in parallel; the relative magnitudes of the rate constants will determine the distribution of various products formed.

6.6. Linear Free-Energy Relationships (LFERs)

We investigate chemical equilibria by examining the change in Gibbs free energy ($\Delta G° = -RT \ln K_{eq}$) that accompanies conversion of reactants to products. As has already been discussed, the ACT approach to chemical kinetics also focuses on changes in Gibbs free energy; $\Delta G°^{\neq}$ is the Gibbs free energy of activation accompanying conversion of reactants to the activated complex, and is related to the reaction rate constant k using Eq. 23.

Consider, for the moment, a reactant that participates in an equilibrium reaction, characterized by K_{eq_0}, and an elementary reaction, characterized by k_0. Replacing one substituent of the reactant for another changes corresponding Gibbs free energies; K_{eq_0} is changed to K_{eq}, and k_0 is changed to k. It is reasonable to postulate that these changes might somehow be related to one another. If changes in Gibbs free energies are proportional to one another, the following relationship should apply:

$$\log(k/k_0) = \rho \cdot \log(K_{eq}/K_{eq_0}) \tag{108}$$

where ρ is a constant that relates to the sensitivity of the two reactions toward substituent replacement. One reaction (such as the one relating to K_{eq_0}) can be taken as a reference point; ρ then becomes a "reaction constant," indicating the

relative sensitivity of other reactions (such as the elementary reaction characterized by k) toward substituent effects. If ρ is greater than one, the reaction is more
sensitive to substituent effects than the reference reaction, while a value of ρ less
than one is indicative of a reaction less sensitive to substituent effects.

Similarly, the reference reaction can be used to define "substituent constants"
(σ), which characterize the effect of particular substituents on K_{eq} or k values. In
the equation presented below, K_{eq} is the equilibrium constant of the reference
reaction for a new substituent, while K_{eq_0} is the equilibrium constant of the
reference reaction for a reference substituent (such as —H):

$$\log(K_{eq}/K_{eq_0}) = (\sigma) \tag{109}$$

Values of the substituent constants can then be used to predict equilibrium
constants or rates for other reactions. Equation 108, for example, becomes:

$$\log(k/k_0) = (\rho) \cdot (\sigma) \tag{110}$$

The existence of LFERs can be understood in the following manner. Small
changes in structure that alter the free energies of reactants relative to products
cause proportional changes in the energy of the transition state (Moore and
Pearson, 1981). The fact that LFERs are *linear* means that replacing one
functional group for another or making a small change in structure has not
altered the structure or form of the transition state in an important way, only its
energy (Jones, 1979). Several forms of LFERs can be constructed, and used to
predict, in a semiempirical way, rate constants for unexplored chemical reactions
(Moore and Pearson, 1981; Hoffmann, 1981; Benson, 1982; Brezonik, Chapter 4,
this volume; Wehrli, Chapter 11, this volume).

It is important to point out that LFERs are often observed for only restricted
sets of chemical reactions. When structural changes become large enough, the
structure and form of the transition state is altered, changing the reaction rate in
ways not accounted for in the LFER equations. Changes may even be great
enough that other competing reactions may have elevated importance, altering
the overall course of the reaction. Used with caution, however, LFERs are more
and more frequently an important means of predicting rates of unexplored
reactions.

7.1. CONCLUSION

Kinetics as a discipline includes a set of concepts and approaches that are useful
in solving problems. Kinetics can be used to explore naturally occurring
processes, to assess the impact of human intervention and to develop new
practices.

It is essential that environmental questions be stated as carefully and
accurately as possible. Spatial and temporal scales need to be known; once time

scales of observation have been defined, for example, processes that are too slow to be significant can be identified and discarded from consideration. The level of certainty required must be decided on. Qualitative information can be useful in deciding whether a certain transformation might take place, but is not sufficient to represent the dynamic behavior of chemical concentrations over time. For a precise model, details concerning spatial heterogeneities (e.g., fissure distribution and sediment size in an aquifer,) and temporal heterogeneities (such as frequency and duration of rainfall or drought) must be known. A synthesis of chemical kinetic information, equilibrium information, and transport information is required to answer important questions concerning the protection and enhancement of natural water quality. The challenge is to integrate a molecular understanding of key chemical reactions with the specific characteristics of actual environmental systems in order to evolve a dynamic picture of the chemistry of natural waters.

REFERENCES

Adam, G., and M. Delbruck (1968), "Reduction of Dimensionality in Biological Diffusion Processes," in A. Rich and N. Davidson, Eds., *Structural Chemistry and Molecular Biology*, Freeman, San Francisco.

Astumian, R. D. and Z. A. Schelly (1984), "Geometric Effects of Reduction of Dimensionality in Interfacial Reactions," *J. Am. Chem. Soc.* **106**, 304–308.

Benson, S. W. (1982), *The Foundations of Chemical Kinetics*, Krieger, Huntington, NY.

Berg, H. C. (1983), *Random Walks in Biology*, Princeton University Press, Princeton, NJ.

Berner, R. A. (1980), *Early Diagenesis: A Theoretical Approach*, Princeton University Press, Princeton, NJ.

Burgess, J. A. (1978), *Metal Ions in Solution*, Ellis Horwood, Sussex, England.

Capellos, C. and B. H. J. Bielski (1980), *Kinetic Systems: Mathematical Description of Chemical Kinetics in Solution*, Krieger, Huntington, NY.

Cornell, R. M., A. M. Posner, and J. P. Quirk (1974), *J. Inorg. Nucl. Chem.* **36**, 1937–1946.

Cotton, F. A., and G. Wilkinson (1980), *Advanced Organic Chemistry*, 4th ed., Wiley-Interscience, New York.

Davis, J. A., R. O. James, and J. O. Leckie (1978), "Surface Ionization and Complexation at the Oxide/Water Interface. I. Computation of Electrical Double Layer Properties in Simple Electrolytes," *J. Colloid Interface Sci.* **63**, 480–499.

Denbigh, K. G., and J. C. R. Turner (1984), *Chemical Reactor Theory*, 3rd ed., Cambridge University Press, Cambridge, England.

Fischer, H. B., E. J. List, R. C. Y. Koh, J. Imberger, and N. H. Brooks (1979), *Mixing in Inland and Coastal Waters*, Academic Press, New York.

Furrer, G., J. Westall, and P. Sollins (1989), "The Study of Soil Chemistry through Quasi-Steady-State Models: I. Mathematical Definition of Model", *Geochim. Cosmochim. Acta* **53**, 595–601.

Gardiner, W. C. (1969), *Rates and Mechanisms of Chemical Reactions*, Benjamin, Menlo Park, CA.

Hardt, S. L. (1979), "Rates of Diffusion Controlled Reactions in One, Two, and Three Dimensions," *Biophys. Chem.* **10**, 239–243.

Hayes, K. F., A. L. Roe, G. E. Brown, K. O. Hodgson, J. O. Leckie, and G. A. Parks (1987), "In Situ X-Ray Absorption Study of Surface Complexes: Selenium Oxyanions on α-FeOOH," *Science* **238**, 783–786.

Hoffmann, M. R. (1981), "Thermodynamic, Kinetic, and Extrathermodynamic Considerations in the Development of Equilibrium Models for Aquatic Systems," *Environ. Sci. Technol.* **15**, 345–353.

Imboden, D. M., and A. Lerman (1978), "Chemical Models of Lakes," in A. Lerman, Ed., *Lakes: Chemistry, Geology, Physics*, Springer, New York.

Imboden, D. M., and R. P. Schwarzenbach (1985), "Spatial and Temporal Distribution of Chemical Substances in Lakes: Modelling Concepts," in W. Stumm, Ed., *Chemical Processes in Lakes*, Wiley-Interscience, New York.

Jennings, A. A. (1987), "Critical Chemical Reaction Rates for Multicomponent Groundwater Contamination Models," *Water Resources Res.* **23**, 1775–1784.

Jones, R. A. Y. (1979), *Physical and Mechanistic Organic Chemistry*, Cambridge University Press, Cambridge, England.

Keck, J. C. (1978), "Rate-Controlled Constrained Equilibrium Method for Treating Reactions in Complex Systems," in R. D. Levine and M. Tribus, Eds., *Maximum Entropy Formalism*, MIT Press, Cambridge, MA.

Laidler, K. J. (1987), *Chemical Kinetics*, Harper, New York.

Lasaga, A. C. (1983), "Rate Laws of Chemical Reactions," Chapter 1, and "Transition State Theory," Chapter 4, in *Reviews in Mineralogy* 8, *Kinetics of Geochemical Processes*, Mineralogical Society, Washington.

Liu, C. W., and T. N. Narashiman (1989), "Redox-Controlled Multiple-Species Reactive Chemical Transport. 1. Model Development," *Water Resources Res.* **25**, 869–882.

March, J. (1985), *Advanced Organic Chemistry*, 3rd ed., Wiley-Interscience, New York.

Moore, J. W., and R. G. Pearson (1981), *Kinetics and Mechanism*, 3rd ed., Wiley-Interscience, New York.

Morel, F. M. M. (1983), *Principles of Aquatic Chemistry*, Wiley-Interscience, New York.

Morgan, J. J., and A. T. Stone (1985), "Kinetics of Chemical Processes of Importance in Lacustrine Environments," in W. Stumm, Ed., *Chemical Processes in Lakes*, Wiley-Interscience, New York.

Motschi, H. (1987), "Aspect of the Molecular Structure in Surface Complexes; Spectroscopic Investigations," in W. Stumm, Ed., *Aquatic Surface Chemistry*, Wiley-Interscience, New York.

Motschi, H., and M. Rudin (1984), "^{27}Al ENDOR Study of VO^{2+} Adsorbed on delta-Alumina," *Colloid Polym. Sci.* **262**, 579–583.

Newman, J. S. (1973), *Electrochemical Systems*, Prentice-Hall, Englewood Cliffs, NJ.

Pankow, J. F., and J. J. Morgan (1981), "Kinetics for the Aquatic Environment," *Environ. Sci. Technol.* **15**, 1155–1164, 1306–1313.

Santschi, P. H. (1988), "Factors Controlling the Biogeochemical Cycles of Trace Elements in Fresh and Coastal Marine Waters as Revealed by Artificial Radioisotopes," *Limnol. Oceanogr.* **33**, 848–866.

Schindler, P. W., and W. Stumm (1987), "The Surface Chemistry of Oxides, Hydroxides and Oxide Minerals," in W. Stumm, Ed., *Aquatic Surface Chemistry*, Wiley-Interscience, New York, pp. 83–110.

Schwarzenbach, R. P., and D. M. Imboden (1984), "Modelling Concepts for Hydrophobic Pollutants in Lakes," *Ecol. Model.* **22**, 171.

Stone, A. T. (1986), "Adsorption of Organic Reductants and Subsequent Electron Transfer on Metal Oxide Surfaces," in *Geochemical Processes at Mineral Surfaces*, J. A. Davis and K. J. Hayes, Eds., American Chemical Society, Washington, DC, pp. 446–461.

Stone, A. T. (1987), "Reductive Dissolution of Manganese(III, IV) Oxides by substituted phenols," *Environ. Sci. Technol.* **21**, 979–988.

Stone, A. T., and J. J. Morgan (1984), "Reduction and Dissolution of Manganese(III) and Manganese(IV) Oxides by Organics," *Environ. Sci. Technol.* **18**, 450–456; *Environ. Sci. Technol.* **18**, 617–624.

Stumm, W., and J. J. Morgan (1981), *Aquatic Chemistry*, 2nd ed., Wiley-Interscience, New York.

Szabo, Z. G. (1969), "Kinetic Characterization of Complex Reaction System," Chapter 1, in C. H. Bamford and C. F. H. Tipper, Eds., *Comprehensive Chemical Kinetics*, Vol. 2, *Theory of Kinetics*, Elsevier, New York.

Tanford, C. (1980), *The Hydrophobic Effect: Formation of Micelles and Biological Membranes*, 2nd ed., Wiley-Interscience, New York.

Tinsley, I. J. (1979), *Chemical Concepts in Pollutant Behavior*, Wiley-Interscience, New York.

Valverde, N., and C. Wagner (1976), "Considerations on the Kinetics and the Mechanism of the Dissolution of Metal Oxides in Acidic Solutions," *Bunsenges. Phys. Chem.* **80**, 330–340.

Walling, C. (1975), "Fenton's Reagent Revisited", *Acc. Chem. Res.* **8**, 125–131.

Westall, J. C. (1987), "Adsorption Mechanisms in Aquatic Surface Chemistry," in W. Stumm, Ed., *Aquatic Surface Chemistry*, Wiley-Interscience, New York.

2

FORMULATION AND CALIBRATION OF ENVIRONMENTAL REACTION KINETICS; OXIDATIONS BY AQUEOUS PHOTOOXIDANTS AS AN EXAMPLE

Jürg Hoigné

Institute for Water Resources and Water Pollution Control (EAWAG), Dübendorf, Switzerland; Swiss Federal Institute of Technology (ETH), Zürich, Switzerland

1. KINETIC CONCEPTS

Within an environmental compartment physical and chemical transformations of specified chemical compounds such as "pollutants" or "probe compounds" or any other chemical species P, are generally controlled both by different environmental factors E_j, such as the activities of environmental reactants acting on them ("driving force"), and the compound-specific rate constant, $k_{j,P}$, with which the specific chemical structures of P respond to such factors j (Smith et al., 1977). Only a strict separation between the terms corresponding to environmental parameters and the chemical constants describing the chemical compound allows for easy generalization of the rate laws and for structuring of kinetic environmental models:

$$P \xrightarrow[k_{j,P}]{E_j} P_{trans} \qquad (1)$$

(where subscript "trans" = transformed).

Figure 1 presents some types of physical and chemical transformations occurring in the environment along with the elementary environmental factors acting as

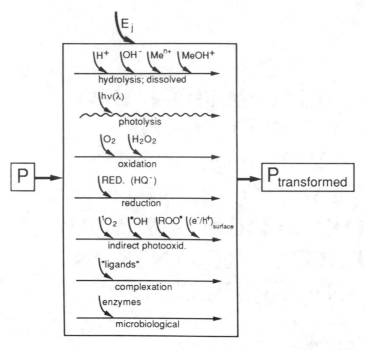

Figure 1. Examples of environmental factors, E_j, mediating chemical transformations.

driving forces. Many such reactions are discussed in this and other chapters of this volume. The general approach of separating the relevant environmental factors and the chemical constants can be applied in each of these cases. For example, the environmental factors controlling the hydrolysis rates are often the activities of H^+, OH^-, or transition-metal species. Detailed formulations of such hydrolytic reactions occurring in homogeneous solutions have recently been reviewed by Mill and Mabey (1986) and are discussed considering environmental processes by Schwarzenbach and Gschwend (Chapter 7, this volume) and Hoffmann (Chapter 3, this volume). In addition, hydrolytic reactions occurring at interfaces or leading to the dissolution of solid surfaces have been reviewed in the previous volume of this series (Stumm and Furrer, 1987; Stone and Morgan, 1987) and are now discussed for other systems by Schnoor (Chapter 17, this volume). The intensity of the solar light spectrum is an additional important environmental factor. It can lead to direct and sensitized photochemical reactions as has been shown in the previous volume (Zepp et al., 1977) and is also discussed in this volume in this chapter and that by Sulzberger (Chapter 14). Other pathways of transformations listed in Figure 1 are the oxidation reactions for which the concentration of oxygen or hydrogen peroxide act as relevant environmental factors (Hoffmann, Chapter 3, this volume). Sometimes also the photoinduced steady-state concentrations of aqueous singlet oxygen ($^1O_2$) or OH radicals (OH·) or peroxyradicals (ROO·) or solvated electrons (e_{aq}^-) can be separated as controlling reactants (this chapter). In the presence of particles of semiconducting surfaces also the light-induced band-gap excitations leading to conduction band electrons and valence band holes that can mediate photoredox reactions occurring at the surface must be considered [for examples, see Hoffmann (Chapter 3, this volume), Pichat and Fox (1989), and Sulzberger (Chapter 14, this volume)]. For thin films of water exposed to the atmosphere, even the dry deposition of ozone, the most abundant atmospheric photooxydant, must sometimes be accounted for. On the other hand, in anaerobic spheres abiotic reducing reactions may occur. Schwarzenbach (Schwarzenbach and Gschwend, Chapter 7, this volume) presents examples where reduced species such as sulfur(II) compounds in sediments form hydroquinone-type species acting as the mediators or environmental factors. In the chapter by Hering and Morel (Chapter 5, this volume) and those by Sulzberger (Chapter 14, this volume) examples are presented where complexing agents act as complementary factors driving the dissolution of solids. Finally, as indicated by the last entry in Figure 1, in microbiological reactions the activity of enzymes can be considered the environmental driving force (see Price and Morel, Chapter 5, this volume).

In natural environments some of the pathways listed in Figure 1 can be interrelated. This complicates all phenomenological predictions, although most specific cases are simplified because of the kinetic predominance of only one or two of the pathways. Therefore, we aim at separating complex reaction pathways into subsets, each of which allows for kinetic quantification and predictions of product formations. Only if these subsets of pathways correspond to the systematics of classical chemistry can the wealth of information on reaction kinetics, mechanisms, and product formations accumulated by the community of fundamental chemists and taught in classical chemistry be used. Furthermore, estimation methods must also account for separated pathways for which known structure–activity or linear free-energy relationships can be applied (see Brezonik, Chapter 4, this volume). Accordingly, the formulations of test guidelines for chemicals should require unambiguous separation between different environmental factors and the corresponding chemical data of the compound [see, e.g., Smith et al. (1977), OECD Test Guidelines up to 1987, and Lyman et al. (1982)]. Such an approach reduces the required data to a two-dimensional matrix.

A strict separation between environmental factors and chemical constants also allows better definition of the role and responsibility of "the environmental chemist" as compared with that of the "producer" or "emitter" of a chemical. As visualized in Scheme 1, it is the physical and analytical chemist working for the "producer of chemicals" or for a "potential emitter' who should provide all chemical data required to predict reaction rate constants for the different pathways of transformations and to determine the types of products produced during such transformations. The "community of environmental chemists" has the responsibility of calibrating the environmental factors and their variations with environmental changes in the different compartments; and formulating and testing the kinetic laws to be considered for describing the chemical transformations.

RESPONSIBILITIES:

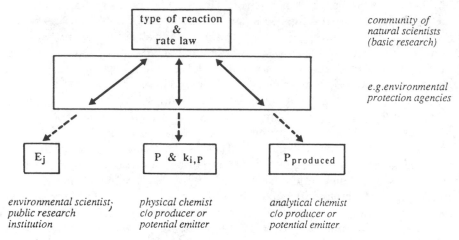

Scheme 1. Responsibilities of the environmental chemist.

Based on the model presented in Eq. 1 and Figure 1, we may formulate the rate of transformation of a specified chemical P converted by pathway j using the kinetic expression:

$$-(d[P]/dt)_j = k_{j,P}[E_j][P] \tag{2}$$

If several pathways occur simultaneously, it is the sum of the rates of all individual reactions that controls the overall loss rate of P:

$$-(d[P]/dt)_{tot} = \sum_j (k_{j,P} \cdot [E_j] \cdot [P]) \tag{3}$$

In natural systems, a pollutant compound P is generally present at a very low concentration relative to the environmental factors (or their reservoirs) considered to mediate its transformation. Thus it does not itself change the environmental factor over the course of the reaction. In this case the time function for the elimination of P versus time t in a well-mixed sphere that is closed for P (semi-batch-type or plug-flow reactor) can be characterized by the pseudo-first order expression:

$$-\ln([P]_t/[P]_0) = k_{j,P}[E_j]^1 t \tag{4}$$

Thus the logarithm of the relative residual concentration of P declines linearly with time t. The slope of the line is $k_{j,P}[E_j]^1$. If $[E_j]$ assumes a steady state, the half-life of P due to reactions in presence of E_j is

$$t_{1/2,P,j} = \ln(2)/(k_{j,P}[E_j]) \tag{5}$$

and, according to Eq. 4, the amount of P transformed by pathway j at reaction time t, $\Delta_j P(t)$, becomes:

$$\Delta_j [P]_t = [P_0] \{1 - \exp(-k_{j,P} [E_j] t)\} \tag{6}$$

If, however, the compartment is open for P (i.e., the concentration of P remains constant), then the amount of transformed P is of interest, and this increases with time such that

$$[P_{\mathrm{trans},\,j}]_t = k_{j,P} [E_j]^1 [P]_0^1 t \tag{7}$$

For example, this is the case when P is a surface of a solid that is steadily renewed during dissolution. In all these rate expressions the chemical constant $k_{j,P}$ and the environmental factor E_j are well separated.

In principle, the rate of formation of individual products can be formulated similarly by Eqs. 6 and 7 when the corresponding rate constants for the specific products formed, instead of that for the loss of the parent molecule, are considered.

1.1   The Reaction-Rate Constant $k_{j,P}$

The reaction-rate constant $k_{j,P}$ is a "chemical constant" characteristic of a compound P with general validity. It can be measured in laboratory experiments designed to isolate the effect of a single environmental factor j. Often, for practical reasons, it is determined only *relative* to that of a well-studied model compound with an absolute rate constant known for the same reaction. In case of slow reactions it is generally easy to measure absolute rate constants directly. For the study of fast reactions, sophisticated short-time measurements, such as pulse radiolysis or flash photolysis, typically combined with kinetic absorption spectroscopy or kinetic phosphorescent measurements, must be applied.

Classical chemistry literature provides comprehensive lists of rate data that can also be applied for predicting abiotic reaction rates of environmental importance when the pathways are well identified. For some reactions, such as hydrolysis or OH radical reactions, enough data are available to determine structure–reactivity correlations that enable one to interpolate within series of chemically related compounds of known reactivity [for review, see Brezonik (Chapter 4, this volume) and Lyman et al. (1982)]. The aquatic chemist is, however, more often confronted with the fact that many classical studies have been performed in organic solvents, but that the speciation of many dissolved chemicals in aquatic systems may vary with pH (caboxylic acids, phenols, etc.) and with ligand concentrations (e.g., all dissolved heavy metal species) or that the chemical species may be adsorbed on surfaces or absorbed by colloidal organic materials. In all these cases, reaction-rate constants must be determined for each individual aqueous species contributing to the overall kinetics. The equilibrium distribution of these species must also be accounted for. Reductions of the complexities often occur for real environmental situations because only one (or two) of the possible species generally dominates the apparent rate of reaction and the product formation.

1.2. The Environmental Factor E_j

Only a few types of environmental factors can be quantified by direct measurement of the concentrations of the reactants mediating the considered reactions (e.g., pH or concentration of a complexing compound). Most factors cannot be easily measured in a direct way, and a calibration assay procedure is more effective. For this the particular enrivonmental factor is calibrated by observing the rate of loss of a reference probe compound P, which must be proved to react exclusively with the considered environmental factor j. Any reference probe P chosen must have a $k_{j,P}$ value known in absolute terms or at least relative to that of other chemical compounds of interest and therefore also be a unique compound. Mixtures of chemicals quantified only by composite (group) parameters, such as total organic carbon (TOC), chemical oxygen demand (COD), total organic halogen compounds (TOX), total color-forming properties, or total infrared (IR) absorption band, contain a sum of individual compounds of different reactivities. The kinetics of the transformation of mixtures therefore depends on the actual composition of these mixtures, which varies with each environmental sample and even changes during the course of the reaction. This results in an undefined shift in the apparent rate constant, specifically, in an "aging of the apparent kinetics." Thus, environmental factors cannot be calibrated by observing the effects on composite parameters, unless in exceptional cases when zero-order kinetics apply in respect to [P] (i.e., systems kinetically "saturated" considering P). Such strict criteria for calibrating chemical process kinetics are not different from those encountered in disinfection studies, where parametrisations must also be based on observations of unique, highly specified "reference species," such as a homogeneous strain of *Escherichia coli*.

Often, $E_{j,P}$ values of secondary parameters in the sense that their concentrations reflect other, still more primary environmental factors. Such dependencies must be well analysed and accounted for. A few examples for such "secondary environmental factors" are given in Table 1 and discussed in Section 2.

An efficient development of a reaction kinetic concept requires that the environmental factor $[E_j]$ used in the rate expression only quantifies the factor directly mediating the loss of a specified P. Parameters that control the speciation of P (preequilibrium) should be separated. Therefore, in the ideal case $[E_j]$ is only a concentration (or activity) of a reactant so that the $k_{j,P}$ values in Eqs. 1–7 are second-order rate constants. To achieve such a clear separation between factors controlling the speciation of P and the kinetic constant, it is necessary that the reaction mechanism be sufficiently well understood so that the

TABLE 1. Examples of Processes and their Primary and Secondary or Composite Environmental Factors

Type of Process	Primary E_j	Secondary E_j or Composite E_j
Hydrolysis	$[H_2O]$	
	$[H^+]$	
	$[OH^-]$	
	$[Me^{n+}]$	
Direct photolysis	$I^0(\lambda)^{(a)}$; Abs. (λ); $z^{(b)}$	$I^z(\lambda)^{(c)}$
Oxidation	$[O_2]$; $[OH^-]^2$; $\cdots$	$[O_2]\cdot[OH^-]^{2\,(d)}$
Sensitized oxidations	$I^0(\lambda)$; $S^z(\lambda)$; $[DOM]$; $\cdots$	$[^1O^2]_{ss}$
		$[ROO\cdot]_{ss}$
	$[NO_2^-]$, or	
	$[NO_3^-]$, or $[Fe(III)]$;	
	$I^0(\lambda)$; $S^z(\lambda)$; $[DOM]$	$[OH\cdot]_{ss}$
Enzymatic reactions	Biomass	Enzymatic activity

(a) $I^0(\lambda)$: light intensity at surface (0 meters).
(b) z: depth of water body (meters).
(c) $I^z(\lambda)$: light intensity averaged over depth z.
(d) Environmental factor for oxidation of iron by oxygen.

crucial rate-limiting step can be identified. There still remain, however, situations in which more complex (composite) environmental factors must be worked with.

1.3.   Composite (Empirical) Environmental Factors

Type 1:   Environmental Factors Involved in the Speciation of P

Example a. The oxidation of aqueous sulfur dioxide [S(IV)] in atmospheric waters (equilibrium system of the aqueous SO_2 species H_2SO_3/HSO_3^-/SO_3^{2-}/ . . .) has been characterized by the empirical rate law

$$r_{-S(IV)} = k_{app}\,[E]\,[SO_2]_{gas} \qquad (8)$$

with

$$[E] = f([H^+]^a, [O_3])$$

Thereby $[H^+]$ controls the speciation of the aqueous S(IV), but $[O_3]$ is a factor that mediates the reaction. The resulting k_{app} based on this composite environmental factor is not a second-order rate constant anymore. A more fundamental and powerful rate law has therefore been achieved by using a kinetic expression in which the environmental factor describing the kinetics is separated from that describing the speciation of S(IV):

$$r_{-S(IV)} = \{k_{H_2SO_3} [H_2SO_3] + k_{HSO_3^-} [HSO^-] + k_{SO_3^{2-}} [SO_3^{2-}] \cdots + \cdots \} [O_3] \tag{9}$$

In this expression the speciation of S(IV) is separately calculated from equilibrium constants characterizing the solubility of SO_2 and the first and second dissociation of H_2SO_3. Thus the three individual k values in Eq. 9 become second-order reaction rate constants that can be compared with those occurring in other systems (Erickson et al., 1977; Maahs, 1983; Hoigné et al., 1985; Hoffmann, 1986). Similar separations between the kinetic environmental factor, and the environmental factor controlling the speciations of P will be applied in Section 2.

Example b. A similar case, cited in a few other chapters, is the empirical kinetic law describing the oxidation of iron(II) with oxygen (Stumm and Lee, 1961; Stumm and Morgan, 1981):

$$r_{-Fe(II)} = k_{app} [E] [Fe(II)] \tag{10}$$

which, for practical reasons, has been formulated by using the composite environmental factor

$$E = p_{O_2} [OH^-]^2 \tag{11}$$

This lumped parameter contains a factor that might be due to the pH-dependent speciation of Fe(II). For easier comparison with other systems, it would therefore be more convenient if that part of the pH effect that controls only the speciation of aqueous Fe(II) could be separated.

Example c. For surface reactions, the absorption of a chemical compound P on surface sites may be controlling the reaction rate (for reviews, see Chapter 14 by Sulzberger and Chapter 7 by Schwarzenbach and Gschwend in this volume). In such cases a separation of an adsorption parameter based on a Langmuir isotherm from that describing the rate constant of the adsorbed species will also allow for an easier comparison of kinetic data for series of different compounds.

Type 2: Environmental Factors Involved in Precursor Reactions.

In a sunlit natural surface water receiving incident light with intensity $I(\lambda)$, dissolved organic material (DOM) acts as a sensitizer for producing the singlet oxygen ($^1O_2$) (for details, see Section 2.2). In this case, the rate of photooxidation of a pollutant P with this photooxydant can be described by the kinetic expression:

$$r_P = k I(\lambda) [DOM] [P] \tag{12}$$

In this empirical kinetic expression the composite environmental factor is

$$E_P = I(\lambda) \, [DOM] \tag{13}$$

Such a composite factor does not, however, allow for any generalizations or comparisons with the second-order rate constants found in compilations for $^1O_2$ and applied in classical chemistry. Therefore, we prefer a rate expression such as

$$r_P = k[^1O_2]_{ss} \, [P] \tag{14a}$$

using

$$[^1O_2]_{ss} = \int_\lambda \{k(\lambda) \, I(\lambda)\} \, [DOM] \ldots \tag{14b}$$

This formulation has the advantage that it is based on the steady-state concentration (subscript ss) of an environmental factor that is also applied in other fields of chemistry and for which fairly comprehensive lists of rate data are or will become available (see Section 2.2).

Type 3: Environmental Factors Including (Autocatalytic) Chain Reactions.

The oxidation induced by ozone is often controlled by a preceding chain reaction that leads to the decomposition of ozone to a more reactive secondary oxidant, OH•. This chain reaction, in which radicals act as chain carriers, is promoted by certain types of solutes but inhibited by others. Therefore, the overall oxidation rate often increases with the ratio of the concentration of the promoter relative to that of the inhibitor. However, a more generally useful treatment would involve treating each reaction step separately and relating it to individual and known reaction steps of OH• (Staehelin and Hoigné, 1985).

2. EXAMPLES: PHOTONS AND PHOTOOXIDANTS AS ENVIRONMENTAL FACTORS

In this section a few examples are given to demonstrate how the kinetic concepts introduced above can be applied and where complications must be accounted for.

During a cloudless summer noon hour, surface waters receive approximately $1 \, kW \, m^{-2}$ of sunlight, or about 2 mol of photons per square meter within the wavelength region of 300–500 nm of interest for photochemical reactions (Fig. 2). Within 1 year about 1300 times this dose is accumulated (Haag and Hoigné, 1986). This irradiation can be considered as a primary environmental constant. A large portion of these photons is absorbed by dissolved organic material (DOM) present in natural water. In addition, a portion can interact with organic and inorganic surfaces of particles and a small fraction of short-wavelength light with

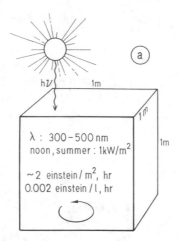

Figure 2. Solar radiation. Mean dose intensity in a mixed 1-m water column in which all light is absorbed. (1 einstein = 1 mol photons). [Adapted from Hoigné et al. (1989).]

nitrate or nitrite or even with "micropollutants" and, mainly in atmospheric waters, also with Fe(III) species.

The resulting rate of interactions between photons and absorbers is very high: assuming that most of the photons of all wavelength ranges are absorbed in a well-mixed 1-m water column, about $2 \, \text{mmol} \, L^{-1} h^{-1}$ of interactions occur between photons and absorbing substrates. In cases where DOM is the main absorber, and assuming an average chromophore unit weight of $120 \, \text{g} \, \text{mol}^{-1}$ in water containing 4 mg of dissolved organic carbon (DOC) per liter, the concentration of absorber unit is 0.033 mM. Thus, each chromophore of DOM is excited about once a minute. Even if only a few of these interactions lead to a chemical reaction, high rates of transformations of chemicals are to be expected. Compared with this, chromophores of other compounds may exhibit a lower overlap with the spectrum of the solar light and absorb a correspondingly smaller fraction of the spectrum and therefore exhibit a correspondingly lower rate of excitations.

As discussed below, some of the excitations of the light absorbers will lead to direct photolysis. In addition, as shown in Eq. 15, dissolved organic material, nitrate (or nitrite), Fe(III) species, and some minerals act as sensitizers or precursors for the production of highly reactive intermediates (so-called 'photoreactants') such as singlet oxygen ($^1O_2$), OH radicals (OH•), DOM-derived peroxy radicals (ROO•), solvated electrons (e_{aq}^-), superoxide anions (O_2^-), triplet states of humic compounds, electron–hole pairs on semiconducting surfaces, reduced transition-metal species, and hydrogen peroxide. In chlorine containing waters even Cl• and OH• would have to be considered as photoreactants. Of these photoreactants only reduced transition-metal species and H_2O_2 accumulate (hours), because of their relatively low reactivity. All other species are highly reactive and short-lived (microsecond range). In spite of the low steady-state concentrations they assume during irradiation, however, they can still be important secondary environmental factors indirectly causing photochemical transformations. [For example, any one standing in a heavy rain become, wet quickly, although this atmospheric environment contains in the order of only 1 ppm (vol/vol) of water.]

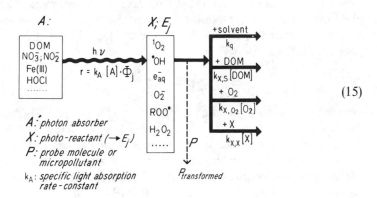

$$(15)$$

A.: *photon absorber*
X: *photo-reactant ($\rightarrow E_j$)*
P: *probe molecule or micropollutant*
k_A: *specific light absorption rate-constant*

2.1. Direct Photolysis

The rate of the primary (direct) photolysis for a species P in an irradiated water can be described by combining the photolytic factors characteristic of the chemical compound itself $(\Phi_\lambda \varepsilon_\lambda)$ and the relevant environmental factor, which in this case is the spectral light intensity (I_λ):

$$-(d[P]/dt)_{\text{direct}} = \int_\lambda \{c\, I_\lambda \Phi_\lambda \varepsilon_\lambda\} d\lambda \, [P] \qquad (16)$$

where

[P] = concentration of species P (M)

Φ_λ = primary quantum yield at wavelength λ

I_λ = spectral light intensity at wavelength λ (mole photons $\text{cm}^{-2} \, \text{nm}^{-1} \, \text{s}^{-1}$)

ε_λ = decadic molar absorption coefficient of P at λ $(M^{-1} \, \text{cm}^{-1})$

$c = 2.3 \times 1000 \, \text{cm}^3 \, \text{L}^{-1}$ for accounting that ε is a $\log_{10}$ unit and I_λ is expressed in cm^{-3} units, but concentration is in L^{-1} units.

From Eq. 16, the rate of the photochemical transformation of P can also be calculated using Eqs. 4–7 considering that

$$E_{\text{dir photol}} = I_\lambda$$

$$k_{\text{dir photol,}} \; P = c \; \Phi_\lambda \varepsilon_\lambda$$

Rates of direct photochemical reactions of aqueous species P as measured in small (thin-layered) samples and corrected for flat-layer exposure correspond to surface values or rates that would occur in the top few centimeters of a water column (Haag and Hoigné, 1986). For mixed water columns of greater depths (z), the mean rate of direct phototransformation must be corrected considering a screening factor (S_λ^z) that accounts for the light attenuation within the water column of depth z:

$$S_\lambda^z = \left(\frac{1 - 10^{-\alpha z}}{2.3} \right) \alpha_\lambda z \tag{17}$$

where α_λ is a diffuse decadic attenuation coefficient and z is depth of the mixed water columns (m).

For water columns that are more rapidly mixed than relevant depletions of P occur, this factor can be estimated for given water depth and measured diffuse attenuation coefficients (Zepp, 1980). The screening factor calculated using the extinction coefficient measured in a 10-cm UV cell has been compared with *in situ* field measurements performed in Vierwaldstättersee before spring algae growth (Bossard, EAWAG laboratories in Kastanienbaum, 1989). Based on such tests, a screening factor can be estimated by approximating α with the absorbance measured in a 10-cm UV spectrophotometer cell (vertical light path).

2.2   Singlet Oxygen Concentration as an Environmental Factor

In surface waters the photoinduced electronic excitation of ground-state (triplet) oxygen molecules ($^3O_2$) to the short-lived singlet oxygen ($^1O_2$) is sensitized by the dissolved organic material (DOM); $^1O_2$ has a chemical reactivity quite different from that of ground-state oxygen (see Schwarzenbach and Gschwend, Chapter 7, this volume). DOM is the dominant dissolved absorber of light in all natural surface waters. Even the photon energy of light of a wavelength of up to 700 nm is sufficient to be absorbed by it and produce the excitation required for the energy transfer to oxygen. The overall reaction can be formulated as

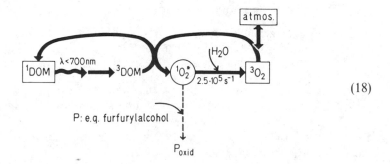

$$\tag{18}$$

Although $^1O_2$ in water is quenched quickly (half-life of 3 μs), a significant steady-state concentration can be achieved during sunshine. Using a model compound that reacts specifically with $^1O_2$, such as furfuryl alcohol (FFA), laboratory experiments have shown that the steady-state concentration of singlet oxygen increases proportionally to the amount of light absorbed by the DOM. Therefore, light in the wavelength region up to $\lambda < 500$ nm is photochemically active (Haag et al., 1984). In natural surface waters the DOM itself is not the sink for $^1O_2$. Instead, $^1O_2$ is quenched by water. In summer noon sunlit surface water of the Greifensee the rate constant for the phototransformation of furfuryl alcohol is $0.03\,h^{-1}$ (half-life of 20 h) (see Table 2). The environmental factor mediating this reaction, $[^1O_2]_{ss}$, can be calculated from this value and the second-order rate constant known for FFA (Eq. 5). Its mean value for water columns of greater depths can be estimated from this surface value using the screening factor accounting for the wavelength dependence of both the absorption of light in the water and the wavelength-dependent yield factor for $^1O_2$ production. Table 2 compares results of such estimates for one meter depth with the surface values. Because most chemically effective light is absorbed within the top meter, any further depth of the mixed water columns would only have to be considered as a "dilution" of what occurs in the top meter. Compared to samples of water from Greifensee, other waters containing higher concentrations of DOM yield a corresponding higher surface value of $[^1O_2]_{ss}$. The $[^1O_2]_{ss}$ averaged over 1 m $[^1O_2]^{1\,m}$ should, however, not vary with the concentration of DOM for such types of water because the higher light absorbance at the surface due to higher DOM is just compensated by lower light penetration into the water body. Alternatively, the same conclusion is reached when assuming that all light is absorbed by absorbers that produce $^1O_2$ with the same quantum efficiency. The experimental finding that the 1-m averages for $[^1O_2]_{ss}$ differ for different lakes and rivers is due to the fact that these quantum efficiencies vary with the type of the DOM (Haag et al., 1984) and that those types of DOM that

TABLE 2. Half-Lives of Added Furfuryl Alcohol and Corresponding Steady-State Concentrations of $^1O_2$

Water Source	DOC (mg L^{-1})	$t_{1/2}$FFA [a] (h)	$[^1O_2]_{ss}^{0m}$ [a] [b] ($M \times 10^{14}$)	$[^1O_2]_{ss}^{1m}$ [a] [c] ($M \times 10^{14}$)
Greifensee	3.5	20	8	4.6
Etang de la Gruyère	13	6	28	1.5
Rhine, Basel	3.2	27	6	3.6
Secondary effluent[d]	15	14	11	2.2

Data from Haag and Hoigné (1986).
[a] Summer midday sunlight, 1 kw m^{-2} h^{-1}.
[b] Concentration at the surface.
[c] Concentration averaged over 1-m depth.
[d] Communal wastewater.

exhibit higher specific light absorptions seem to have somewhat lower quantum efficiencies for singlet oxygen formations (Frimmel et al., 1987; Haag and Hoigné, 1986; Zepp et al., 1977).

In laboratory experiments, the rate constants for series of potential chemical pollutants reacting with $^1O_2$ could be determined relative to that of FFA or comparable reference compounds (Scully and Hoigné, 1987). This list, containing a few reaction rate constants, can be extended by correctly interpreting further rate constants tabulated in comprehensive lists on singlet oxygen reactions known from other fields of research (Wilkinson and Brummer, 1981). However, some limitations must be carefully considered: (1) the quoted rate constants often include physical quenching as well as the chemical reaction of singlet oxygen, and (2) most data have been measured in organic solvents. As discussed in Section 1.1, such compilations therefore do not include all chemical species of relevance in water. They do not, for instance, consider deprotonated acids and phenols, which are even more easily attacked by the electrophilic photooxidants than are the nondeprotonated species (Haag and Hoigné, 1986).

The rate constants $k_{1O_2,P}$ (left scale) for a few model pollutants have been plotted against pH in Figure 3. The right scale gives the half-life of these pollutants in presence of a singlet oxygen concentration determined for summer sunlit Greifensee Lake, used here as a reference. The scales at the right have been

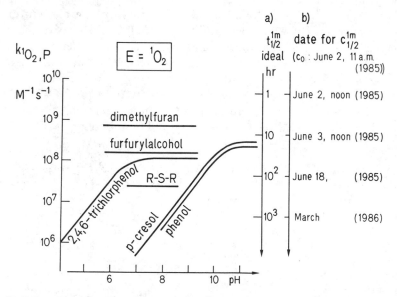

Figure 3. Comparison of rate constants for reactions of $^1O_2$ (left scale) and sunlight irradiation times required for solute eliminations (right scales, 1 m average depth) versus pH. Data are from Scully and Hoigné (1987). (*a*) Half-life of selected pollutants in Greifensee water during exposure to June midday sunlight (1 kW m^{-2}) yielding $[^1O_2]_{ss}^{1m} = 4 \times 10^{-14}$ M (Haag and Hoigné, 1986). (*b*) Scale of dates when the concentration of the chemicals becomes reduced to 50% of its initial value. These dates are an estimate based on the real sum curve of measured solar irradiations in Dübendorf, Switzerland, starting on a clear summer day (i.e., June 2, 1985, 11 A.M.). [Adapted from Hoigné et al. (1989).]

calibrated with the rate data for the transformation of dimethylfurane or FFA determined in sample tubes and evaluated according to Eq. 6. The examples show that in those cases where $^1O_2$ reacts only with deprotonated species (e.g., phenolate anions), the apparent rate constants decrease in the pH region below the pK_a of the chemical by one order of magnitude per pH unit decrease. This corresponds to the increased degree of masking the reactive anions by protonation. Comparing this pH dependence of the apparent rate constant with the pH-dependent degree of dissociation of the chemical, a reaction-rate constant for both the deprotonated and nondeprotonated species can be deduced (Scully and Hoigné, 1987).

A thorough analysis of all the products formed in reactions of singlet oxygen with organic molecules is known for only a few cases. However, singlet oxygen is so selective that it reacts only with very special functional chemical structures such as are present in 1,3 dienes or polycondensed aromatic hydrocarbons (with delocalized π-electron bonds) or in sulfides or trialkylamines. Even in the primary oxidation products, further reactive sites are mostly missing. These primary products are therefore generally resistant to further singlet oxygen reactions. The expected types of products can thus be predicted from chemical experience.

We may conclude that all concepts given in Section 1 can be successfully applied for analyzing and generalizing the effects of singlet oxygen.

2.3. OH Radical Concentration as an Environmental Factor

Hydroxyl radicals are produced in the aqueous environment by photolysis of nitrite, nitrate, and aqueous iron complexes or, in water treatment, from photolysis of hypochlorous acid and catalytic decay of aqueous ozone and, very slowly, from the UV photolysis of hydrogen peroxide. [For recent publications giving comprehensive reviews on these different pathways of OH radical formations, see Zepp et al. (1987a), Faust and Hoigné (1990), Nowell and Hoigné (1989), Hoigné (1989) and Sturzenegger (1989). The kinetically dominant sink for OH• in most surface waters is DOM. Only in cases of very pure waters ($DOC/[HCO_3^-] < 0.3$ mg mmol^{-1}) does bicarbonate contribute significantly to scavenging OH radicals (Hoigné and Bader, 1978, 1979)]. The scheme presented in reaction 19 gives a model for the case where the nitrate is the precursor:

$$\boxed{NO_3^-} \xrightarrow{\lambda \sim 320nm} \overset{NO_2}{\wedge} O^- \xrightarrow{H^+} \overset{\textstyle\boxed{DOM}}{\underset{\sim 10^5 \text{ s}^{-1} \ *}{\odot \dot{O}H}}$$

P: e.g. butylchloride

P_{oxid}

(19)

* for Greifenseewater (DOC ~ 4 mg/l)

Although in most surface water OH• is consumed quickly (half-life of a few microseconds), kinetically significant concentrations can still occur during sunshine, provided the concentration of a suitable precursor substance (such as nitrate) is high relative to scavenger concentrations. Reference compounds P used for determining the steady-state concentration of OH radicals in laboratory model experiments, which are not photodegraded or reactive with singlet oxygen, or peroxy radicals, or ozone include alkanes, chloroalkanes, or even benzene, toluene, and nitrobenzene (Hoigné and Bader, 1978, 1979; Haag and Hoigné, 1985; Zepp et al., 1987a; Nowell and Hoigné, 1990). The choice of a suitable probe compound was made in each case according to analytical convenience and system conditions. From the quantum yield of OH radical

TABLE 3. Photoreactants as Environmental Factors E_j in a Surface Water

A. Dependencies of E_j on Water Composition and Depth

E_j	Functionalities of E_j	
	at Surface	1-m Layer[a]
$[^1O_2]_{ss}$	$\propto [DOM]$	Independent[b,c]
$[OH \cdot]_{ss}$	$\propto [NO_3^-]/[DOM]$	$\propto [NO_3^-]/[DOM]^2$
$[ROO\cdot]_{ss}$	$\propto [DOM]$	Independent[b,c]
$[e_{aq}^-]_{ss}$	$\propto [DOM]/[O_2]$	$\propto 1/[O_2]$

[a] Light screening by suspended particles is neglected.
[b] Independent of [DOM].
[c] For $DOC < 5\ mg\,L^{-1}$.

B. Numerical Examples for Lake Greifensee, June Noon Sunlight ($1\ kW\,m^{-2}$)

E_j	Probe or Reference P	$k_{j,P}$ (M^{-1} s^{-1})	$k_{j,P}[E_j]_{ss}^{0m}$ (% h^{-1})	$[E_j]_{ss}^{0m}$ (M)	$[E_j]_{ss}^{1m}$ (M)	
$[^1O_2]_{ss}$	Furfuryl alcohol	1.2×10^8	3	8×10^{-14}	5×10^{-14}	(e)
$[OH\cdot]_{ss}$	Butyl chloride	3×10^9	0.2	2×10^{-16}	4×10^{-17}	(f)
$[ROO\cdot]_{ss}$	Trimethyl-phenol	(?)[d]	15	(?)[d]	(?)[d]	(g)
$[e_{aq}^-]_{ss}$	CCl$_4$	3×10^{10}	0.13	1.2×10^{-17}	5.2×10^{-18}	(h)

[d] Assumed relevant wavelength region = 355 nm for corresponding screening factor.
[e] Data from Haag and Hoigné (1986).
[f] Data from Zepp et al. (1987a).
[g] Data from Faust and Hoigné (1987).
[h] Data from Zepp et al. (1987b).

formation from nitrate photolysis, the solar terrestrial light spectrum, and the chemical composition of the water, the transformation of butyl chloride (reference) during noon summer sunshine at the surface of Greifensee Lake used as a reference ($[NO_3^- -N]/DOC = 0.4$ mg mg^{-1}) becomes 0.002 h^{-1} [see Table 3B and Zepp et al. (1987a)]. From this the steady-state concentration of the environmental factor, $[OH\cdot]_{ss}$ can be estimated using Eq. 5.

The kinetics of reaction of OH radicals with interfering sinks has been extensively calibrated for waters in which OH$\cdot$ was produced from decomposed ozone or UV-photolyzed hydrogen peroxide (Hoigné, 1988; Hoigné and Bader, 1978). Combining such information, we can predict that the surface value of $[OH]_{ss}$ should increase with increasing concentration of the precursors, [e.g., nitrate, aqueous chlorine, or Fe(III)] relative to the concentration of DOM (DOC). Only in case of waters that are low in DOC must bicarbonate also be considered as an additional scavenger. If such exceptionally clean waters are neglected, the 1-m value of $[OH\cdot]_{ss}$ would decrease proportionally to the concentration of DOM to the second power, because both the fraction of the light that is screened by the DOM and the rate of OH radical scavenging are linear functions of the concentration of DOM (see entry in Table 3A).

Comprehensive lists of absolute rate constants of OH radicals reacting with hundreds of aqueous solutes are given in the literature (see e.g., Buxton et al., 1988). From these the rate of transformation relative to that of the reference applied can be estimated. A few examples are presented in Figure 4.

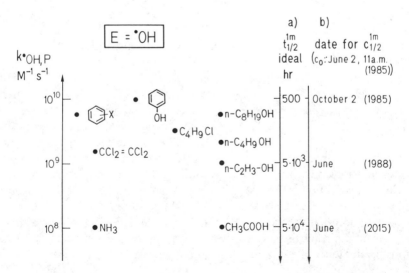

Figure 4. Compilation of rate constants for reacting of OH radicals (left scale) and sunlight irradiation times required for solute elimination (right scales). Definitions as in Figure 3. Rate constants for tetrachloroethylene and butylchloride are from our own measurements of competition kinetics (Hoigné and Bader, 1979); other data are from Farhataziz and Ross (1972) and Buxton et al. (1988). The mean half-life at the surface would be about 20 times shorter because of the absence of light screening by DOM. Assumption $[\cdot OH]_{ss}^{1m} = 1 \times 10^{-17}$ M (Zepp et al., 1987).

The concept applied here corresponds to the one that is so successfully applied for quantifying the oxidizing role of atmospheric $[OH\cdot]_{ss}$ acting in the troposphere. In contrast to the minor occurrence in the aqueous phase, OH radicals are the dominant tropospheric photooxidants. In the troposphere OH is predominantly produced from photolyzed ozone followed by $O(^1D)$ reacting with H_2O. It is scavenged by CH_4 (the main organic material of the atmosphere) and CO. The $[OH\cdot]_{ss}$ in this sphere has also been determined by investigating the kinetics of the disappearance of a reference probe compound. There, generally the ubiquitous 1,1,1-trichloroethane has been used as a reference as this is refractory with respect to other tropospheric environmental factors and reacts relatively slowly with $OH\cdot$. Rate constants of large series of atmospheric pollutants reacting with $OH\cdot$ are known or can be estimated from well-developed structure–reactivity relationships (Atkinson, 1986).

The products formed by reactions of OH radicals with aqueous organic compounds can hardly ever be analyzed in detail. $OH\cdot$ is highly nonselective and can therefore abstract an H atom from many sites of an organic molecule or add to any C–C double bond or accept an electron. From all primary reactions a free radical is formed, which reacts at a nearly diffusion-controlled rate with a second radical. This partner is mostly oxygen, a biradical, present in much higher concentrations than any other radical in natural water. From an addition of this biradical yet another radical (peroxyradical) is formed. The succeeding radical transformation reactions and radical–radical reactions then lead to a wide spectrum of different products. The case where OH reacts primarily with biocarbonate, which then acts as a secondary weakly reacting oxidant, could also be of relevance for the transformation of some exceptionally easily oxidizable compounds (Larson and Zepp, 1988; Neta et al., 1988).

Many examples of product formations due to OH radical reactions have been analyzed by radiation chemists who produce OH radicals during high-energy radiolysis of water. However, because of the nonselectivity of the primary reactions and the large number of bimolecular and unimolecular secondary transformations of the radicals formed, only specific examples can be analyzed, which may serve for only general information and instruction (von Sonntag, 1986; Schuchmann and von Sonntag, 1987). Complementary to basic research, such as questions regarding the acceptability of high-energy irradiation applied for food conservation, many studies have been undertaken to determine the wholesomeness of aqueous solutions in which radiolytically produced OH radicals are dominant reactants. Results from such studies could be applied to environmental problems if the different groups of the chemical community would formulate their results in terms of comparable systems of unit reactions.

2.4. Solvated Electron Concentration as an Environmental Factor

A hypothesis describing the source and sink of photolytically produced solvated electrons (e_{aq}^-) is given in Eq. 20:

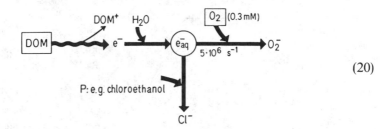

$$(20)$$

Flash-photolysis experiments in natural water samples have shown significant formation of a transient exhibiting spectroscopic and kinetic characteristics of e_{aq}^-. However, no evidence for such concentrations of e_{aq}^- was found in the transformation rate of typical e_{aq}^- probe compounds applying continuous solar irradiation of natural surface waters [for review and discussion, see Zepp et al. (1987b) and Cooper et al. (1989)]. For these experiments, chlorinated organic compounds have been applied for selectively probing for e_{aq}^-. These probes scavenge e_{aq}^- in a diffusion-controlled reaction to produce chloride anion that can be measured directly. However, the steady-state concentration of e_{aq}^- is low because the formation rate of e_{aq}^- is low and e_{aq}^- is quickly scavenged by dissolved oxygen. Therefore for such measurements a high enough concentration of the probe compound was applied to scavenge all electrons released during the illumination. The steady-state concentration of solvated electrons was then estimated from the measured formation of chloride anion considering that the lifetime of e_{aq}^- in natural waters and in absence of such scavengers is controlled by the rate with which the electrons are scavenged by dissolved oxygen (half-life $\sim 0.1\ \mu s$).

From such results we could deduce that the value of $[e_{aq}^-]_{ss}$ for Lake Greifensee water (reference) during June midday sunshine ($1\ kW\ m^{-2}$) is about $1.2 \times 10^{-17}\ M$. The effect of such an environmental factor on reactive micropollutants can be estimated by combining this value with second-order reaction-rate constants found in comprehensive tables (Buxton et al., 1988). Illustrations are given in Figure 5. and in Table 3B. Trichloroacetic acid and chloroform, both of which react at nearly diffusion controlled rates, would be degraded at a rate of only $0.13\%\ h^{-1}$ even when at the surface. Estimates for deeper water columns are not possible until the wavelength dependence of the quantum efficiency of the production of solvated electrons is also known. Such information is difficult to achieve and might not be too relevant when considering the small effect this photooxidant has for the transformation of pollutants.

Although DOMs are presumably the main precursors for photolytically produced solvated electrons, reactions with oxygen constitute the major sink. At the surface and constant oxygen concentration the value of $[e_{aq}^-]_{ss}$ will therefore increase with the concentration of DOM. Correspondingly, as discussed for the case of singlet oxygen, the mean concentration of this environmental factor, when averaged over a water depth in which most photoactive light is absorbed by DOM, becomes independent of the DOM (see Table 3A).

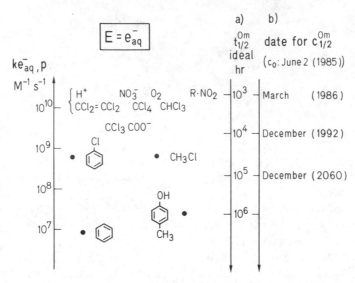

Figure 5. Comparison of rate constants for surface-level reactions of solvated electrons (left scale), (Anbar et al., 1973 and Buxton et al., 1988) and sunlight irradiation times required for solute elimination (right scales); sunlight intensity and definitions as in Figure 3. The mean half-life within a steadily mixed 1-m layer would be about five times longer because of light screening by DOM. Assumption: $[e_{aq}^{-}]_{ss}^{0m} = 1.2 \times 10^{-17}\ M$ (Zepp et al., 1987b).

Solvated electrons combining with oxygen form superoxide anion (HO_2/O_2^{-}). Because added electron scavengers, even when added in high concentrations, did not affect the formation rate of hydrogen peroxide in illuminated Greifensee Lake water, the solvated electron itself is not considered a main precursor for hydrogen peroxide formation in this water (Sturzenegger, 1989).

2.5.   Organic Peroxy Radical Concentration as an Environmental Factor

For measuring the steady-state concentration of organic peroxy radicals (ROO•) produced in sunlit natural waters series of "antioxidants," such as poly-methylphenols, have been successfully applied as selective probe compounds (Faust and Hoigné, 1987). The rates of transformation have shown that the steady-state concentration of the apparent photooxidants increases with the amount of light absorbed by the DOM. The sink for ROO• has not been identified, but kinetic evidence is that DOM, even when the DOC amounts up to $5\ mg\,L^{-1}$, does not control the lifetime of the peroxy radicals. The following reaction scheme summarizes the results of the kinetic analysis for peroxy radical formation:

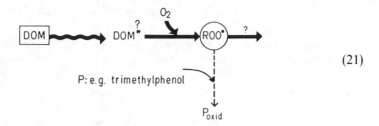

$$(21)$$

Because the environmental factor determined by the loss rate of antioxidants seems to be a mixture of DOM-derived peroxy radicals of different reactivity, absolute concentrations and absolute rate constants must be estimated with caution.

The steady-state concentration of the DOM-derived peroxy radicals present at the surface of Lake Greifensee water during noon summer sunshine (reference) transforms 2,4,6-trimethylphenol, for example, with a rate constant of 15% h^{-1}.

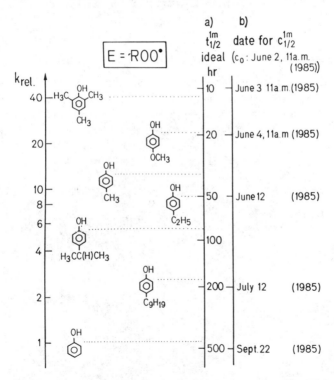

Figure 6. Comparison of experimentally determined relative rate constants for ROO· radicals (left scale) and sunlight irradiation times required for solute eliminations in Greifensee Lake (right scales); see Figure 3 for sunlight intensity and definitions. Half-life scales are depth-averaged values for a well-mixed 1-m water column, calculated by using Eq. 17 with a screening factor based on α_{366} $=0.01$ cm^{-1}. Calculations using $\alpha_{313}=0.022$ cm^{-1} predict values to be ~ 1.8 times the half-lives given. [All data from Faust and Hoigné (1987).]

The rates of reaction of other compounds were also measured relative to this reference compound. Results of such comparisons are presented in Figure 6. Empirical correlations between the kinetic effects of fulvic acid derived peroxy radicals and those of unique reference peroxy radicals indicate that even for this case structure–reactivity relationships can be applied for estimations. However, correlations established for waters containing different types of DOM appear to be somewhat different (Faust and Hoigné, 1987). In addition, because the light screening effects depend on the wavelength, depth functions for different water bodies would require information also on the wavelength dependences of the formation of this environmental factor.

We conclude that this environmental factor consists of a mixture of different types of DOM derived peroxyradicals. Therefore the absolute second-order reaction rate constant and the absolute concentration of ROO• molecules cannot be separated from each other for this group of photooxidants. This is different to the cases of $^1O_2$ or OH or e_{aq}^-, where the photoreactants are unique entities for which absolute calibrations are possible. This conclusion must also be considered when structuring modelling programs.

2.6. Dry Deposition of Atmospheric Photooxidants as Environmental Factors

If we assume a sunlit summer atmosphere containing 10^{12} ozone molecules per cubic centimeter (~ 100 μg ozone m^{-3}), but only 1000 times fewer hydroperoxy radicals (HO$_2$•) and even 1,000,000 times fewer OH radicals, we may conclude that the flux of atmospheric oxidants to a liquid surface is due mainly to dry

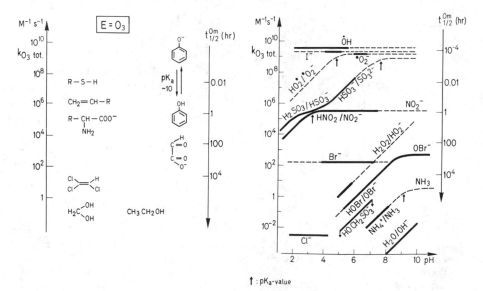

Figure 7. Comparison of rate constants for reactions of ozone in aqueous films (left scales) and hours of reactions required and half-life scales (right) for the given pollutants in presence of 10^{-9} M ozone (aqueous equilibrium conc. for gas-phase conc. of 10^{12} molecules cm^{-3}) [Data from Hoigné and Bader (1983) and Hoigné et al. (1985).]

deposition of ozone, although the deposition velocity is expected to be low because of the low solubility of ozone in water. If this, for example, should occur with a rate of only $0.05\,cm\,s^{-1}$ (Galbally and Roy, 1980; Broder and Gygax, 1985; Walcek, 1987), then the flux of ozone to the surface would still achieve $0.1\,mg\,m^{-2}\,h^{-1}$. In this case, a film of water of 0.1-mm depth would receive a mean dose rate of up to $1\,mg\,L^{-1}$ of ozone per hour $(20\,\mu M L^{-1}\,h^{-1})$.

The reactions controlling the lifetime of ozone at the surface depend on the chemical composition of the aqueous film. If this is not specified, it is therefore not possible to estimate the steady-state concentration for this photooxidant in the aqueous phase. However, the chemical effect of ozone in films or atmospheric droplets can be estimated by assuming that the steady-state concentration of ozone is still in equilibrium with the concentration of the ozone in the atmosphere. An atmospheric concentration of ozone of 10^{12} molecules per cubic centimeter would result in an aqueous equilibrium concentration of about 1 nanomolar (20°C) as the environmental factor to be considered in the aqueous film. Figure 7 gives examples for the rate constants for different types of compounds and a scale for the corresponding half-lives of these compounds exposed to the estimated concentration of ozone. (For further rate constants see Neta et al., 1988 or Hoigné and Bader, 1983 and Hoigné et al., 1985).

2.7.   Other Photoreactants as Environmental Factors

For modeling the transformation of special types of aqueous solutes in surface waters, even the steady-state concentrations of DOM-derived excited triplet states could be important (Zepp et al., 1985). Hydrogen peroxide accumulates in many photolyzed aquatic systems [for review, see Cooper et al. (1988, 1989) and Sturzenegger (1989)]. It is, however, of very low reactivity toward most chemicals appearing in aerobic environments, unless transition metals or enzymes, such as peroxidase, initiate an activation. So far, we have not found evidence for significant effects of these and further photoreactants on the elimination of aqueous solutes when these are present in filtered waters. In unfiltered waters the possibility of photoinduced surface reactions occurring in presence of algae or inorganic particles should, however, also be considered (Sulzberger, Chapter 14, this volume; Zepp and Schlotzhauer, 1983). Critical testing for all factors is necessary if the behaviors of chemicals in further types of waters or fates of other types of pollutants of less common structures are to be predicted.

2.8.   Relative Roles of Different Photooxidants

The dependences of the environmental factors on DOC or other relevant water parameters are given in Table 3A. Because these vary for the different photo-reactants (even the dimensions of the functions vary!), only specified cases can be compared with each other. Therefore, surface values of rate constants with which appropriate probe compounds are transformed by the various photoreactants present under summer sunshine in Lake Greifensee Water (reference) are summarized in Table 3B primarily for comparison. The corresponding surface steady-state concentrations of the photoreactants are deduced by considering the rate constants of the list references. Taking into account the depth dependence of light penetration and the specific relationships of the environmental factors with water composition, an average value for the 1-m depth average is also estimated. Greater depths would only contribute to a further dilution effect.

On the basis of the calibrated steady-state concentrations (given in Table 3B) and known lifetimes (given in the text) of the specified photoreactants, we can estimate the total production of photoreactant per kilowatt-hour of absorbed light or per unit area and year (about $1300 \, kW \, h \, m^{-2}$). The resulting values listed in Table 4 are remarkably high. Comparison of the yearly total production of photoreactants with the amount of DOM present per square meter of the water column indicates that these reactants may also be important in the aging of DOM. For example, a lake containing $4 \, mg \, L^{-1}$ of DOC as "DOM molecular units" with a molecular weight of $120 \, g \, C \, mol^{-1}$ contains $0.3 \, mol \, m^{-2}$ of such DOM units in a 10-m-deep water column. Comparison of this value with the amounts of photoreactants produced shows that 10% of the DOM molecular units would react, for example, with •OH radicals in one year. If, in addition, we assume a dry deposition of atmospheric ozone ($100 \, \mu g \, m^{-3}$) with a deposition velocity of $0.05 \, cm \, s^{-1}$ during 5000 "ozone deposition-effective hours" per year, we can conclude that this oxidant will produce nearly as much oxidations as the OH radicals produced in situ. Aging effects of DOM induced by singlet oxygen or peroxide radicals cannot be quantified as easily, because these do not significantly react with DOM and corresponding rate constants cannot be measured. However, for each photoproduced solvated electron a corresponding DOM cation radical formation must occur. We expect the DOM cation radical to undergo molecular rearrangements that may either split out a cationic entity

TABLE 4. Estimates for Production Rates of Photoreactants in Greifensee Water (Reference Water) and Dry Deposition Rates of Tropospheric Ozone

Reactant	mol/kW	mol/(m²·yr)
$^1O_2$ produced in water	50×10^{-3}	60
•OH produced in water	20×10^{-6}	0.025
ROO• produced in water	(?)[a]	(?)[a]
e_{aq}^- produced in water	70×10^{-6}	0.1
H_2O_2 produced in water	?	0.2
O_3 dry deposition		0.02
Comparison: primary biomass production		120[b]

[a] Cannot be estimated because of reasons explained in the text.
[b] Expressed as mole reduction. Estimate for mesotrophic lake (Switzerland) based on a primary production of $360 \, g \, C \cdot m^{-2} \, y^{-1}$ (Data based on Uehlinger and Bloesch, 1989).

(possibly a proton) or add an anion (OH^-) and O_2. Therefore, DOM might also undergo significant transformations by such reactions. The formation rate of H_2O_2 can easily be twice that of solvated electrons (Petasne and Zika, 1987; Sturzenegger, 1989). This must also be compensated by a corresponding oxidation effect on a precursor substrate such as DOM (Hoigné, 1988). Although highly speculative, such estimates show that aqueous and atmospheric photo-reactants could have large effects not only on specified micropollutants but also on the DOM present in natural waters. In addition, direct photolytic reactions of DOM might proceed.

3. LIMITS OF THE CONCEPT

Relevant environmental factors and their chemical effects have been successfully calibrated for homogeneous or "quasi-homogeneous" environmental compartments, where inhomogeneities are statistically averaged. However, many environmental factors and accumulations of pollutants relate to poorly defined discrete heterogeneous compartments of nonrandom distribution, such as surfaces of particles and heterogeneities on solid surfaces, surface films, oil droplets, biological tissues, and structured sediments. In all these cases the description of the role of the heterogeneity of natural systems for reaction kinetics remains a challenge to the community of environmental scientists. Such advanced approaches will lead from well-defined deterministic systems into the real world of nondeterministic systems. This significantly limits extrapolations from specific examples studied. For the time being, therefore, we must often accept compromises and account for uncertainties. However, for many relevant cases a much better basis for generalizable predictions and for formulating improved research projects or monitoring programs could be achieved if the established basic kinetic concepts and mechanistic understanding were applied in the critical way they deserve.

Acknowledgments

Parts of Section 2 in this chapter are directly based on a recent review by Hoigné et al. (1989) published in the ACS series '*Advances of Chemistry*', Volume 219. We thank ACS for the permission to reproduce adapted figures.

Section 2 summarizes a series of studies carried out during the last years at EAWAG and is based on earlier publications and discussions. The author thanks H. Bader, B. Faust, W. Haag, L. Nowell, F. Scully, and R. Zepp for their participation in these studies and P. Tratnyek and Janet Hering for stimulating discussion and all their reviewing efforts. Financial support of part of the studies by the Foundation of the President of the Swiss Federal Institute of Technology is acknowledged.

REFERENCES

Anbar, M., M. Bambenek, and A. B. Ross (1973), "Selected Specific Rates of Reactions of Transients from Water in Aqueous Solution, I. Hydrated Electron," National Bureau of Standards, NSRDS-NBS 43, Washington, DC, Government Printing Office.

Atkinson, R. (1986), "Kinetics and Mechanisms of the Gas-Phase Reactions of the Hydroxyl Compounds under Atmospheric Conditions," *Chem. Rev.* **86**, 69–201.

Brock Neely, W. (1985), "Environmental Exposure from Chemicals", in W. Brock Neely and G. E. Blau, Eds. *Hydrolysis*, Vol. I, CRC Press, Boca Raton, FL, pp. 157–173.

Broder, B. and H. A. Gygax (1985), "The Influence of Locally Induced Wind Systems on the Effectiveness of Nocturnal Dry Deposition of Ozone," *Atmos. Environ.* **19**, 1627–1637.

Buxton, G. V., C. L. Greenstock, W. P. Helman, and A. B. Ross (1988), "Critical Review of Rate Constants for Reactions of Hydrated Eletrons, Hydrogen Atoms and Hydroxyl Radicals in Aqueous Solution", *J. Phys. Chem. Ref. Data*, **17**, 513–886.

Cooper, W. J., R. G. Zika, R. G. Petasne, and J. M. C. Plane (1988), "Photochemical Formation of H_2O_2 in Natural Waters Exposed to Sunlight," *Environ. Sci. Technol.* **22**, 1156–1160.

Cooper, W. C., R. G. Zika, R. G. Petasne, and A. M. Fischer (1989), "Sunlight-Induced Photochemistry of Humic Substances in Natural Waters," in *Major Reactive Species*, I. H. Suffet and P. MacCarthy, Eds. *Aquatic Humic Substances, Advances in Chemistry Series*, Vol. 219, American Chemical Society, Washington, DC, pp. 334–362.

Erickson, R. E., L. M. Yates, R. L. Clark and D. McEwen (1977), "The Reaction of Sulfur Dioxide with Ozone in Water and Its Possible Atmospheric Significance," *Atmos. Environ.* **11**, 813–817.

Farhataziz., and A. B. Ross (1972), "Selected Specific Rates of Reactions of Transients from Water in Aqueous Solution. III. Hydroxyl Radical and Perhydroxyl Radical and Their Radical Ions," National Bureau of Standards, NSRDS-NBS 59, Washington, DC, Government Printing Office.

Faust, B. C., and J. Hoigné, (1987), "Sensitized Photooxidation of Phenols by Fulvic Acid and in Natural Waters," *Environ. Sci. Technol.* **21** (10), 957–964.

Faust, B. C., and J. Hoigné, (in press), "Photolysis of Fe (III)-Hydroxy Complexes as Sources of OH Radicals in Clouds, Fog and Rain," *Atmos. Environ.*

Frimmel, F. H., H. Bauer, J. Putzlen, P. Murasecco, and A. M. Braun (1987), "Laser Flash Photolysis of Dissolved Aquatic Humic Material and the Sensitized Production of Singlet Oxygen," *Environ. Sci. Technol.* **21**, 541–545.

Galbally, I. E., and C. R. Roy (1980), "Destruction of Ozone at the Earth's Surface," *Quart. J. R. Met. Soc.* **106**, 599–620.

Haag, W., and J. Hoigné (1985), "Photo-sensitized Oxidation in Natural Water via OH Radicals," *Chemosphere* **14**, 1659–1671.

Haag, W., and J. Hoigné (1986), "Singlet Oxygen in Surface Waters. 3. Photochemical Formation and Steady-State Concentrations in Various Types of Waters," *Environ. Sci. Technol.* **20** (4), 341–348.

Haag, W. R., J. Hoigné, E. Gassmann, and A. M. Braun (1984), "Singlet Oxygen in Surface Waters. Part I: Furfuryl Alcohol as a Trapping Agent. Part II: Quantum Yields of Its Production by Some Natural Humic Materials as a Function of Wavelength," *Chemosphere* **13** (5/6), 631–640, 641–650.

Hoffmann, M. R. (1986), "On the Kinetics and Mechanism of Oxidation of Aquated Sulfur Dioxide by Ozone," *Atmos. Environ.* **20** (6), 1145–1154.

Hoigné, J. (1988), "The Chemistry of Ozone in Water" in S. Stuchi, Ed. *Process Technologies for Water Treatment*, Plenum, New York, pp. 121–143.

Hoigné, J. (1988), "Effect of Abiotic Photochemical Processes on the Chemistry of Aquatic Solutes and on Global Cycles," *Appl. Geochem.* **3**, 63.

Hoigné, J., and H. Bader (1978), "Ozone and Hydroxyl Radical-Initiated Oxidations of Organic and Organometallic Trace Impurities in Water," *Organometals and Organomettaloids; Occurrence and Fate in the Environment*, F. E. Brinckman and J. M. Bellama, Eds. (ACS Symposium Series No. 92), (American Chemical Society), Washington, DC, 292–313.

Hoigné, J., and H. Bader (1979), "Ozonation of Water: 'Oxidation-Competition Values' for OH Radicals of Different Types of Waters Used in Switzerland," *Ozone: Sci. Eng.* **1**, 357–372.

Hoigné, J., and H. Bader (1983), "Rate Constants of Reactions of Ozone with Organic and Inorganic Compounds in Water. I. Non-dissociating Organic Compounds. II. Dissociating Organic Compounds," *Water Res.* **17**, 173–183, 185–194.

Hoigné, J., H. Bader, W. R. Haag, and J. Staehelin (1985), "Rate Constants of Reactions of Ozone with Organic and Inorganic Compounds in Water-III, Inorganic Compounds" *Water Res.* **19**, 993–1004.

Hoigné, J., B. C. Faust, W. Haag, F. Scully and R. Zepp (1989), "Aquatic Humic Substances as Sources and Sinks of Photochemically Produced Transient Reactants", in *Aquatic Humic Substances*, I. H. Suffet and P. McCarthy, Eds., (Advances in Chemistry Series No. 219), American Chemical Society Washington, DC, pp. 363–381.

Larson, R. A., and R. G. Zepp (1988), "Reactivity of the Carbonate Radical with Aniline Derivatives," *Environ. Toxicol. Chem.* **7**, 265–274.

Leifer, A. (1988), "The Kinetics of Environmental Aquatic Photochemistry; Theory and Practice," ACS professional reference book, Maple Press, York, PA.

Lyman, W. J., W. F. Reehl, and D. H. Rosenblatt (1982), *Handbook of Chemical Property Estimation Methods; Environmental behaviour of Organic Compounds*, McGraw-Hill, New York.

Maahs, H. G. (1983), "Measurements of the Oxidation Rate of Sulfur (IV) by Ozone in Aqueous Solution and Their Relevance to SO_2 Conversion in Nonurban Tropospheric Clouds," *Atmos. Environ.* **17**, 341–345.

Mill, T., and W. R. Mabey (1986), "Hydrolysis of Organic Chemicals," in O. Hutzinger, Ed., *Handbook of Environmental Chemistry*, Vol. 2, Part D, *Reactions and Processes*, Springer-Verlag, Berlin, pp. 71–111.

Neta, P., R. E. Huie, and A. B. Ross (1988), "Rate Constants of Inorganic Radicals in Aqueous Solution," *J. Phys. Chem. Ref. Data*, **17**, 1027–1284.

Nowell, L., and J. Hoigné (1990), "Photolysis of Aqueous Chlorine and Hydroxyl Radical Production at Sunlight and UV Wavelength," *Water Res.*

OECD (1987), *Guidelines for Testing of Chemicals* OECD, Paris (1981–1987).

Petasne, R. G., and R. G. Zika (1987), "Fate of Superoxide in Coastal Sea Water," *Nature (London)* **325**, 516–518.

Pichat, P., and M. A. Fox (1989), "Photocatalysis on Semiconductors," Chapter G. 1 in M. A. Fox and M. Chanon, Eds., *Photoinduced Electron Transfer*, Part D, Elsevier, Amsterdam, 1988.

Roof, A. A. M. (1982), "Aquatic Photochemistry," in O. Hutzinger, Ed., *Handbook of Environmental Chemistry*, Vol. 2B, Springer-Verlag, Amsterdam, 43–72.

Schuchmann, M. N., and C. von Sonntag (1987), "Hydroxyl Radical-Induced Oxidation of Diisopropyl Ether in Oxygenated Aqueous Solution. A Product and Pulse Radiolysis Study," *Z. Naturforsch.* **42b**, 495–502.

Scully, F., and J. Hoigné (1987), "Rate Constants for Reactions of Singlet Oxygen with Phenols and Other Compounds in Water," *Chemosphere* **16**, 681–694.

Smith, J. H., W. R. Mabey, N. Bohonos, B. R. Holt, S. S. Lees. T.-W. Chou, D. C. Bomberger, and T. Mill (1977), "Environmental Pathways of Selected Chemicals in Freshwater Systems. Part I: Background and Experimental Procedures," SRI International, Contract No. 68-03-2227.

von Sonntag, C. (1986), "Disinfection by Free Radicals and UV-Radiation," *Water Supply* **4**, 11–18.

Staehelin, J., and J. Hoigné (1985), "Decomposition of Ozone in Water in the Presence of Organic Solutes Acting as Promoters and Inhibitors of Radical Chain Reactions," *Environ. Sci. Technol.* **19**, 1206–1213.

Stone, A. T., and J. J. Morgan (1987), "Reductive Dissolution of Metal Oxides," in W. Stumm, Ed., *Aquatic Surface Chemistry*, Wiley, New York, pp. 221–252.

Stumm, W., and G. Furrer (1987), "The Dissolution of Oxides and Aluminium Silicates; Examples of Surface-coordination-controlled Kinetics", in W. Stumm, Ed., *Aquatic Surface Chemistry*, Wiley, New York, pp. 197–217.

Stumm, W. and G. F. Lee (1961). "Oxygenation of Ferrous Iron," *Ind. Eng. Chem.* **53**, 143–146.

Sturzenegger, V. (1989), "Wassers Hyperoxid in Oberflächengewässern: Photochemische Prochilition und Abban," Ph.D. Thesis, Swiss Federal Institute of Technology Zürich, Thesis ETH Nr. 9004.

Uehlinger, U., and J. Bloesch (1989), "Primary Production of Different Phytoplankton Size Classes in an Oligo-mesotrophic Swiss Lake," *Arch. Hydrobiol.* **116**, 1–21.

Walcek, C. J. (1987), "A Theoretical Estimate of O_3 and H_2O_2 Dry Deposition over the Northeast United States", *Atmos. Environ.* **21**, 2649–2659.

Wilkinson, F., and J. G. Brummer (1981), "Rate Constants for the Decay and Reactions of the Lowest Electronically Excited Singlet State of Molecular Oxygen in Solution", *J. Phys. Chem. Ref. Data* **10**, 809–999.

Zepp, R. G. (1980), "Experimental Approaches to Environmental Photochemistry," in O. Hutzinger, Ed., *Handbook of Environmental Chemistry*, Vol. 2B, Springer-Verlag, Amsterdam, pp. 19–41.

Zepp, R. G., and P. F. Schlotzhauer (1981), "Effects of Equilibration on Photoreactivity of the Pollutant DDE Sorbed on Natural Sediments," *Chemosphere* **10**(5), 453–460.

Zepp, R. G. and P. F. Schlotzhauer (1983), "Influence of Algae on Photolysis Rates of Chemicals in Water," *Environ. Sci. Technol.* **17**, 462–468.

Zepp, R. G., N. L. Wolfe, G. L. Baughman, and R. C. Hollis (1977), "Singlet Oxygen in Natural Waters," *Nature* **267**, 421–423.

Zepp, R. G., P. F. Schlotzhauer, and R. M. Sink (1985), "Photosensitized Transformations Involving Electronic Energy Transfer in Natural Waters: Role of Humic Substances," *Environ. Sci. Technol.* **19**, 74–81.

Zepp, R. G., J. Hoigné, and H. Bader (1987a), "Nitrate-induced Photooxidation of Trace Organic Chemicals in Water," *Environ. Sci. Technol.* **21** (5), 443–450.

Zepp, R. G., A. M. Braun, J. Hoigné, and J. A. Leenheer (1987b), "Photoproduction of Hydrated Electrons from Natural Organic Solutes in Aquatic Environments," *Environ. Sci. Technol.* **21**, 485–490.

3

CATALYSIS IN AQUATIC ENVIRONMENTS

Michael R. Hoffmann

Department of Environmental Engineering Science, California Institute of Technology, Pasadena, California

1. GENERAL CONSIDERATIONS OF CHEMICAL CATALYSIS

1.1. Introduction

1.1.1. Definitions. Early in the history of chemical kinetics a *catalyst* was defined as *a chemical species that changes the rate of a reaction without undergoing an irreversible change itself* (Ostwald, 1902). Subsequent definitions of a catalyst included (1) *a catalyst is a chemical species that may be chemically altered but is not involved in a whole number stoichiometric relationship among reactants and products* and (2) *a catalyst is a chemical species that appears in the rate law with a reaction order greater than its stoichiometric coefficient.* In the latter case it was realized that either a product of the reaction (*autocatalysis*) or a reactant may also function as a catalyst. From a practical perspective, a catalyst is a chemical species that influences the rate of a chemical reaction regardless of the fate of the catalytic species. However, a catalyst has no influence on the thermodynamics of a reaction. In other words, the concentration of a catalyst is reflected in the rate law but is not reflected in the equilibrium constant. This latter definition was modified and approved by the International Union of Pure and Applied Chemistry (IUPAC, 1981) to read as follows:

> A *catalyst* is a substance that increases the rate of reaction without modifying the overall standard Gibbs energy change in the reaction; the process is called *catalysis*, and a reaction in which a catalyst is involved is known as a *catalyzed reaction.*

1.1.2. Environmental Chemical Catalysis. "Catalysis" in a single phase is normally referred to as homogeneous catalysis, while catalysis that takes place at the interface of two phases is classified as heterogeneous catalysis. In environmental systems, several categories of catalyzed reactions play an important role in the transformation of inorganic and organic chemical species.

In the realm of homogeneous catalysis we often encounter examples of acid- and base-catalyzed hydration–dehydration and hydrolysis, metal-catalyzed hydrolysis and autoxidation, photocatalytic oxidation and reduction, metal-catalyzed electron transfer, acid-catalyzed decarboxylation, photocatalytic decarboxylation, metal-catalyzed free-radical chain reactions, acid-catalyzed nucleophilic substitutions, and enzymatic catalysis.

1.2.   Categories of Catalysis

1.2.1. Homogeneous Catalysis.

Homogeneous catalysis in the context of aquatic chemistry implies that the reactants and the catalyst are present as dissolved species in water. In a catalyzed reaction the reactants follow a reaction pathway that is accelerated (i.e., the rate of reaction is increased) relative to the noncatalyzed pathway. This feature implies that the mechanism (i.e., the set of elementary reactions that constitute the overall reaction) of the stoichiometric reaction has been changed. The change in mechanism that occurs during catalysis yields a reaction coordinate over a potential-energy surface with a lower potential-energy barrier. In other words, the potential-energy surface for the mechanism involving the catalyst has a lower activation-energy barrier ($\Delta G_{cat}^{\neq}$) at the saddlepoint than the mechanism for the noncatalyzed pathway ($\Delta G_{rxn}^{\neq}$). In transition-state theory, the rate constant k for the rate-determining step of the mechanism is given by

$$k = (C^{\varnothing})^{1-n}\left(\frac{RT}{Lh}\right)e^{-\Delta G^{\neq}/RT} \tag{1}$$

and where k has units given by $(C^{\varnothing})^{1-n}\,s^{-1}$, $C^{\varnothing}$ is a standard concentration (i.e., mol L^{-1} or M), L is Avogadro's number, h is Planck's constant, R is the Universal Gas Constant, and $\Delta G^{\neq} = \Delta H^{\neq} - T\Delta S^{\neq}$. Thus, if $\Delta G_{cat}^{\neq} < \Delta G_{rxn}^{\neq}$, then $k_{cat} > k_{rxn}$.

An example of a catalyzed reaction that has an obvious shift in mechanism is the oxidation of thiourea by H_2O_2 (Hoffmann and Edwards, 1977). The stoichiometry for this reaction is

$$2(NH_2)_2C{=}S + H_2O_2 + 2\,H^+ \longrightarrow (NH_2)_2CS-SC(NH_2)_2^{2+} + 2\,H_2O \tag{2}$$

The experimentally determined rate law was found to be of the form

$$-\frac{d[(NH_2)_2C{=}S]}{dt} = k_2[H_2O_2][(NH_2)_2C{=}S]$$

$$+ k_3[H_2O_2][(NH_2)_2C{=}S][H^+] \tag{3}$$

Hypothetical potential-energy–reaction coordinate profiles for each term of the rate law of Eq. 3 are shown in Figure 1. The first term of the rate law of Eq. 3

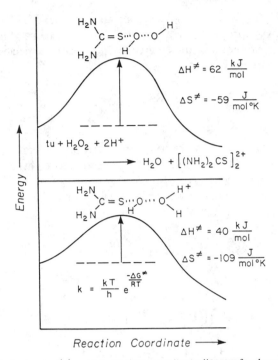

Figure 1. A hypothetical potential-energy–reaction coordinate diagram for the oxidation of thiourea by hydrogen peroxide. The upper diagram shows the noncatalyzed pathway and its associated activation parameters, while the lower diagram shows the reduction in the potential energy barrier due to the acid-catalyzed pathway. The acid-catalyzed pathway has a lower $\Delta H^{\neq}$ (40 kJ mol^{-1}) than the noncatalyzed pathway ($\Delta H^{\neq} = 62$ kJ mol^{-1}). The lowering of the activation enthalpy is compensated somewhat by the decrease $\Delta S^{\neq}$ in going from the noncatalyzed to the catalyzed pathway.

gives the elemental composition of the transition state (Edwards et al., 1968; Bunnett, 1986) for the noncatalyzed pathway (Fig. 1a; $\Delta G^{\neq}_{rxn} = 79.6$ kJ mol^{-1}; $E_a = 64.5$ kJ mol^{-1}), and the second term of the rate law gives the elemental composition of the transition state (located at the top of the energy barrier) for the acid-catalyzed pathway (Fig. 1b; $\Delta G^{\neq}_{cat} = 72.5$ kJ mol^{-1}; $E_a = 42.5$ kJ mol^{-1}). In the noncatalyzed pathway, the reaction proceeds via the nucleophilic attack by a nonbonded electron pair of

$$\underset{R}{\overset{R}{>}}C=S:$$

on the linear

$$:\ddot{O} \quad \overset{/}{\underset{..}{O}}:$$

bond of the H_2O_2 with the subsequent displacement of OH^-. The acid-catalyzed pathway proceeds in a similar manner except that the H_2O_2 is protonated in a prior equilibrium to give an intermediate species, $H_3O_2^+$, which is attacked subsequently by

$$\underset{R}{\overset{R}{\diagdown}}C=S:$$

with the displacement of a water group. In nucleophilic displacement reactions, displacement of H_2O molecule will proceed more favorably than will displacement of OH^-, which is more strongly bound in the transition state. The stepwise sequence of elementary reactions (i.e., the mechanism) for the second term of Eq. 3 can be written as

$$H_2O_2 + H^+ \underset{}{\overset{K_{a0}^{-1}}{\rightleftharpoons}} H_3O_2^+ \tag{4}$$

The rate–determining step is

$$(NH_2)_2C=S + H_3O_2^+ \xrightarrow{k_3} (NH_2)_2C\text{–}S\text{–}OH^+ + H_2O \tag{5}$$

$$(NH_2)_2C\text{–}S\text{–}OH^+ + H^+ \underset{}{\overset{K_{a1}^{-1}}{\rightleftharpoons}} (NH_2)_2C\text{–}S\text{–}OH_2^{2+} \tag{6}$$

$$(NH_2)_2C=S + (NH_2)_2C\text{–}S\text{–}OH_2^{2+} \xrightarrow{k_4} (NH_2)_2CS\text{–}SC(NH_2)_2^{2+} \tag{7}$$

The mechanism of the noncatalyzed pathway is similar except that the protonation step of Eq. 4 is missing and the rate-determining step is a simple bimolecular elementary reaction:

$$(NH_2)_2C=S + H_2O_2 \xrightarrow{k_2} (NH_2)_2C\text{–}S\text{–}OH^+ + OH^- \tag{8}$$

In this particular case, the more favorable energetics of the catalyzed pathway are apparent. This reaction is not unique; a number of nucleophiles are oxidized by H_2O_2 via a similar mechanism. They include Cl^-, Br^-, SCN^-, CN^-, I^-, $S_2O_3^{2-}$, and RSH (Leung and Hoffmann, 1985).

Specific acid catalysis as illustrated in the mechanism of Eqs. 4–7 is the most common form of catalysis encountered in environmental systems. However, we may ask whether the reaction of hydrogen peroxide with thiourea is truly acid-catalyzed. According to some of the early definitions of chemical catalysis, this reaction would not be categorized as a catalytic, since the proton is a stoichiometric reactant, but in accord with the IUPAC definition the pathway involving H^+

$$v_{cat} = k_3 \left[\underset{NH_2}{\overset{NH_2}{\diagdown}}C=S \right] [H_2O_2][H^+]$$

has an energetically more favorable potential energy surface, and thus we may call this a *catalyzed reaction*.

1.2.2. Heterogeneous Catalysis.

In heterogeneous catalysis, the catalytic species is present in a phase different from that of the reactants in an overall chemical reaction (Boudart and Burwell, 1974; Haller and Degless, 1986). In aquatic systems the heterogeneous catalyst is often in the solid phase and the reactants are dissolved in the aqueous phase. Furthermore, the reaction takes place at specific sites (i.e., reactive sites) on the surface of the solid. As in the case of homogeneous catalysis, the heterogeneous catalyst provides an alternative mechanism for the reaction that is energetically more favorable. Heterogeneous catalysis of gas-phase reactions at high temperatures plays an extremely important role in commercial scale chemical synthesis and in pollution-control processes such as flue-gas scrubbing. In aquatic systems, a number of surfaces may provide active catalytic sites for reactions such as ester hydrolysis and the redox reactions of organic and inorganic compounds. Potential catalytic surfaces include metal oxide particles, aluminium silicate minerals, organic (humic substances) colloids, inorganic (polymeric forms of iron and aluminum) colloids, soot particles (graphitic carbon), organic micelles, and surface organic microlayers.

The rate of a heterogeneous reaction is defined normally with respect to either the reactive surface area or the mass of the solid catalyst as

$$v = \frac{1}{A} \frac{d\xi}{dt} \tag{9}$$

where ξ is known as the degree of advancement of the reaction, which is defined as

$$n_i = n_i^0 + \gamma_i \xi \tag{10}$$

where n_i and n_i^0 are the number of moles of chemical component i at any time and at $t = 0$, respectively, γ_i is the stoichiometric coefficient of component i in the stoichiometric equation for the reaction that is being catalyzed, and A is the total reactive surface area (or the total number of moles of surface sites per unit area).

For the simple case of the formation of a product in a catalyzed reaction, at $t = 0$, $n_i^0 = 0$, and Eq. 9 can be recast as

$$v_{\text{het}} = \frac{1}{A} \frac{1}{\gamma_i} \frac{dn_i}{dt} = \frac{V}{A} \frac{1}{\gamma_i} \frac{dC_i}{dt} \tag{11}$$

where $C_i = n_i/V$. Thus we expect to find experimentally that the rate of a heterogeneous reaction will be dependent of the total reactive surface area (or proportional to the concentration of reactive sites in moles per square meter).

$$v_{\text{het}} = \frac{dC_i}{dt} = \frac{k_{\text{het}} \gamma_i A}{V} \prod_{i=1}^{N} C_i^{\beta_i} \tag{12}$$

where the term $\prod_{i=1}^{N} C_i^{\beta_i}$ represents a function that is a product of the concentrations of chemical components i involved in the rate-determining step of the mechanism raised to their appropriate kinetic orders β_i.

Several examples of the catalytic effect of surfaces in aquatic systems have been provided in the recent literature. Sung and Morgan (1981) studied the catalytic effect of γ–FeOOH particulate surfaces on the rate of autoxidation of Mn(II) to Mn(IV), which has the following stoichiometry:

$$\text{Mn(II)} + \text{O}_2 \xrightarrow{\gamma - \text{FeOOH}} \text{MnO}_{2(s)} \tag{13}$$

Earlier Morgan (1967) showed that the autoxidation of Mn(II) in water was an autocatalytic process in that the solid products of the reaction were shown to accelerate the rate of reaction, which was described by the following rate law:

$$-\frac{d[\text{Mn(II)}]}{dt} = k_1 [\text{Mn(II)}][\text{O}_2][\text{OH}^-]^2 + k_2 [\text{MnO}_2][\text{Mn(II)}] \tag{14}$$

where $k_2 \approx k_2' [\text{O}_2][\text{OH}^-]^2$ and MnO_2 is actually a nonstoichiometric product that contains several mineral forms (e.g., γ–MnOOH, Mn_3O_4, δ–MnO$_2$, {MnO$_{1.3}$ to MnO$_{1.9}$}). A similar multiterm rate law was observed by Sung and Morgan (1981); others have also noted the catalytic effect of surfaces on the autoxidation of Mn(II) (Davies, 1985; Brewer, 1975; Emerson et al., 1979).

More recently, Wehrli and Stumm (1988) reported on the catalytic effect of coordinated surface hydroxyl groups of δ–Al$_2$O$_3$ and TiO$_2$ on the rate of autoxidation of VO^{2+} over the pH range of 4–9. They formulated their kinetic observations in terms of a simple rate law that took into account the surface speciation (reactive-site-specific) of the metal oxides:

$$\frac{d[\text{V(IV)}]}{dt} = k \left\{ \begin{array}{c} \text{structure} \end{array} \right\} [\text{O}_2] \tag{15}$$

where

is the actual surface concentration (in moles per square meter) of the mononuclear bidentate surface complex of VO^{2+} with either δ-Al_2O_3 or TiO_2 surfaces

$$\begin{array}{c} \rule[-1.5em]{0.15em}{2.5em}\!\!-OH \\[-0.3em] \!\!-OH \end{array}$$

and $[O_2]$ is the bulk solution-phase concentration of oxygen.

Stone (1989) has recently reported on the catalytic effect of Al_2O_3 suspensions on the rate of hydrolysis of monoterephthalate (MPT), which is normally sensitive to specific-base catalysis:

$$(16)$$

He found rate enhancements of more than an order of magnitude compared to the homogeneous base-catalyzed reaction rate and attributed these enhancements to the reaction of specifically adsorbed MPT^- reacting with excess OH^- that had accumulated in the electrical double layer (in the outer Helmholz layer) as a counterion to the positively charged Al_2O_3 surface. He gives the following rate law to account for his experimental observations:

$$-\frac{d[MPT^-]}{dt} = k_x K^s [OH^-]_d [\blacksquare\!-\!Al\!-\!MPT]$$

$$+ k_b [OH^-]_{bulk} [MPT]_{bulk} \qquad (17)$$

1.2.3. Biochemical–Enzymatic Catalysis.   In any natural water, the role of biochemical and microbial catalysis of a wide variety of chemical reactions must be recognized (Arnold et al., 1986, 1988). Enzymes are proteins that may be either intracellular (either membrane-bound or freely diffusing within the cytoplasm) or extracellular. For a number of enzymes the catalytic activity is restricted to a confined region of the protein molecule; this region normally is called the *active site*. In certain cases, the catalytic activity of enzymes can be mimicked by specific transition-metal–organic (macromolecular) complexes (Boyce and Hoffmann, 1983; Hoffmann and Hong, 1987; Leung and Hoffmann, 1988; 1989a,b). Virtually all categories of chemical reactions are sensitive to the catalytic influences of enzymes. Reactions sensitive to enzymatic catalysis include oxidations, reductions, electron transfers, proton transfers, hydrolyses, disproportionations, atom transfers, group transfers, decarboxylations, ring hydroxylations, ether cleavages, dehydrogenation, dehalogenation, dehydrohalogenation, and demethylation. The kinetics of enzymatic catalysis has been treated thoroughly by many authors (Cleland, 1986; Cornish-Bowden, 1976; Laidler and Bunting, 1973).

1.2.4. Photochemical catalysis.

The term "photocatalysis" may be misleading. Moore and Pearson (1981) contend that "catalysis by light" is an improper phrase; they argue that when the rate of a reaction is accelerated by a means other than by a chemical species, it should not be classified as catalytic. However, it is clear that some reactions, which are thermodynamically favorable (i.e., $\Delta G^{\circ}_{rxn} < 0$), can be assisted by the interaction with light in the UV–visible range. These reactions are sometimes called photoassisted reactions. They can occur either homogeneously or heterogeneously.

Photoassisted reactions can be further categorized as being either photosynthetic (i.e., $\Delta G^{\circ}_{rxn} > 0$) or photocatalytic (i.e., $\Delta G^{\circ}_{rxn} < 0$). An example of a photosynthetic reaction would be the photoassisted elimination–rearrangement reactions of aldehydes (March, 1977):

$$R_2CH-CR_2CR_2-\underset{O}{\overset{\|}{C}}-R' \xrightarrow{h\nu} R_2C{=}CR_2 \;+\; R_2CH-\underset{O}{\overset{\|}{C}}-R'$$

(18)

Another example of a photoassisted reaction is provided by the irradiation of o-nitrobenzaldehydes to give the corresponding benzoic acids:

(19)

In the category of photocatalytic reactions (a "malaprop" according to Moore and Pearson) we could include examples of a variety of two-electron oxidations and reductions that are very slow in the absence of catalytic influences. A primary example for this situation is the oxidation of S(IV) $\{SO_2 \cdot H_2O, HSO_3^-, SO_3^{2-}\}$ by oxygen to S(VI) $\{SO_4^{2-}\}$, which has the following stoichiometry:

$$2\,SO_3^{2-} + O_2 \longrightarrow 2\,SO_4^{2-}$$

(20)

In the absence of any catalytic influence, this reaction is exceedingly slow (Hoffmann and Boyce, 1983; Hoffmann and Jacob, 1984; Boyce et al., 1983). However, in the presence of light with $\lambda \leq 285$ nm this reaction is accelerated manyfold; thus in a liberal definition of catalysis, wherein we consider the wave particle duality of light and matter (Hawking, 1988), this reaction is catalyzed nominally by light. The autoxidation of S(IV) (Eq. 20) is also sensitive to trace metal catalysis, enzymatic catalysis (*sulfite oxidase*) and to heterogeneous photocatalysis (Faust et al., 1989).

1.2.5. Free-Radical Chain Reactions. Free-radical chain reactions are sensitive to two types of catalysis; a catalyst may increase the rate of initiation of chains by introducing an additional initiation pathway or may lead to new chain propagation steps. For example a chain reaction involving S(IV) and O_2 can be initiated by a number of alternative pathways:

$$SO_3^{2-} \xrightarrow{hv < 285 \text{ nm}} SO_3^{-\bullet} + e_{aq} \tag{21}$$

$$SO_3^{2-} + Co^{3+} \longrightarrow SO_3^{-\bullet} + Co^{2+} \tag{22}$$

$$SO_3^{2-} + {}^{\bullet}OH \longrightarrow SO_3^{-\bullet} + OH^- \tag{23}$$

However, the catalyst in a free-radical reaction is often irreversibly utilized. For example, in the case by of Eq. 22, the nominal catalyst, Co^{3+}, is reduced in the initiation step to Co^{2+}. Thus Co^{3+} would be better classified as an initiator than a catalyst, although in this particular case, Co^{2+} is readily oxidized back to Co^{3+} either directly by O_2 or by an intermediate produced in the chain propagation steps, peroxymonosulfate:

$$Co^{2+} + O_2 \longrightarrow Co^{3+} + O_2^- \tag{24}$$

$$2Co^{2+} + HSO_5^- + H^+ \longrightarrow 2Co^{3+} + SO_4^{2-} + H_2O \tag{25}$$

Thus the net effect of the sequence of Eqs. 22 and 24 would be one of apparent catalysis. Chain reactions are less likely in aquatic systems than they are in atmospheric systems, but there is strong evidence that free chain reactions involving S(IV) may be taking place in cloudwater droplets, especially in the remote troposphere where H_2O_2 is in short supply.

1.3. General Considerations of Catalytic Mechanisms

Even though there are a variety of mechanisms leading to catalysis, many catalyzed reactions for a single substrate and a single catalyst can be considered according to the following general scheme (Laidler, 1987):

$$C + S \underset{k_{-1}}{\overset{k_1}{\rightleftharpoons}} A + B \tag{26}$$

$$A + D \xrightarrow{k_2} P_1 + E \tag{27}$$

$$B + E \xrightarrow[\text{rapid}]{} C + P_2 \tag{28}$$

where C represents the catalyst; S, the substrate, A, B, and E, intermediates; and

P_1 and P_2, reaction products. In surface catalysis A would be the surface complex and E would be H^+; in acid catalysis, C would transfer a proton to S to form a protonated intermediate, A, while B would be the conjugate base of the acid. In the latter case, D may be a water molecule. In selected cases, the reaction of A with D is sufficiently slow such that Eq. 26 can be considered to be a rapidly attained prior equilibrium. Under these conditions $k_2[A][D] \ll k_{-1}[A][B]$, and we can consider the intermediates to be in actual equilibrium with the reactants; intermediates of this type are often called *Arrhenius intermediates*. The intermediate, $H_3O_2^+$, formed in Eq. 4 can be classified as an Arrhenius intermediate. In the converse situation, $k_{-1}[A][B] \ll k_2[A][D]$, with the concentration of the A very small, the steady-state approximation can be applied around A; intermediates under these condition have been called *van't Hoff intermediates* (Laidler, 1987).

If we apply the notion of a prior equilibrium to the above mechanism and if $[S]_0 \gg [C]_0$ where $[S]_0 = [S] + [A]$ and $[C]_0 = [C] + [A]$ ($[S]_0$ and $[C]_0$ can be considered to be the total concentrations of substrate and catalyst at time $= t$ or the initial concentrations of each at $t = 0$), then the rate of the reaction (valid for consideration of initial rates) can be expressed as

$$v_0 = \frac{k_2 K_1 [C]_0 [S]_0 [D]}{K_1 [S]_0 + [B]} \tag{29}$$

This rate expression for the *Arrhenius intermediate* is characteristic of surficial and enzymatic catalysis with single substrates. On the other hand, the steady-state treatment of the *van't Hoff intermediate* for the condition of $k_2[A] \gg k_{-1}[B]$ yields a rate expression of the following form:

$$v_0 = \frac{k_2 k_1 [C]_0 [S]_0 [D]}{k_1 ([C]_0 + [S]_0) + k_{-1}[B] + k_2[D]} \tag{30}$$

1.4. Activation Energies for Catalyzed Reactions

The overall experimentally determined activation energy and the activation energies for the individual elementary reactions in a mechanism may have a complicated relationship; thus, unless the mechanism is fairly well understood, the information obtained from determination of an apparent activation energy, E_a (or apparent activation parameters: $\Delta G^{\neq}$, $\Delta H^{\neq}$, and $\Delta S^{\neq}$) may be of limited value. We can consider several cases to illustrate this point. For the general mechanism of Eqs. 26–28, if we have a simple catalyst–substrate complex forming in the first step of the mechanism (i.e., $C + S \rightleftharpoons C–S$, where $C–S = A$ of Eq. 26; B *and* D *are not present in this limiting case*), then we obtain the following rate expression:

$$v_0 = \frac{k_2 K_1 [C]_0 [S]_0}{K_1 [S]_0 + 1} \tag{31}$$

When $K_1[S]_0 \ll 1$ (i.e., at low substrate concentrations), we obtain

$$v_0 \simeq k_2 \frac{k_1}{k_{-1}}[C]_0[S]_0 \tag{32}$$

A hypothetical potential-energy–reaction coordinate diagram for this situation $(k_{-1} \gg k_2$; an Arrhenius intermediate) is presented in Figure 2a. For this limiting case, the overall activation energy E_{a1}, for the k_2 step as the rate-determining step is given by

$$E_{a1} = E_1 + E_2 - E_{-1} \tag{33}$$

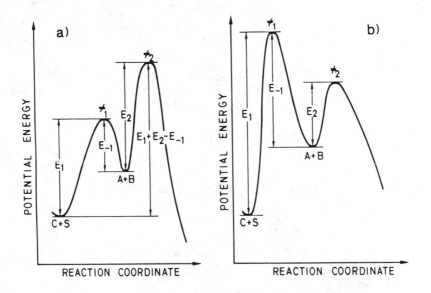

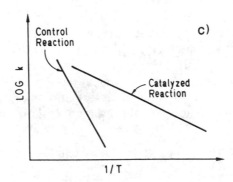

Figure 2. Hypothetical potential energy–reaction-coordinate diagrams for the $k_{-1} \gg k_2$ case.

For the case of high substrate concentrations (i.e., $K_1[S]_0 \gg 1$), where the rate expression reduces to $v_0 \simeq k_2[C]_0$, the overall activation energy is given by

$$E_{a2} = E_2 \tag{34}$$

If we now consider the formation of a van't Hoff intermediate (the steady-state case), the rate expression for the single catalyst–substrate complex is

$$v_0 = \frac{k_2 k_1 [C]_0 [S]_0}{k_1([C]_0 + [S]_0) + k_{-1} + k_2} \tag{35}$$

If we again consider, first, the case of high substrate concentrations ($[S]_0 \gg [C]_0$), then Eq. 35 can be reduced to

$$v_0 \simeq k_2[C]_0 \tag{36}$$

with a corresponding activation energy of $E_{a2}^{ss} = E_2$; however, at low substrate concentrations ($\{k_1[S]_0 + k_1[C]_0\} \ll \{k_{-1} + k_2\}$) the rate expression derived from the steady-state approximation around [A] reduces to

$$v_0 \simeq \frac{k_2 k_1 [C]_0 [S]_0}{k_{-1} + k_2} \tag{37}$$

which for $k_2 \gg k_{-1}$ yields $v_0 \approx k_1 [C]_0 [S]_0$ and an activation energy (Fig. 2b) of

$$E_{a1}^{ss} = E_1 \tag{38}$$

In this latter case, crossing of the initial energy barrier will determine the observed rate of the reaction. For the converse case of $k_2 \ll k_{-1}$, the steady-state treatment reduces to

$$v_0 \simeq \frac{k_2 k_1}{k_{-1}} [C]_0 [S]_0 \tag{39}$$

which is equivalent to the result obtained with the prior equilibrium assumption and, thus, the activation energy, E_{a1}^{ss} is again given by the right-hand side of Eq. 33.

With respect to the empirical determination of overall activation energy, we expect to find a lesser slope for the E_a / T^{-1} relationship of the catalyzed reaction when compared to the noncatalyzed reaction as shown in Figure 1c.

1.5. Principle of Microscopic Reversibility and Free-Energy Profiles

On one hand, the presence of a catalyst in a reaction system has the net effect of accelerating the rate of attainment of equilibrium by providing an alternative mechanism with $\Delta G_{cat}^{\neq} < G_{rxn}^{\neq}$; on the other hand, it has no ground-state energetic effect on the equilibrium state as defined by ΔG_{rxn}^0. This conclusion follows directly from basic thermodynamic arguments in that ΔG_{rxn}^0 depends only on the initial and final states of the system and is independent of the path (i.e., *mechanism*) of going from one state to another.

We can prove this assertion by using several alternative reaction mechanisms and by invoking the *principle of microscopic reversibility*, which states that when an overall reaction is at equilibrium, each step in the mechanistic sequence must also be at equilibrium (i.e., the rate of forward reaction equals the rate of the reverse reaction). Three alternative mechanisms for the hydration of an aldehyde can be considered as follows:

$$RCHO + H_2O \underset{k_{-1}}{\overset{k_1}{\rightleftharpoons}} RCH(OH)_2 \tag{40}$$

$$RCHO + H_2O + H^+ \underset{k_{-2}}{\overset{k_2}{\rightleftharpoons}} RCHO(H_3O)^+ \underset{k_{-3}}{\overset{k_3}{\rightleftharpoons}} RCH(OH)_2 + H^+ \tag{41}$$

$$RCHO + H_2O + OH^- \underset{k_{-4}}{\overset{k_4}{\rightleftharpoons}} \underset{HO}{\overset{R}{>}}\!\!\!C\!\!-\!\!O\cdot H_2O \underset{k_{-5}}{\overset{k_5}{\rightleftharpoons}} RCH(OH)_2 + OH^- \tag{42}$$

At equilibrium, where the rate of the forward reaction is equal to the rate of the reverse reaction ($v_f = v_r$), we can write the following equality for Eq. 40:

$$k_1[RCHO]_{eq}[H_2O]_{eq} = k_{-1}[RCH(OH)_2]_{eq} \tag{43}$$

or

$$K_1 = K_{eq} = \frac{[RCH(OH)_2]_{eq}}{[RCHO]_{eq}[H_2O]_{eq}} = \frac{k_1}{k_{-1}}. \tag{44}$$

Application of the kinetic definition of equilibrium to the mechanisms of Eqs. 40–42 yields the following relationship:

$$K_{eq} = \frac{k_3 k_4}{k_{-3} k_{-4}} = \frac{k_5 k_6}{k_{-5} k_{-6}} = \frac{k_1}{k_{-1}} \tag{45}$$

This relationship indicates that the value of the equilibrium constant will be the same regardless of the mechanism and that the magnitudes of the various rate constants will be restricted to certain ranges.

The significance of the principle of microscopic reversibility is that it allows the mechanism for reverse reaction to be determined with an accuracy similar to that for the mechanism for the forward reaction. This means that if a mechanism for a reaction is followed in the forward direction the same mechanism is followed in reverse order for the back reaction. Using the reaction coordinate diagrams of Figure 2 to further illustrate this point, we note that if the energy barriers (E_1, E_{-1}, E_2) and the intermediates (A) are known in the forward direction, then the principle of microscopic reversibility allows us to predict the mechaism of the back reaction by following the energy profile from right to left in Figures 2a and 2b.

2. KINETICS AND MECHANISMS OF SELECTED REACTIONS OF ENVIRONMENTAL INTEREST

2.1. Homogeneous Catalysis

2.1.1. Acid and Base Catalysis

a. *Acid Catalysis of S(IV) Reactions.*   Perhaps the least complicated form of catalysis in aquatic systems is specific acid–base catalysis. Given the limited concentrations of weak organic and inorganic acids and bases (general acids and bases) in natural waters (although there are a few notable exceptions, e.g., Los Angeles cloudwater), we find that catalysis by acids and bases is dominated by specific-acid or specific-base catalysis (i.e., catalysis by H^+ and OH^-, respectively). Several reactions classes exhibit a strong tendency toward acid and/or base catalysis. They include aldehyde and ketone hydration reactions, ester hydrolyses, nucleophilic substitutions, and electrophilic substitutions. The oxidation of thiourea by hydrogen peroxide (Eqs. 4–7) represents a nucleophilic substitution that is sensitive to specific-acid catalysis. On the other hand, the reaction of hydrogen peroxide with S(IV) is as an example of an electrophilic substition (redox reaction) that is sensitive to both specific-acid and general-acid catalysis. This particular reaction, which occurs in atmospheric water droplets (e.g., clouds, fogs, and rain), provides a very important pathway for the oxidation of SO_2 on a global basis and for the production of acidic atmospheres in the near-urban troposphere (Hoffmann and Jacob, 1984). Acid catalysis of the $S(IV)$–H_2O_2 reaction is a major molecular-level reason for the importance of this siagle reaction in the conversion of S(IV) in the atmosphere. The subtleties of this assertion will be explained below.

The stoichiometry for the reaction of S(IV) with H_2O_2 is quite simple:

$$H_2O_2 + HSO_3^- \longrightarrow H_2O + H^+ + SO_4^{2-} \tag{46}$$

The oxidation of aquated sulfur dioxide proceeds via a nucleophilic displacement of HSO_3^- by H_2O_2 to form peroxymonosulfurous acid as an intermediate that, in turn, undergoes an acid-catalyzed (both specific-acid and general-acid) rearrangement to give the products as described by the stoichiometry of Eq. 46 (McArdle and Hoffmann, 1983). The mechanism can be written as follows:

$$H_2O \cdot SO_2 \overset{k_{a1}}{=\!=\!=} H^+ + HSO_3^- . \tag{47}$$

$$HSO_3^- + H_2O_2 \underset{k_{-1}}{\overset{k_1}{\rightleftharpoons}} \begin{array}{c} ^-O \\ O \end{array}\!\!\!\!>S\!\!-\!\!OOH + H_2O \qquad (48)$$

peroxymonosulfurous acid

The rate-determining step for specific acid catalysis is

$$\begin{array}{c} ^-O \\ O \end{array}\!\!\!\!>S\!\!-\!\!OOH + H^+ \xrightarrow{k_2} HSO_4^- + H^+ \qquad (49)$$

The rate-determining step for general acid catalysis is

$$\begin{array}{c} ^-O \\ O \end{array}\!\!\!\!>S\!\!-\!\!OOH + HA \xrightarrow{k_3} HSO_4^- + HA \qquad (50)$$

The mechanism of Eqs. 47–50 were originally proposed by Hoffmann and Edwards (1975) and reaffirmed by many other investigators (Overton, 1985). The rate expression that results from this mechanism is as follows:

$$v = \frac{d[S(VI)]}{dt} = \frac{k_1 K_{a1}[H_2O_2][S(IV)]}{k_{-1} + k_2[H^+] + k_3[HA])(K_{a1} + [H^+]}(k_2[H^+] + k_3[HA]) \qquad (51)$$

where HA represents any general acid (i.e., weak acid or proton donor), $[S(IV)]$ $= [SO_2 \cdot H_2O] + [HSO_3^-] + [SO_3^{2-}]$.

In addition to the $S(IV) - H_2O_2$ reaction, the reactions of other peroxides such as peroxymonosulfate, peroxyacetic acid, and methyl hydroperoxide with $S(IV)$ are also sensitive to specific-acid catalysis (Hoffmann and Calvert, 1985). The rate of oxidation of $S(IV)$ by HSO_5^- is comparable to the rate of oxidation of $S(IV)$ by hydrogen peroxide (Betterton and Hoffmann, 1988). We have proposed a general mechanism for the ROOH–S(IV) reaction in which the rate-determining step involves the acid-catalyzed decomposition of a peroxide–bisulfite intermediate. The rate expression applicable for this mechanism is

$$-\frac{d[HSO_3^-]}{dt} = \frac{k_1(k_2/k_{-1})K_{a1}\{H^+\}[ROOH][S(IV)]}{(1+(k_2/k_{-1})\{H^+\})(K_{a1} + \{H^+\})} \qquad (52)$$

where for HSO_5^- $k_1 = 1.21 \times 10^6 \, M^{-1}s^{-1}$, $k_2/k_{-1} = 5.9 \, M^{-1}$, $k_1 k_2/k_{-1} = 7.14 \times 10^6 \, M^{-2} s^{-1}$, $K_{a1} = 2.64 \times 10^{-2} \, M$ at 5°C and $\mu = 0.2 \, M$. For oxidation of $S(IV)$ by H_2O_2 (at 15°C, $\mu = 1.0 \, M$), the equivalent parameters are $k_1 = 2.6 \times 10^6 \, M^{-1}s^{-1}$, $k_2/k_{-1} = 16 \, M^{-1}$, and $K_1 k_2 = 2.4 \times 10^7 \, M^{-2}s^{-1}$. The rate constants for the two oxidants are thus very similar when allowance is made for the difference in temperature, but the activation parameters, particularly $\Delta S_{k1}^{\neq}$ (the entropy of activation for the k_1 step), are markedly different: $\Delta H_{k1}^{\neq} = (25.74 \pm 0.77) \, kJ \, mol^{-1}$; and $\Delta S_{k1}^{\neq} = (-88.1 \pm 2.7) \, J \, mol^{-1} \, K^{-1}$ for HSO_5^-; but $\Delta H_{k1}^{\neq} = 37 \, kJ \, mol^{-1}$; and $\Delta S_{k1}^{\neq} = 4 \, J \, mol^{-1} \, K^{-1}$ for H_2O_2.

By analogy to the reactions of H_2O_2, CH_3OOH, and CH_3CO_3H with HSO_3^-, the following acid-catalyzed nucleophilic substitution is proposed as the favored mechanism for the oxidation of S(IV) by HSO_5^-. In this mechanism an intermediate, containing two sulfur atoms in two formal oxidation states [S(IV) and S(VI)], and with a molecular formula $S_2O_7^{2-}$, undergoes an acid-catalyzed (i.e., the acid can be either H^+ or a weak acid HA) rearrangement to yield the disulfate ion, $S_2O_7^{2-}$, which, in turn, undergoes an acid-catalyzed hydrolysis to yield HSO_4^-;

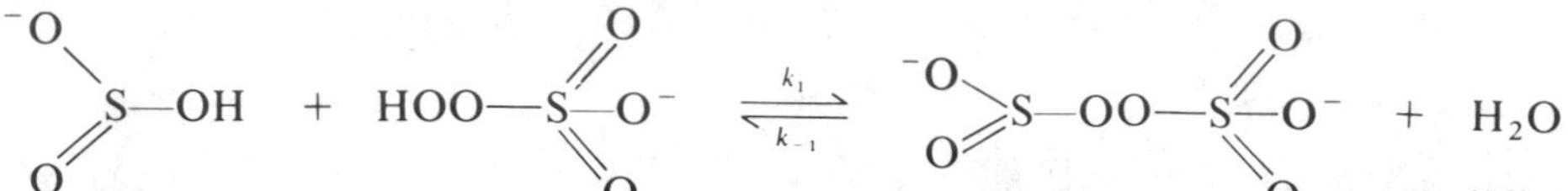

$$(53)$$

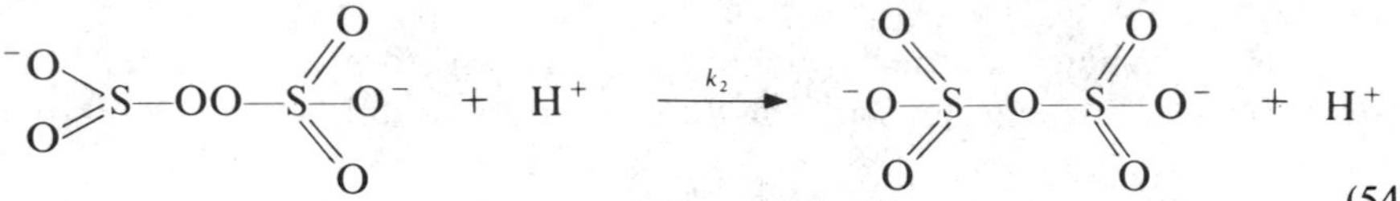

$$(54)$$

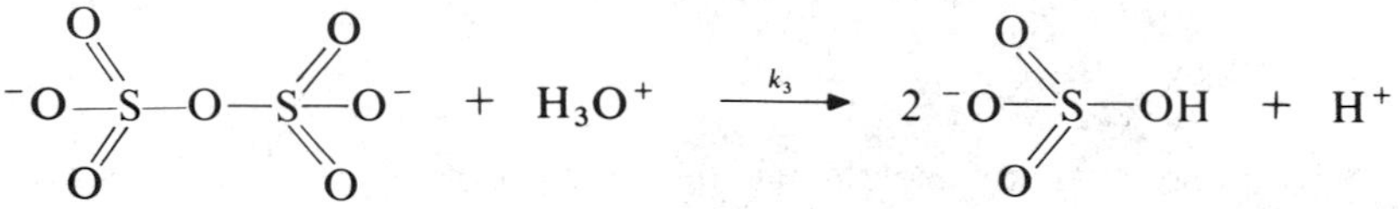

$$(55)$$

The similar rates of oxidation of S(IV) by HSO_5^- and by H_2O_2 suggests that the respective reactions may lead to similar transition-state intermediates. If the intermediate has the following structure:

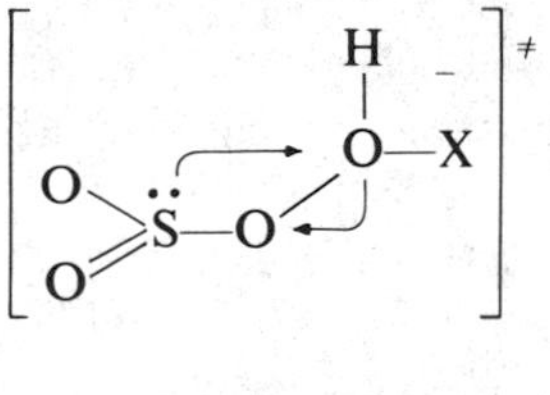

$$(56)$$

where $X = SO_3^-$ (for HSO_5^-), H (for H_2O_2), CH_3 (for CH_3OOH), and CH_3C (for CH_3CO_3H), then the ability of the substituent X to enhance the protonation of O and thus to facilitate S–O bond formation may be compensated by the tendency of X to stabilize the O–O bond.

Although peroxymonosulfate is potentially highly reactive from a thermodynamic point of view, there may be a kinetic barrier for reaction with HSO_3^-. This, in turn, suggests that under certain conditions in tropospheric water droplets where S(IV) is the dominant reductant the concentration of HSO_5^- may build up to unexpectedly high levels.

In an open atmospheric system, the oxidation of S(IV) in a unbuffered water droplet quickly lowers the pH. This has the dual effect of shutting down the oxidation of S(IV) by other oxidants such as O_3, O_2, and $\cdot OH$ (v_{O_3}, v_{O_2}, and v_{OH} are proportional to $[H^+]^{-1}$) and lowering the total concentration of S(IV) in the droplet according to the following relationship

$$[S(IV)] = H_{SO_2} P_{SO_2} \alpha_0^{-1} \tag{57}$$

where

$$\alpha_0 = \frac{[H^+]^2}{[H^+]^2 + K_{a1}[H^+] + K_{a1}K_{a2}}$$

In the case of the H_2O_2–S(IV) reaction the increase in the reaction rate with a decrease in pH offsets the lowering in the S(IV) concentration with a decrease in pH. Thus, for open-phase conditions found in a cloudwater droplet, the H_2O_2–S(IV) reaction essentially is independent of pH (except when $K[H^+] \gg 1$) as shown in Eq. 58:

$$\frac{d[S(VI)]}{dt} = v_{open-phase} = \frac{k' P_{H_2O_2} H_{H_2O_2} P_{SO_2} H_{SO_2}}{1 + K[H^+]} \tag{58}$$

where $K - 16\ M^{-1}$, $k' = 7.5 \times 10^5\ M^{-1}s^{-1}$, $P_{H_2O_2}$ and P_{SO_2} are the respective partial pressures, H_{SO_2} and $H_{H_2O_2}$ are the respective Henry's law constants, and S(VI) represents total sulfate $\{[S(VI)] = [SO_4^{2-}] + [HSO_4^-]\}$. In addition to this unique feature among the oxidants of S(IV), H_2O_2 is highly soluble in the aqueous phase with a Henry's law constant of $H_{H_2O_2} = 10^5\ M\ atm^{-1}$ ($P_{H_2O_2} \simeq 1 - 4 \times 10^{-9}$ atm). For comparison, $H_{O_3} = 10^{-2}$ and $H_{SO_2} = 1.24\ M\ atm^{-1}$, respectively.

b. Ester Hydrolysis. In addition to the nucleophilic and electrophilic substitutions discussed above, we find that most hydrolytic reactions in natural systems are sensitive to specific-acid or specific-base catalysis. In acid-catalyzed ester hydrolysis, the role of the proton is often thought to provide a reaction pathway of lower energy by withdrawing electrons for the carbonyl group of the ester and therefore weakening the –C–O– bond to be broken. In base-catalyzed ester hydrolysis the role of the base (OH⁻) is to act as the active nucleophile in the initial attack on the carbonyl carbon of the ester. Since OH⁻ is a much more effective nucleophile (nucleophilicity is a function of basicity and polarizability) than H_2O (the conjugate acid of OH⁻), we anticipate that a hydrolysis reaction that proceeds via a mechanism involving the nucleophilic attack of OH⁻ will occur at a faster rate the pathway involving attack by H_2O alone.

Hydrolysis represents a significant pathway for the abiotic transformation of organic esters such as carboxylates, amides, carbamates, and organophosphates (Mabey and Mill, 1978; Mill and Mabey, 1987; Macalady and Wolfe, 1983) in various aquatic environments. A mechanism for acid catalysis has been suggested by Bender (1971), as follows:

$$
\underset{R\quad OR'}{\overset{O}{\underset{\|}{C}}} \;+\; H^+ \quad \overset{\text{fast}}{\underset{k_{-1}}{\overset{k_1}{\rightleftharpoons}}} \quad \underset{R\quad OR'}{\overset{OH^+}{\underset{\|}{C}}}
\tag{59}
$$

$$
\underset{R\quad OR'}{\overset{OH^+}{\underset{\|}{C}}} \;+\; H_2O \quad \overset{\text{slow}}{\underset{k_{-2}}{\overset{k_2}{\rightleftharpoons}}} \quad \underset{\substack{R\;\;O\;\;OR'\\ \;\;\;H}}{\overset{OH}{\underset{\|}{C}}} \;+\; H^+
\tag{60}
$$

$$
\underset{\substack{R\;\;O\;\;O-R'\\ \;\;\;H}}{\overset{OH}{C}} \;+\; H^+ \quad \underset{k_{-3}}{\overset{k_3}{\rightleftharpoons}} \quad \underset{\substack{R\;\;O\;\;O\\ \;\;\;H\;\;\;R'}}{\overset{OH}{C}}H^+ \quad \rightleftharpoons \quad R'OH \;+\; R-\underset{OH}{\overset{OH}{C}}{}^+
\tag{61}
$$

$$
R-\underset{OH}{\overset{OH}{C}} \quad \rightleftharpoons \quad R-\underset{OH}{\overset{O}{C}} \;+\; H^+
\tag{62}
$$

The base-catalyzed pathway proceeds via the direct nucleophilic addition of OH^- to the carbonyl carbon as follows:

$$
\underset{R\quad OR'}{\overset{O}{\underset{\|}{C}}} \;+\; OH^- \quad \overset{\text{slow}}{\underset{k_{-4}}{\overset{k_4}{\rightleftharpoons}}} \quad \underset{\substack{R\;\;O\;\;OR'\\ \;\;\;H}}{\overset{O^-}{\underset{|}{C}}}
\tag{63}
$$

$$
\underset{\substack{R\;\;O\;\;OR'\\ \;\;\;H}}{\overset{O^-}{\underset{|}{C}}} \quad \underset{k_{-5}}{\overset{k_5}{\rightleftharpoons}} \quad R'O^- \;+\; R-\underset{OH}{\overset{O}{C}}
\tag{64}
$$

$$
R'O^- \;+\; H_2O \quad \underset{k_{-6}}{\overset{k_6}{\rightleftharpoons}} \quad R'OH \;+\; OH^-
\tag{65}
$$

The overall rate of hydrolysis of esters is also sensitive to the effects of general-acid catalysis and general-base catalysis, although in most natural waters these effects are not significant (Perdue and Wolfe, 1983). The total rate of hydrolysis for a particular ester is then the summation of terms for the various pathways.

$$v_T = -\frac{d[\text{R--C(=O)--OR}']}{dt} = k_{\text{obs}}[\text{R--C(=O)--OR}'] \tag{66}$$

where k_{obs} is a pseudo-first-order rate constant containing multiple terms.

$$k_{\text{obs}} = k_n[\text{H}_2\text{O}] + k_2 K_1[\text{H}^+] + k_4[\text{OH}^-] + \sum_i k_i[\text{HA}_i] + \sum_j k_j[\text{B}_j] \tag{67}$$

The k_n term reflects the rate of hydrolysis due to a water molecule alone; the $k_2 K_1$ and k_4 terms account for specific-acid and specific-base catalysis, respectively; and the k_i and k_j terms reflect the contributions made by general-acid and general-base catalysis, respectively.

  c. *Reversible Nucleophilic Additions.*   The nucleophilic addition of S(IV) to aldchydes to form α-hydroxysulfonates has been shown to be an important pathway for the apparent stabilization of S(IV) in the atmospheric aqueous phase (Munger et al., 1986). Below pH $\sim$ 3, the rate of adduct formation is governed by the nucleophilic attack of bisulfite or sulfite on the carbonyl carbon of the electrophile. As with the addition of other strong nucleophiles to the carbonyl group, sulfite addition is not strongly catalyzed by general acids or bases. However, specific-acid catalysis is observable at pH ≤ 1 (a pH typical of highly acidic haze aerosol). The mechanism for the formation of hydrox-yalkylsulfonates in the simplest case is given by Eqs. 68–75 (Boyce and Hoffmann, 1984; Olson et al., 1986, 1988; Betterton and Hoffmann, 1987, 1988; Betterton et al., 1988; Olson and Hoffmann, 1988, 1989).

$$\text{RCH(OH)}_2 \underset{}{\overset{K_d}{\rightleftharpoons}} \text{RCHO} + \text{H}_2\text{O} \qquad\qquad \text{fast} \tag{68}$$

$$\text{H}_2\text{O}\cdot\text{SO}_2 \underset{}{\overset{K_{a1}}{\rightleftharpoons}} \text{HSO}_3^- + \text{H}^+ \qquad\qquad \text{fast} \tag{69}$$

$$\text{HSO}_3^- \underset{}{\overset{K_{a2}}{\rightleftharpoons}} \text{SO}_3^{2-} + \text{H}^+ \qquad\qquad \text{fast} \tag{70}$$

$$\text{RCHO} + \text{HSO}_3^- \underset{k_{-1}}{\overset{k_1}{\rightleftharpoons}} \text{RCH(OH)SO}_3^- \qquad\qquad \text{slow} \tag{71}$$

$$\text{RCHO} + \text{SO}_3^{2-} \underset{k_{-2}}{\overset{k_2}{\rightleftharpoons}} \text{RCH(O}^-)\text{SO}_3^- \qquad\qquad \text{slow} \tag{72}$$

$$\text{RCHO} + \text{H}^+ \underset{}{\overset{1/K_{a0}}{\rightleftharpoons}} \text{RCHOH}^+ \qquad\qquad \text{fast} \tag{73}$$

$$RCHOH^+ + HSO_3^- \underset{k_{-0}}{\overset{k_0}{\rightleftharpoons}} RCH(OH)SO_3^- + H^+ \quad \text{slow} \tag{74}$$

$$RCH(OH)SO_3^- \overset{K_{a4}}{\rightleftharpoons} RCH(O^-)SO_3^- + H^+ \qquad \text{fast} \tag{75}$$

This mechanism is appropriate for the addition of S(IV) to formaldehyde, benzaldehyde, methylglyoxal, and hydroxyacetaldehyde, where Eqs. 73–74 represent the acid-catalyzed pathway. The corresponding three-term rate expression is

$$v_1 = \frac{d[RCH(OH)SO_3^-]}{dt} = \left(\frac{k_0 \{H^+\} \alpha_1}{K_{a0}} + k_1 \alpha_1 + k_2 \alpha_2 \right)$$
$$\times \left(\frac{K_d}{1 + K_d} \right) [S(IV)][RCHO]_T \tag{76}$$

where $[S(IV)] = [H_2O \cdot SO_2] + [HSO_3^-] + [SO_3^{2-}]$. This mechanism is based on the supposition that the dehydration of the carbonyl form of the aldehyde is rapid. However, the assumption is critically dependent on the pH range of interest.

   d. Hydration–Dehydration Reactions.   The dehydration of the *gem*-diol form of the aldehydes is sensitive to both specific-acid and specific-base catalysis and general-acid and general-base catalysis where the macroscopic dehydration rate constant is given by

$$k_d = k_W + K_{H+}[H^+] + k_{OH-}[OH^-] + k_{HA}[HA] + k_{B-}[B^-] \tag{77}$$

where in the specific case of methyl glyoxal, Betterton and Hoffmann (1987) have determined the following values: $k_W = 0.09 \text{ s}^{-1}$, $K_{H+} = 28 \ M^{-1} \text{s}^{-1}$, $k_{OH-} = 3.1 \times 10^4 \ M^{-1} \text{s}^{-1}$, $k_{HA}(H_2PO_4^-) = 0.08 \ M^{-1} \text{s}^{-1}$, and $k_{B-}(HPO_4^-) = 0.57 \ M^{-1} \text{s}^{-1}$.

2.1.2.   *Metal Ion and Metal-Complex Catalysis*

   a. Metal-Catalyzed Ester Hydrolysis.   In selected cases an additional term must be added to Eq. 67 to account for the role of metal ions and their complexes as catalysts for ester hydrolysis (Plastourgou and Hoffmann, 1984). A metal ion or metal- ion complex can mediate or facilitate hydrolysis or nucleophilic displacement in two ways: either by direct polarization of the substrate and subsequent external attack by the nucleophile (i.e., OH^-/H_2O) or by the *in situ* generation of a reactive basic reagent such as MOH^+. In the first case, the metal ion functions as a Lewis acid that directly polarizes the substrate and enhances displacement reactions:

$$\text{OH}^-, \text{H}_2\text{O}, \text{N:} \quad + \quad \underset{R \quad\quad Y}{\overset{\displaystyle \overset{\big|}{\underset{\displaystyle \text{O}}{\text{Zn}^{2+}}}}{\underset{\displaystyle \text{C}}{\|}}} \quad\longrightarrow \tag{78}$$

where N: is a nucleophile. Polarization can also be achieved by coordination to the leaving group as in the case of phosphate esters. Divalent metal ions can increase the rate of hydrolysis by a factor of $\sim 10^4$ when a direct polarization mechanism is operative.

In the second case, the aquated metal ion hydrolyzes to give a coordinated reactive nucleophile. The bimolecular nucleophilic displacement can be represented as follows:

$$
\begin{array}{ccc}
\overset{+}{\text{Zn}}\!\!-\!\!\text{OH} & & \overset{+}{\text{Zn}}\!\!-\!\!\text{N} \\
\quad\vdots & \text{or} & \text{HO:} \quad\to\quad \text{C=O} \\
R\text{—C=O} & & \quad Y \\
\quad Y & &
\end{array} \tag{79}
$$

A general mechanism for catalysis by a metal–ligand complex is as follows:

$$\text{M(OH)}_n^{m-n+} + \text{R}''\text{COOR}' \xrightarrow{k_m} (\text{OH})_{n-1}\,\text{M—O—}\underset{\overset{\displaystyle |}{\text{OR}''}}{\overset{\overset{\displaystyle \text{OR}'}{|}}{\text{C}}}\text{—OH]}^{m-n+} \tag{80}$$

$$(\text{OH})_{n-1}\,\text{M—O—}\underset{\overset{\displaystyle |}{\text{OR}''}}{\overset{\overset{\displaystyle \text{OR}'}{|}}{\text{C}}}\text{—OH]}^{m-n+} \xrightarrow{k_p} (\text{OH})_{n-1}\,\text{M—O—}\overset{\overset{\displaystyle \text{O}}{\|}}{\text{C}}\text{—R}'^{m-n+} + \text{R}''\text{OH} \tag{81}$$

with a corresponding overall rate expression

$$-\frac{d[\text{R}'\overset{\overset{\displaystyle \text{O}}{\|}}{\text{C}}\text{OR}'']}{dt} = (k_m[\text{M(OH)}_n^{m-n+}] + k_n + k_2 K_1[\text{H}^+] + k_4[\text{OH}^-])[\text{R}'\overset{\overset{\displaystyle \text{O}}{\|}}{\text{C}}\text{OR}''] \tag{82}$$

Kinetic data for the base-catalyzed hydrolysis of substituted p-nitrophenyl acetates is shown in Figure 3, while in Figure 4, the relative catalytic effect of a variety of metal hydroxy complexes is compared in a linear free–energy relationship (LFER) of $\log k_{\text{MeOH}}$ versus $pK_{\text{MeH}_2\text{O}}$. The LFER of Figure 4 establishes

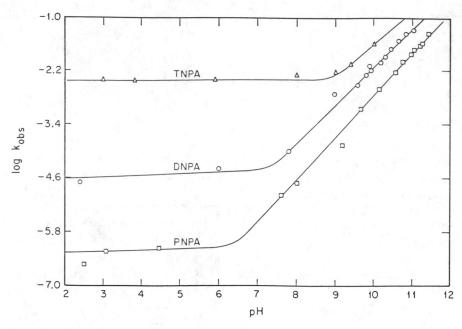

Figure 3. Kinetic plots of log k_{obs} versus pH for the base-catalyzed hydrolysis of p-nitrophenyl acetate (NPA), dinitrophenyl acetate (DNPA), and trinitrophenyl acetate (TNPA).

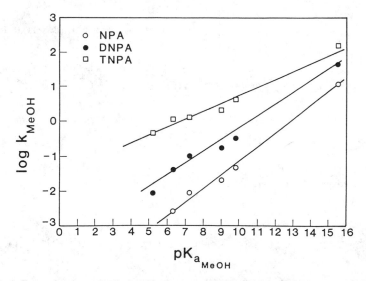

Figure 4. A linear free-energy relationship between the logarithm of rate constant for metal-catalyzed hydrolysis (log k_{MeOH}) of p-nitrophenyl acetate (NPA), dinitrophenyl acetate (DNPA), and trinitrophenyl acetate (TNPA) and the hydrolysis constant ($MeH_2O \rightleftharpoons MeOH + H^+$) for the formation of the corresponding hydroxy complex. This relationship provides support for the argument that the MeOH species is involved in a direct nucleophilic attack on the carbonyl carbon of the nitrophenyl acetate esters.

that the value of k_m increases as the first hydrolysis constant (K_{MOH}) for the complexes decreases (e.g., as pK_{MOH} increases, the basicity of the MOH functionality increases). It also provides supporting evidence for the mechanism of Eqs. 79–81.

b. Metal-Catalyzed Autoxidation.

Autoxidation reactions of many inorganic and organic compounds are accelerated in the presence of first-row transition-metal ions and complexes in which the metal center has access to more than one stable oxidation state. A considerable amount of research has been directed toward determining the mechanisms of transition-metal catalysis in the reactions of O_2 with a wide variety of reduced sulfur compounds such as SO_2, H_2S, and mercaptans, RSH in aqueous solution (Hoffmann and Lim, 1979; Boyce et al., 1983; Hoffmann and Hong, 1987; Leung and Hoffmann, 1988, 1989).

The noncatalytic autoxidation of mercaptans may occur as follows:

$$^3\Sigma_g^+ O_2 + RSH_{(3p_u \to 3d_g)} \to O_2^- + RS^\bullet + H^+ \tag{83}$$

or

$$^1\Delta_g O_2 + RSH_{(3p_u)} \to O_2^- + RS^\bullet + H^+ \tag{84}$$

In the first reaction, the half-filled π_g^* orbital of ground-state oxygen overlaps with an excited sulfur atom ($3p_u \to 3d_g$ transition) on the mercaptan, whereas in the second reaction, a direct overlap of empty π_g^* orbital of an excited-singlet-state oxygen with the filled $3p_u$ orbital of the sulfur atom is required. However, both reactions have relatively large activation energies. These large activation-energy barriers for autoxidation can be overcome by metal-complex catalysis in which both the substrate, RSH, and the oxidant, O_2, are simultaneously bound to the metal complex.

Cobalt(II)-tetrasulfophthalocyanine, Co(II)TSP, has been found to be an effective catalyst for the autoxidation of 2-mercaptoethanol and cysteine. In our study of the autoxidation of 2-mercaptoethanol, ethanethiol, and 2-aminoethanethiol as catalyzed by Co(II)TSP, we provided evidence for a mechanism that proceeds via a catalytic center that is bridged by the 2-mercaptoethanol anion (Leung and Hoffmann, 1988, 1989). In the proposed mechanism, electron transfer from the Co(II) metal center to bound dioxygen was considered to be the rate-determining step, which was followed by the release of hydrogen peroxide and a mercaptan radical. These intermediates react further to produce disulfide, RSSR, as the final product of the autoxidation. Two additional 2-mercaptoethanol molecules are oxidized by H_2O_2 to give the disulfide.

The following general rate law was found to hold for the substrates over pH range of 8.8–13.5:

$$v = \frac{-d[RS^-]}{dt} = k_{obs}[CoTSP]_T[RS^-] \tag{85}$$

where k_{obs} is of the following form:

$$k_{obs} = \frac{k_{31}K_{21} + (k_{32}K_{22}K'_1/a_{H^+}) + k_{33}K_{23}K'_1K'_2/a^2_{H^+}}{[1 + (K'_1/a_{H^+}) + (K'_1K'_2/a^2_{H^+})][1 + (a_{H^+}/K'^2_{a1})]} \qquad (86)$$

where k_{3i} and K_{2i} are the rate constants for the electron transfer and the equilibrium constants for substrate complexation of the ith catalytic center, respectively; and K'_2, K'_2, and K'_{a1} are the apparent acid dissociation constants of the pyrrole groups of Co(II)TSP–RS$=$Co(II)TSP and Co(II)TSP–RS$=$Co(II) TSP$^-$, and RCH$_2$CH$_2$SH, respectively.

Hydrogen peroxide and mercaptan radicals, RS$^\bullet$, have been identified as reaction intermediates in the autoxidation of both substrates. These intermediates react further to produce the corresponding disulfide, RSSR, as the final product of the autoxidation.

Transition-metal ions such as Fe(III), Cu(II), Co(II), Co(III), and Mn(II) have been shown to be effective homogeneous catalysts for the autoxidation of sulfur dioxide in aqueous solution. Hoffmann and coworkers have shown that Fe(III) and Mn(II) are the most effective catalysts at ambient concentrations for the catalytic autoxidation of S(IV) to S(VI) in cloudwater and fogwater (Jacob and Hoffmann, 1983; Hoffmann and Jacob, 1984; Hoffmann and Calvert, 1985). Mechanisms for the homogeneous catalysis by Fe(III) and Mn(II) that have been proposed include a free-radical chain mechanism, a polar mechanism involving inner-sphere complexation followed by a two-electron transfer from S(IV) to bound dioxygen, and photoassisted electron transfer.

The reaction of ground-state molecular oxygen with S(IV) proceeds at very slow rates in the absence of light or specific catalysts. Transition metals, such as Fe(III), which are capable of coordinating S(IV) and/or O$_2$ accelerate the rate of S(IV) autoxidation. The mechanisms invoked to explain transition-metal-catalyzed S(IV) autoxidations can be classified into two categories: (1) one-electron free-radical reactions and (2) two-electron polar reactions involving ternary complexes comprised of the transition metal, S(IV), and molecular oxygen.

FREE-RADICAL CHAIN MECHANISMS.   A frequently postulated initiation step for the free-radical autoxidation of S(IV) catalyzed by Fe(III) is as follows:

$$Fe^{3+} + SO_3^{2-} \rightarrow Fe^{2+} + SO_3^{-\bullet}. \qquad (87)$$

Most researchers have assumed that electron transfer between Fe^{3+} and SO$_3^{2-}$ is an outersphere process. However, formation of an inner-sphere complex appears to be a necessary step in this mechanism when open coordination sites are available in the octahedral coordination sphere of Fe(III) (Conklin and Hoffmann, 1988a–c). On the other hand, Fe(2,2'-bipyridine)$_3^{3+}$, which has a fully occupied coordination sphere, oxidizes SO$_3^{2-}$ via an outer-sphere electron transfer to give SO$_3^{-\bullet}$ with the following rate law:

$$v = k''[Fe(bipy)_3^{3+}][SO_3^{2-}] \qquad (88)$$

where $k'' = 2.1 \times 10^8\ M^{-1}\,s^{-1}$ (Wilmarth et al., 1983). When O_2 is introduced to this system, the consumption of SO_3^{2-} is enhanced; this enhanced reaction can be attributed to the involvement of a free-radical chain reaction as shown below (Hoffmann and Boyce, 1983):

$$SO_3^{-\bullet} + O_2 \xrightarrow{k_1} SO_5^{-\bullet} \tag{89}$$

$$SO_5^{-\bullet} + HSO_3^{-} \xrightarrow{k_2} SO_4^{-\bullet} + SO_4^{2-} + H^+ \tag{90}$$

$$SO_5^{-\bullet} + SO_3^{2-} \xrightarrow{k_3} SO_4^{-\bullet} + SO_4^{2-} \tag{91}$$

$$SO_4^{-\bullet} + HSO_3^{-} \xrightarrow{k_4} SO_3^{-\bullet} + SO_4^{2-} + H^+ \tag{92}$$

$$SO_4^{-\bullet} + SO_3^{2-} \xrightarrow{k_5} SO_3^{-\bullet} + SO_4^{2-} \tag{93}$$

$$2SO_3^{-\bullet} \xrightarrow{k_6} S_2O_6^{2-} \tag{94}$$

$$2SO_4^{-\bullet} \xrightarrow{k_7} S_2O_8^{2-} \tag{95}$$

$$2SO_5^{-\bullet} \xrightarrow{k_8} S_2O_8^{2-} + O_2 \tag{96}$$

However, when several open coordination sites (sites normally occupied by water) on the metal catalyst are available the following mechanism appears to be operative (Hoffmann and Jacob, 1984; Conklin and Hoffmann, 1988a–c):

$$HSO_3^{-} \underset{}{\overset{K_{a2}}{\rightleftharpoons}} SO_3^{2-} + H^+ \tag{97}$$

$$Fe^{3+} + H_2O \underset{}{\overset{K_b^*}{\rightleftharpoons}} FeOH^{2+} + H^+ \tag{98}$$

$$FeOH^{2+} + SO_3^{2-} \underset{}{\overset{K_0}{\rightleftharpoons}} HOFeOSO_2$$
$$\text{red complex} \tag{99}$$

$$HOFe^{III}OS^{IV}O_2 \underset{k_{-1}}{\overset{k_1}{\rightleftharpoons}} HOFe^{II}OS^{V}O_2$$
$$\text{red complex} \qquad\qquad \text{orange complex} \tag{100}$$

$$HOFe^{II}OS^{V}O_2 + O_2 \underset{k_{-2}}{\overset{k_2}{\rightleftharpoons}} HOFe^{III}OS^{V}O_2\!\!\underset{}{\overset{\displaystyle O^{\cdot\,\cdot}}{\diagup}}\!\!\overset{}{\underset{}{O}}$$
$$\text{orange complex} \tag{101}$$

$$\underset{\substack{\text{HOFe}^{\text{III}}\text{OS}^{\text{V}}\text{O}_2}}{\overset{\displaystyle O{=}O^{\cdot-}}{|}} \xrightarrow{k^3} \underset{\substack{\text{HOFe}^{\text{III}}\text{OS}^{\text{VI}}\text{O}_2}}{\overset{\displaystyle O{=}O^{2-}}{|}} \xrightarrow{+2^+} \text{HOFe}^{2+} + \text{SO}_3 + \text{H}_2\text{O}_2 \quad (102)$$

$$\text{SO}_3 + \text{H}_2\text{O} \longrightarrow 2\text{H}^+ + \text{SO}_4^{2-} \quad\quad\quad (103)$$

$$\text{H}_2\text{O}_2 + \text{HSO}_3^- \longrightarrow \text{H}_2\text{O} + \text{H}^+ + \text{SO}_4^{2-} \quad\quad (104)$$

The rate expression obtained from the above mechanism is

$$\frac{d[\text{SO}_4^{2-}]}{dt} = \left[\frac{k_2}{k_3 + k_{-2}}\right]\left[\frac{k_1}{k_2[\text{O}_2] + k_{-1}}\right] K_0 \alpha_1 \beta_1 [\text{Fe(III)}][\text{S(IV)}][\text{O}_2]$$

$$(105)$$

If $k_3 \gg k_{-2}$ and $k_2[\text{O}_2] \gg k_{-1}$, this rate expression can be reduced to

$$\frac{d[\text{SO}_4^{2-}]}{dt} = k_1 K_0 \alpha_2 \beta_1 [\text{Fe(III)}][\text{S(IV)}] \quad\quad (106)$$

The predicted rate expression closely approximates the empirical rate law as observed by many investigators (Hoffmann and Calvert, 1985):

$$-\frac{d[\text{S(IV)}]}{dt} = k[\text{Fe(III)}][\text{S(IV)}]\alpha_2 \quad\quad (107)$$

One possible structure for the active catalytic intermediate, which takes into account the known geometry for octahedrally coordinated Fe(III) and the nature of the Fe–O–S bonding as determined by Raman measurements, is as follows (Conklin and Hoffmann, 1988a–c):

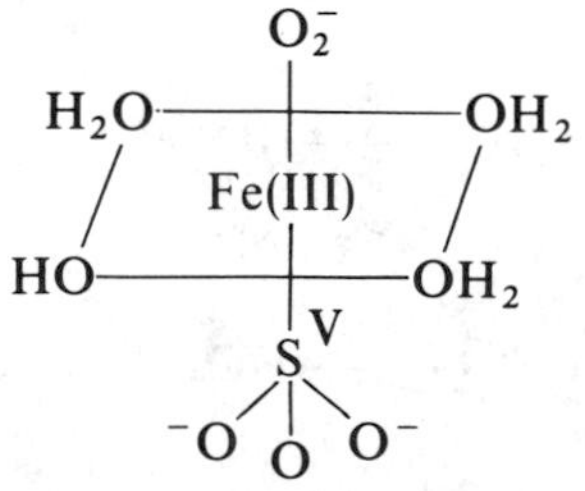

Recent model calculations by a number of investigators (Pandis and Seinfeld, 1989; Jacob, in press; Graedel et al., 1986; Jacob and Hoffmann, 1983; Seigneur and Saxena, 1988) have shown that the Fe(III)-catalyzed autoxidation of S(IV) represent a major pathway for the production of acidity in clouds and fogs.

2.2. Heterogeneous Catalysis

2.2.1. Surface-Catalyzed Ester Hydrolysis. Mineral surfaces (e.g., metal oxides) may accelerate the rate of ester hydrolysis by similar mechanisms at the solid–liquid interface as reported by Stone (1989), who has argued that accelerated ester hydrolysis is due to a combination of the enhancement of $[OH^-]$ in the electrical double layer relative to the bulk solution and specific adsorption of the ester to the surface. These heterogeneous reactions can be interpreted in terms of a mechanism involving the direct nucleophilic additions of the ▮–SOH and ▮–SiO$^-$ surface groups to the carbonyl carbon atom of the carboxylic acid esters followed by the dissociation of the surface complexes. An intermediate surface complex in the specific case of the *p*-nitrophenyl acetates (Gonzalez and Hoffmann, 1989a), which is similar to the one formed in the homogeneous reactions (Eqs. 80–81), is postulated.

$$\text{▮—SOH} + \underset{\underset{CH_3}{|}}{\overset{\overset{O}{\|}}{C}}\text{—O—}\bigcirc\text{—NO}_2 \rightleftharpoons \text{▮—S—O—}\underset{\underset{CH_3}{|}}{\overset{\overset{OH}{|}}{C}}\text{—O—}\bigcirc\text{—NO}_2 \tag{108}$$

$$\text{▮—S—O—}\underset{\underset{CH_3}{|}}{\overset{\overset{OH}{|}}{C}}\text{—O—}\bigcirc\text{—NO}_2 \;\blacktriangleright\; \text{▮ S O } \underset{CH_3}{\overset{O}{C}}$$

$$\text{HO—}\bigcirc\text{—NO}_2 \tag{109}$$

where ▮–SOH represent a surface hydroxyl group of a generic metal oxide particle in aqueous suspension. The corresponding rate expression is as follows:

$$\frac{d[P]}{dt} = \left\{ k_1 \left[\frac{S_T}{([H^+]/K_{a1}^s) + 1 + (K_{a2}^s/[H^+])} \right] \right.$$

$$+ k_2 \left[\frac{S_T}{([H^+]^2/K_{a1}^s K_{a2}^s) + ([H^+]/K_{a2}^s) + 1} \right]$$

$$\left. + k_3 [OH^-] \right\} [NPA] \tag{110}$$

where the k_1 term represents the reaction involving ▮–SOH surface sites, the k_2 term represents the reaction involving ▮–SiO$^-$ surface sites, S_T is the total site concentration, and the k_3 term represents the base-catalyzed hydrolysis term.

Other aspects of surface catalysis, including electrical double-layer effects, must also be taken into account for a complete kinetic description (Stone, 1989).

Results of these studies indicate that surfaces may have a pronounced influence on the rates of ester hydrolysis, especially in aquifers and at the sediment–water interface.

2.2.2. Heterogeneous Photocatalysis on the Surface of Metal Oxide Semiconductors

a. General Considerations.   Semiconductors such as TiO_2, Fe_2O_3, CdS, and ZnS are known to be photochemical catalysts for a wide variety of reactions. When a photon with an energy of hv matches or exceeds the band-gap energy, E_g, of the semiconductor, an electron, e^-, is promoted from the valence band, VB, into the conduction band, CB, leaving a hole, h^+, behind. Electrons and holes can then recombine (and dissipate the input energy as heat) or migrate to the surface and be trapped in surface states of the material. Surface-site trapping becomes the predominant process if redox catalysts such as Pt (an efficient e^- trap) or RuO_2 (an h^+ trap) are in contact with the semiconductor's surface. Thus several electrons and holes from consecutive photon absorption processes can be stored on a single particle that acts as a reservoir for redox reactions with solute molecules.

The redox potentials of both e^- and h^+ are determined by the relative position of the conduction and velence band, respectively. Band-gap positions are material constants that are known for a variety of semiconductors. Most materials show "Nernstian" behavior, which results in a shift of the surface potential by 59 mV in the negative direction with a $\Delta pH = +1$. Thus, electrons are better reductants in the basic pH range, while holes have a higher oxidation potential in the acidic pH range. With the right choice of semiconductor and pH, the redox potential of the e_{cb}^- can be varied from $+0.5$ to -1.5 V and that of the h_{vb}^+, from $+1.0$ to more than $+3.5$ V. If $\Delta G° < 0$ for the overall reaction, then photoassisted reactions can be classified as photocatalytic, whereas if $\Delta G° > 0$, then photoassisted reactions can be classified as photosynthetic.

b. Photocatalytic Oxidation of S(IV) by O_2.   As mentioned above, the catalytic autoxidation of SO_2 in deliquescent haze aerosol, clouds, fogs, and hydrometeors appears to be a viable pathway for the rapid formation of sulfuric acid in humid atmospheres. Jacob and Hoffmann (1983) have shown that Fe(III) and Mn(II) are the most effective catalysts at ambient concentrations for the autoxidation of S(IV) to S(VI) in cloudwater and fogwater. Although Fe(III) has been found in high ($\simeq 1$ mM) concentrations in atmospheric water droplets, much of this iron is predicted to be in form of solid particles or colloids based on thermodynamic considerations (Stumm and Morgan, 1981). Ferric oxides and oxyhydroxides (α-Fe_2O_3, Fe_3O_4, α-FeOOH, γ-FeOOH) have been identified as components of airborne particles; iron-containing particles such as these are likely to serve as cloud and fog condensation nuclei and to be suspended and/or dissolved in the liquid droplet that results.

Metal oxide semiconductors such as α-Fe_2O_3 (hematite) can function as either photosensitizers or photocatalysts. Absorption of a photon with an energy equal to or greater than the band-gap energy E_g of a semiconductor results in the transient formation of an electron–hole pair:

$$\alpha\text{-}Fe_2O_3 \xrightarrow{h\nu} e_{cb}^- + h_{vb}^+ \tag{111}$$

In the absence of suitable electron and hole scavengers adsorbed to the surface of a semiconductor particle, recombination occurs within 1 ns. However, when appropriate scavengers are present, the valence-band holes, h_{vb}^+, ($E_H^7 = 2.3$ V {oxidation potential} function as powerful oxidants while the conduction-band electrons, e_{cb}^-, ($E_H^7 = 0.0$ V {reduction potential} function as moderately powerful reductants.

The kinetics and mechanism of the photoassisted oxidation of S(IV) in the presence of suspensions of α-Fe_2O_3 have been studied over the pH range of 2–10.5 (Faust et al., 1989). Similar kinetic behavior toward S(IV) was observed for colloidal suspensions of TiO_2. Quantum yields, ϕ, ranged from 0.08 to 0.3 with a maximum yield found at pH 5.7. On band-gap illumination, conduction-band electrons and valence-band holes are separated; the trapped electrons e_{tr}^-, are transferred either to surface-bound dioxygen or to Fe(III) sites on or near the surface, while the trapped holes h_{tr}^+ accept electrons from adsorbed S(IV) to produce S(VI). The formation of S(V) radicals indicates that the reaction proceeds via successive one-electron transfers. The relatively high quantum yields can be attributed in part to the desorption of SO_3^- from the α-Fe_2O_3 surface and subsequent initiation of a homogeneous free-radical chain reaction.

The photoassisted heterogeneous oxidation of S(IV) appears to proceed via rapid formation of Fe(III)–S(IV) surface complexes formed by ligand exchange with the surface hydroxyl groups. For example, two possible surface complexes are as follows:

$$\blacksquare\text{-}FeOH + HSO_3^- \;\underset{}{\overset{K_1^s}{=\!=}}\; \blacksquare\text{-}FeOSO_2^- + H_2O \tag{112}$$

$$\blacksquare\text{-}Fe\begin{smallmatrix}OH\\\\OH\end{smallmatrix} + H^+ + HSO_3^- \;\underset{}{\overset{K_2^s}{\rightleftharpoons}}\; \blacksquare\text{-}Fe\begin{smallmatrix}O\\\\O\end{smallmatrix}S\!-\!O + 2H_2O \tag{113}$$

In the first case, a mononuclear monodentate surface complex is formed, while in the second case a mononuclear bidentate complex is formed. In the pH range of 1–3, the overall rate of reaction is limited by $[\blacksquare\text{-}FeOSO_2^-]$ or $[\blacksquare\text{-}FeO_2SO]$ up to the saturation limit imposed by the total number of available surface sites (~ 10 sites nm^{-2}). Enhanced reactivity in the pH range of 5–7 can be attributed either to the more favorable surface complexation of SO_3^{2-} (e.g., $\blacksquare\text{-Fe}\underset{\text{OH}_2}{\overset{\text{OH}}{<}}$

$+ SO_3^{2-} \rightleftharpoons \blacksquare\text{-Fe}\underset{\text{OSO}_2^-}{\overset{\text{OH}}{<}} + H_2O$) or to the increased contribution of reaction

pathways that occur independently of the surface coordination sites. For example, $SO_3^{-\bullet}$ generated at a surface site may diffuse into the bulk phase and initiate a free-radical chain reaction involving O_2. Likewise, Fe(II) can be oxidized readily above pH 5 to give Fe(III), which can, in turn, catalyze the autoxidation of S(IV) in the aqueous phase.

Light absorption with photon energies exceeding the band-gap energy of α-Fe_2O_3 leads to the formation of electron–hole pairs (Eq. 111) followed by the one-electron oxidation of adsorbed sulfite yielding the $SO_3^{3-\bullet}$ radical anion:

$$\blacksquare\text{-FeOSO}_2^- \xrightarrow[e_{cb}^-]{\overset{h\vec{v}b \,\leftarrow\, e^-}{hv}} \blacksquare\text{-FeOS(IV)O}_2^- \xrightarrow[e_{cb}^-]{} \blacksquare\text{-FeOS(V)O}_2^\bullet \quad (114)$$

The formation of sulfite radical anion (i.e., $SO_3^{-\bullet}$) as an intermediate is indicated by the fact that $S_2O_6^{2-}$ ($2SO_3^{2-\bullet} \rightarrow S_2O_6^{2-}$) is one of the end products when the illuminations are carried out under N_2.

In the absence of a suitable electron acceptor bound to the surface the $\blacksquare\text{-Fe(III)OH}$ surface site (e.g., $\blacksquare\text{-Fe(III)O=O}$), the $\blacksquare\text{-Fe(III)OH} + e_{cb}^-$ reaction produces $\blacksquare\text{-Fe(II)OH}$, which leads to the subsequent release of Fe_{aq}^{2+} to the solution phase and the progressive dissolution of the particle as shown below:

$$\begin{matrix} \blacksquare\text{-Fe(III)OS(IV)O}_2^- & & \blacksquare\text{-Fe(III)OS(V)O}_2^\bullet \\ \blacksquare\text{-Fe(III)OH} & \xrightarrow{\quad\quad} & \blacksquare\text{-Fe(II)OH} \\ e_{cb}^- & & \end{matrix} \quad (115)$$

$$\begin{matrix} \blacksquare\text{-Fe(III)OS(V)O}_2^\bullet & & \blacksquare\text{-Fe(III)OS(V)O}_2 \\ \blacksquare\text{-Fe(II)OH} & \xrightarrow{\quad Fe_{aq}^{2+} +} & \blacksquare\text{-OH} \end{matrix} \quad (116)$$

Even though the trapping of the conduction-band electron should be rapid, the subsequent dissolution step may be slow.

Hong et al. (1987a) have mathematically analyzed the reaction system involving the generation of radicals on the surface of either Fe_2O_3 or TiO_2, diffusion of SO_3^- to the bulk phase, and a free-radical chain reaction according to Eqs. 89–96. An additional free-radical termination step involves the surficial reaction of $SO_5^{-\bullet}$ as follows:

$$\begin{matrix} \blacksquare\text{-Fe(III)OSO}_4^\bullet & & \blacksquare\text{-Fe(III)OSO}_2^+ \\ \blacksquare\text{-Fe(III)OH} & \xrightarrow{\ + h_{vb}^+ \ \overset{k_{112}}{\quad}\ } & \blacksquare\text{-Fe(III)OH} + O_2 \end{matrix} \quad (117)$$

Given this mechanism, the following rate expression has been derived in the case of photocatalysis by metal oxide particle suspensions for pH < 5:

$$-\frac{d[S(IV)]}{dt} = -\frac{k_2 k_{112}}{4k_8}\{M_x O_y\}[HSO_3^-]$$

$$+[HSO_3^-]\sqrt{\left[\frac{k_2 k_{112}}{4k_8}\right]^2 \{M_x O_y\}^2 + \frac{k_2^2}{4k_8}\phi I_0 (1-10^{-\epsilon L\{M_x O_y\}})}\{\blacksquare-MOSO_2^-\}$$

$$(118)$$

In the case of TiO$_2$, ϕ ranged from 0.5 to 300. Quantum yields in excess of one provide strong evidence for a free-radical chain mechanism. These quantum yields are relatively high compared to tther photoassisted semiconductor-catalyzed redox reactions; thus some contribution to the overall rate may, in fact, be due to a homogeneous free-radical chain pathway with a very low concentration of freely diffusing initiator radicals (i.e., SO$_3^-$, SO$_4^{-\bullet}$, $^\bullet$OH).

    c. Photocatalytic Production of Hydrogen Peroxide. The formation of H$_2$O$_2$ and organic peroxides in illuminated aqueous suspensions of a variety of metal oxides and desert sands in the presence of O$_2$ and organic electron donors has been studied by Hong et al. (1987b) and Korman et al. (1988a,b). The photocatalytic rate of formation of H$_2$O$_2$ on metal oxides surfaces was shown to depend on the O$_2$ partial pressure, on the concentration of organic electron donors, and on the concentration of H$_2$O$_2$. During the initial phase of irradiation the rate production of H$_2$O$_2$ is given by

$$\frac{d[H_2O_2]}{dt} = (\phi_0 - \phi_1[H_2O_2])\frac{d[hv]_{abs}}{dt} \qquad (119)$$

where ϕ_0 is the quantum yield for H$_2$O$_2$ formation, ϕ_1 is the quantum yield for H$_2$O$_2$ degradation, and $d[hv]_{abs}/dt$ is the photon flux. The steady-state concentration of H$_2$O$_2$ was shown to be independent of the adsorbed photon flux as predicted by the above equation; $[H_2O_2]_{ss} = \phi_0/\phi_1$. Steady-state concentrations of H$_2$O$_2$ in excess of 100 μM were obtained in illuminated suspensions of ZnO. In the case of ZnO, the quantum yield for H$_2$O$_2$ formation, ϕ_0, approached 15% at 330 nm. whereas the photodegradation reactions were relatively inefficient. On the other hand, ϕ_1 for TiO$_2$ (1% at 330 nm) was found to be an order of magnitude less than the value for ZnO. Photooxidation of acetate and a wide variety other organic electron donors on colloidal ZnO has been shown to produce substantial concentrations of organic peroxides, ROOH. Illumination of ZnO suspensions containing acetate produced greater than 40% of the measured total peroxide (H$_2$O$_2$ and ROOH) as ROOH.

    Hydrogen peroxide can be formed by either the reduction of O$_2$ by e$_{cb}^-$ or the oxidation of H$_2$O by h$_{vb}^+$ as follows:

$$O_2 + 2e_{cb}^- + 2H_{aq}^+ \longrightarrow H_2O_2 \qquad (120)$$

$$2H_2O + 2h_{vb}^+ \longrightarrow H_2O_2 + 2H_{aq}^+ \qquad (121)$$

Appreciable yields of hydrogen peroxide are detected only in the presence of appropriate electron donors, D. The electron donor, D, which must be adsorbed on the particle surface, reacts with the a valence-band hole as follows:

$$D + h_{vb}^+ \longrightarrow D^{+\bullet} \tag{122}$$

Electron donors bound to the surface of the semiconductor particles interfere with e_{cb}^+/h_{vb}^+ recombination allowing e_{cb}^- (conduction-band electrons) to react with molecular oxygen via Eq. 120. Isotopic labeling studies have shown that H_2O_2 formed in irradiated suspensions of ZnO contained oxygen atoms derived exclusively from O_2. Titanium dioxide is much less effective for H_2O_2 production (Kormann et al., 1988b).

On the other hand, H_2O_2 may also be photodegraded. It can be reduced to water or oxidized to O_2 as follows:

$$H_2O_2 + 2e_{cb}^- \xrightarrow{\text{H}^+} 2H_2O \tag{123}$$

$$H_2O_2 + 2h_{vb}^+ \longrightarrow O_2 + 2H^+ \tag{124}$$

Both processes are first order in $[H_2O_2]$ for low concentrations of H_2O_2. The combination of these reactions will result in the attainment of a steady-state concentration of hydrogen peroxide during continuous irradiation. The rate of degradation of H_2O_2 can be described by the following relationship:

$$-\frac{d[H_2O_2]}{dt} = f_2([D], [O_2], [H_2O_2]) = \Phi_1 [H_2O_2]\frac{d[hv]_{abs}}{dt} \tag{125}$$

The oxidation of organic molecules as electron donors D by photogenerated valence-band holes leads to organic radicals $D^{\bullet+}$ (or $D^\bullet$). Carbon centered radicals add oxygen at diffusion-controlled rates to form the corresponding peroxy radical:

$$R^1R^2R^3C^\bullet + O_2 \longrightarrow R^1R^2R^3C(O_2)^\bullet \tag{126}$$

Under environmental conditions low steady-state concentrations are likely to favor first-order decay processes. If one of the substituents is OH, O^-, or NR_2 (e.g., $CH_3\dot{C}HOH$, $\dot{C}O_2^-$, $\dot{C}H_2N(CH_3)_2$ (i.e., a heteroatom with a nonbonding or antibonding electron pair), the peroxy radical is unstable and decays within a few milliseconds to yield the corresponding aldehyde, ketone, acid, or imine and a superoxide radical ($O_2^{-\bullet}$):

$$R^1R^2(OH)C(O_2)^\bullet \longrightarrow R^1R^2CO + O_2^{-\bullet} + H^+ \tag{127}$$

Specific- and general-base catalysis further accelerates the elimination of $O_2^{-\bullet}$ from α-hydroxyalkylperoxy radicals. Superoxide ($O_2^{-\bullet}$) ultimately forms hydrogen peroxide in a bimolecular process. No organic peroxides are formed according to this route:

$$O_2^- \cdot + HO_2^* \xrightarrow{\text{H}+} H_2O_2 + O_2 \tag{128}$$

If none of the substituents R^1, R^2, R^3 provides a nonbonding electron pair to the carbon-centered radical, the corresponding peroxy radical (e.g., $\cdot O_2CH_2CO_2^-$) is stable toward unimolecular or base-catalyzed elimination of O_2^-.

2.2.3. Solid-Supported Surface Catalysis by Metal Complexes.

Hong et al. (1987a, b, in press) have prepared a variety of hybrid catalysts between Co(II) phthalocyanine complexes and the surfaces of silica gel, polystyrene-divinylbenzene, and TiO_2 and tested these hybrids for catalytic activity with respect to the autoxidation of hydrogen sulfide, sulfur dioxide, 2-mercaptoethanol, cysteine, and hydrazine:

$$HS^- + 2O_2 \longrightarrow H^+ + SO_4^{2-} \tag{129}$$

$$2SO_3^{2-} + O_2 \longrightarrow 2SO_4^{2-} \tag{130}$$

$$2RSH + O_2 \longrightarrow RSSR + H_2O_2 \tag{131}$$

$$N_2H_4 + O_2 \longrightarrow N_2 + 2H_2O, \tag{132}$$

They found the most efficient hybrid catalyst to be the a Co(II)-tetraaminophthalocyanine ($Co^{II}TAP$), which was formed on the complexation of an imidazole group on a functionalized silica gel directly to the Co(II) center on the $Co^{II}TAP$. The next most efficient catalyst appeared to be a $Co^{II}TAP$ derivative on silica gel that was bound through the peripheral amino substituent of the phthalocyanine ligand. The Co(II)–tetrasulfophthalocyanine ($Co^{II}TSP$) derivatives bound to either polystyrene–divinylbenzene or TiO_2 directly through the central metal appeared to have lower catalytic activity.

Using the autoxidation of S(IV) as a specific example, we can break down surface catalysis in a hybrid heterogeneous system into the following processes:

Diffusion of reactants from the bulk solution to the solid surface

Diffusion of the reactants to the active catalytic centers located either on the surface or in the pores of the solid

Chemical reactions at the catalytic centers

Diffusion of reactions products out from the pores to the bulk solution

Diffusion of products away from the surface into the bulk solution

Based on the previous work of Boyce et al. (1983) and Hoffmann and Hong (1987), Hong et al. (in press) postulated the following rate expression for the surface-catalyzed reaction:

$$v = -\frac{d[S(IV)]}{dt} = \frac{k'[\blacksquare-Co^{II}TSP]_T[O_2][SO_3^{2-}]}{K_A + K_B[O_2] + K_C[SO_3^{2-}] + [O_2][SO_3^{2-}]} \tag{133}$$

where $[\blacksquare-Co(II)TSP]_T$ represents the concentration of surface-bound catalyst; k' is a composite rate constant; and K_A, K_B, and K_C are ratios of intrinsic rate constants.

Assuming that the autoxidation of S(IV) in the pores of the porous hybrids is isothermal, we can write the convective-diffusion equation in spherical geometry for O_2 depletion as

$$D_{e,O_2}\left(\frac{\partial^2[O_2]}{\partial r^2} + \frac{2}{r}\frac{\partial[O_2]}{\partial r}\right) + R_{O_2} = \varepsilon_p \frac{\partial[O_2]}{\partial t} \tag{134}$$

where the boundary conditions are $[O_2] = [O_2]_{bulk}$ at $r = a$ and $\partial[O_2]/\partial t = 0$ at $r = 0$. The initial condition is $[O_2] = [O_2]_{0, bulk}$ at $t = 0$; $[O_2]$ denotes the concentration of O_2 in the pore; R_{O_2} the depletion rate· of O_2; ε_p is the porosity; and a is the radius of the solid. The first boundary condition implies that the external resistance to diffusion is eliminated by stirring, and the second boundary condition arises from the symmetry of the particle. The initial condition corresponds to a system saturated with oxygen at equilibrium.

A similar continuity equation can be written for SO_3^{2-}:

$$D_{e,SO_3^{2-}}\left(\frac{\partial^2[SO_3^{2-}]}{\partial r^2} + \frac{2}{r}\frac{\partial[SO_3^{2-}]}{\partial r}\right) + R_{SO_3^{2-}} = \varepsilon_p \frac{\partial[SO_3^{2-}]}{\partial t} \tag{135}$$

where the boundary conditions are also similar: $[SO_3^{2-}] = [SO_3^{2-}]_{bulk}$ at $r = a$ and $\partial[SO_3^{2-}]/dt = 0$ at $r = 0$ with the initial conditions of $[SO_3^{2-}] = 0$ at $t = 0$ for $r < a$ and $[SO_3^{2-}] = [SO_3^{2-}]_{0, bulk}$ at $t = 0$ for $r = a$. These initial conditions correspond to the initiation of the reaction between S(IV) and O_2.

The solution to the preceding coupled differential equations above can be simplified by assuming a steady-state condition. Furthermore under the pseudo-first-order conditions of $[O_2]_0 > [SO_3^{2-}]$, we can express the rate of S(IV) depletion as

$$R_{SO_3^{2-}} = -\frac{K_E[SO_3^{2-}]}{K_F + K_G[SO_3^{2-}]} \tag{136}$$

where $K_E = k'[\blacksquare-Co(II)TSP]_T[O_2]$, $K_F = K_A + K_B[O_2]$, and $K_G = K_C + [O_2]$. We can assume that K_E, K_F, and K_G are essentially constant since a small neutral molecule such as O_2 will diffuse through the porous matrix faster the SO_3^{2-}. With this approximation, Eq. 135 can be rewritten as

$$D_e\left(\frac{\partial^2 C_r}{\partial r^2} + \frac{2}{r}\frac{\partial C_r}{\partial r}\right) = \frac{K_E C_r}{K_F + K_G C_r} \tag{137}$$

where we now write $C_r = [SO_3^{2-}]_{pore}$ and $C = [SO_3^{2-}]_{bulk}$; the boundary conditions remain the same.

The effectiveness factor, η, is defined as the ratio of the reaction rate with pore diffusion resistance to the reaction rate without pore diffusion resistance (i.e., all of the active catalytic sites are restricted to the external surface of the particle). In mathematical terms

$$\eta = \frac{\int R(C_r)dV_p}{V_p R(C)} \tag{138}$$

where V_p represents volume of the particles. In general, the effectiveness factor can be derived by first solving the continuity equation to obtain a concentration profile of reactants in the particle; this profile is then used in the rate expression, and the latter is integrated over the particle volume to yield the actual rate of reaction. The actual rate can be used in Eq. 138 to evaluate η. The value of η is useful as an indicator of the extent of the diffusion limitation to the overall reaction and for expressing the observed actual rate in terms of measurable bulk concentrations.

For a first-order reaction ($R = kC$) within a rectangular slab of width $2L$, η is given by (Aris, 1969):

$$\eta = \frac{\tanh \varphi}{\varphi} \tag{139}$$

where $\varphi = L(k/D_e)^{1/2}$ {i.e., the Thiele modulus}. The simple form of η can be extended to arbitrary geometries by modifying the Thiele modulus as follows:

$$\varphi = \frac{V_p}{S_x}\left(\frac{k}{D_e}\right)^{1/2} \tag{140}$$

where S_x is the external surface area of the catalyst bead. We can modify φ further for a general rate expression and an arbitrary geometry.

$$\varphi = \frac{V_p R(C)}{S_x(2D)^{1/2}}\left(\int R(C_r')dC_r'\right)^{-1/2} \tag{141}$$

Incorporation of the rate expression given in Eq. 136, gives a Thiele modulus of the following form:

$$\varphi = \frac{aK_E^{1/2}C}{3(2D_e)^{1/2}(K_F + K_G C)}\left\{\frac{C}{K_G} - \frac{K_F}{K_G^2}\ln\left[\frac{K_G C}{K_F} + 1\right]\right\}^{-1/2} \tag{142}$$

and thus the actual rate of reaction is given by

$$R_{actual} = \eta R(C) = \frac{\tanh \varphi}{\varphi}\frac{K_E C}{K_F + K_G C} \tag{143}$$

Kinetic analysis of the heterogeneous catalytic systems suggests that the mechanism of oxidation involves activation of molecular oxygen and complexation of the substrate as a prelude to electron transfer. In the case of S(IV), this mechanism is analogous to that proposed for the corresponding homogeneous system (Boyce et al., 1983). In general, reactions on solid matrices occur more slowly because access of the reactants to the active site may be influenced by mass-transfer effects such as pore diffusion. In the case of S(IV) oxidation at pH 6.7, attachment of homogeneous catalysts to solid supports results in an enhancement of catalytic activity due, in part, to the initiation of a free-radical chain reaction and to the suppression of the formation of μ-peroxo Co(III) dimers, which are catalytically inactive (Hong et al., 1989).

The hybrid metal phthalocyanine complexes may find application for specific pollution control processes such as SO_2 stack-gas scrubbing, post-Klaus plant scrubbing of H_2S and SO_2, sweetening of sour oil-refinery wastes, odor and corrosion control in wastewater facilities, and the elimination of excess rocket fuel wastes.

3. CONCLUSIONS

Chemical catalysis has been found to play an important role in a variety of chemical reactions that take place in the natural aquatic environment. The phenomenon of catalysis can be treated rigorously with respect to kinetic formulations and mechanistic descriptions. The predominant form of catalysis encountered in natural waters is specific-acid and specific-base catalysis. However, in several situations trace metal catalysis and surface catalysis of reactions can play an important role in the transformation of chemical species in the aquatic environment. The autoxidation of S(IV) in clouds is highly sensitive to catalysis by Fe(III) and Mn(II), while the autoxidation of Mn(II) to Mn(III)/Mn(IV) and V(IV) to V(V) is sensitive to surface catalysis by metal oxides and oxyhydroxides. However, surface-catalyzed reactions may be limited by diffusion through porous matrices in certain systems. Homogeneous and heterogeneous photocatalysis in aquatic systems exposed to light can play significant roles in the transformation of chemical compounds. The experimentalist should be cognizant of these latter phenomena when performing experiments pertinent to the aquatic environment (i.e. light as a potential reaction variable is often ignored).

Acknowledgments

I am grateful to Werner Stumm of EAWAG/ETH for providing the time, support, office space, and living arrangements that allowed me to write this chapter with a single focus. My appreciation is also extended to the California Institute of Technology for granting me a sabbatical leave. Support for the individual research projects originating within my group at Caltech, which are described in this chapter, has been provided by a number of funding agencies that include the U.S. Environmental Protection Agency, the National

Science Foundation, the U.S. Department of Energy, the U.S. Public Health Service, and the Electric Power Research Institute. Without the generous support of these agencies our progress in the study of chemical catalysis applicable to environmental systems would have been severely limited.

REFERENCES

Aris, R. (1969), *Elementary Chemical Reactor Analysis*, Prentice-Hall, Englewood Cliffs, NJ, 352 pp.

Arnold, R. G., T. M. Olson, and M. R. Hoffmann (1986), "Kinetics and Mechanism of the Dissimilative Reduction of Fe(III) by *Pseudomonas* sp. 200," *Biotech. Bioeng.* **28**, 1657–1671.

Arnold, R. G., T. J. DiChristina, and M. R. Hoffmann (1988), "Reductive Dissolution of Iron Oxides by *Pseudomonas* sp. 200," *Biotech. Bioeng.*, **32**, 1081–1096.

Bahnemann, D. W., M. R. Hoffmann, A. P. Hong, and C. Kormann (1987), "Photocatalytic Formation of Hydrogen Peroxide," in R. Johnson, Ed., *The Chemistry of Acid Rain: Sources and Atmospheric Processes*, (ACS Symposium Series No. 239), American Chemical Society, Washington, DC, pp. 120–132.

Bender, M. I. (1971), *Mechanism of Homogeneous Catalysis from Protons to Proteins*, Wiley-Interscience, New York.

Betterton, E. A., and M. R. Hoffmann (1987), "Kinetics, Mechanism, and Thermodynamics of the Formation of the S(IV) Adduct of Methylglyoxal in Aqueous Solution," *J. Phys. Chem.* **91**, 3011–3020.

Betterton, E. A., and M. R. Hoffmann (1988), "Rapid Oxidation of Dissolved SO_2 by Peroxymonosulfate," *J. Phys. Chem.*, **92**, 5962–5965 (1988).

Betterton, E. A., Y. Erel, and M. R. Hoffmann (1988), "Aldehyde–Bisulfite Adducts: Prediction of Some of Their Thermodynamic and Kinetic Properties," *Environ. Technol.* **22**, 92–97.

Boudart, M., and R. L. Burwell, Jr. (1974), "Mechanism in Heterogeneous Catalysis", in E. S. Lewis, Ed., *Techniques of Chemistry Vol. VI: Investigation of Rates and Mechanisms of Reactions*, Wiley-Interscience, New York, pp. 693–740.

Boyce, S. D., and M. R. Hoffmann (1984), , "Kinetics and Mechanism of the Formation of Hydroxymethanesulfonic Acid at Low pH," *J. Phys. Chem.* **88**, 4740–4746.

Boyce, S. D., M. R. Hoffmann, P. A. Hong, and L. M. Moberly (1983), "Catalysis of the Autoxidation of Aquated Sulfur Dioxide by Homogeneous Metal–Phthalocyanine Complexes," *Environ. Sci. Technol.* **17**, 602–611.

Brewer, P. G. (1975), "Minor elements in Seawater," in J. P. Riley, and G. Skirrow, Eds., *Chemical Oceanography*, Vol. 1, Academic Press, New York, pp. 445–496.

Bunnett, J. F. (1986) "From Kinetic Data to Reaction Mechanism," in C. F. Bernasconi, Ed., *Techniques of Chemistry Vol. VI: Investigation of Rates and Mechanisms of Reactions*, 4th ed., Wiley-Interscience, New York, pp. 251–372.

Cleland, W. W. (1986), "Enzyme Kinetics as a Tool for Determination of Enzyme Mechanisms," in C. F. Bernasconi, Ed., *Techniques of Chemistry Vol. VI: Investigation of Rates and Mechanisms of Reactions*, 4th ed., Wiley-Interscience, New York, pp. 791–870.

Conklin, M. C., and M. R. Hoffmann (1988a), "Metal Ion-Sulfur(IV) Chemistry. 1. Structure and Thermodynamics of Transient Cu(II)–S(IV) Complexes," *Environ. Sci. Technol.* **22**, 883–891.

Conklin, M. C., and M. R. Hoffmann (1988b), "Metal Ion–Sulfur(IV) Chemistry. 2. Kinetic Studies of the Redox Chemistry of Cu(II)–S(IV) Complexes," *Environ. Sci. Technol.* **22**, 891–898.

Conklin, M. C., and M. R. Hoffmann (1988c), "Metal Ion–Sulfur(IV) Chemistry. 3. Thermodynamics, Structure, and Kinetics of Transient Fe(III)–S(IV) Complexes," *Environ. Sci. Technol.* **22**, 899–907.

Cornish-Bowden, A. J. (1976), *Principles of Enzyme Kinetics*, Butterworth, London.

Davies, S. R. H. (1987), in *Geochemical Processes at Mineral Surfaces*, J. A. Davis, and K. F. Hayes, Eds., American Chemical Society, Washington, DC, pp. 487–   .

Edwards. J. O., E. F. Greene, and J. Ross (1968), "From Stoichiometry and Rate Law to Mechanism," *J. Chem. Educ.* **45**, 381–385.

Emerson, S., R. E. Cranston, and P. S. Liss (1979), "Redox Species in a Reducing Fjord: Equilibrium and Kinetic Considerations," *Deep-Sea Res.* **26**, 859–878.

Faust, B. C., M. R. Hoffmann, and D. W. Bahnemann (1989), "Photocatalytic Oxidation of Sulfur Dioxide in Aqueous Suspensions of α-Fe_2O_3," *J. Phys. Chem.*, **93**, 6371–6381.

Gonzalez, A. C., and M. R. Hoffmann (1989a), "The Kinetics and Mechanism of the Catalytic Hydrolysis of Nitrophenyl Acetates by Metal Oxide Surfaces," *J. Phys. Chem.*, submitted.

Gonzalez, A. C., and M. R. Hoffmann (1989b), "Kinetics and Mechanism of the Catalytic Hydrolysis of Parathion by Metal Hydroxo Complexes," *Environ. Sci. Technol.*, submitted.

Graedel, T. E., M. L. Mandich, and C. J. Weschler (1986), "Kinetic Model Studies of Atmospheric Droplet Chemistry. 2. Homogeneous Transition Metal Chemistry in Raindrops," *J. Geophys. Res.*, **91**, 5205–5221.

Haller. G. L. (1986), "Mechanism in Heterogeneous Catalysis," in C. F. Bernasconi, Ed., *Techniques of Chemistry Vol. VI: Investigation of Rates and Mechanisms of Reactions*, 4th ed., Wiley-Interscience, New York, pp. 951–980.

Hawking, S. W. (1988), *A Breif History of Time*, Bantam, New York, 198 pp.

Hoffmann, M. R., and S. D. Boyce (1983), "Theoretical and Experimental Considerations of the Catalytic Autoxidation of Aqueous Sulfur Dioxide in Relationship to Atmospheric Systems," *Adv. Environ. Sci. Technol.* **12**, 149–148. (Wiley-Interscience, New York).

Hoffmann, M. R., and J. G. Calvert (1985), "Chemical Transformation Modules for Eulerian Acid Deposition Models, Vol. 2, The Aqueous-Phase Chemistry," EPA/600/3-85/017, U.S. Environmental Protection Agency, Research Triangle Park, NC.

Hoffmann, M. R., and J. O. Edwards (1975), "Kinetics and Mechanism of the Oxidation of Sulfur Dioxide by Hydrogen Peroxide in Acidic Solution," *J. Phys. Chem.* **79**, 2096–2098.

Hoffmann, M. R., and J. O. Edwards (1977), "Kinetics and Mechanism of the Oxidation of Thiourea and N,N,-Dialkylthioureas by Hydrogen Peroxide," *Inorg. Chem.* **16**, 3333–3338.

Hoffmann, M. R., and D. J. Jacob (1984), "Kinetics and Mechanism of the Catalytic

Oxidation of Dissolved SO_2 in Atmospheric Droplets: Free Radical, Polar and Photoassisted Pathways," in J. G. Calvert, Ed., *Acid Precipitation: SO_2, NO, NO_x Oxidation Mechanisms: Atmospheric Considerations*, Butterworth, Boston-London, pp. 101–172.

Hoffmann, M. R., and A. P. Hong (1987), "Catalytic Oxidation of Reduced Sulfur Compounds by Homogeneous and Heterogeneous Co(II) Phthalocyanine Complexes," *Sci. Total Environ.* **64**, 99–115.

Hoffmann, M. R., and B. C. H. Lim (1979), "Kinetics and Mechanism of the Oxidation of Sulfide by Oxygen: Catalysis by Homogeneous Metal Phthalocyanine Complexes," *Environ. Sci. Technol.* **13**, 1406–1414.

Hong, A. P., D. W. Bahnemann, and M. R. Hoffmann (1987a), "Co(II) Tetrasulfophthalocyanine on Titanium Dioxide: A New Efficient Relay for the Photocatalytic Formation and Depletion of Hydrogen Peroxide in Aqueous Suspensions," *J. Phys. Chem.* **91**, 2109–2116.

Hong, A. P., D. W. Bahnemann, and M. R. Hoffmann (1987b), "Co(II) Tetrasulfophthalocyanine on Titanium Dioxide. II. Photocatalytic Oxidation of Aqueous Sulfur Dioxide," *J. Phys. Chem.* **91**, 6245–6251.

Hong, A. P., S. D. Boyce, and M. R. Hoffmann (in press), "Catalytic Autoxidation of Chemical Contaminants by Hybrid Complexes of Co(II) Phthalocyanine," *Environ. Sci. Technol.* **23**.

IUPAC (1981), "Manual of Symbols and Terminology for Physicochemical Quantities and Units, Appendix V, Symbolism and Terminology in Chemical Kinetics," *Pure Appl. Chem.* **53**, 753.

Jacob, D. J., and M. R. Hoffmann (1983), "A Dynamic Model for the Production of H^+, NO_3^- and SO_4^{2-} in Urban Fog," *J. Geophys. Res.* **88**, 6611–6621.

Jacob, D. J., E. W. Gottlaeb, and M. J. Prather (1989), "Chemistry of a Polluted Boundary Layer," *J. Geophys. Res.* **94**, 12,975–13,002.

Kormann, C., D. W. Bahnemann, and M. R. Hoffmann (1988a), "Preparation and Characterization of Quantum-Size Titanium Dioxide," *J. Phys. Chem.* **92**, 5196–5201.

Kormann, C., D. W. Bahnemann, and M. R. Hoffmann (1988b), "Photocatalytic Production of H_2O_2 and Organic Peroxides in Aqueous Suspensions of TiO_2, ZnO, and Desert Sand, " *Environ. Sci. Technol.* **22**, 798–806.

Kormann, C., D. W. Bahnemann, and M. R. Hoffmann (1989), "Environmental Photochemistry: Is Iron Oxide (α-Fe_2O_3) an Active Photocatalyst? A Comparative Study of α-Fe_2O_3, ZnO and TiO_2," *J. Photochem.*, **48**, 161–169.

Laidler, K. J. (1987), *Chemical Kinetics*, 3rd. ed., Harper, New York, 531 pp.

Laidler, K. J., and P. S. Bunting (1973), *The Chemical Kinetics of Enzyme Action*, 2nd. ed., Clarendon, Oxford.

Leung, K., and M. R. Hoffmann (1985), "Kinetics and Mechanism of the Oxidation of 2-Mercaptoethanol by Hydrogen Peroxide," *J. Phys. Chem.* **89**, 5267–5271.

Leung, K., and M. R. Hoffmann (1988), "Kinetics and Mechanism of Autoxidation of 2-Mercaptoethanol Catalyzed by Co(II)–4,4′,4″,4‴-Tetrasulfophthalocyanine in Aqueous Solution," *Environ. Sci. Technol.* **22**, 275–282.

Leung, K., and M. R. Hoffmann (1989), "Kinetics and Mechanism of Autoxidation of 2-Aminoethanethiol and Ethanethiol Catalyzed by Co(II)–4,4′,4″,4‴-Tetrasulfophthalocyanine in Aqueous Solution," *J. Phys. Chem.*, **93**, 431–433, 434–441.

Mabey, W. R., and T. J. Mill (1978), *J. Phys. Chem. Ref. Data* **7**, 383–415.

McArdle, J. V., and M. R. Hoffmann (1983), "Kinetics and Mechanism of the Oxidation of Aquated Sulfur Dioxide by Hydrogen Peroxide at Low pH," *J. Phys. Chem.* **87**, 5425–5429.

Macalady, D. L., and N. L. Wolfe (1983), "New Perspectives on the Hydrolytic Degradation of the Organophosphorothioate Insecticide Chloropyrifos," *J. Agric. Food Chem.* 1139–1147.

March, J. (1977), *Advanced Organic Chemistry*, 2nd. ed., McGraw-Hill, New York, 1328 pp.

Mill, T. J., and Mabey, W. R. (1987), "Hydrolysis of Organic Chemicals," in O. Hutzinger, Ed., *Handbook of Environmental Chemistry, Vol. 2 Part D, Reactions and Processes*, Springer-Verlag, Berlin, pp. 72–111.

Moore, J. W., and R. G. Pearson (1981), *Kinetics and Mechanism*, 3rd. ed., Wiley, New York, 455 pp.

Morgan, J. J. (1967), "Chemical Equilibria and Kinetic Properties of Manganese in Natural Waters," in S. D. Faust and J. V. Hunter, Eds., *Principles and Applications of Water Chemistry*, Wiley, New York.

Munger, J. W., C. Tiller, and M. R. Hoffmann (1986), "Identification and Quantification of Hydroxymethanesulfonic Acid in Atmospheric Water Droplets," *Science* **231**, 247–249.

Olson, T. M., and M. R. Hoffmann (1986), "On the Kinetics of Formaldehyde–S(IV) Adduct Formation in Slightly Acidic Solution," *Atmos. Environ.* **20**, 2277–2278.

Olson, T. M., and M. R. Hoffmann (1988), "The Kinetics, Mechanism, and Thermodynamics of Glyoxal–S(IV) Adduct Formation," *J. Phys. Chem.* **92**, 533–540, 4246–4253.

Olson, T. M., L. A. Torry, and M. R. Hoffmann (1988), "Kinetics of the Formation of Hydroxyacetaldehyde–S(IV) Adducts at Low pH," *Environ. Sci. Technol.* **22**, 1284–1289.

Olson, T. M., S. D. Boyce, and M. R. Hoffmann (1986), "Kinetics, Thermodynamics, and Mechanism of the Formation of Benzaldehyde–S(IV) Adducts," *J. Phys. Chem.* **90**, 2482–2488.

Olson, T. M., and M. R. Hoffmann (1989), "Hydroxyalkylsulfonate Formation: Its Role as a S(IV) Reservoir in Atmospheric Water Droplets", *Atmos. Environ.* **23**, 985–997.

Ostwald, W. (1902), *Lehrbuch de Allgemeinen Chemie*, 2nd. ed., Vol. 2, Part 2, Akademische Verlagsgesellschaft, Leipzig, pp. 249–262.

Overton, J. H., Jr. (1985), "Validation of the Hoffmann and Edwards' S(IV)–H_2O_2 Mechanism," *Atmos. Environ.* **19**, 687–690.

Pandis, S. N., and J. H. Seinfeld (1989), "Sensitivity Analysis of a Chemical Mechanism for Aqueous-Phase Atmospheric Chemistry," *J. Geophys. Res.* **94**, 1105–1126.

Perdue, E. M., and N. L. Wolfe (1983), "Prediction of Buffer Catalysis in Field and Laboratory Studies of Pollutant Hydrolysis Reactions," *Environ. Sci. Technol.* **17**, 635–642.

Plastourgou, M., and M. R. Hoffmann (1984), "Transformation and Fate of Organic Esters in Layered-Flow Systems: The Role of Trace Metal Catalysis," *Environ. Sci. Technol.* **18**, 756–764.

Seigneur, C., and P. Saxena (1988), "A Theoretical Investigation of Sulfate Formation in Clouds," *Atmos. Environ.* **18**, 101–115.

Stone, A. T. (1989), "Enhanced Rates of Monophenyl Terephthalate Hydrolysis in Aluminium Oxide Suspensions", *J. Colloid Interface Sci.*, **127**, 429–440.

Stumm, W., and J. J. Morgan (1981), *Aquatic Chemistry*, 2nd. ed., Wiley, New York, 780 pp.

Sung, W., and J. J. Morgan (1981), "Oxidative Removal of Mn(II) from Solution Catalyzed by the γ-FeOOH (Lepidocrocite) Surface," *Geochim. Cosmochim. Acta* **45**, 2377–2383.

Wehrli, B., and W. Stumm (1988), "Oxygenation of Vanadyl (IV). Effect of Coordinated Surface Hydroxyl Groups and OH$^-$," *Langmuir* **4**, 753–758.

Wilmarth, W. K., D. M. Stanbury, J. E. Byrd, H. N. Po, C.-P. Chua (1983), "Electron-Transfer Reactions Involving Simple Free Radicals," *Coord. Chem. Rev.* **51**, 155–179.

4

PRINCIPLES OF LINEAR FREE-ENERGY AND STRUCTURE–ACTIVITY RELATIONSHIPS AND THEIR APPLICATIONS TO THE FATE OF CHEMICALS IN AQUATIC SYSTEMS

Patrick L. Brezonik

Department of Civil and Mineral Engineering, University of Minnesota, Minneapolis, Minnesota

1. INTRODUCTIONS

Modern chemical kinetics has been described as having three levels of inquiry: phenomenological, mechanistic, and statistical mechanical. These phases deal respectively with the following aspects: (1) fitting data to rate expressions and evaluating rate constants as functions of physical conditions, (2) elucidating the elementary steps that make up the stoichiometric reaction, and (3) studying the energetics of elementary steps, details of bond breaking and making, and the nature of transition states. Based on the discussion in this chapter, we may add a fourth level to these: the correlational level. In contrast to the first three levels, which are concerned with individual reactions, the correlational level focuses on relationships among rates and equilibria of *related* reactions and *series* of reactants. It should be realized that the four phases of kinetics often are closely interrelated and represent a continuum of inquiry rather than discrete levels.

Correlational approaches also are important in analyzing and/or predicting the chemical and biological fate of chemical contaminants in aquatic systems. Of interest in this regard is the development of correlations between measures of a compound's chemical reactivity, bioactivity or "environmental behavior", and measure(s) of the compound's physicochemical properties or strucutral charac-teristics. For convenience, we will refer generically to such correlations as "attribute relationships."

This chapter describes applications of attribute relationships in aquatic chemistry. The underlying theoretical basis for these methods is explored, along with their strengths and limitations as analytical and predictive tools for contaminant fate in aquatic systems.

2. NATURE OF CORRELATIONAL ANALYSES IN AQUATIC CHEMISTRY

The search for relationships among the dynamic and equilibrium properties of related series of compounds has been a paradigm of chemists for many years. This search is a manifestation of three major goals of chemists: (1) to categorize existing information on the behavior of chemicals, (2) to explain their behavior in terms of fundamental physicochemical principles, and (3) to predict unmeasured properties from measured properties of the same chemicals or from the behavior of related compounds. The discovery of such unifying principles and predictive relationships is, of course, a source of intellectual satisfaction, but such relation-ships also have practical benefits (e.g., to water-quality managers faced with predicting the ecological and toxicological effects and fate of organic contami-nants in natural waters).

Numerous relationships exist among the structural characteristics, physico-chemical properties, and/or biological qualities of classes of related compounds. Simple examples include bivariate correlations between physicochemical prop-erties such as aqueous solubility and octanol–water partition coefficients (K_{ow}) and correlations between equilibrium constants of related sets of compounds. Perhaps the best-known attribute relationships to chemists are the correlations between reaction rate constants and equilibrium constants for related reactions commonly known as *linear free-energy relationships* or LFERs. The LFER concept also leads to the broader concepts of property–activity and structure–activity relationships (PARs and SARs), which seek to predict the environmental fate of related compounds or their bioactivity (bioaccumulation, biodegradation, toxicity) based on correlations with physicochemical properties or structural features of the compounds. Table 1 summarizes the types of attribute relationships that have been used in chemical fate studies and defines some important terms used in these relationships.

Development of quantitative relationships to predict the fate and effects of aquatic contaminants from easily measured properties is important for three reasons: (1) the variety of organic compounds potentially present in aquatic systems; (2) the number and complexity of physical and chemical processes involved in contaminant transport and transformation, and (3) the diversity of organisms that may be affected by contaminants or involved in their bio-degradation. Most attention has focused on organic compounds, but some

TABLE 1. Characteristic Types of Attribute Relationships

Type of Predictor Variable[a]	Type of Predicted Variable	Name[b]	Example[c]
Structure	Activity	SAR, QSAR	Bioaccumulation vs. molecular connectivity index (e.g., $^1X^v$)
Property	Activity	PAR	Bioaccumulation vs. octanol–water partition coefficient, log K_{ow} (LFERs are a type of PAR: predictor property = equilibrium constant; predicted activity = reaction rate constant)
Property	Property	PPR	log K_{ow} vs. log S_w (aqueous solubility) or vs. log t_r (chromatographic retention time)
Structure	Property	SPR	log S_w vs. TSA (total molecular surface area)
Molecular	Activity	SAR, MAR	Biodegradability vs. $\Delta\|\delta\|_{x-y}$ (difference in atomic charge across key bond, x–y)

[a] For purposes of this discussion, these terms are defined as follows: *structure*— geometric attribute of molecule (based on size, shape, arrangement of atoms component molecule); *property*— physical-chemical attribute of molecule that can be measured directly or calculated readily from measurable variables; *molecular property*—attribute not measured directly but calculated from molecular structure based on fundamental (theoretical) relationships such as molecular orbital theory, statistical, or quantum mechanics; *activity*—attribute of molecule related to its biological behavior.

[b] PAR and (Q) SAR are the only commonly accepted terms in the literature.

[c] Examples are discussed later in this chapter.

relationships have been reported for metal ions. Although attribute relationships are inherently empirical, most have an underlying theoretical basis. A goal of research in this field is to develop generalizations from empirical relationships so that a few fundamental principles can explain many observations (Agmon, 1981).

3. LINEAR FREE-ENERGY RELATIONSHIPS

It is well known that no *general* relationship exists between the thermodynamics of reactions and the rates at which they proceed toward equilibrium. Many reactions have favorable energetics but proceed imperceptibly, if at all, and numerous environmental examples exist of such "kinetically hindered" reactions. The absence of a general relationship is understandable from transition-state theory, from which we know that reaction rates are controlled by the energy difference between reactants and a transition-state complex, rather than by the energy difference between reactants and products.

In spite of the lack of a universal relationship between reaction rates and their equilibria, correlations do exist between the rates and energetics of reactions for sets of related compounds. Correlations among equilibrium constants for two sets of compounds also are common. For example, logarithms of stability constants for Al(III) and Fe(III) complexes with various ligands are closely correlated (Langmuir, 1979); the same is true for many other pairwise combinations of metal complexes. Because log K_{eq} is proportional to the free energy of reaction ($\Delta G°$), these are simple examples of linear free-energy relationships (LFERs). LFERs are useful in kinetics because they enable us to predict reaction rates from more easily measured (or more readily available) equilibrium properties, and they are equally valuable in improving our understanding of reaction mechanisms and rate controlling steps.

The term "LFER" is used for such correlations because they usually are linear correlations between logs of rate constants and logs of equilibrium constants for reactions of the compounds. According to transition-state theory, rate constants are exponentially related to the free energy of activation ($\Delta G^{\neq}$), and thermodynamics tells us that equilibrium constants are similarly related to $\Delta G°$. If two reactions exhibit a LFER we can write

$$\ln k_2 - \ln k_1 = \alpha\{\ln K_2 - \ln K_1)\tag{1}$$

or

$$\{-\Delta G_2^{\neq} + \Delta G_1^{\neq}\}/RT = \alpha\{-\Delta G_2° + \Delta G_1°\}/RT\tag{2}$$

For series of i reactants undergoing a given reaction

$$\ln k_i = \alpha \ln K_i + C, \text{ or } \Delta G_i^{\neq} = \alpha \Delta G_i° + C\tag{3}$$

The reactions on the kinetic and equilibrium sides of the equation do not have to be the same. For a series of i reactants and related reactions j and k:

$$\ln k_{ij} = \alpha \ln K_{ik} + C, \text{ or } \Delta G_{ij}^{\neq} = \alpha \Delta G_{ik}° + C\tag{4}$$

The free energy of activation is composed of an entropy activation ($\Delta S^{\neq}$) and an enthalpy of activation ($\Delta H^{\neq}$). The former is associated with the pre-exponential factor A of the Arrhenius equation and the latter with the experimental E_{act}, which defines the sensitivity of reaction rate to temperature. In some reactions, substituents affect E_{act} (or $\Delta H^{\neq}$) and $\Delta H°$ primarily, while A (or $\Delta S^{\neq}$) and $\Delta S°$ change only slightly. In these cases, $\Delta G^{\neq}$ and $\ln k$ vary in the same way as E_{act}, and $\ln k$ varies linearly with $\Delta G°$. Frequently, however, $\ln k$ and $\ln K$ are correlated, even though both A and E_{act} vary as the substituent changes. This can happen if $\Delta H^{\neq}$ (or E_{act}) and $\Delta S^{\neq}$ are correlated, and similar relationships are observed between the values of $\Delta H°$ and $\Delta S°$ for the reactions; this is called the *compensation effect*. The net effect is a simple LFER in which the change in k is less than if E_{act} or A changed alone. It is easy to show that this situation leads to

Eq. (5):

$$k = A' \exp\left(\frac{\Delta H^{\neq}}{R}\left[\frac{1}{T} - \frac{1}{\theta}\right]\right) \tag{5}$$

This is called the *isokinetic relationship*, and θ is the isokinetic temperature, where all k values for related series of reactions are the same. Although θ can be obtained as the slope of plot of $\Delta S^{\neq}$ versus $\Delta H^{\neq}$, significant statistical problems may be encountered (Exner, 1972). At $T < \theta$, reactions with smaller E_{act} are faster. At $T > \theta$, reactions with larger E_{act} are faster.

The existence of an LFER for a set of reactants is equivalent to saying that the free energy of activation is a constant fraction of the free energy of reaction for the reactants ($\Delta G^{\neq}/\Delta G^{\circ} = \alpha$). Although an exact theory is not available to explain why this should be, the basis for the relationships can be understood qualitatively from diagrams of energy profiles along a reaction coordinate (Fig. 1). Such diagrams show that if the energy curves for related reactions have similar shapes (a reasonable assumption), then by simple geometry the change in E_{act} must be proportional to the overall change in reaction energetics (Adamson, 1979). If $\Delta S^{\neq}$ and ΔS° are constant for reactions in a series, then the change in $\Delta G^{\neq}$ is proportional to the change in ΔG°, and $\log k$ is proportional to $\log K$. If the entropy terms are not constant but the entropies and enthalpies are correlated (the compensation effect), the LFER still holds.

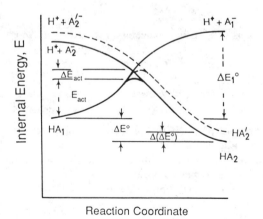

Reaction Coordinate

Net Reaction: $\mathrm{HA_1 + A_2^- \rightarrow HA_2 + A_1^-}$

or $\mathrm{HA_1 + A_2'^- \rightarrow HA_2' + A_1^-}$

Figure 1. Existence of LFERs can be understood from geometric similarity of energy profiles along the reaction coordinate for related reactants and reactans. Modified from Adamson (1979).

3.1. Major Categories of LFERS

Linear free-energy reactions are well known in organic chemistry, and the effects of substituent groups on the kinetics and equilibria of reactions for related organic compounds have been studied since the 1920s. Substituents influence reactivity by altering electron density at the reaction site and by steric effects. Electronic effects are easiest to evaluate for aromatic compounds with substituents in meta and para positions, where steric influences (which are more difficult to predict) are minimized. Reactions favored by high electron density are accelerated by electron-donor substituents; reactions favored by low electron density are favored by electron-withdrawing groups. Substituents can induce two categories of electronic effects: inductive (polar) and resonance. The former are transmitted along a chain of atoms without reorganizing chemical bonds and decrease rapidly with distance from a substituent group. The latter occur in compounds with conjugated double bonds and produce changes in electron densities at certain positions in conjugated systems.

Four major categories of LFERs have been developed over the past 60 years or so (Table 2). These relationships apply to a wide variety of classes of organic and inorganic compounds and a wide range of reactions—coordinative reactions (dissociation–association) of acids and metal complexes, hydrolysis, hydration, substitution, substituent group oxidations, electron exchange between metal ions, and so on. This section describes the basis for these categories of correlative relationships and the types of reactions to which they apply.

TABLE 2. Major Classes of LFERs Applicable to Reactions in Aquatic Systems

Relationship	Types of Reaction or Reactants	Basis of LFER
Brønsted	Acid- and base-catalyzed reactions: hydrolysis, dissociation, association	Rate related to K_a or K_b of product or catalyst
Hammett (sigma)	Reactions of para- or meta-substituted aromatic compounds: hydrolysis, hydration of alkenes, substitution, oxidation, enzyme-catalyzed oxidations; some Type II photooxidations	Electron withdrawal and/or donation from/to reaction site by substituents on aromatic rings via resonance effects
Taft	Hydrolysis and many other reactions of aliphatic organic compounds	Steric and polar effects of substituents
Marcus	Outer-sphere electron exchange reactions of metal ions, chelated metals, and metal ion oxidation by organic oxidants such as pyridines and quinones	The three components of energy needed to produce transition state; for related redox reactions, $\ln k$ proportional to E°

Brønsted LFER.   Brønsted and Pederson (1923) were the first to describe a relationship between rates and equilibria for a series of compounds. They found that $\log k_B$ for base-catalyzed decomposition of nitramide, $H_2N_2O_2$, varies linearly with $\log K_{HB^+}$, the acidity constant of the conjugate acid of the catalyst. Rate constants for many other acid- or base-catalyzed reactions (including the hydrolysis of amides, esters, carbamates, and organophosphates, and dissociation of acids and metal-ion complexes) are log–log related to the acid (or base) dissociation constant of the catalyst and follow either of the equations

$$\log k_{HB} = A + \alpha \log K_{HB} \tag{6a}$$

$$\log k_B = A + \alpha \log K_B \tag{6b}$$

Proportionality constant α ranges between 0 and 1.

A commonly cited Brønsted LFER is the correlation between k_f and K_a for acid dissociation, $HA + H_2O \rightarrow A^- + H_3O^+$, which is linear over many orders of magnitude, with $\alpha \approx 1.0$. This is a trivial example, however. Rate constants for acid recombination are nearly uniform because they are at the diffusion-controlled limit (Adamson, 1979). Because $K_a = k_{for}/k_{rev}$, k_{for} *must* be proportional to K_a and α_f must be unity. Conversely, α_r must be zero for diffusion-controlled reactions. The linear relationship between k_{for} and K_a thus eventually must break down. When $K_a \gg 1$, $k_{for} \rightarrow 10^{11} \, M^{-1} s^{-1}$ (the diffusion-controlled limit), and $\alpha_f \rightarrow 0$. A slope change in Brønsted plots often is interpreted as signifying a change in the rate-limiting step from chemical to diffusion control, but it also could arise from changes in the relationship between $\Delta G°$ and $\Delta G^{\neq}$ (Agmon, 1981). Proton transfer is the rate-limiting step in many acid-catalyzed reactions, and linear correlations between $\log k$ and $\log K$ (with $\alpha = 1$) thus are not surprising. Brønsted LFERs with $\alpha < 1$ also are common; these are explainable only by qualitative arguments (e.g., Fig. 1). The value of α is used to make inferences about transition states. If an acid catalyst were completely dissociated before reacting, α would be 1. If $\alpha \approx 1$, the leaving group (conjugate base) is assumed to be loosely bound in the transition state.

Hammett LFER.   An LFER with broad implications was developed in the 1930s by Hammett (1937) to explain substituent effects on reactions of meta- or para-substituted benzene compounds. Hammett found that hydrolysis of substituted ethyl benzoates, ionization of substituted benzoic acids, and many related reactions are affected by nature and position of aromatic substituents, and that the effect could be predicted quantitatively. Using rate and equilibrium constants for ionization of benzoic acid as references, Hammett defined the LFER:

$$\log\left(\frac{k}{k_0}\right) = \rho \log\left(\frac{K}{K_0}\right) = \rho\sigma \tag{7}$$

where 0 denotes unsubstituted benzoic acid and ρ is a constant that depends on the reaction and solvent. Hammett defined the log of the ratio of the ionization constants as a substituent parameter, σ, which is a characteristic of a given substituent and its position on the ring. The effects of meta and para substituents (as expressed by σ) are thought to be additive. For a given reaction, ρ is determined by least-squares fit of normalized rate data for substituted benzene compounds to the corresponding σ. Because acid recombination reactions are diffusion-controlled, k for ionization of benzoic acids is directly proportional to K_a, and $\rho = 1.00$ for this reaction. The basic Hammett relationship does not apply to ortho substituents because they may exert steric as well as electronic effects on reaction centers.

The predictive importance of the Hammett relationship is impressive. It applies not only to hydrolysis reactions but also to substitution and oxidation reactions of aromatic compounds and even to enzyme-catalyzed reactions like the oxidation of phenols and aromatic amines by peroxidase (Job and Dunford, 1976). According to Exner (1972), data available in 1953 allowed prediction of 42,000 rate or equilibrium constants, of which only 3180 had been measured at the time.

The fact that σ applies to such a variety of reactions implies that it measures a fundamental property like electron density at the reactive site (Weston and Schwarz, 1972). The substituent parameter σ decreases with decreasing K_a, and the extent of ionization decreases as electron density increases at the O–H bond in a carboxylic acid group. Thus σ and electron density must be inversely related. Electron-withdrawing groups such as –Cl and –NO_2 have positive σ values; electron donors such as –CH_3 and –NH_2 have negative σ values. Similarly, ρ measures a reaction's sensitivity to electron density. Whereas ρ is positive for nucleophilic reactions that are hindered by high electron density, electrophilic reactions that are accelerated by high electron density have negative values. Although the Hammett relationship was derived empirically, both σ and ρ have been calculated from quantum-mechanical indices, and the form of the equation has been deduced theoretically from molecular-orbital methods (Exner, 1972).

Many other substituent parameters have been developed to improve correlations for specific types of reactions. Brown and Okamoto (1958) developed substituent constants (σ^+) for electrophilic reactions based on hydrolysis rates of meta- and para-substituted 2-chloro-2-phenylpropanes (CPP), which react by electrophilic carbonium ion intermediates. Formation of these intermediates is facilitated by high electron density at the reactive carbon (i.e., by meta- or para-electron donors). The parameter σ^+ is defined by $-4.54\sigma^+ = \log(k/k_0)$, where -4.54 is the best estimate of ρ for hydrolysis of the CPPs (based on Hammett's σ constants for meta substituents), k_0 is the hydrolysis rate constant for unsubstituted CPP, and k is the rate constant for a substituted CPP.

Taft LFERs.   Both σ and σ^+ measure resonance effects on electron distributions. Relationships using these parameters generally do not work very well for ortho substituents and for substituents on aliphatic compounds because they influence reaction rates by polar (inductive) and steric effects. Separation of substituent constants into polar, resonance and steric terms has been the subject of many investigations, but the primary achievement was by Taft in the 1950s (Taft, 1956). He theorized that these terms contribute additively to σ and evaluated polar contributions (called σ^*) from rates of acid- and base-catalyzed hydrolysis of esters of general formula (RCOOR′), where R is the substituent being evaluated. This led to the equation

$$\sigma^* = \frac{\log(k/k_0)_{\text{base}} - \log(k/k_0)_{\text{acid}}}{2.48} = \frac{\log(k/k_0)_{\text{base}} - E_s}{2.48} \tag{8}$$

The factor 2.48 puts σ^* on the same scale as Hammett's σ, and the k_0 values are rate constants for acid and base hydrolysis of acetic acid esters (i.e., R is a methyl group in the reference compound). Usually R′ is an ethyl or methyl group, but in many cases the rate constants do not depend on the nature of R′. Equation 8 is based on the fact that acid hydrolysis rates of substituted benzoic acid esters are only slightly affected by the nature of the substituent, but acid hydrolysis rates of aliphatic esters are strongly affected by substituents. These effects were taken to be caused by steric factors; thus $\log(k/k_0)_{\text{acid}}$ defines E_s. It is reasonable to assume that steric factors affect base-catalyzed rates in the same way. Substituent effects on base hydrolysis of aliphatic compounds are composed of both polar and steric effects, and subtraction of the latter yields a measure of the former. The parameter σ^* is important because it allows one to evaluate substituent effects on aliphatic reaction rates by a formula analogous to the Hammett equation, or by a bivariate relationship, the Taft–Pavelich equation (Pavelich and Taft, 1957):

$$\log(k/k_0) = \rho^*\sigma^* + \delta E_s \tag{9}$$

where ρ^* is analogous to Hammett's ρ, and δ is a measure of the sensitivity of the reaction to steric effects, as measured by E_s (Hancock et al., 1961).

   Other σ constants have been developed, and several books and reviews (Exner, 1972; Weston and Schwarz, 1972; Wells, 1968; Hansch and Leo, 1979) describe the Hammett and related LFERs in detail, along with their uses and limitations in assessing substituent effects on rates of organic reactions. Although sigma relationships have obvious uses in predicting reaction rates, they are probably more important to organic chemists in unraveling mechanisms and developing reaction theories (Weston and Schwarz, 1972).

   Marcus LFER.   Oxidation–reduction reactions involving metal ions occur by two types of mechanisms: inner- and outer-sphere electron transfer. In the former, the oxidant and reductant approach intimately and share a common primary hydration sphere so that the activated complex has a bridging ligand between the two metal ions (M–L–M′). Inner-sphere redox reactions thus involve bond forming and breaking processes like other group transfer and substitution reactions, and transition-state theory applies directly to them. In outer-sphere electron transfer, the primary hydration spheres remain intact. The metal ions are separated by at least two water molecules (or other ligands), and only the electron moves between them. Such mechanisms are explained by the Franck–Condon principle.

Electrons move much more rapidly than atoms and nuclei. For example, vibrational periods of atoms are on the order of 10^{-13} s, but electron transitions occur on a time scale of 10^{-15} s. If electronic transitions are so rapid, the question arises as to why many redox reactions are so slow. For metal ions in solution the answer lies in the fact that charge distributions in the ions induce structural configurations in the ligands and surrounding solvent. The charge distributions and thus structural configurations differ between the reactants and products in an electron-exchange reaction, and energy is required to distort the ions to a shape (the transition state) from which they can form the product ions spontaneously.

Several theories have been developed to explain rates of electron transfer reactions. The most widely used model was derived by Marcus (1963), which views $\Delta G^{\neq}$ for such reactions as the sum of three terms: (1) electrostatic work in bringing two charged species together, (2) energy required to modify the solvent structure, and (3) energy required to distort the metal–ligand bond lengths in the reactants. The rate constant can be formulated in terms of the free energy required for these tasks:

$$k_{AB} = Z_{AB} \exp(-\Delta G_{AB}^{\neq}/RT) \qquad (10)$$

where Z_{AB} is the collision frequency of A and B, and $\Delta G_{AB}^{\neq} = \omega_{elec} + \Delta G_{solv}^{\neq} + \Delta G_{lig}^{\neq}$. Electrostatic work ω can be calculated readily from Coulomb's law. The solvent and ligand energy terms were quantified by Marcus (Weston and Schwarz, 1972; Marcus, 1963), and his theoretical model leads to the following equation:

$$k_{AB} = (k_{AA} k_{BB} K_{AB} f)^{0.5} \qquad (11a)$$

where

$$\ln f = (0.25 \ln K_{AB})^2 / \ln(k_{AA} k_{BB} / Z^2) \qquad (11b)$$

If $\Delta G_{AB}^{\circ} \approx 0$, f approaches 1. While k_{AA} and k_{BB} are rate constants for electron exchange between the oxidized and reduced states of an element, K_{AB} is related to the standard reduction potentials (E°) or $p\varepsilon^{\circ}$ values of the half-reactions.

Under limiting conditions, the Marcus theory (Eq. 11) leads to a simple LFER. For outer-sphere redox reactions of two metal ions and a series of ligands complexing one of the metals (e.g., $A_{ox} L_i + B_{red} = A_{red} L_i + B_{ox}$), a plot of $\ln k$ (or $\Delta G_{AB}^{\neq}$) versus ΔG_{AB}° (or $\ln K$ or $p\varepsilon_{AB}^{\circ}$ for one-electron transfers) will have a slope of ~ 0.5 for small values of ΔG_{AB}°. In such cases k_{BB} is constant, but K_{AB} varies. Exchange rate constants (k_{AA}) for the various complexes of A will vary somewhat, causing scatter about the line. An *average* value of k_{AA} can be estimated from the y-intercept of the plot (where $\Delta G_{AB}^{\circ} = 0$); in a plot of $\ln k$ versus

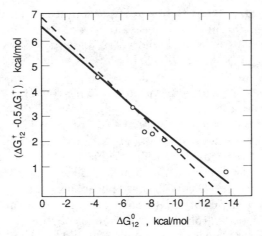

Figure 2. LFER for oxidation of various substituted phenanthroline Fe(II) complexes. Dashed line represents theoretical slope of 0.5 from Marcus LFER. Data redrawn from Dulz and Sutin (1963).

$\ln K$, the intercept is one-half the sum of exchange rate constants k_{AA} and k_{BB} (see Eq. 11a). The slope in such plots is 0.5 only when ΔG°_{AB} is near zero because at large values (more negative than ~ -5 kcal mol^{-1}) f no longer is ~ 1.

Figure 2 shows this for oxidation of various substituted phenanthroline Fe(II) complexes by Ce(IV). An average rate constant of 2×10^3 M^{-1} s^{-1} was found for the phenanthroline–Fe(II)–FE(III) exchanges by this approach (Dulz and Sutin, 1963); this compares with a value of 4 M^{-1} s^{-1} for the free ions. Many studies have verified the Marcus relationship for metal ion redox reactions, and large deviations are assumed to indicate that the reaction occurs by an inner-sphere mechanism. (*Note*: An outer-sphere mechanism can be inferred if the redox reaction is faster than the rates of ligand exchange for the metal ions.)

Linear plots of $\ln k$ versus ΔG° are found in some electron-transfer reactions that proceed by atom transfer, but the slopes are not 0.5 as predicted for outer-sphere reactions by the Marcus theory. Care thus needs to be taken in inferring mechanisms for redox reactions from LFERs. Other LFERs, such as the Hammett and Brønsted relationships, also have been applied to redox reactions of organically complexed metal ions and similarly lead to linear plots of $\log k$ versus E°_{AB} (see next section).

3.2. Applications of LFERs to Reactions in Natural Waters

Linear Free-energy reactions have been applied to the reactions of many contaminants in natural waters, and they are especially useful in implementing simple transport and fate models for organic chemicals. If experimental values of rate constants are not available and if they can be estimated by LFERs, the applicability of such models can be extended. Hydrolysis reactions have received the most attention in developing LFERs for organic contaminants. In most cases, only base-catalyzed hydrolysis is important at natural water pH values, and Brønsted relationships have been fit to such data for several sets of compounds (Table 3). For example, plots of second order $\log k$ values versus pK_a of the product alcohols showed good correlations within classes of carbamates, but slopes varied among classes Wolfe et al. (1978). Hydrolysis rates of these pesticides thus are correlated with the acidity of the constituent alcohol, but the rates are very slow, and hydrolysis is an important degradation pathway for these compounds only when the alcohol has a $pK_a < \sim 12$.

Alkaline hydrolysis rates of organophosphate and organophosphorothionate esters also follow Brønsted relationships (Table 3). These classes of compounds include pesticides, plasticizers, and hydraulic fluid additives. Rate constants for O,O-dimethyl-O- and O,O-diethyl-O-(aryl) phosphates and phosphorothionates are correlated with the acidity of the conjugate acid of the aryl leaving group, which is preferentially hydrolyzed under alkaline conditions, leaving the dialkyl ester anions. When the leaving-group conjugate acid has a $pK_a \sim 10$, the half-life

TABLE 3. Examples of LFERs for Aquatic Contaminants

Alkaline Hydrolysis	Reference
Organophosphates (Brønsted) O,O-Diethyl-P $\log k = 0.28 \log K_a - 0.22$; $r^2 = 0.93$, $n = 4$ O,O-Dimethyl-P $\log k = 0.28 \log K_a + 0.50$; $r^2 = 0.97$, $n = 4$	Wolfe, 1980a
Organophosphorothionates (Brønsted) O,O-Diethyl-P $\log k = 0.21 \log K_a - 1.6$; $r^2 = 0.95$, $n = 4$ O,O-Dimethyl-P $\log k = 0.25 \log K_a + 0.34$; $r^2 = 0.97$, $n = 5$	Wolfe, 1980a
Triaryl phosphate esters (Hammett) $\log k = 1.4 \, \Sigma\sigma + \log k_o$; $r^2 = 0.99$, $n = 4$	Wolfe, 1980a
Aliphatic primary amides (Taft) $\log k = 1.6\sigma^* - 1.37$; $r^2 = 0.948$, $n = 11$	Wolfe, 1980b
Diphthalate esters (Taft–Pavelich) $\log k = 4.59\sigma^* + 1.52E_s - 1.02$; $R^2 = 0.975$, $n = 5$	Wolfe, 1980b
Other Reactions	
Acid hydrolysis of 2,2-substituted alkenes (Brown–Okamoto) $\log k_H^- = 12.3 \, \Sigma\sigma^+ - 8.5$; $r^2 = 0.97$, $n = 24$	Wolfe, 1980b
pK_a values of substituted 2-nitrophenols (*not* a kinetic LFER) $pK_a = pK_{aH} - 2.59 \, \Sigma\sigma_i$; $r^2 = 0.98$, $n = 17$	Schwarzenbach et al., 1988
pK_a values of aldehyde–bisulfite adducts (Taft): $pK_a = 12.1 - 1.29 \, \Sigma\sigma^*$; $n = 6$	Betterton et al., 1988

for hydrolysis at pH 8 approaches 1 yr. Alkaline hydrolysis of the phosphate esters thus is rapid only when the aryl group has strong electron-withdrawing substituents, which provide conjugate acid $pK_a \ll 10$. The O,O,O-trialkyl esters are all very slow to hydrolyze ($t_{1/2} > 10^4$ days at pH 8) because alcohols are very weak acids. The phosphorothionates are less reactive than the phosphate esters; hydrolysis of the former compounds is slow to very slow under environmental conditions.

Brønsted LFERs also apply to reactions of metal ions (Lewis acids). Dissociation rates of Ni(II) complexes are correlated with corresponding dissociation equilibrium constants. This suggests that the reactions occur by dissociative interchange, in which breakage of the Ni(II)–ligand bond predominates over formation of the Ni(II)–water bond in the rate-determining step (Hoffmann, 1981). In addition, rates of metal-catalyzed decarboxylation of malonic acid are correlated with the stability constants for the metal–malonate complexes (Prue, 1952).

Hammett (and related sigma) relationships have been applied to aquatic reactions of several classes of aromatic contaminants. For example, alkaline hydrolysis of triaryl phosphate esters fits a Hammett relationship (Table 3); $\Sigma\sigma$ is the sum of the substituent constants for the aromatic groups and k_0 is the hydrolysis rate constant for triphenyl phosphate ($0.27 \ M^{-1} s^{-1}$; $t_{1/2} = 30$ days at pH 8). Triaryl esters thus hydrolyze much more rapidly than trialkyl or dialkyl–monoaryl esters under alkaline conditions. Rates of photooxidation of deprotonated substituted phenols by singlet oxygen ($^1O_2$) have been found to be correlated with Hammett σ constants (Scully and Hoigné, 1987). The electronic effects of substituents on pK_a values of substituted 2-nitrophenols also fit a Hammett relationship; this, of course, is not a *kinetic* LFER. Two compounds (4-phenyl-2-NP and 3-methyl-2-NP) did not fit the relationship and were not included in the regression. Steric effects may account for the discrepancy for the latter compound. Nitrophenols are used as intermediates in synthesis of dyes and pesticides and also used directly as herbicides and insecticides.

Success in applying LFERs to aliphatic compounds has been mixed, reflecting the complications of both steric and polar effects of substituent groups on reaction rates. Rates of acid-catalyzed hydration of 2,2-substituted alkenes ($RR'C=CH_2$) to form secondary alcohols [$RR'C(OH)CH_3$] fit a σ^+ correlation for electrophilic reactions over almost 16 orders of magnitude in k (Fig. 3a). This class of compounds includes vinyl chloride, propylene, styrene, and other compounds used as starting materials in polymer, pesticide, and dye manufacturing. The correlation, although impressively wide, is not precise; predicted and measured values of rate constants agree within about a factor of only 10. Note that σ^+ constants were developed to describe the effects of *aromatic* substituents on electrophilic reactions; that they work at all for aliphatic alkenes may be considered an unexpected bonus. Base-catalyzed hydrolysis rates of aliphatic primary amides fit an LFER with Taft's σ^* constants for polar effects (Fig. 3b). Rate constants for primary amides such as acetamide, acrylamide, and propionamide are about 0.7–$2.7 \times 10^{-3} \ M^{-1} s^{-1}$ ($t_{1/2}$ at pH 8 $\sim 3 \times 10^3$ to 10^4 days).

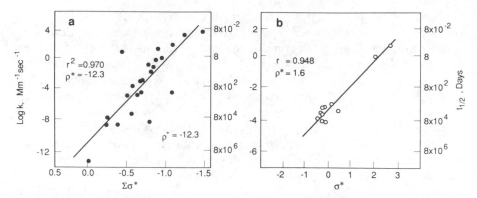

Figure 3. (a) Electrophilic σ^+ LFER for hydration of 2,2-substituted alkenes, and (b) Taft σ^* LFER for alkaline hydrolysis of primary amides. From Wolfe (1980b) with permission of Butterworth Publishers.

The high r^2 for this LFER (Table 3) is somewhat deceptive in that the regression is driven by two outlying data points.

The alkaline hydrolysis of phthalate diesters has been fit to the Taft–Pavelich equation (Eq. 9). Dimethyl phthalate (DMP) hydrolyzes to phthalic acid (PA) in two steps: $DMP + H_2O \rightarrow MMP + CH_3OH$ and $MMP + H_2O \rightarrow PA + CH_3OH$. The first step is about 12 times faster than the second, and nearly all the diester is converted to the monoester before product PA is formed. Other diesters are assumed to behave similarly. An LFER was obtained from rate measurements on five phthalate esters (Wolfe et al., 1980b). The reaction constants, ρ^* and δ, were determined by multiple regression analysis of the measured rate constants and reported values of σ^* and E_s for the alkyl substituents. The fitted intercept compares favorably with the measured rate constant (log $k_{OH} = -1.16 \pm 0.02$) for the dimethyl ester (for which σ^* and $E_s = 0$ by definition). Calculated half-lives under pseudo-first-order conditions (pH 8.0, 30°C) range from about 4 months for DMP to over 100 years for di-2-ethylhexyl phthalate.

Finally, several equilibrium and kinetic properties of aldehyde–bisulfite adducts were found to be linearly related Taft's σ^* parameter (Betterton et al., 1988). These compounds, which include α-hydroxymethane sulfonate and other α-hydroxyalkyl sulfonates, may be important reservoirs of S(IV) species in clouds, fog, and rain. Fairly good relationships were found between equilibrium properties (e.g. acidity constants) and $\Sigma\sigma^*$ values, but rates constants for nucleophilic addition of SO_3^{2-} to the aldehydes showed only a crude fit. Similarly, poor results were found in applying σ^* to hydrolysis reactions of volatile alkyl chlorides (T. Vogel, University of Michigan, personal communication, 1989), and this has been shown to be a general characteristic of reactions of alkyl halides with nucleophiles (Okamoto et al., 1967).

Various LFERs have been applied to redox reactions of potential interest in aquatic systems, including reactions of dissolved metal ion oxidants and metal oxides with both organic and inorganic reductants. A Marcus LFER has been shown to fit the autoxidation of several metal ions, including Fe^{2+} and its hydroxy complexes (see Chapter 11, this volume). Oxidation rates of some pyridine and bipyridyl derivatives, such as biologically important nicotinamide and the herbicides diquat and paraquat, by Co(III), were found to fit a Marcus relationship (Fanchiang, 1982). Oxidation rates of various substituted phenols by soluble complexes of Fe(III) (with cyanide, phenanthroline, or bipyridyl as ligands) are correlated with σ constants of the phenol substituent groups. Similar trends have been found for the reductive dissolution of Mn(III, IV) oxides by substituted phenols (Stone, 1987). Electron-donating groups ($\sigma < 0$) decrease reduction potentials, and electron-withdrawing groups ($\sigma > 0$) increase reduction potentials of substituted phenols. Correlation of oxidation rates with σ constants implies that rates are correlated with reduction potentials of the phenols (thus with the overall $E°$ and $\Delta G°$ values of reaction; see Fig. 4).

Although the above discussion is not intended to be an exhaustive review of LFER applications in aquatic chemistry, it is illustrative of the literature on this topic. Several comments and conclusions are pertinent. First, the huge database regarding LFERs in the chemistry literature has hardly been tapped by existing aquatic studies. Several reasons can be cited for this, including the fact that many compounds studied by chemists in laboratory situations are not important as environmental contaminants. Also, some reactions occurring in nonaqueous systems are not relevant to the aquatic behavior of organic contaminants. Second, most of the reported LFERs for aquatic contaminants are based on correlations with very small sample sizes ($n = 4$ or 5 for many regression equations in Table 3). Consequently, a fair amount of uncertainty exists about

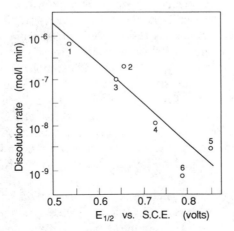

Figure 4. LFER for reductive dissolution of Mn(III, IV) oxides by substituted phenols. Rate of Mn^{2+} formation versus $E_{\frac{1}{2}}$ for reduction of phenols: 1 = p-methylphenol, 2 = p-chlorophenol, 3 = phenol, 4 = p-hydroxybenzoic acid, 5 = o-hydroxybenzoic acid, 6 = 4'-hydroxyacetophenone. From (Stone, 1987) with permission of the American Chemical Society.

the broader applicability of these relationships. Third, few of the reported correlations are very precise. It is unclear whether this primarily reflects measurement uncertainties or inherent lack of fit to the supposed linear relationships. Even when the coefficient of explanation is high (e.g., $r^2 > 0.95$), the log–log nature of the relationships, which often span many orders of magnitude, leads to large uncertainties in estimating k (easily $\pm 5 \times$ or $\pm 10 \times$). Nonetheless, such LFERs still can be useful in screening groups of compounds for further study. Even crude estimates (factor of 10) may be sufficient for compounds with (predicted) rates that are very fast or very slow. Attention then can be focused on obtaining more reliable rates for compounds of intermediate reactivity (characteristic times of days to a few years). Finally, further developmental work is needed on some important classes of aquatic contaminants such as volatile chlorinated alkanes and alkenes, for which none of the existing LFERs seems to apply.

4. PROPERTY–ACTIVITY AND STRUCTURE–ACTIVITY RELATIONSHIPS (PARs AND SARs)

The LFERs described in preceding sections are but a subset of the attribute relationships that have been developed to relate the environmental behavior of compounds (in particular their bioactivity) to their structural or equilibrium properties (e.g., PARs and SARs; see Table 1). The original impetus for research on PARs and SARs was the need to predict the effectiveness of new drugs and medicines, but in recent years this subject has become an important area of environmental chemistry.

Variables commonly used in PARs and SARs are summarized in Table 4. The main processes of interest relative to the bioactivity of aquatic contaminants are bioaccumulation, biodegradation, and acute toxicity (LC_{50}), but inhibition of key biological processes such as respiration rate and photosynthesis also are used in some PARs and SARs as measures of a compound's toxicity. The physicochemical properties listed in Table 4 reflect molecular structure, but they are not structural characteristics themselves. Relationships based on these properties thus should be called *property–activity relationships* (PARs), and the term (quantitative) "structure–activity relationship," "(Q)SAR" should be restricted to relationships based on structural or topological parameters. However, the literature is not consistent in this terminology, and the line between structural characteristics and properties resulting from structure is not always clear.

Aqueous solubility (S_w) and lipophilicity (as measured by partition coefficients between water and octanol, K_{ow}) are among the most common physicochemical properties used to predict a compound's bioactivity or chemical reactivity. In turn, these properties often are predicted by correlations with another phsyco-chemical property; these may be called *property–property relationships* (PPRs). For example, K_{ow} is correlated with S_w (Fig. 5a) and with easily measured indices such as HPLC retention times (t_r). Alternatively, some properties used to predict

TABLE 4. Common Attributes Used in Environmental Correlations[a]

Structural Variables	Physicochemical Properties	Biological Activity
Molar volume	Molecular weight	Bioaccumulation, BF
Surface area	Lipophilicity (hydrophobicity)	Acute toxicity, LC_{50}
Chain length	K_{ow} K_p (sediment–water partition coefficient) K_{oc} (K_p normalized to organic content)	Inhibition of a function or process such as photosynthesis, respiration, or bioluminescence, EC_{50}
Topological properties of atom-bond arrangements (degree of branching): X and X^v, molecular connectivity indices	Aqueous solubility, S_w Chromatographic retention time, t_r	Biodegradation rate, K_{deg}, (BOD_5 as fraction of theoretical BOD)
Molecular fragment approaches (group additivity) of fundamental atomic properties	Molar refractivity, R_M LFER parameters, e.g., σ, σ^+, σ^* LSER parameters: solvatochromic properties, π^*, β, α_m Hard–soft acid parameters, σ_p, σ_k, for metals	

[a] Attributes and symbols are defined in the text

bioactivity can be predicted from structural characteristics of molecules; these may be termed structure-property relationships (SPRs). Examples include (1) the prediction of K_{ow} (Fig. 5b) or S_w from a compound's molar volume or molecular surface area (perhaps the simplest measures of structure), (2) prediction of K_{ow} from substituent lipophilicity parameters or lipophilic terms for molecular fragments, and (3) prediction of S_w or K_{ow} from topological indices such as the molecular connectivity indices. These properties generally reflect the aqueous thermodynamic activity of a compound (solvation free energy) rather than specific chemical reactivity.

4.1. Property–Activity Relationships

K_{ow}. Many PARs are based on partition coefficients or related parameters. The partition coefficient of a compound between octanol (o) and (w), $K_{ow} = C_o/C_w$, is the standard measure of a compound's lipophilicity, and this

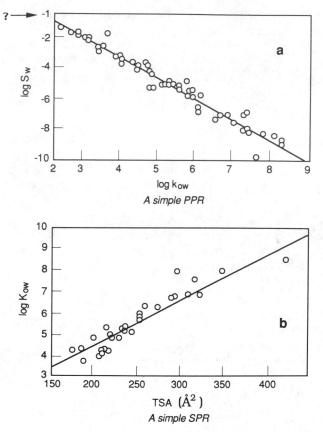

Figure 5. (a) simple PPR: measured log K_{ow} versus log S_w for 55 aromatic hydrocarbons; from (Andren et al., 1987). (b) A simple SPR: log K_{ow} versus TSA for 32 aromatic hydrocarbons; from (Doucette and Andren, 1987). With permission of the American Chemical Society.

characteristic has important implications for biological transport of molecules. Interest in this property dates back to the classic studies of Meyer and Overton near the turn of the twentieth century, which showed a relationship between olive oil–water partition coefficients and the narcotic action of organic compounds. The parameter K_{ow} is probably the most widely used predictor variable in PARs and other attribute relationships. It can be measured directly, but such methods are tedious, and difficulties arise for highly lipophilic compounds [e.g., polynuclear aromatic hydrocarbons (PAHs) and polychlorinated biphenyls (PCBs) with log $K_{ow} > 5$] because their low aqueous solubility makes measurement of equilibrium concentrations in the aqueous phase problematic. Direct measurement of K_{ow} by conventional shaker-flask methods is not practical for most such compounds, but a generator-column method has been described (Woodburn et al., 1984) that extends measurement to log $K_{ow} > 8$. However, substantial variations (0.5–2.0 log units) can be found in the recent literature even for *measured* values at $\log K_{ow} < 5$ for many common aquatic contaminants, including some chlorophenols, PAHs, and PCBs (Sabljic, 1987).

A variety of indirect methods have been developed to estimate K_{ow} (Table 5). These include simple PPRs and SPRs (Fig. 5), as well as group additivity methods in which the lipophilic contributions of molecular fragments have been estimated (e.g., Rekker's fragment technique and UNIFAC). The existence of so many methods to estimate K_{ow} suggests there are some fundamental difficulties in accurately determining this important parameter. As a minimum, we can expect that there will be differences in the values of a compound's K_{ow} determined by the different techniques. This is the case, and a large literature has arisen on this subject (Andren et al., 1987; Sabljic, 1987; Eadsworth, 1986; De Voogt et al., 1988; Veith et al., 1979).

A Bioactivity Model.   The biological processes for which PARs and SARs are sought are much more complicated than the chemical reactions for which LFERs are developed. Biological processes consist of many physicochemical steps, and their rate-controlling steps are poorly defined compared with chemical reactions. Given this complexity, it is surprising that simple PARs and SARs exist, and even more surprising that they often work quite well. The applicability

TABLE 5. Methods of Determining K_{ow}

Method	Reference
Direct measurement	Woodburn et al., 1984
Hansch–Leo π approach for substituted aromatic compounds $\pi_X = \log K_{ow,X} - \log K_{ow,H}$	Hansch and Leo, 1979
For chlorinated PCBs and PAHs $\log K_{ow}^{\cdot} = (n+1)^b \log K_{ow}^0$ where n = number of Cl atoms $b \log K_{ow}^0$, (K_{ow} of parent compound)	Kaiser, 1983
Fragment constant approach $\log K_{ow} = \Sigma n_i f_i$ where n_i = number of fragments of type i in molecule f_i = contribution of ith fragment to $\log K_{ow}$	Rekker, 1977; Hansch and Leo, 1979
Estimation from UNIFAC $K_{ow} = 0.115 \gamma_w / \gamma_o$. with γ_w and γ_o calculated by UNIFAC	Arbuckle, 1983
Correlation with aqueous solubility, S_w	Andren et al., 1987
Correlation with $\log k'$, HPLC retention coefficient $k' = (t_r - t_o)/t_o$	Eadsworth, 1986

of simple relationships over a range of organisms, processes, and organic compounds suggests that there are a few common *key steps*. These likely are transport across cell membranes and chemical reaction involving a key enzyme. Ease of transport across membranes is related to the lipophilicity of molecules because membranes have lipid layers. Chemical reactions are subject to the same laws whether they occur *in vitro* or *in vivo*; consequently, many biotic reactions are amenable to LFERs such as Brønsted or Hammett relationships.

These considerations can be used to derive models relating the bioactivity and physicochemical properties of organic compounds. Hansch and Fujita (1964) were the first to develop models of this type. Their model (Table 6) is called the *linear free-energy relationship model*; it describes a compound's bioactivity as an additive combination (on a log–log basis) of lipophilic and electronic (substituent) effects. In theory, other factors such as steric effects can be added to the model. As the derivation in Table 6 shows, the model can be simplified in a number of ways, depending on the biological process and nature of the organic compound being considered. For example, if an enzymatic reaction is not involved or never becomes rate-limiting (within a series of compounds), bioactivity is related only to lipophilicity (K_{ow}) in either a linear or curvilinear fashion.

Bioaccumulation.   The PARs described in Table 7 for the three major biotic processes (bioaccumulation, biodegradation, toxicity) fit the Hansch–Fujita model types given in Table 6. Most of the PARs involve K_{ow} or a correlate such as t_r. Bioaccumulation (BF), which can be viewed as a simple partitioning process, has been correlated with K_{ow} for a wide range of compounds and organisms. In most cases, the slope of a log–log plot is close to one, and Mackay (1982) argued that the slope *should be* one, implying that BF and K_{ow} have a linear relationship. He found a good linear fit for a set of 51 compounds, but arguments (based on both equilibrium and kinetic reasons) can be made that the relationship is *not* necessarily linear over a broad range of K_{ow}.

Other molecular attributes besides K_{ow} affect BF for some classes of compounds. Bioaccumulation of PCBs depends on congener shape, as well as lipophilicity. The BF values of the congeners displayed a rather scattered bell-shaped relationship with K_{ow}, but linear correlations were obtained when $\log K_{ow}$ was multiplied by an empirical steric coefficient or by t_r values for congeners on HPLC activated carbon columns (Shaw and Connell, 1984). Planar congeners have the largest t_r values (are most strongly adsorbed) and highest BF values. Polychlorinated biphenyls with chlorine atoms ortho to the bond joining the two rings are *not* planar (because of steric factors); instead, the rings are perpendicular to each other. Thus PCB bioaccumulation is affected by both lipophilicity and molecular shape.

Biodegradation.   This is much more complicated process than bioaccumulation, and a given one-parameter relationship typically applies only to a narrow range of compounds. Within a class of compounds, transport across bio-

TABLE 6. Property–Activity Model of Hansch and Fujita (1964)

General scheme of rate-limiting processes for biological response to organic compound X within a series of related compounds:

$$\text{Compound in extra cellular phase} \xrightarrow{} \text{Site of action in cellular phase} \xrightarrow[\substack{\text{Critical} \\ \text{reaction}}]{k_X} \cdots \longrightarrow \text{Biological response} \qquad (1)$$

$$\text{Response rate} = d(\text{biotic condition})/dt = d(\text{bc})/dt = Ak_X[C_X] \qquad (2)$$

where A = probability X will reach active site in some time interval, $f(K_{ow})$; k_X = rate constant for critical step; and $[C_X]$ = external concentration of X; if we assume that A vs. $\log K_{ow}$ fits a normal distribution and k_X is related to k for a parent compound by the Hammett relationship, Eq. (2) here becomes:

$$d(\text{bc})/dt = ak_H[C]\exp(\rho\sigma - [\pi - \pi_o]^2/b) \qquad (3)$$

where π is a substituent constant for lipophilicity; for a fixed response and assay time (e.g., 96-h LC_{50}), $d(\text{bc})/dt$ is replaced by a constant, and this leads to

$$\log(1/[C]) = -c(\pi^2 + \pi_o^2) + c'\pi\pi_o + c''\pi + \rho\sigma + c''' + \log(k_H) \qquad (4)$$

For a given parent compound and organism, π_o and k_H are subsumed in the other constants, giving Eq. (5), which can be simplified under four conditions:

$$\log(1/[C]) = -c\pi^2 + c'\pi + \rho\sigma + c'' \qquad (5)$$

where $\pi_o \gg \pi$; σ small or zero

Type I: $\log(1/C) = a\pi + b$ [bioaccumulation; nonspecific toxins (narcosis)]
where $\pi \approx \text{or} > \pi_o$; σ still unimportant

Type II: $\log(1/C) = -a\pi^2 + b\pi + c$ (curvilinear)
where π is unimportant (membrane transport not rate-limiting)

Type III: $\log(1/C) = \rho\sigma + a$ (Hammett-like behavior)
$\pi_o \gg \pi$; σ important

Type IV: $\log(1/C) = a\pi + \rho\sigma + b$ (activity related to both K_{ow} and σ)

---

membranes may be rate-controlling for some compounds, while enzymatic reaction (or some other process) may limit the degradation rate for others. A change from transport limitation to reaction rate limitation often is inferred from a change in the slope of K_{deg} versus K_{ow}. Parabolic relationships (Type II in Table 6) have been reported between k_{deg} and K_{ow} for phenols, ketones, alcohols, and phthalate esters (Urushigawa and Yonezawa, 1979); k_{deg} typically reaches a maximum at a fairly low value of K_{ow} (~ 3), and then declines at larger K_{ow} values.

TABLE 7. Simple Property–Activity Relationships for Organic Compounds

		Reference
Bioaccumulation		

Bioaccumulation

Mussels	$\log BF = 0.86 \log K_{ow} - 0.81$; $n = 16$, $r^2 = 0.91$	Geyer et al., 1982
Fish	$\log BF = 0.85 \log K_{ow} - 0.70$; $n = 59$, $r^2 = 0.90$	Veith et al., 1979
Chlorella	$\log BF = 0.68 \log K_{ow} + 0.16$; $n = 41$, $r^2 = 0.81$	Grimes and Morrison, 1975

Linear fit of most of same data   Mackay, 1982
$\log BF = \log K_{ow} - 1.32$
or BF $= 0.048 K_{ow}$; $n = 51$, $r^2 = 0.95$

Uptake and clearance rates in fish   Hawker and Connell, 1985

Uptake $\log k_1 = 0.337 \log K_{ow} - 0.373$
Clearance $-\log k_2 = 0.663 \log K_{ow} - 0.947$

Biodegradation of diphthalates
$\log k_h = -2.09 \log t_r^2 + 1.19 \log t_r - 1.15$, $r^2 = 0.99$ Urushigawa and
$\log k_b = 2.1 \log k_{OH} - 6$; $n = 4$, $r^2 = 0.93$ Yonezawa, 1979
$\log k_b = 9.6\sigma^* + 3.2E_s - 8.1$ Wolfe et al., 1980a

Narcosis (unreactive compounds)
Acute Toxicity to guppies
$\log LC_{50}(\mu M) = -0.87 \log K_{ow} + 4.87$; $n = 50$, $r^2 = 0.974$ Veith et al., 1983
Bilinear model:
$\log LC_{50} = -0.941 \log K_{ow} + 0.94$ Veith et al., 1983
$\qquad \log(1 + 0.000068 \log K_{ow}] - 1.25$
Chlorobenzene inhibition of primary production
in green alga
$\qquad -\log EC_{50} = 0.985 \log K_{ow} - 2.626$ $n = 12$ Wong et al., 1984
$\qquad \log EC_{50} = 0.587 \log S_w - 2,419$ $n = 12$ Wong et al., 1984

Toxicity of reactive organic halides to guppies

$\log LC_{50}(\mu M) = 1.30 \log(1604 + 1/k_{NBP}) - 4.35$;   Hermens, 1986
$n = 15$, $r^2 = 0.88$
$k_{NBP} =$ rate constant for reaction of compound with
nucelophilic reagent, 4-nitrobenzylpyridine

Because of covariance among predictor variables, PARs may not always provide an unambiguous way to infer the rate-limiting step for a given reaction. For example, k_{deg} is correlated with t_r for phthalate di-esters (Urushigawa and Yonezawa, 1979), which suggests that their biodegradation is limited by transport into cells (since t_r is strongly correlated with K_{ow}). However, k_{deg} also is correlated with the rate constant (k_{OH}) for chemical (alkaline) hydrolysis for these compounds, and a Taft–Pavelich LFER has been derived for this reaction (Table

3). Thus k_{deg} is closely correlated with measures of both transport and substituent (electronic and steric) properties. Because these properties probably are intercorrelated and because the PARs are based on so few data points, it is not possible to infer which is more important from existing information.

Fit of decomposition data to PARs does not necessarily prove that the mechanism is biological. For example, degradation of the herbicide propyzamide and nine analogs in soil fit a Type IV model: $\log k = -2.74 - 1.22\sigma + 0.58\log K_{ow}$ (Cantier et al., 1986), but other evidence (effect of temperature, sterilization) supported chemical hydrolysis rather than microbial degradation as the primary loss mechanism. The negative sign on σ implies an electrophilic reaction (one accelerated by high electron density). The influence of K_{ow} on the rate suggests that compound degradation in soils occurs in the adsorbed phase since K_{ow} is correlated with the strength of adsorption to soil particles.

Toxicity. The toxicities of organic compounds that behave as narcotics are inversely correlated with K_{ow}. "Narcosis" is defined as a nonspecific, reversible, physiological effect independent of chemical structure. Compounds behaving in this manner include many representatives in the following classes: aliphatic, aromatic, and chlorinated hydrocarbons; alcohols; aldehydes; ketones; and ethers. In general, narcotic compounds are unreactive molecules. According to the theory of narcotic action, the thermodynamic activities of narcotics in blood are approximately the same at the same level of physiological effects. For nonvolatile chemicals in aquatic systems, Veith et al. (1983) showed that the thermodynamic activity (a_{nar}) needed to produce narcosis can be estimated by the expression $a_{nar} \approx C_{nar}/S_w$, where C_{nar} is the aqueous concentration of chemical producing the narcotic effect. If a_{nar} is constant among narcotic compounds, C_{nar} is proportional to S_w. Since S_w and K_{ow} are inversely correlated, we can expect an inverse correlation between LC_{50} and K_{ow} for narcotic compounds. This has been verified for several data sets (see Table 7).

Linear solvation energy relationships (LSERs) also have been used to predict the toxicities of compounds that act by narcosis (Kamlet et al., 1987). The LSER approach explains the solubility of compounds in terms of three types of solute–solvent interactions: cavity formation (to make room for the solute), solvent–solute dipolar interactions, and hydrogen-bonding interactions, which are modeled as linear combinations of the resulting free-energy terms. The solute characteristic that affects the cavity term is molar volume, and the exergonic effects of solute–solvent dipole–dipole interactions are described in terms of several "solvatochromic" parameters. The LSER approach provides a closer fit to toxicity data than single variable ($\log K_{ow}$) relationships and helps explain why some compounds with similar K_{ow} values have different toxicities. Its disadvantage is that the solvatochromic variables needed to compute LSERs are not available for many compounds and must be experimentally measured or computed from other molecular properties.

Chemically reactive compounds generally are more toxic than unreactive (narcotic) compounds, and the toxicity of reactive compounds is affected by electronic structure, as well as lipophilicity. A major toxic mechanism of reactive compounds is nucleophilic displacement (SN) reactions (Hermens, 1986). These are important in biological systems because nucleophilic sites (–SH, –NH$_2$, –OH groups) are common on biomacromolecules. Also, as SN reactivity is influenced by inductive and resonance effects, substituent parameters should be useful in developing predictive equations for the toxicity of such compounds. For example, Hermens (1986) found that reactive organic halides like allyl and benzyl chloride were much more toxic than predicted from K_{ow} relationships for narcotic compounds, and LC$_{50}$ values were not strongly correlated with K_{ow}. A better correlation was found between their toxicity and their reactivity with a standard nucleophilic reagent, 4-nitrobenzylpyridine (Table 7).

4.2. Structure–Activity Relationships

The structural parameters used to develop SARs and SPRs can be grouped into four categories: (1) simple numerical indices such as carbon number or chain length; (2) measures of molecular size such as molar volume and molecular surface area; (3) group additivity (fragment) parameters based on fundamental molecular or atomic properties; and (4) topological parameters based on molecular shape, such as branching indices. Good correlations with carbon number have been found for restricted sets of compounds, but the limitations of the first category are obvious. Molar volume and surface area have been used primarily to predict physicochemical properties such as S_w.

A fragment-based SAR derived from fundamental molecular characteristics has been described to predict biodegradability of several classes of organic compounds. The percent biodegradation in five days was correlated with the difference in modulus of absolute atomic charge $\Delta|\delta|_{x-y}$ across a key bond common to a class of molecules (Dearden and Nicholson, 1987). The value $\Delta|\delta|_{x-y}$ was calculated by obtaining the minimum energy configuration of a molecule by a molecular mechanics program; such computed parameters, which are not quite structural nor physicochemical attributes (as defined in Table 1), may be termed "molecular attributes". Provided the correct bond is chosen for the calculation, impressive correlations can be achieved. For example, a correlation based on $\Delta|\delta|_{c-o}$ for eight amino acids yielded an r^2 of 0.995, but a correlation based on $\Delta|\delta|_{N-o}$ for the same compounds yielded an r^2 of 0.15. The biodegradation of amino acids thus is controlled by the charge difference at the C–O bond (and presumably a reaction involving that bond) and not the C–N bond. A single $\Delta|\delta|_{x-y}$ biodegradation correlation involving 112 compounds in six classes (phenols, amines, aldehydes, carboxylic acids, amino acids, halogenated compounds) yielded an r^2 of 0.978. In addition, for a subgroup of 19 alcohols, electrophilic superdelocalizability on the carbon atom to which the hydroxyl group is attached was highly correlated with biodegradability (r^2), implying that the reactivity of that carbon atom is the controlling factor in alcohol biodegradation (Dearden and Nicholson, 1987).

The idea that the geometric structure of a molecule encodes information on its chemical properties and biological activity has led to the analysis of molecular

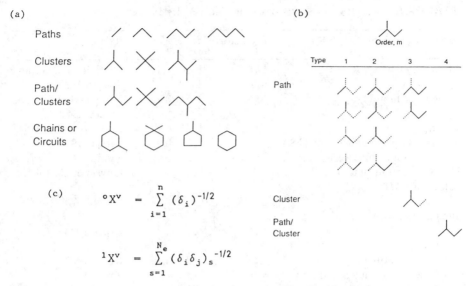

Figure 6. (a) Types of molecular connectivity subgraphs; (b) subgraphs for isopentane; order m = number of contiguous bonds in subgraph; (c) definition of valence-based zero and first-order MC indices (χ^v) in terms of $\delta_i = V_n\text{-}n_{Hi}$, where V_n = valence of nonhydrogen atoms (4,5,6 for C,N,O, respectively) and n_{Hi} = number of hydrogens attached to nonhydrogen atom; N_e = number of edges (bonds) in reduced carbon skeleton.

data by topographical methods and the development of topological indices. The Randić Branching Index forms the basis for molecular connectivity (MC) indices (Kier and Hall, 1976) that are widely used as predictor variables in SARs. The MC approach is a bond counting method that describes the "connectedness" of nonhydrogen atoms in organic compounds, and it considers four types of bonding patterns (Fig. 6). A variety of MC indices can be calculated from the reduced representation (carbon skeleton) of a molecule.

Many SARs and SPRs have been derived from MC indices. Molecular conductivity indices are correlated with physical properties such as boiling point and density; for a review, see Keir and Hall (1986). Examples relevant to aquatic systems are given in Table 8. Although such relationships are empirical, a theoretical basis can be described for some. The fact that first-order MC indices (^1X and $^1X^v$) are closely correlated with so many physicochemical properties of organic compounds suggests that they measure some underlying fundamental molecular property. Most probably that is total molecular surface area; Sabljic (1987) found the following correlation between TSA and ^1X: TSA $= 24.6^1X$ $+ 57.7$; $r^2 = 0.91$ ($n = 72$). The ^1X index also is assumed to reflect the "branchedness" of compounds, but the structural characteristics encoded in other MC indices are not well defined. In general, the compounds used to develop MC SARs should be sufficiently similar that their metabolic fates and mechanisms of attack are the same or closely related; for example, compare the correlation

TABLE 8. Some SPRS and SARS Based on Connectivity Indices

Predicted Variable	Predictor	n	r^2	Reference
Molar refractivity, R_M	$^1X^v$	101	0.99	Nirmalakhandan and Speece, 1988
Aqueous solubility, S_w	$^1X^v$	315	0.95	Nirmalakhandan and Speece, 1988
Log K_{ow}	^1X	138	0.97	Murray et al., 1975
Total molecular surface area (TSA)	^1X	72	0.91	Sabljic, 1987
Soil sorption coefficient K_{ow}	^1X	72	0.95	Sablic, 1987
HPLC retention times and TLC R_f values of PAHs	$^1X^v$			Govers et al., 1984
SARS				
Bioaccumulation				
	$^1X^v$	21	0.92	Koch, 1983
Toxicity				
Chlorophenols to guppies	$^1X^v$	10	0.96	Koch, 1982
Diverse halogenated and non-aliphatic compounds to *Daphnia magna*	$^1X^v$	13	0.70	Koch, 1982
		31	0.82	Koch, 1983
Chlorophenols to protozoan	$^1X^v$	10	0.79	Koch, 1982
Chlorophenols to protozoan	$^2X^v$	10	0.82	Koch, 1982

Multivariable SARs

Chlorosis induced in *Lemna minor* by substituted phenols

$-\log EC_{50} = -0.55 + 1.116\,^1X^v + 0.603\Sigma\sigma$ $R^2 = 0.92, n = 25$ Kier and Hall, 1976

$-\log EC_{50} = 2.12 + 1.114\,^1X^v - 0.273\,pK_a$ $R^2 = 0.93, n = 25$ Kier and Hall, 1976

Acute toxicity of organotin compounds to *Daphnia magna*

$-\log LC_{50} = 0.75\,^1X^v - 5.63$ $r^2 = 0.86; n = 12$ Vighi and Calamari, 1985

$-\log LC_{50} = 0.41\log K_{ow} + 0.52\,pK_a + 0.10$ $r^2 = 0.96; n = 12$ Vighi and Calamari, 1985

$-\log LC_{50} = 0.21\log K_{ow} + 0.51\,pK_a + 0.21\,^1X^v - 0.82$ $r^2 = 0.98; n = 12$ Vighi and Calamari, 1985

coefficients (Table 8) for toxicity of chlorophenols to guppies and the toxicity of a more diverse group of chemicals to *Daphnia*.

In spite of their apparently nonmechanistic basis, SARs derived from MC indices have many attractive features, including simplicity of calculation (by computer) and applicability to a wide variety of compounds, organisms, and biological activities. In addition, MC indices are deterministic; there is no uncertainty in the value of a given index for a particular compound. In contrast, considerable uncertainties may exist in values of physicochemical properties or structural attributes (K_{ow}, S_w, TSA) that are measured directly or calculated by methods described earlier in this chapter. These uncertainties violate a fundamental assumption of conventional regression analysis that the dependent variable contains all the uncertainty in each data pair; this problem does not occur with MC-based SARs.

Finally, it should be noted that SARs involving $^1X^v$, which is essentially a steric parameter, are improved by including explicit measures of electronic and lipophilic characteristics (e.g., σ, K_{ow}) in multiple regression models. For example, chlorosis induced in *Lemna minor* by substituted phenols is correlated with $^1X^v$ ($r^2 = 0.84$), and addition of a σ term or the pK_a of the phenol increases the r^2 to 0.93 and reduces the standard error by about 40% (Table 8). This reinforces the idea that no single variable captures all the factors that affect the bioactivity of organic compounds. While steric, electronic, and lipophilic parameters, among other important molecular properties, are interrelated, they are not equivalent measures of a single underlying controlling variable. Consequently, further improvements in the development of SARs and PARs probably will involve multivariable models. However, caution must be observed in developing such multivariable models to avoid problems resulting from multicollinearity (correlations among predictor variables).

5. SUMMARY AND CONCLUSIONS

Effects of substituents on reaction rates of related organic compounds generally reflect their electron-donating and/or withdrawing and steric properties. These effects can be quantified and used to produce LFERs that predict rates of reactions from the equilibrium properties of the reactants and products. However, because of scatter in relationships, LFER predictions of organic transformation rates in natural waters generally are useful only as "first approximations". Simple property–property correlations (e.g., $\log K_{ow}$ vs. $\log S_w$) and property–activity relationships (e.g., bioaccumulation vs. $\log K_{ow}$) are essentially LFERs. Although such relationships are empirical, they work because they express underlying fundamental relationships. Their accuracy is limited, however, by both measurement problems and theoretical constraints, and caution must be observed in extending such equations (for predictive purposes) beyond the compounds on which they are based. Simple attribute–reactivity relationships work best when limited to compounds that "behave" similarly (i.e., induce response by the same mechanism).

Correlations based on fundamental molecular (or atomic) properties (LSERs, predictions based on molecular mechanics) and molecular topology (MC indices) show much promise. In spite of their empiricism, MC indices are attractive because of their high correlations with many physical, chemical, and biological variables. Their deterministic nature is an advantage relative to variables such as K_{ow}, which can have large measurement uncertainties.

Although numerous PARs and SARs have been reported in the past 10 years, much work remains to be done. The validity of relationships beyond the few species used to develop them is uncertain, and few relationships are available for subacute effects. Finally, analysis of fit and *lack of fit* of compounds to a given LFER, PAR, or SAR can yield useful mechanistic information; this benefit probably exceeds the predictive values of correlational analyses.

REFERENCES

Adamson, A. W. (1979), *A Textbook of Physical Chemistry*, 2nd ed., Academic Press, New York.

Agmon, N. (1981), "From Energy Profiles to Structure–Reactivity Correlations," *Int. J. Chem. Kinet.* **13**, 333–365.

Andren, A. W., W. J. Doucette, and R. M. Dickhut (1987), "Methods for Estimating Solubilities of Hydrophobic Organic Compounds: Environmental Modeling Efforts," in R. A. Hites and S. J. Eisenreich, Eds., *Sources and Fates of Aquatic Pollutants*, (ACS Symposium Series 216), American Chemical Society, Washington, D.C., pp. 3–26.

Arbuckle, W. B. (1983), "Estimating Activity Coefficients for Use in Calculating Environmental Parameters," *Environ. Sci. Technol.* **17**, 537–542.

Betterton, E. A., Y. Erel, and M. R. Hoffmann (1988), "Aldehyde–Bisulfite Adducts: Prediction of Some of Their Thermodynamic and Kinetic Properties," *Environ. Sci. Technol.* **22**, 92–99.

Brønsted, J. N., and K. Pederson (1923), "Die Katalytische Zersetzung des Nitramids und ihre Physikalisch-Chemische Bedeutung," *Z. Phys. Chem.* **108**, 185.

Brown, H. C., and Y. Okamoto (1958), "Electrophilic Substituent Constants," *J. Am. Chem. Soc.* **80**, 4979.

Cantier, J. M., J. Bastide, and C. Coste (1986), "Structure–Degradability Relationships for Propyzamide Analogues in Soil," *Pest. Sci.*, **17**, 235–241.

Dearden J. C., and R. M. Nicholson (1987), "Correlation of Biodegradability with Atomic Charge Difference and Superdelocalizability," in K. L. E. Kaiser, Ed., *QSAR in Environmental Toxicology*—II, Reidel Amsterdam, pp. 83–89.

De Voogt, P., J. W. M. Wegener, J. C. Klamer, G. van Zijl, and H. Govers (1988), "Prediction of Environmental Fate and Effects of Heteroatomic Polycyclic Aromatics by QSARs: The Position of *n*-Octanol/water Partition Coefficients," *Biomed. Environ. Sci.* **1**, 194–209.

Doucette, W. J., and A. Andren (1987), "Correlation of Octanol/Water Partition Coefficients and Total Molecular Surface Area for Highly Hydrophobic Aromatic Compounds," *Environ. Sci. Technol.* **21**, 821–824.

Dulz, G., and N. Sutin (1963), "The Kinetics of the Oxidation of Iron(II) and Its Substituted *tris*-(1,10-phenanthroline) Complexes by Cerium(IV)," *Inorg. Chem.* **2**, 917–921.

Eadsworth, C. V. (1986), "Application of Reverse-Phase H. P. L. C. for the Determination of Partition Coefficients," *Pest. Sci.* **17**, 311–325.

Exner, O. (1972), "The Hammett Equation—The Present Position," in N. B. Chapman and J. Shorter, Eds., *Advances in Linear Free Energy Relationships*, Plenum Press, London, pp. 1–69.

Fanchiang, Y.-T. (1982), "Free-Energy Correlations for Catalysis of Outer Sphere Electron-Transfer Reactions by Noncoordinated Pyridine Derivatives," *Int. J. Chem. Kinet.* **14**, 1305.

Geyer, H. P., Sheehan, D. Kotzias, D. Freitag, and F. Korte (1982), "Prediction of Ecotoxicological Behaviour of Chemicals: Relationships between Physicochemical Properties and Bioaccumulation of Organic Chemicals in the Mussel *Mytilus edulis*," *Chemosphere* **11**, 1121–1134.

Govers, H. C., Ruepert, and H. Aiking (1984), "Quantitative Structure–Activity Relationships for Polycyclic Aromatic Hydrocarbons," *Chemosphere* **13**, 227.

Grimes, D. J., and S. M. Morrison (1975), "Bacterial Bioconcentration of Chlorinated Insecticides from Aqueous Systems," *Microb. Ecol.* **2**, 43–59.

Hammett, L. P. (1937), "The Effect of Structure upon the Reactions of Organic Compounds. Benzene Derivatives," *J. Am. Chem. Soc.* **59**, 96–103.

Hancock, C. K., E. A. Meyers, and B. J. Yager (1961), "Quantitative Separation of Hyperconjugation Effects from Steric Substituent Constants," *J. Am. Chem. Soc.* **83**, 4211–4213.

Hansch, C., and T. Fujita (1964), "ρ-σ-π Analysis. A Method for the Correlation of Biological Activity and Chemical Structure," *J. Am. Chem. Soc.* **86**, 1616–1626.

Hansch, C., and A. Leo (1979), *Substituent Constants for Correlation Analysis in Chemistry and Biology*, Wiley, New York.

Hawker, D. W., and D. W. Connell (1985), "Prediction of Bioconcentration Factors under Non-equilibrium Conditions," *Chemosphere* **14**, 1835–1843.

Hermens, J. L. M. (1986), "Quantitative Structure–Activity Relationships in Aquatic Toxicology," *Pest. Sci.* **17**, 287–296.

Hoffmann, M. R. (1981), "Thermodynamic, Kinetic, and Extrathermodynamic Considerations in the Development of Equilibrium Models for Aquatic Systems," *Environ. Sci. Technol.* **15**, 345–353.

Job, D. and H. B. Dunford (1976), "Substituent Effect on the Oxidation of Phenols and Aromatic Amines by Horseradish Peroxidase Compound I," *Eur. J. Biochem.* **66**, 607–614.

Kaiser, K. L. E. (1983), "A Non-linear Function for the Approximation of Octanol/Water Partition Coefficients of Aromatic Compounds with Multiple Chlorine Substitution," *Chemosphere* **12**, 1159.

Kamlet, M. J., R. M. Doherty, R. W. Taft, M. H. Abraham, G. D. Veith, and D. J. Abraham (1987), "Solubility Properties in Polymers and Biological Media. 8. An Analysis of the Factors that Influence the Toxicities of Organic Nonelectrolytes to the Golden Orfe Fish (*Leuciscus idus melanotus*)," *Environ. Sci. Technol.* **21**, 149–155.

Kier, L. B., and L. H. Hall (1976), *Molecular Connectivity in Chemistry and Drug Research*, Academic Press, New York.

Kier, L. B., and L. H. Hall (1986), *Molecular Connectivity in Structure–Activity Analysis*, Research Studies, London.

Koch, R. (1982), "Molecular Connectivity and Acute Toxicity of Environmental Pollutants," *Chemosphere*, **11**, 925–931.

Koch, R. (1983) *Toxicol. Environ. Chem.* **6**, 87.

Langmuir, D. (1979), "Techniques of Estimating Thermodynamic Properties for Some Aqueous Complexes of Geochemical Interest," in E. Jenne, Ed., *Chemical Models*,

(ACS Symposium Series No. 93), American Chemical Society, Washington, D.C., pp. 353–387.

Mackay, D. (1982), "Correlation of Bioconcentration Factors," *Environ. Sci. Technol.* **16**, 274–278.

Marcus, R. A. (1963), "On the Theory of Oxidation–Reduction Reactions Involving Electron Transfer. V. Comparison and Properties of Electrochemical and Chemical Rate Constants," *J. Phys. Chem.*, **67**, 853–857.

Murray, W. J., L. H. Hall, and L. B. Kier (1975), *J. Pharm. Sci.* **75**, 1978.

Nirmalakhandan, N., and R. E. Speece (1988), "Structure–Activity Relationships," *Environ. Sci. Technol.* **22**, 606–615.

Okamoto, K., I. Nitta, T. Imoto, and H. Shingu (1967), "Kinetic Studies of Bimolecular Nucleophilic Substitution. II. Structural Effects of Alkyl Halides on the Rate of S_N2 Reactions—A Reinvestigation of the Linear Free-energy Relationships for the Structural Variation of the Alkyl Groups," *Bull. Chem. Soc. Jpn.*, **40**, 1905–1908.

Pavelich, W. A. and R. W. Taft (1957), "The Evaluation of Inductive and Steric Effects on Reactivity. The Methoxide Ion-Catalyzed Rates of Methanolysis of *l*-Menthyl Esters in Methanol," *J. Am. Chem. Soc.* **79**, 4935–4940.

Prue, J. E. (1952), "The Kinetics of the Metal-Ion Catalyzed Decarboxylation of Acetonedicarboxylic Acid," *J. Chem. Soc.*, 2331.

Rekker, R. F. (1977), *The Hydrophobic Fragmental Constant*, Elsevier, New York.

Sabljic, A. (1987), "On the Prediction of Soil Sorption Coefficients of Organic Pollutants from the Molecular Structure: Application of Molecular Topology Model," *Environ. Sci. Technol.* **21**, 358–366.

Schwarzenbach, R. P., R. Stierli, B. R. Folsom, and J. Zeyer (1988), "Compound Properties Relevant for Assessing the Environmental Partitioning of Nitrophenols," *Environ. Sci. Technol.* **22**, 83–92.

Scully, F. E., and J. Hoigné (1987), "Rate Constants for Reactions of Singlet Oxygen with Phenols and Other Compounds in Water," *Chemosphere* **16**, 681–694.

Shaw, G. R., and D. W. Connell (1984), "Physicochemical Properties Controlling Polychlorinated Biphenyl (PCB) Concentrations in Aquatic Organisms," *Environ. Sci. Technol.* **18**, 18–23.

Stone, A. T. (1987), "Reductive Dissolution of Manganese (III/IV) Oxides by Substituted Phenols," *Environ. Sci. Technol.* **21**, 979–987.

Taft, R. W., Jr. (1956), "Separation of Polar, Steric, and Resonance Effects in Reactivity," in M. S. Newman, Ed., *Steric Effects in Organic Chemistry*, Wiley, New York, pp. 556–675.

Urushigawa, Y., and Y. Yonezawa (1979), "Chemicobiological Interactions in Biological Purification System VI. Relation between Biodegradation Rate Constants of Di-*n*-alkyl Phthalate Esters and Their Retention Times in Reverse Phase Partition Chromatography," *Chemosphere* **5**, 317–320.

Veith, G. D., D. L. DeFoe, and B. V. Bergstedt (1979), "Measuring and Estimating the Bioconcentration Factor of Chemicals in Fish," *J. Fish. Res. Board Can* **36**, 1040–1048.

Veith, G. D., D. J. Call, and L. T. Brooke (1983), "Structure–Toxicity Relationships for the Fathead Minnow, *Pimephales promelas*: Narcotic Industrial Chemicals," *Can. J. Fish. Aquat. Sci.*, **30**, 743–748.

Vighi, M., and D. Calamari (1985), "QSARs for Organotin Compounds on *Daphnia magna*," *Chemosphere* **14**, 1925–1932.

Wells, P. R. (1968), *Linear Free Energy Relationships*, Academic Press, New York.

Weston R. E., Jr., and H. A. Schwarz (1972), *Chemical Kinetics*, Prentice-Hall, Englewood Cliffs, NJ.

Wolfe, N. L. (1980a), "Organophosphate and Organophosphorothionate Esters: Application of Linear Free Energy Relationships to Estimate Hydrolysis Rate Constants for Use in Environmental Fate Assessment," *Chemosphere* **9**, 571–579.

Wolfe, N. L. (1980b), "Determining the Role of Hydrolysis in the Fate of Organics in Natural Waters," in R. Haque, Ed., *Dynamics, Exposure and Hazard Assessment of Toxic Chemicals*, Ann Arbor Science Publ., Ann Arbor, MI, 1980. pp. 163–178.

Wolfe, N. L., D. F. Paris, W. C. Steen, and G. L. Baughman (1980a), "Correlation of Microbial Degradation Rates with Chemical Structure," *Environ. Sci. Technol.* **14**, 1143–1144.

Wolfe, N. L., W. C. Steen, and L. A. Burns (1980b), "Phthalate Ester Hydrolysis: Linear Free Energy Relationships," *Chemosphere* **9**, 403–408.

Wolfe, N. L., R. G. Zepp, and D. F. Paris (1978), "Use of Structure-Reactivity Relationships to Estimate Hydrolytic Persistence of Carbamate Pesticides," *Water Research* **12**, 561–563.

Wong, P. T. S., Y. K. Chau, J. S. Rhamey, and M. Docker (1984), "Relationship between Water Solubility of Chlorobenzenes and their Effects on a Freshwater Green Alga," *Chemosphere* **13**, 991–996.

Woodburn, K. B., W. J. Doucette and A. W. Andren (1984), "Generator Column Determination of Octanol/Water Partition Coefficients for Selected Polychlorinated Biphenyl Congeners," *Environ. Sci. Technol.* **18**, 457–459.

5

THE KINETICS OF TRACE METAL COMPLEXATION: IMPLICATIONS FOR METAL REACTIVITY IN NATURAL WATERS

Janet G. Hering

Institute for Water Resources and Water Pollution Control (EAWAG), Dübendorf, Switzerland; Swiss Federal Institute of Technology, (ETH), Zürich, Switzerland

and

François M. M. Morel

Ralph M. Parsons Laboratory for Water Resources and Hydrodynamics, Massachusetts Institute of Technology, Cambridge, Massachusetts

1. INTRODUCTION

The biogeochemical reactivity of trace metals must be examined within the context of natural waters as dynamic systems (Hoffmann, 1981; Pankow and Morgan, 1981a, b; Morgan and Stone, 1985). The concentrations of trace metals and organic complexing agents in natural waters may be increased by inputs or decreased by removal or transformation processes (as outlined in Fig. 1). The biological effects and biogeochemical cycling of trace metals are profoundly influenced by their chemical speciation as has been demonstrated for the interactions of metals with the biota (Anderson and Morel, 1982), solid surfaces (Davis and Leckie, 1978; Dalang et al., 1984), light (Waite and Morel, 1984), and reactive solutes (Moffett and Zika, 1987a). In modeling the effect of complexation by organic ligands on processes of biogeochemical interest, the organic and inorganic complexation reactions are generally considered to be at pseudoequilibrium. That is, the complexation reactions are taken to be fast relative to other processes, and an equilibrium distribution among dissolved organic and in-

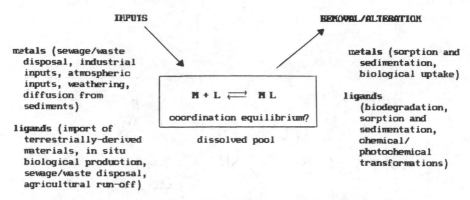

Figure 1. Outline of metal and ligand cycling in natural waters.

organic metal species is assumed even as the total dissolved metal concentration varies as a result of some controlling biogeochemical process. This approach has met with considerable success in modeling the reactivity of metals in natural and laboratory systems. The pseudoequilibrium assumption is not necessarily valid under natural water conditions, however, and, in some instances, the rates of biogeochemical reactions of metals may be influenced, or even controlled, by the rates of metal complexation reactions. Here we describe complexation reactions as slow if the characteristic time of the reaction is long compared with that of controlling biogeochemical or hydrologic processes; in practice, this refers to a time scale of hours to years.

1.1. Metal Speciation in Natural Waters

The question of the extent and nature of the interactions of metals with naturally occurring ligands is fundamental to the consideration of coordination kinetics in natural waters. The kinetics of complexation reactions can significantly influence metal reactivity only for those metals for which the equilibrium speciation is dominated by organic complexes. The determination of metal speciation in natural waters is based on the nonreactivity of dissolved metals in chemical or biological assays, which is generally attributed to complexation of metals by dissolved organic complexing agents (Duinker and Kramer, 1977, and references cited therein).

Complexing ligands in natural waters include strong, low molecular weight chelators either of anthropogenic origin, such as EDTA (see Table 1 for symbols and abbreviations), or produced by biological activity, such as siderophores. Because of the variability in the sources of such ligands, their concentration in natural waters is difficult to estimate. In polluted rivers, concentrations of EDTA from 30 to 150 nM have been reported (Brauch and Schullerer, 1987; Dietz, 1987). EDTA has also been measured in contaminated groundwaters (see Section 3.4). Production of metal complexing agents in algal cultures has been

TABLE 1. Glossary of Symbols and Abbreviations

EDTA	ethylenediaminetetraacetate
CDTA	cyclohexylenediaminetetraacetate
HEPES	N-2-hydroxyethylpiperazine-N'-2-ethanesulfonic acid
k_{-w}	rate constant for water loss from metal ion (s^{-1})
k_f	rate constant for formation of metal complex
k_d	rate constant for dissociation of metal complex
k_{in}	rate constant for internalization of metal by a cell
K_{os}	stability constant for outer-sphere complex (M^{-1})
K_{eq}	overall equilibrium constant for a given reaction
K_{ML} or K_{HL}	equilibrium constant for formation of complex ML or HL
K_D	distribution coefficient [sorbed concentration (mass g^{-1})/dissolved concentration (mass L^{-1} or mL^{-1})]
K_s	Michaelis–Menten constant
a	distance of closest approach of aquated ions (m)
N	Avogadro's number
Z_M or Z_L	charge on metal or ligand
μ	ionic strength
k_B	Boltzmann constant [$J\ K^{-1}$]
T	temperature (K)
D	dielectric constant ($J^{-1}\ C^2\ m^{-1}$)
e	proton charge (C)
r_i	ionic radius (Å)
S_T	total surface binding sites (mol g^{-1})
ρ or ρ_{max}	cellular uptake or maximal cellular uptake rate
μ'	steady-state growth rate (s^{-1})
Q	cellular quota of substrate (mol $cell^{-1}$)
τ	residence time (h)

widely observed (Zhou and Wangersky, 1989, and references cited therein) but the ligands have, at best, been only partially characterized. Based on chemical assays for bound hydroxamates, estimated siderophore concentrations in the media of iron-limited, stationary-phase cultures are approximately 1–10 μM (McKnight and Morel, 1980; Trick et al., 1983a, b). The majority of metal complexation, however, can be attributed to metal interactions with humic substances, a heterogeneous mixture of ill-defined composition (McKnight et al., 1983; Cabaniss and Shuman, 1988b). The nature of metal binding by humates, particularly the extent of competition between metals, is not known. The physical chemical nature of humate–metal interactions has kinetic as well as thermodynamic implications (Cacheris and Choppin, 1987; Lavigne et al., 1987).

The extent of organic complexation varies from close to 100% for Cu(II) in surface waters to negligible organic complexation for Mn(II) and Cd (Table 2). For Cu(II), the concentration of inorganic (or reactive) Cu in surface water is usually below the detection limit of the analytical technique. The concentration

TABLE 2. Extent of Organic Complexation of Metals in Natural Waters

Metal	Percent Complexed	Water Type	Sample Treatment	Method	Reference[a]
Cu	100	N. Pacific	Filtered (0.3 μm)	DPASV	(a)
	89–99.8	Atlantic coastal	Filtered (0.45 μm)	DPCSV/MnO$_2$	(b)
	>99.9	Atlantic coastal	Filtered (0.45 μm)	SEP-PAK	(c)
	>99.9	Atlantic coastal	Unfiltered	Solvent extraction	(d)
	>95	N.Y. Bight	Unfiltered	ASV	(e)
	>99	N.Y. Bight	Unfiltered	Amperometry and/or bioassay	(f)
Zn	60–95	Scheldt estuary	Filtered (0.45 μm)	DPCSV	(g)
	42	N. Pacific	Filtered (0.45 μm)	EDTA and/XAD column	(h)
	14	river	Filtered (0.45 μm)	MnO$_2$	(i)
	0	Gulf Stream	Unfiltered	Chelex	(j)
Cd	0	North Sea	Filtered (0.45 μm)	ASV	(k)
	0	Gulf Stream	Unfiltered	Chelex	(j)
Mn	0	Gulf Stream	Unfiltered	Chelex	(j)
Al	39	lake	Centrifuged	Ion-exchange chromato-graphy and colorimetry	(l)
	42	stream	Centrifuged		(l)

[a] (a) Coale and Bruland (1988), (b) Buckley and van den Berg (1986), (c) Sunda and Hanson (1987), (d) Moffett and Zika (1987b), (e) Huizenga and Kester (1983), (f) Hering et al. (1987), (g) van den Berg et al. (1987), (h) Hirose et al. (1982), (i) van den Berg and Dharmvanij (1984), (j) Sunda (1984), (k) Duinker and Kramer (1977), (l) Driscoll et al. (1984).

of organic ligands is determined by titration of the water sample with Cu(II); the free cupric ion concentration is estimated from model calculations of the complexation of Cu(II) at ambient concentration. Although different models have been used to describe metal–organic interactions [reviewed by Perdue and Lytle (1983), Cabaniss et al. (1984), Dzombak et al. (1986), Fish et al. (1986), Turner et al. (1986)], several assumptions common to all are involved in the estimation of free cupric ion concentration:

1. The estimation relies on extrapolation outside the range of the analytical technique. The inherent difficulties of such extrapolation have been discussed by Cabaniss et al. (1984).
2. The total filterable metal is considered as dissolved (in a true chemical sense) and the possibility that some of the observed nonreactivity of filterable metal might be due to the presence of inorganic colloidal species is usually disregarded. The importance of filterable (or nonsettleable) colloidal species has been discussed elsewhere (Gschwend and Wu, 1985; Leppard et al., 1986; Morel and Gschwend, 1987).
3. Equilibrium between (dissolved) metal and ligand species on both the geochemical time scale of natural waters and the analytical time scale of titration experiments is assumed.

In this chapter, we examine the validity of the assumption of coordination equilibrium between metals and organic ligands for both natural and analytical conditions. In so doing, we consider several factors: the initial distribution of metal and ligand species, the nature of the perturbation, the rates and mechanisms of the reactions involved in the reestablishment of coordination equilibrium, and the rates of competing biogeochemical processes. We do not consider here the issues of catalysis or surface, biological, and photochemical reactions that may be important in natural waters. For discussion of these issues, the reader is referred to other chapters in this volume.

2.   METAL COMPLEXATION REACTIONS: MECHANISMS AND RATES

The kinetics of complexation reactions (such as complex formation and dissociation, metal-, ligand-, and double-exchange reactions) of well-defined ligands has been comprehensively reviewed by Margerum et al. (1978). On the basis of such studies, a conceptual framework for estimating the rates of complexation reactions in natural waters can be developed. Within such a framework, the principles governing the coordination kinetics of well-defined ligands may be applied, with due caution, to reactions of metals with heterogeneous and ill-defined naturally occurring ligands under environmentally relevant conditions (Hoffmann, 1981; Morgan and Stone, 1985). The following discussion, which is based largely on the previous work of Margerum et al. (1978), examines the parameters controlling the rates and mechanisms of complexation reactions, such as the nature and relative and absolute concentrations of reacting species and the presence of competing species.

2.1. Complex Formation

In the Eigen mechanism for metal complex formation, formation of an outer-sphere complex between a metal and a ligand is followed by rate-limiting loss of water from the inner coordination sphere of the metal (Eigen and Wilkins, 1965, and references cited therein; Burgess, 1978). Thus

$$[M(H_2O)_6]^{2+} + L^{n-} \underset{K_{os}}{\rightleftarrows} [M(H_2O)_6, L]^{(2-n)+}$$

$$[M(H_2O)_6, L]^{(2-n)+} \xrightarrow{k_{-w}} [M(H_2O)_5 L]^{(2-n)+} + H_2O$$

The stability constant for the outer-sphere complex, K_{os}, is primarily dependent on the charges on the reacting species and the ionic strength of the medium. The rate constant for water loss, k_{-w}, is characteristic of the reacting metal. Water loss from the ligand, not included here explicitly, is considered to be fast relative to metal water loss and to precede formation of the outer-sphere complex (Margerum et al., 1978, and references cited therein). Then the rate of complex formation (omitting coordinated waters) can be written as

$$\frac{d[ML^{(2-n)+}]}{dt} = k_f [M^{2+}][L^{n-}] \tag{1}$$

where the formation rate constant $k_f = K_{os} k_{-w}$. Values for K_{os} can be calculated from the expression [from Wilkins (1970)]

$$K_{os} = \frac{4\pi N a^3}{0.003} \exp\left[\frac{-Z_M Z_L e^2}{a D k_B T}\right] \exp\left[\frac{Z_M Z_L e^2 \kappa}{D k_B T(1 + \kappa a)}\right] \tag{2}$$

where

$$\kappa^2 = \frac{8\pi N e^2 \mu}{1000 D k_B T} \tag{3}$$

(see Table 1 for definition of symbols). This mechanism can also be applied to the reaction of multidentate ligands if the rate-limiting step for complexation is formation of the first metal–ligand bond. The effect of ionic strength on the calculated rate constant for complex formation (due to the dependence of K_{os} on ionic strength) can be seen in Figure 2, which shows the comparison between calculated and observed rate constants for formation of Ni(II) complexes as a function of ligand charge. Some of the variability in the rate constants for

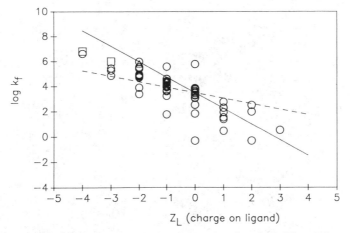

Figure 2. Rate constants for formation of Ni complexes as a function of ligand charge. Observed formation rate constants for ionic strength $\mu \rightarrow 0$ ($\triangle$), 0.01 ($\square$), and 0.1–0.3 ($\bigcirc$) from Margerum et al. (1978) (ionic strength data obtained from cited references). Calculated rate constants for ionic strength 0 (——) and 0.1 (– – –) determined as described in the text for $k_{-w} = 10^4\,s^{-1}$, $a = 5 \times 10^{-10}$ m, $Z_M = 2$, $D = 6.95 \times 10^{-10}\,J^{-1}\,C^2\,m^{-1}$ at 298 K.

reactions of ligands of the same charge can be rationalized on the basis of ligand structure. Decelerating factors include ligand protonation, particularly at the site of formation of the first metal–ligand bond, and chelation-controlled substitution, in which the rate-limiting step is shifted to subsequent bond formation or ring-closure steps. The markedly slower rate of reaction of zwitterionic amino acids as compared with other neutral ligands is an example of chelation-controlled substitution. Formation reactions may be accelerated by an internal conjugate base mechanism in which the K_{os} term is effectively increased by hydrogen bonding between a basic donor atom of the ligand and a coordinated water of the reacting metal.

Despite the obvious difficulties in quantitatively predicting the rates of complex formation based on the Eigen mechanism, it remains a useful interpretation of observed rates of complex formation. In particular, the relative kinetic reactivity of different metals can be estimated on the basis of the rate constants for water loss. As shown in Figure 3, for many metals these rate constants can be related to the ratio of the charge to the ionic radius of the metal cation, which provides a measure of the cation–water dipole electrostatic interaction. Metal hydrolysis can also significantly accelerate the rate of complex formation because of the increased rate of water loss when one or more coordinated waters are replaced by OH^-. This effect is especially significant for slow-reacting metals such as Cr(III), Al(III), and Fe(III) (Crumbliss and Garrison, 1988).

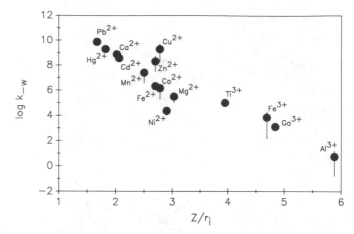

Figure 3. Rate of water loss from metal cations as a function of the ratio of the charge to the radius of the metal ion. Rate constants from Margerum et al. (1978), Z/r_i ratios calculated from ionic radii tabulated in the *CRC Handbook of Chemistry and Physics*. [Adapted with permission from Crumbliss and Garrison *Comments Inorg. Chem.* 8: 1–26 (1988), Gordon and Breach Science Publishers, Inc., New York.]

2.2. Complex Dissociation

Rate constants for complex dissociation can be related to formation rate constants through the principle of microreversibility. Thus for

$$M + L \underset{k_d}{\overset{k_f}{\rightleftharpoons}} ML$$

the dissociation rate constant $k_d = k_f / K_{eq}$. This method provides a reasonable estimate for dissociation rate constants even for multidentate ligands such as CDTA as shown in Table 3. Consistency between kinetic rate constants and the equilibrium stability constant for the zinc-glycine complex has also been noted by Correia dos Santos and Goncalves (1986).

2.3. Metal-Exchange Reactions

The overall reaction for metal exchange

$$M' + ML \longrightarrow M'L + M$$

involves three reacting species and may proceed by two general pathways, disjunctive and adjunctive. Omitting protonated species and charges for convenience, these mechanisms may be written as follows:

Disjunctive Mechanism	**Adjunctive Mechanism**
$ML \rightleftharpoons M + L$	$M' + ML \rightleftharpoons M'LM$
$M' + L \longrightarrow M'L$	$M'LM \longrightarrow M'L + M$

TABLE 3. Comparison between Observed Rate Constants for Complex Dissociation[a] and Rate Constants Predicted from Measured Formation Rate Constants[b] and Equilibrium Constants[c] for

$$M^{2+} + HL^{3-} \underset{k_d}{\overset{k_f}{\rightleftharpoons}} ML^{2-} + H^+ \qquad (L = CDTA^{4-})^d$$

Metal	k_f $(M^{-1} s^{-1})$	$\log(K_{ML}/K_{HL})$ $(\mu = 0.1)$	k_d (pred.) $(M^{-1} s^{-1})$	k_d(obs.) $(M^{-1} s^{-1})$
Ni	1.9×10^5 $(\mu = 0.3)$	7.9	2.4×10^{-3}	3.4×10^{-4} $(\mu = 0.1)$
Pb	8×10^8 $(\mu = 0.2)$	7.9	10	23 $(\mu = 0.1)$

[a] Dissociation rate constants from Margerum et al. (1978).
[b] Formation rate constants for Pb from Margerum et al. (1978); for Ni, from Cassatt and Wilkins (1968).
[c] Equilibrium constants from Martell and Smith (1974).
[d] CDTA = cyclohexylenediaminetetraacetate.

Disjunctive metal exchange involves dissociation of the initial complex and reaction of the incoming metal with the free (or protonated) ligand intermediate as compared with the direct attack of the incoming metal on the initial complex in the adjunctive pathway. We have introduced this terminology to denote these as stoichiometric mechanisms which may be distinguished by the dependence of

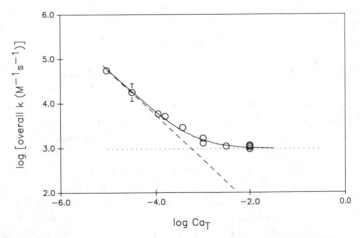

Figure 4. Overall rate constants for metal-exchange reaction of Cu(II) with CaEDTA at pH = 8.2 as a function of (excess) Ca for rate = $-d(Cu)/dt = k[\text{CaEDTA}][\text{Cu}]$, where $k = [k_{disj}/[\text{Ca}]) + k_{adj}$. Data (O) is shown with empirically derived contributions of the adjunctive pathway ($\cdots$) and disjunctive pathway ($---$) to the overall rate constant (———). [Adapted with permission from Hering and Morel, *Environ. Sci. Tech.* **22**: 1469–78. Copyright 1988 American Chemical Society.]

the observed rate constant on the concentration of the outgoing metal (Hering and Morel, 1988b). This dependence is observed in the exchange of Cu(II) for Ca for L = EDTA in the presence of excess Ca (Fig. 4). The overall rate constant (i.e., for rate $= -d[ML]/dt = k[M'][ML]$) reflects the decreasing contribution of the disjunctive pathway with increasing Ca (at constant pH) until, at $Ca_T \approx 0.3$ mM (pH = 8.2), the adjunctive pathway becomes predominant (Hering and Morel, 1988b). Thus the prevailing mechanism for the overall reaction depends on the concentration of the outgoing metal as well as the rate constants for the alternative pathways.

These rate constants are dependent on ligand structure and, for the disjunctive pathway, on pH. The rate constant for the disjunctive pathway is determined by the rate constant for reaction of the incoming metal with the steady-state concentration of the free or protonated ligand intermediate. It is thus inversely proportional to the conditional equilibrium stability constant for the initial metal complex and directly proportional to the formation rate constant for the final metal complex. The adjunctive rate constant is more dependent on ligand structure since either formation or dissociation of the intermediate dinuclear complex can be rate-limiting. Specifically:

$$ML \rightleftharpoons M\text{-}L$$

$$M\text{-}L + M' \rightleftharpoons MLM' \rightarrow M + M'L$$

For the reaction of transition metals with alkaline earth metal complexes, the formation of the dinuclear intermediate is more likely to be rate-limiting. The adjunctive rate constant can be rationalized on the basis of the stability of the metal complex with a ligand fragment (corresponding to a partially dissociated complex), the stability of the initial complex, and the rate of reaction of the incoming metal with the partially dissociated complex (Margerum, 1963; Margerum et al., 1978).

In comparing alkaline earth-for-transition metal exchange with transition metal-for-transition metal exchange, two important differences are apparent. Because of the larger stability constants of transition metal complexes, as compared with alkaline earth metal complexes, the disjunctive mechanism is less favorable and the adjunctive exchange pathway predominates. Also, as shown in Table 4, the rate constants for adjunctive exchange are slower for transition metal complexes than for alkaline earth complexes.

This discussion is predicated on competitive binding of the exchanging metals. For well-defined, discrete ligands, competition between metals and its effect on reaction kinetics can be precisely quantified. For humic substances, however, the situation is less clear. In extrapolating these principles from model to natural ligands, the nature of the metal–ligand interactions must be considered. Studies of Cu(II) complexation by humic substances indicate that alkaline earth metals and Cu(II) do not compete for the same binding sites of natural ligands (Sunda and Hanson, 1979; Cabaniss and Shuman, 1988a; Hering and Morel, 1988a) possibly because of the affinities of the metals for different heteroatoms, for

TABLE 4. Rate Constants for Adjunctive Metal-Exchange Reaction of Cu with Metal–EDTA complexes[a]

Metal	$k\ (M^{-1}\,s^{-1})$
Ca	970
Zn	19
Co	15.3
Pb	5.1
Cd	2.2
Ni	0.016

[a] Data from Margerum et al. (1978); for Ca, from Hering and Morel (1988b).

instance, O, N, and S (Niebor and Richardson, 1980). Competitive binding of Cd and Cu has been observed for fulvic acid (Fish, 1984), but Giannissis et al. (1985) have reported no competition between Pb and Cu for Ca-flocculated humic acids.

2.4. Ligand-Exchange Reactions

The pathways for ligand-exchange reactions with the overall stoichiometry and rate

$$ML + L' \longrightarrow L + ML'$$

$$-d[ML]/dt = k[ML][L']$$

are analogous to those described for metal-exchange reactions. The disjunctive pathway involves reaction of the incoming ligand with the (intermediate) free metal, produced by dissociation of the initial complex. The adjunctive pathway involves direct attack of the incoming ligand on the initial complex and formation of a ternary intermediate, LML'. The overall rate constant for an excess of the outgoing ligand, L, reflects the contributions of the adjunctive pathway (which is independent of [L]) and of the disjunctive pathway (which decreases with increasing [L]), as shown in Figure 5. This interpretation is also consistent with the observed dependence of the rate of ligand-exchange reactions of humate-bound Cu on the concentration of free humate (Hering and Morel, 1990). As for metal-exchange reactions, the disjunctive rate constant for ligand exchange can be related to the stability constant of the initial complex.

For heterogeneous materials, intramolecular redistribution of bound metal among binding sites (effectively, intramolecular ligand exchange) may be significant. Such a mechanism has been proposed to explain the observed effect of preequilibration time on the rate of dissociation of thorium from humate complexes (Cacheris and Choppin, 1987).

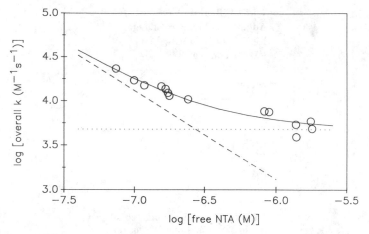

Figure 5. Overall rate constants for ligand-exchange reaction of Calcein with CuNTA at pH = 7.4 as a function of (excess) free NTA concentration for rate = k[CuNTA][Calcein] where k = (k_{disj}/[NTA]) + k_{adj}. Data (○) are shown with empirically derived contributions of the adjunctive pathway ($\cdots$) and disjunctive pathway ($---$) to the overall rate constant (——). [Data from Hering and Morel (1990).]

For both metal- and ligand-exchange reactions, the dependence of the overall rate of reaction on the *relative* concentrations of the reacting species must be considered in extrapolating from laboratory to natural water conditions. Thus for the exchange reaction of the initial complex, ML, with either an incoming metal or ligand, X, the rate of the overall reaction

$$XL + M \text{ (metal exchange)}$$

$$ML + X \longrightarrow \qquad \text{or}$$

$$MX + L \text{ (ligand exchange)}$$

can be described by the rate expression

$$\frac{-d[ML]}{dt} = k_{overall}[ML][X]$$

Under conditions where the outgoing species, specifically, M for metal exchange or L for ligand exchange, is in excess

$$k_{overall} = \frac{k_{disj}}{[\text{outgoing ligand or metal}]} + k_{adj}$$

The disjunctive pathway is inhibited at increasing concentrations of the outgoing species. Thus the predominant pathway for metal- or ligand-exchange reactions depends on the relative reactant concentrations and particularly on the excess concentration of the outgoing metal or ligand.

2.5. Double-Exchange Reactions

In double-exchange reactions, four reacting species are involved in simultaneous metal- and ligand-exchange reactions, where the overall reaction

$$ML + M'L' \longrightarrow ML' + M'L$$

may proceed through two types of mechanisms, "ligand-initiated" and "metal-initiated":

Ligand-initiated:

$$ML \rightleftharpoons M + L$$

$$L + M'L' \longrightarrow M'L + L'$$

Metal-initiated

$$ML \rightleftharpoons M + L$$

$$M + M'L' \longrightarrow ML' + M'$$

Both of these pathways are disjunctive with respect to the complex ML and may be either adjunctive or disjunctive with respect to the reaction of the intermediate free M or L with the complex M'L'. The complete dissociation of both initial complexes is an unlikely pathway unless both are weak as may be the case for alkaline earth metal complexes. Double-exchange reactions of transition metal complexes often involve coordination chain mechanisms (for example, further reaction of L', produced by a ligand-initiated pathway, with the initial complex ML) and the overall rates may be strongly dependent on trace concentrations of reactants that promote or terminate coordination chain reactions (Margerum et al., 1978).

An example of double-exchange reactions is the reaction of added Cu(II) with a Ca-equilibrated mixture of a stronger ligand, S, and a weaker ligand, W. Thus

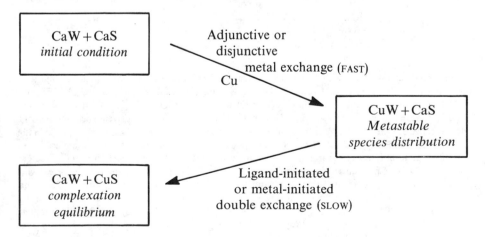

As discussed previously, the rate of reaction with the added Cu is faster for the Ca complex of the weaker ligand, CaW, than for the Ca complex of the stronger ligand, CaS. In the resulting distribution of metal and ligand species, Cu speciation is dominated by the weaker ligand. Subsequent equilibration to give the thermodynamically favored complex with the stronger ligand proceeds slowly through a double-exchange reaction. As shown in Figure 6, the reaction of added Cu with a mixture of EDTA and humic acid was followed by observing the quenching of the natural fluorescence of the humic acid due to formation of Cu–humate complexes (Hering and Morel, 1989). In the absence of Ca, no quenching of the humate fluorescence is observed indicating immediate formation of the CuEDTA complex. In the presence of 0.01 M Ca, however, rapid quenching of the humate fluorescence is observed followed by a slow recovery of fluorescence as the Cu is exchanged between humate binding sites and EDTA. This reaction can be modeled assuming a metal-initiated double exchange in which inorganic Cu (in pseudoequilibrium with humate binding sites) reacts with CaEDTA (solid line in Fig. 6). At lower Cu: humate ratios, as prevalent in natural waters (ca. 10 μmol Cu g^{-1} humic), the formation of CuEDTA is predicted to proceed over a period of about a year at 0.01 M Ca (Hering and Morel, 1989).

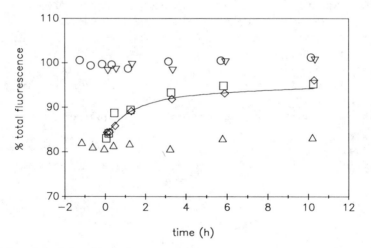

Figure 6. Reaction of Cu with a mixture of ligands, EDTA and Suwannee Stream humic acid (4.94 mg L^{-1} humic acid, 2 μM total EDTA, 2 μM total Cu in 0.1 M NaCl, 5 mM HEPES, pH = 7.4). Percent fluorescence of the humic acid is shown over time after addition of Cu at $t = 0$ to the ligand mixture in the absence of Ca (∇) or with 0.01 M Ca ($\square$, $\diamond$). For reference, the percent fluorescence over time of the ligand mixture in the absence of Cu ($\bigcirc$) and of humic acid and Cu in the absence of EDTA ($\triangle$ with 0.01 M Ca) are shown. (———) calculated percent free humate assuming pseudoequilibrium between inorganic Cu and humate binding sites and reaction of inorganic Cu with CaEDTA. Cu—humate interactions are modeled with three humate binding sites with conditional stability constants $10^{10.2}$, $10^{8.4}$, and $10^{5.8}$ and total ligand concentrations of 0.25, 1, and 9 μM. Reaction of inorganic Cu with CaEDTA is described with a single pH- and Ca-dependent rate constant (7.2×10^4 M^{-1} min^{-1}). [Adapted with permission from Hering and Morel, *Geochim. Cosmochim. Acta* 53: 611–8 (1989), Pergamon Press plc].

Thus, in systems containing at least two competing metals and two competing ligands, slow double-exchange reactions are possible. Depending on the initial speciation of both metals and ligands, perturbation of the system may result in a metastable distribution of metal and ligand species that is far from the equilibrium species distribution.

3. COORDINATION KINETICS IN NATURAL WATERS

3.1. Analytical Measurements

Analytical determinations of metal speciation in natural waters involve additions of metals or ligands to water samples. Rapid coordination equilibrium is then assumed, and, certainly, measurements of metal speciation by metal titrations demonstrate that some complexation reactions are fast on an analytical time scale. In some cases the effect of equilibration time (Kramer, 1986; Coale and Bruland, 1988) or the possible error due to incomplete equilibration (Sunda and Hanson, 1987) have been examined.

In practice, metal titration experiments provide information only on the concentration of the inorganically complexed and free metal species. Based on our study of double-exchange kinetics, we can imagine the concentration of the various organic complexes during the course of a metal titration of a mixture of ligands (both natural and anthropogenic) in seawater. This exercise has been performed from a *thermodynamic* perspective by Dzombak et al. (1986). If coordination equilibrium is attained for each addition of metal to the sample, the metal complexes will be formed in the order of decreasing stability (or, more rigorously, of the product of the stability constant and the ligand concentration); that is, the strongest complexes are formed first. Because of the slower reaction of alkaline earth complexes of strong ligands, however, we would predict, from a *kinetic* perspective, that the weaker rather than the stronger ligands would react initially with the added metal. Reequilibration of the metal species on the analytical time scale is unlikely. (*Note*: Depending on the sensitivity of the analytical technique for the inorganic metal, such redistribution of metal species, even if it occurs, might not result in any detectable change in the concentration of measurable forms of the metal.) As the titration progresses, the weaker ligands can no longer complex all the added metal and significant concentrations of inorganic (or free) metal are available to react with alkaline earth complexes of the strong ligands. This reaction should give a detectable change over several hours in the concentration of measurable metal. In practice, however, the detection of such a change may be complicated by the difficulty in relating the observed instrumental response to the concentration of free and inorganic metal species in the presence of incompletely titrated weak ligands.

Thus for mixtures of competing ligands in seawater the concentration of strong ligands (such as EDTA) may be underestimated in metal titrations. The problem will be exacerbated if the strong ligands are initially complexed with transition metals rather than alkaline earth metals. Thus the multimetal complexation model for metal titrations, in which rapid equilibration of Cu with competing metals is assumed (Hirose, 1988), is, on kinetic grounds, an unlikely model for observed titration curves. The question of incomplete equilibration of competing metals also arises in techniques that involve additions of radiotracers and assume complete equilibration of the trace concentration of the radioisotope with natural ligands in the presence of excess concentrations of stable isotopes of the metal. A much lower estimate of Zn complexation has been obtained by such methods than by electrochemical techniques (compare values from references g, i, and j in Table 2).

It should be noted, however, that siderophores, biogenic strong complexing agents, undergo fairly rapid exchanges both of Mg for Cu or Fe(III) (Hering, unpub.; Hudson, 1989) and of Cu for Fe(III) (McKnight and Morel, 1980). Formation of iron complexes under environmental conditions is prerequisite for siderophore-mediated iron transport by microorganisms. Thus we might expect the overall rates for metal-exchange reactions of biogenic strong complexing agents, in general, to differ from that of anthropogenic ligands such as EDTA.

There are two major consequences of slow kinetics for the analytical determinations of metal speciation. The first is that the interactions of metals with strong complexing agents may not be accessible through metal titrations. The problems of incomplete equilibration cannot be resolved simply by extending the equilibration period because of the possibility of sample alteration (e.g., real changes in ligand concentration due to biological activity in unfiltered samples). Conversely, the kinetic inertness of strong metal complexes could be exploited in alternative investigative techniques such as the chromatographic separation of metal complexes. In any case, coordination kinetics must at least be considered in assessing analytical determinations of metal speciation in natural waters.

3.2.   Biogeochemical Processes

Slow coordination kinetics may also contribute to the overall rates of biogeochemical processes such as biological metal uptake, sorption, or chemical or photochemical transformations. Coordination kinetics should be considered if the reacting species (either metal or ligand) exists predominantly in complexed form and the complex itself does not undergo the biogeochemical process of interest. That is, the complex itself is not sorbed or transported into a cell or otherwise transformed. A major class of exceptions are electron transfer reactions; for example, in reduction of Cu(II) at cell surfaces the rate of Cu(II) reduction is a function of the total (complexed) Cu(II) concentration rather than the free cupric ion concentration (Jones et al., 1987; Price and Morel, Chapter 8, this volume). In contrast, biological uptake or sorption reactions are proportional to the concentration of free or inorganic species. In the study of Cd sorption on calcite in the presence of EDTA (Davis et al., 1987), direct sorption of the CdEDTA complex is not observed. In this case, contribution of the coordination kinetics of CdEDTA dissociation to the overall sorption rate is suggested by the observation that the decrease in the sorption rate for 1 μM Cd_T (from $\sim 100\%$ Cd sorption in 5 min in the absence of EDTA to > 100 h at 1.5 μM EDTA) is greater than the corresponding decrease in the $[Cd^{2+}]$ as a result of the presence of EDTA complexes (10–100-fold decrease depending on pH). Similar factors may be important in controlling the overall rates of biogeochemical processes in natural waters.

Clearly, coordination kinetics do not necessarily influence the rates of biogeochemical processes. The importance of coordination kinetics will depend on the relative rates of the complexation reactions as compared to the rates of competing biogeochemical processes. The following examples are chosen, however, to illustrate the conditions for which coordination kinetics may be important and the possible implications of coordination kinetics for metal reactivity in natural waters.

3.3. Example I: Oxidation of Fe(II) in the Presence of Complexing Ligands

The influence of complexing ligands on the rate of oxidation of Fe(II) due to the decrease in the pseudoequilibrium concentration of Fe^{2+} [where the Fe(II)L complex is not directly oxidized] has been discussed by Pankow and Morgan (1981a). In this example, we analyze the influence of coordination kinetics on the overall rate of Fe(III) production when L_T is in excess of Fe(II). The effect of changing the rate constants for formation and dissociation of the Fe(II)L complex is examined over the pH range of 7–8, that is, effectively at different rates for oxidation of Fe(II) since the oxidation rate is pH-dependent. Conditions are given in Table 5. As shown in Figure 7, the effect of changing the formation (and thus dissociation) rate constant by a factor of 10 is negligible at pH 7. At this pH,

TABLE 5. Fe(II) Oxidation in the Presence of Complexing Ligands

Conditions: $Fe_T = 1$ μM, $L_T = 10$ μM, $p_{O_2} = 0.2$ atm, pH 7.0–8.0

Reactions: (1) Fe(II) oxidation

$$Fe^{2+} \xrightarrow[O_2]{k_1} Fe(III)$$

(2) Fe(II) complexation

$$Fe^{2+} + L \underset{k_3}{\overset{k_2}{\rightleftharpoons}} Fe(II)L$$

Rate and equilibrium constants:
  (1) $k_1 = k'[OH^-]^2$ where $k' = 2.6 \times 10^{11}$ M^{-2} s^{-1}
     (for $p_{O_2} = 0.2$ atm)
  (2) $\log K_{Fe(II)L} = 6.0$; $K_{Fe(II)L} = k_2/k_3$

Assumptions:
  (1) No back reaction of Fe(III)
  (2) No oxidation of Fe(II)L

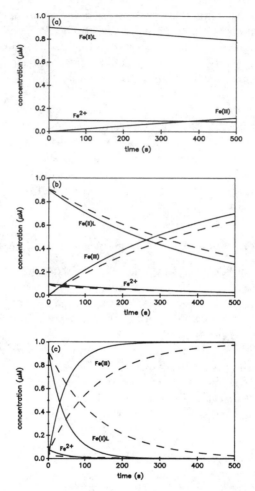

Figure 7. Model of Fe(II) oxidation in the presence of complexing ligands for conditions given in Table 5 for (a) pH = 7.0, (b) pH = 7.5, and (c) pH = 8.0. The influence of coordination kinetics on the overall rate of Fe(II) oxidation is examined by changing the rates of formation (and dissociation) of the Fe(II)L complex for $k_2/k_3 = K_{Fe(II)L} = 10^6$. (———) $k_2 = 10^5 \ M^{-1} \ s^{-1}$, (– – –) $k_2 = 10^4 \ M^{-1} \ s^{-1}$.

the oxidation kinetics are sufficiently slow that the assumption of coordination equilibrium is valid. However, at higher pH (i.e., faster oxidation rates), the influence of coordination kinetics becomes increasingly pronounced. At pH 8, decrease in the rate constants by a factor of 10 results in doubling the half-life of the Fe(II) complex with respect to oxidation. Under these conditions, a model assuming pseudoequilibrium between free and complexed Fe(II) cannot adequately describe the overall oxidation rate. Although the absolute change in the overall reaction half-life due to coordination kinetics is not significant on an environmental time scale, this example illustrates the difficulties that might arise in interpreting the rate of some overall reaction or process without considering explicitly the kinetics of the coordination reactions involved.

3.4. Example II: Equilibration of ^{60}CoEDTA with Soil

The importance of coordination kinetics in soil column experiments even in the *absence* of organic complexing agents has been discussed by Kirkner et al. (1985). A model including only transport and equilibrium partitioning of metals between solution and solid phases was insufficient to describe the desorption of Ni and Cd in column wash-out experiments. The agreement of the model prediction with observed Cd desorption was significantly improved by including desorption kinetics in the model. In the presence of organic complexing agents the assumption of rapid equilibration of sorbed and dissolved metals may be even more questionable.

Field observations of radionuclide migration in groundwater have been attributed to transport of organic complexes for plutonium (Cleveland and Rees, 1981) and ^{60}Co (Means et al., 1978; Killey et al., 1984). Variable "apparent distribution coefficients" were determined on the basis of the observed partitioning of ^{60}Co between solution and solid phases. From electrophoretic measurements, ^{60}Co was found to be transported in groundwater as an anionic species that did not undergo isotopic exchange with ^{57}Co in the absence of light (Killey et al., 1984). Migration of ^{60}Co was attributed to transport of the ^{60}CoEDTA complex; total EDTA concentrations of up to 0.34 μM were measured in groundwater samples.

The time scale for equilibration of ^{60}CoEDTA with soil for the conditions reported by Means et al. (1978) was estimated using the equilibrium and rate constants given in Table 6. The model considers only the acid-catalyzed dissociation of ^{60}CoEDTA and reaction of free ^{60}Co with soil surface binding sites ($\equiv$SO). Direct reactions of CoEDTA with surface binding sites or with dissolved metals in the groundwater are neglected. If such direct reactions are significant the predicted half-life of ^{60}CoEDTA would be decreased because the rate of equilibration of ^{60}CoEDTA with soil in this model is controlled by the pH-dependent rate of CoEDTA dissociation (constants and conditions given in Table 6). This crucial parameter was estimated based on the measured rate constant for reaction of Co^{2+} with HEDTA and the equilibrium constant. With these assumptions, a half-life for ^{60}CoEDTA of $\approx$ 120 days is estimated. It is interesting to note the correspondance of this estimated half-life with the residence time of groundwater ($\approx$ 200–700 days) between one of the ^{60}Co disposal sites and discharge to the surface water (Killey et al., 1984). In such a case, slow coordination kinetics might contribute to observations of variable "apparent distribution coefficients," although other factors such as heterogeneity of aquifer materials cannot be neglected. In comparison to groundwater residence times, the kinetics of complexation reactions on time scales of approximately one year may often be negligible. However, coordination kinetics

TABLE 6. Equilibration of ^{60}CoEDTA with Soil

Conditions: $EDTA_T = 0.34 \ \mu M$, $^{60}Co_T = 10^{-11} \ M$, $[\equiv SO] = 0.35 \ M$,[a]
$[\equiv SOM] = 10^{-3} \ M$, pH = 6.7

Reactions: (1) Dissociation of ^{60}CoEDTA[b]

$$CoEDTA^{2-} + H^+ \underset{k_1}{\overset{k_{-1}}{\rightleftharpoons}} Co^{2+} + HEDTA^{3-}$$

(2) Equilibrium between protonated EDTA species

$$HEDTA^{3-} + H^+ \rightleftharpoons H_2EDTA^{2-}$$

(3) Sink for EDTA

$$HEDTA + \equiv SOM \xrightarrow{k_3} MEDTA + \equiv SOH$$

(4) Sorption of ^{60}Co onto soil

$$Co + \equiv SO \underset{k_{-4}}{\overset{k_4}{\rightleftharpoons}} \equiv SOCo$$

Rate and equilibrium constants:

(1) $\log(K_{CoEDTA}/K_{HEDTA}) = 7.08$[c], $k_1 = 4 \times 10^6 \ M^{-1} \ s^{-1}$,[d]
$k_{-1} = 0.33 \ M^{-1} \ s^{-1}$,[e]

(2) $\log(K_{H_2EDTA}) = \dfrac{[H_2EDTA]}{[HEDTA][H^+]} = 6.32$[c]

(3) $k_3 = 10^5 \ M^{-1} \ s^{-1}$,[f]

(4) $\log K_{=SOCo} = 5.54$,[g] $k_4 = 10^5 \ M^{-1} \ s^{-1}$,[f] $k_{-4} = 0.28 \ s^{-1}$,[e]

---

[a] Calculated from CEC (cation exchange capacity) 20 meq/100 g and bulk density of soil 1.75 kg L^{-1}.

[b] Direct reactions of CoEDTA with $\equiv$ SO or with dissolved metals in groundwater were neglected based on observations of negligible sorption of CdEDTA on calcite (Davis et al., 1987) and of Fe(II)EDTA on magnetite (Blesa et al., 1984) and on measured concentrations of dissolved trace metals less than the total EDTA concentration used in the model (Jacobs et al., 1988).

[c] Martell and Smith (1974).

[d] Margerum et al. (1978).

[e] Calculated from the formation rate constant and equilibrium constant (i.e., k_f/K_{eq}).

[f] Diffusion-limited rate constant for reaction of dissolved species with surface proposed by Astumian and Schnelly (1984).

[g] Calculated from K_D reported by Means et al. (1978) such that

$$K_{=SOCo} \ (\text{in } M^{-1}) = K_D \left(\text{in} \frac{\text{dpm/g}}{\text{dmp/mL}} \right) \div (1000) S_T \ (\text{in mol g}^{-1})$$

where S_T is taken to be equal to the CEC of 20 meq 100 g^{-1}.

may be quite significant in extrapolating the results of column experiments to field conditions.

3.5. Example III: Constraints on Biological Metal Uptake Due to Coordination Kinetics

The availability of a nutrient metal such as iron to phytoplankton is governed by the ambient concentration of dissolved metal in the medium and by the rate of reaction of the metal with metal-binding sites on the cell surface. As described by Hudson (1989), this reaction may be written as follows:

$$Fe(OH)_2^+ + L_{memb} \underset{k_d}{\overset{k_f}{\rightleftharpoons}} FeL_{memb} \xrightarrow{k_{in}} Fe_{intracell},$$

and the minimum number of "carrier" sites (L_{memb}) can be estimated by comparing the rate of complex formation with the Fe uptake rate required to maintain steady-state intracellular Fe concentration (as shown in Table 7). The coordination kinetics (i.e., $k_f [Fe(OH)_2^+]$) effectively determine the minimum number of "carrier" sites needed to supply the cells' iron requirement. This minimum value for L_{memb} has two important consequences for the cell (see Table 7). The internalization of the Fe-"carrier" complex is fairly slow; the calculated residence time for Fe in the membrane is ≈ 2.5 h. And the required density of "carriers" is high compared with membrane proteins; the calculated average distance between "carrier" centers is ≈ 4 nm for a 7-μm-diameter cell. Thus coordination kinetics and diffusion are the chemical and physical bounds that constrain the mechanism of biological metal uptake by phytoplankton.

TABLE 7. Constraints on Biological Metal Uptake Due to Coordination Kinetics (from Hudson, 1989)

$$Fe(OH)_2^+ + L_{memb} \underset{k_d}{\overset{k_f}{\rightleftharpoons}} FeL_{memb} \xrightarrow{k_{in}} Fe_{intracell}$$

where Fe uptake is described by the Michaelis–Menten expression

$$\rho = \rho_{max} \frac{[Fe(OH)_2^+]}{K_s + [Fe(OH)_2^+]} \quad \text{and} \quad K_s = \frac{k_d + k_{in}}{k_f}$$

(1) Minimum number of "carrier" sites (L_{memb})

$$L_{memb} \geqslant \mu'Q/k_f [Fe(OH)_2^+] = 2 \times 10^{-17} \text{ mol cell}^{-1}$$
$$\text{for } k_f = 10^6 \ M^{-1} \text{ s}^{-1}, [Fe(OH)_2^+] = 10^{-10} \ M$$
$$\mu'Q = 2 \times 10^{-21} \text{ mol cell}^{-1} \text{ s}^{-1}$$

(2) Internalization of Fe-"carrier" complex[a]

$$\tau_{FeL-memb} \approx \frac{1}{k_f [Fe(OH)_2^+]} \approx \frac{L_{memb}}{\mu'Q} \approx 2.5 \text{ h}$$

(3) Average distance between "carrier" centers[b]
$$\approx 4 \text{ nm for a spherical cell of radius } r_{cell} = 3.5 \ \mu m$$

---

[a] The expression $\tau = L_{membrane}/\mu'Q$ defines the mean residence time of all porters. This is the same as for the internalization of FeL if almost all "carrier" sites are iron-bound, that is, if $K_s \approx [Fe(OH)_2^+]$.
[b] Calculated for a flat surface with an area equivalent to the area of a spherical cell ($r_{cell} = 3.5 \ \mu m$, $A_{cell} = 150 \ \mu m^2$) and assuming that each "carrier" binds a single iron atom.

3.6.   Example IV: Indirect Aluminum Toxicity

Recently, Rueter et al. (1987) proposed that the toxicity of Al to phytoplankton might be due not to the direct effects of Al itself, to which phytoplankton (unlike higher organisms) are relatively insensitive, but to the displacement of Cu(II) from complexes with naturally occurring ligands and the toxic effect of increased free cupric ion concentration. In this model for indirect Al toxicity, the coordination kinetics of the Al-for-Cu exchange would influence observed toxicity if the rates of metal exchange and biological metal uptake were comparable. The initial rate of Cu(II) release from natural organic complexes as a function of the Al-for-Cu metal-exchange rate constants can be compared with the estimated range for the rate of biological metal uptake. Figure 8 indicates that, at environmentally relevant concentrations of Al and organically complexed Cu, coordination kinetics would be predicted to influence observed (indirect) toxicity if the rate constant for Al-for-Cu metal exchange is less than about 1 $M^{-1} s^{-1}$. This model, of course, presumes direct competitive interaction between Al and Cu with naturally occurring ligands. The possible importance of coordination kinetics in indirect toxicity is difficult to estimate, however, because of the scarcity of information on rates of Al-for-Cu metal exchange even for well-defined ligands.

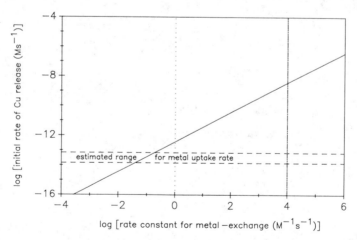

Figure 8. Initial rate of Cu(II) release from an organic complex CuL as a function of the Al-for-Cu metal-exchange rate constant calculated for [Al] = 6 μM, [CuL] = 0.06 μM, where $d[Cu(II)]/dt$ = k[Al][CuL] (assuming negligible change in [Al] and [CuL] over some initial time). This is compared with an estimated range for the metal uptake rate by phytoplankton calculated for 10^6 cells L^{-1} and a cellular metal uptake rate of ≈ 0.5–2.5×10^{-16} mol cell^{-1} h^{-1} (Sunda and Huntsman, 1986; Harrison and Morel, 1986). Also shown are the estimated maximum rate constants for metal-exchange reactions of Al^{3+} ($\cdots$) and of Al(OH)$^{2+}$ ($-\cdots-$) based on the water-loss rate constants for these species (Crumbliss and Garrison, 1988) and a $K_{os} = 1$. [Note that the K_{os} term may be significantly greater than 1 as a result of electrostatic interactions between Al^{3+} or Al(OH)$_2^+$ and Cu complexes with an overall negative charge.]

4. CONCLUSIONS

An understanding of coordination kinetics allows for rationalization (and even prediction) of the mechanisms and rates of coordination reactions of well-defined ligands even in complicated systems. The rates of complexation reactions in natural waters will depend on the initial speciation of both metals and ligands, the type of perturbation, the relative concentrations of reacting species, the strength of metal–ligand interactions, and the ligand structure (particularly the extent of competitive metal binding). The difficulties in applying the principles of coordination kinetics derived from the study of well-defined ligands to naturally occurring ligands arise primarily because of our limited knowledge of their structure and the nature of their interactions with metals. Of particular importance is the extent to which competing metals interact with specific binding sites, the heterogeneity of binding sites, and the contribution of electrostatic effects to the apparent affinity of specific binding sites. Nonetheless, the observed kinetic behavior of strong complexing agents suggests that coordination kinetics may be slow in natural waters, especially in systems containing mixtures of strong and weak ligands and high concentrations of alkaline earth metals or mixtures of competing transition metals.

Acknowledgments

We thank R. Hudson for providing unpublished data, M. Hoffmann for helpful comments, and W. Stumm for his encouragement and support. This work was funded in part by the Swiss National Science Foundation and by the U.S. Office of Naval Research (N00014-86-K-0325), National Science Foundation (8615545-OCE), and Environmental Protection Agency (CR-815293-01).

REFERENCES

Anderson, M. A., and F. M. M. Morel (1982), "The Influence of Aqueous Iron Chemistry on the Uptake of Iron by the Coastal Diatom *Thalassiosira weisflogii*," *Limnol. Oceanogr.* **27**, 789–813.

Astumian, R. D., and Z. A. Schelly (1984), "Geometric Effects of Reduction of Dimensionality in Interfacial Reactions," *J. Am. Chem. Soc.* **106**, 304–308.

Blesa, M. A., E. B. Borghi, A. J. G. Maroto, and A. E. Regazzoni (1984), "Adsorption of EDTA and Iron-EDTA Complexes on Magnetite and the Mechanism of Dissolution of Magnetite by EDTA," *J. Colloid Interface Sci.* **98**, 295–305.

Brauch, H. J., and Schullerer (1987), "Verhalten von Ethylendiamintetracacetat (EDTA) und Nitrilotriacetat (NTA) bei der Trinkwasseraufbereitung," *Vom Wasser* **69**, 155–164.

Buckley, P. J. M., and C. M. G. van den Berg (1986), "Copper Complexation Profiles in the Atlantic Ocean," *Mar. Chem.* **19**, 281–296.

Burgess, J. (1978) *Metal Ions in Solution*, Ellis Horwood Ltd., Chichester, England.

Cabaniss, S. E., and M. S. Shuman (1988a), "Copper Binding by Dissolved Organic Matter: I. Suwannee River Fulvic Acid Equilibria," *Geochim. Cosmochim. Acta* **52**, 185–193.

Cabaniss, S. E., and M. S. Shuman (1988b), "Copper Binding by Dissolved Organic Matter: II. Variation in Type and Source of Organic Matter," *Geochim. Cosmochim. Acta* **52**, 195–200.

Cabaniss, S., M. S. Shuman, and B. J. Collins (1984), "Metal–Organic Binding: A Comparison of Models," in C. J. M. Kramer and J. C. Duinker, Eds., *Complexation of Trace Metals in Natural Waters*, Martinus Nijhoff/Dr. W. Junk, The Hague, Netherlands, pp. 165–179.

Cacheris, W. P., and G. R. Choppin (1987), "Dissociation Kinetics of Thorium-Humate Complex," *Radiochim. Acta* **42**, 185–190.

Cassatt, J. C., R. G. Wilkins (1968), "The Kinetics of Reaction of Nickel(II) Ion with a Variety of Amino Acids and Pyridinecarboxylates," *J. Am. Chem. Soc.* **90**, 6045–6050.

Cleveland, J. M., and T. F. Rees (1981), "Characterization of Plutonium in Maxey Flats Radioactive Trench Leachates," *Science* **212**, 1506–1509.

Coale, K. H., and K. W. Bruland (1988), "Copper Complexation in the Northeast Pacific," *Limnol. Oceanogr.* **33**, 1084–1101.

Correia dos Santos, M. M., and M. L. S. Simoes Goncalves (1986), "Electroanalytical Chemistry of Copper, Lead, and Zinc Complexes of Amino Acids at the Ionic Strength of Seawater (0.70 M $NaClO_4$)," *J. Electroanal. Chem.* **208**, 137–152.

CRC Handbook of Chemistry and Physics, 1st Student ed. (1988), R. C. Weast, Ed., CRC Press, Boca Raton, FL, p. F-105.

Crumbliss, A. L., and J. M. Garrison (1988), "A Comparison of Some Aspects of the Aqueous Coordination Chemistry of Aluminum(III) and Iron(III)," *Comments Inorg. Chem.* **8**, 1–26.

Dalang, F., J. Buffle, and W. Haerdi (1984), "Study of the Influence of Fulvic Substances on the Adsorption of Copper(II) Ions at the Kaolinite Surface," *Environ. Sci. Technol.* **18**, 135–141.

Davis, J. A., and J. O. Leckie (1978), "Effect of Adsorbed Complexing Ligands on Trace Metal Uptake by Hydrous Oxides," *Environ. Sci. Technol.* **12**, 1309–1315.

Davis, J. A., C. C. Fuller, and A. D. Cook (1987), "A Model for Trace Metal Sorption Processes at the Calcite Surface: Adsorption of Cd^{2+} and Subsequent Solid Solution Formation," *Geochim. Cosmochim. Acta* **51**, 1477–1490.

Dietz, F. (1987), "Neue Messergebnisse ueber die Belastung von Trinkwasser mit EDTA," *GWF Wasser/Abwasser* **128**, 286–288.

Driscoll, C. T., J. P. Baker, J. J. Bisogni, and C. L. Schofield (1984), "Aluminum Speciation and Equilibria in Dilute Acidic Surface Waters of the Adirondack Region of New York State," in O. R. Bricker, Ed., *Acid Precipitation: Geological Aspects*, Butterworth, Boston.

Duinker, J. C., and C. J. M. Kramer (1977), "An Experimental Study on the Speciation of Dissolved Zinc, Cadmium, Lead, and Copper in River Rhine and North Sea water, by Differential Pulsed Anodic Stripping Voltammetry," *Marine Chem.* **5**, 207–228.

Dzombak, D. A., W. Fish, and F. M. M. Morel (1986), "Metal–Humate Interactions. 1. Discrete Ligand and Continuous Distribution Models," *Environ. Sci. Technol.* **20**, 669–675.

Eigen, M., and R. G. Wilkins (1965), "The Kinetics and Mechanism of Formation of Metal Complexes", in *Mechanisms of Inorganic Reactions* (ACS Symposium Series No. 49), American Chemical Society, Washington, DC, pp. 55–80.

Estep, M., J. E. Armstrong, and C. van Baalen (1975), "Evidence for the Occurrence of Specific Iron (III)-Binding Compounds in Near-Shore Marine Ecosystems," *Appl. Microbiol.* **30**, 186–188.

Fish, W. (1984), "Modeling the Interactions of Trace Metals and Aquatic Humic Materials," Ph.D. thesis, Massachusetts Institute of Technology, Cambridge, MA.

Fish, W., D. A. Dzombak, and F. M. M. Morel (1986), "Metal–Humate Interactions. 2. Application and Comparison of Models," *Environ. Sci. Technol.* **20**, 676–683.

Giannissis, D., G. Dorange, and M. Guy, (1985), "Association (Complexation and Adsorption) Phenomena of Heavy Metal Cations with Humic Substances (Role of Calcium Ions)," in T. D. Lekkas, Ed., *5th International Conference on Heavy Metals in the Environment*, Vol. 2, pp. 469–471.

Gschwend, P. M., and S. Wu (1985), "On the Constancy of Sediment–Water Partition Coefficients of Hydrophobic Organic Pollutants," *Environ. Sci. Technol.* **19**, 90–96.

Harrison, G. I., and F. M. M. Morel (1986), "Response of the Marine Diatom *Thalassiosira weisflogii* to Iron Stress", *Limnol. Oceanogr.* **31**, 989–997.

Hering, J. G., and F. M. M. Morel (1988a), "Humic Acid Complexation of Calcium and Copper," *Environ. Sci. Technol.* **22**, 1234–1237.

Hering, J. G., and F. M. M. Morel (1988b), "Kinetics of Trace Metal Complexation: Role of Alkaline Earth Metals," *Environ. Sci. Technol.* **22**, 1469–1478.

Hering, J. G., and F. M. M. Morel (1989), "Slow Coordination Reactions in Seawater," *Geochim. Cosmochim. Acta* **53**, 611–618.

Hering, J. G., and F. M. M. Morel (1990), "Kinetics of Trace Metal Complexation: Ligand-Exchange Reactions", *Environ. Sci. Technol.* 1, **24**, 242–252.

Hering, J. G., W. G. Sunda, R. L. Ferguson, and F. M. M. Morel (1987), "A Field Comparison of Two Methods for the Determination of Copper Complexation: Bacterial Bioassay and Fixed-Potential Amperometry," *Marine Chem.* **20**, 299–312.

Hirose, K. (1988), "Metal–Organic Ligand Interactions in Seawater: Multimetal Complexation Model for Metal Titration," *Mar. Chem.* **25**, 39–48.

Hirose, K., Y. Dokiya, and Y. Sugimura (1982), "Determination of Conditional Stability Constants of Organic Copper and Zinc Complexes Dissolved in Seawater Using Ligand Exchange Method with EDTA," *Mar. Chem.* **11**, 343–354.

Hoffmann, M. (1981), "Thermodynamic, Kinetic, and Extrathermodynamic Considerations in the Development of Equilibrium Models for Aquatic Systems," *Environ. Sci. Technol.* **15**, 345–353.

Hudson, R. (1989), "The Chemical Kinetics of Iron Uptake by Marine Phytoplankton", Ph.D. thesis, Massachusetts Institute of Technology, Cambridge, MA.

Huizenga, D. L., and D. R. Kester (1983), "The Distribution of Total and Electrochemically Available Copper in the Northwestern Atlantic Ocean," *Mar. Chem.* **13**, 281–291.

Jacobs, L. A., H. R. von Gunten, R. Keil, and M. Kuslys (1988), "Geochemical Changes along a River-Groundwater Infiltration Flow Path: Glattfelden, Switzerland," *Geochim. Cosmochim. Acta* **52**, 2693–2706.

Jones, G. J., B. P. Palenik, and F. M. M. Morel (1987), "Trace Metal Reduction by Phytoplankton: The Role of Plasmalemma Redox Enzymes," *J. Phycol.* **23**, 237–244.

Killey, R. W. D., J. O. McHugh, D. R. Champ, E. L. Cooper, and J. L. Young (1984), "Subsurface Cobalt-60 Migration from a Low-Level Waste Disposal Site," *Environ. Sci. Technol.* **18**, 148–157.

Kirkner, D. J., A. A. Jennings, and T. L. Theis (1985), "Multisolute Mass Transport with Chemical Interaction Kinetics," *J. Hydrol.* **76**, 107–117.

Kramer, C. J. M. (1986), "Apparent Copper Complexation Capacity and Conditional Stability Constants in North Atlantic Waters," *Mar. Chem.* **18**, 335–349.

Lavigne, J. A., C. H. Langford, and M. K. S. Mak (1987), "Kinetic Study of the Speciation of Nickel(II) Bound to a Fulvic Acid," *Anal. Chem.* **59**, 2616–2620.

Leppard, G. G., J. Buffle, and R. Baudat (1986), "A Description of the Aggregation Properties of Aquatic Pedogenic Fulvic Acid Combining Physico-chemical and Microscopical Observations," *Water Res.* **20**, 185–196.

Margerum, D. W. (1963), "Exchange Reactions of Multidentate Ligand Complexes," *Rec. Chem. Progr.* **24**, 237–251.

Margerum, D. W., G. R. Cayley, D. C. Weatherburn, and G. K. Pagenkopf (1978), "Kinetics and Mechanism of Complex Formation and Ligand Exchange," in *Coordination Chemistry*, Vol. 2, A. Martell, Ed. (ACS Symposium Series No. 174), Washington, DC, pp. 1–220.

Martell, A. E., and R. M. Smith (1974), *Critical Stability Constants*, Vol. 1, Plenum Press, New York.

McKnight, D. M., and F. M. M. Morel (1980), "Copper Complexation by Siderophores from Filamentous Blue–Green Algae," *Limnol. Oceanogr.* **25**, 62–71.

McKnight, D. M., G. L. Feder, E. M. Thurman, R. L. Wershaw, and J. C. Westall (1983), "Complexation of Copper by Aquatic Humic Substances from Different Environments," *Sci. Tot. Environ.* **28**, 65–76.

Means, J. L., D. A. Crerar, and J. O. Duguid (1978), "Migration of Radioactive Wastes: Radionuclide Mobilization by Complexing Agents," *Science* **200**, 1477–1481.

Moffett, J. W., and R. G. Zika (1987a), "Reaction Kinetics of Hydrogen Peroxide with Copper and Iron in Seawater," *Environ. Sci. Technol.* **21**, 804–810.

Moffett, J. W., and R. G. Zika (1987b), "Solvent Extraction of Copper Acetylacetonate in Studies of Copper(II) Speciation in Seawater," *Mar. Chem.* **21**, 301–313.

Morel, F. M. M., and P. M. Gschwend (1987), "The Role of Colloids in the Partitioning of Solutes in Natural Waters," in W. Stumm, Ed., *Aquatic Surface Chemistry*, Wiley-Interscience, New York, Chapter 15.

Morgan, J. J., and A. T. Stone (1985), "Kinetics of Chemical Processes of Importance in Lacustrine Environments," in W. Stumm, Ed., *Chemical Processes in Lakes*, Wiley-Interscience, New York, Chapter 17.

Murphy, T. P., D. R. S. Lean, and C. Nalewajko (1976), "Blue–Green Algae: Their Excretion of Iron-Selective Chelators Enables Them to Dominate Other Algae," *Science* **192**, 900–902.

Niebor, E., and D. H. S. Richardson (1980), "The Replacement of the Nondescript Term 'heavy metals' by a Biologically and Chemically Significant Classification of Metal Ions," *Environ. Poll. (Ser. B)* **1**, 3–26.

Pankow, J. F., and J. J. Morgan (1981a), "Kinetics for the Aquatic Environment," *Environ. Sci. Technol.* **15**, 1155–1164.

Pankow, J. F., and J. J. Morgan (1981b), "Kinetics for the Aquatic Environment," *Environ. Sci. Technol.* **15**, 1206–1313.

Perdue, M. P., and C. R. Lytle (1983), "A Critical Examination of Metal–Ligand Complexation Models: Application to Defined Multiligand Mixtures," in R. F. Christman and E. T. Gjessing, Eds., *Aquatic and Terrestrial Humic Materials*, Ann Arbor Science, Ann Arbor, MI, Chapter 14.

Rueter, J. G., K. T. O'Reilly, and R. R. Petersen (1987), "Indirect Aluminum Toxicity to the Green Alga Scenedesmus through Increased Cupric Ion Activity," *Environ. Sci. Technol.* **21**, 435–438.

Sunda, W. (1984), "Measurement of Manganese, Zinc, and Cadmium Complexation in Seawater Using Chelex Ion Exchange Equilibria," *Mar. Chem.* **14**, 365–378.

Sunda, W. G., and P. J. Hanson (1979), "Chemical Speciation of Copper in River Water," in *Chemical Modeling in Aqueous Systems*, E. A. Jenne, Ed. (ACS Symposium Series No. 93), American Chemical Society, Washington, DC, Chapter 8.

Sunda, W. G., and A. K. Hanson (1987), "Measurement of Free Cupric Ion Concentration in Seawater by a Ligand Competition Technique Involving Copper Sorption onto C_{18} SEP-PAK Cartridges," *Limnol. Oceanogr.* **32**, 537–551.

Sunda, W. G., and S. A. Huntsman (1986), "Relationships among Growth Rate, Cellular Manganese Concentrations and Manganese Transport Kinetics in Estuarine and Oceanic Species of the Diatom Thalassiosira," *J. Phycol.* **22**, 259–270.

Trick, C. G., R. J. Andersen, A. Gillam, and P. J. Harrison (1983a), "Prorocentrin: An Extracellular Siderophore Produced by the Marine Dinoflagellate *Procentrum minimum*," *Science* **219**, 306–308.

Trick, C. G., R. J. Andersen, N. M. Price, A. Gillam, and P. J. Harrison (1983b), "Examination of Hydroxamate Siderophore Production by Neritic Eukaryotic Marine Phytoplankton," *Mar. Biol.* **75**, 9–17.

Turner, D. R., M. S. Varney, M. Whitfield, R. F. C. Mantoura, and J. P. Riley (1986), "Electrochemical Studies of Copper and Lead Complexation by Fulvic Acid. I. Potentiometric Measurements and a Critical Comparison of Metal Binding Models," *Geochim. Cosmochim. Acta* **50**, 289–297.

van den Berg, C. M. G., and S. Dharmvanij (1984), "Organic Complexation of Zinc in Estuarine Interstitial and Surface Water Samples," *Limnol. Oceanogr.* **29**, 1025–1036.

van den Berg, C. M. G., A. G. A. Merks, and E. K. Duursma (1987), "Organic Complexation and Its Control of the Dissolved Concentrations of Copper and Zinc in the Scheldt Estuary," *Est. Coast. Shelf Sci.* **24**, 785–797.

Waite, T. D., and F. M. M. Morel (1984), "Photoreductive Dissolution of Colloidal Iron Oxides in Natural Waters," *Environ. Sci. Technol.* **18**, 860–868.

Wilkins, R. G. (1970), "Mechanisms of Ligand Replacement in Octahedral Nickel(II) Complexes," *Acc. Chem. Res.* **3**, 408–416.

Zhou, X., and P. J. Wangersky (1989), "Production of Copper-Complexing Organic Ligands by the Marine Diatom *Phaeodactylum tricornutum* in a Cage Culture Turbidostat," *Mar. Chem.* **26**, 239–259.

6

THE FRONTIER-MOLECULAR-ORBITAL THEORY APPROACH IN GEOCHEMICAL PROCESSES

George W. Luther, III

College of Marine Studies, University of Delaware, Lewes, Delaware

1. INTRODUCTION

Recently, we have used an inner-sphere-type mechanism combined with a frontier-molecular-orbital theory (MOT) approach to demonstrate electron-transfer processes [both oxidation and reduction] at the pyrite surface. Specifically, in the *initial* step an electron from the S_2^{2-} of FeS_2 is transferred to oxidants (e.g., Fe^{3+}), whereas an electron is transferred from reducing agents (e.g., Cr^{2+}), to S_2^{2-} in FeS_2 (Luther, 1987). In the case of oxidation of FeS_2, a strengthening of the S–S bond is predicted that is consistent with the eventual formation of thiosulfate, whereas for the reduction of FeS_2, a decrease in the S–S bond stability is predicted for the eventual formation of hydrogen sulfide. In addition, this approach explains the rapid reactivity of Fe^{3+} versus the *slower* reactivity of O_2 at low pH with pyrite.

The work herein will address the reactivity of hydrogen sulfide and pyrite with a variety of diatomic inorganic oxidizing reagents [halogens, peroxide, hypochlorite, and oxygen (triplet and singlet)], some of which are found or are transient in environmental ecosystems. In addition, this chapter will demonstrate the principal factors responsible for (1) the relative pH dependent reactivity of H_2S and SH^- with O_2 (including trace metal catalysis), MnO_2 and Fe(III) minerals, (2) the slow oxidation of Mn(II) by O_2, and (3) the relatively faster oxidation of Fe(II) by O_2.

Throughout the chapter, the various reactions will be shown to have similar types of frontier molecular orbitals for describing reactivity. These similarities should aid the reader in understanding the relative speed of reaction and the nature of electron-transfer reactions in the aquatic environment. The frontier MOT approach is based on detailed quantum-mechanical calculations that

Lasaga (Lasaga and Gibbs, Chapter 9, this volume) has recently used to describe silicate dissolution.

2.   MOLECULAR-ORBITAL THEORY PRINCIPLES

It is important to note for the following discussion that in *electron-transfer processes* the reductant's highest occupied molecular orbital (HOMO) should combine with the oxidant's lowest unoccupied molecular orbital (LUMO) of the same symmetry to ensure proper overlap of reductant and oxidant orbitals to initiate electron transfer. That is, electron transfer will occur readily from π^* to π^* orbitals on different species or from σ^* to σ^* but not π^* to σ^* in a *linear* arrangement of atoms [e.g., A–B–C in Appendix I (following references at the end of this chapter)]. In the case of outer-sphere electron-transfer processes, π- to π-electron transfers are favored over σ to σ because (1) such transfers do not require major changes in bond lengths in the precursor complex (lower activation energy) and (2) the π orbitals are more diffuse or better exposed than σ orbitals. This process is well documented for transition metals. For inner-sphere electron-transfer processes, both π- to π- and σ- to σ-electron transfers are most favored (Purcell and Kotz, 1980).

In *chemical reactions*, a Lewis base HOMO (donor) must combine with a Lewis acid LUMO (acceptor) with electron density flowing from the HOMO to the LUMO. The orbitals must have the same symmetry with respect to the bond axis as above so that they can overlap in some way (linear, angular, etc.) according to the principles of MOT as described by Pearson (1976) (see also Appendix I). The major principles to remember are listed below. If all of these criteria are met, then a reaction is termed "symmetry-allowed":

1. The molecular orbitals must be positioned for good overlap.
2. The energies of the HOMO and the LUMO must be comparable for net positive overlap. The energy of the LUMO should be lower than that of the HOMO or about 6 eV above that of the HOMO.
3. The net effect of the HOMO to LUMO electron transfer will correspond to bonds being made and broken. The bonds being made or broken must be consistent with the expected end products of the reaction.

The first principle is illustrated best in Appendix I. The second principle is best approximated from the ionization potential (IP; IE) of the Lewis base HOMO and from the electron affinity (EA) of the Lewis acid LUMO. The third principle will be illustrated in the specific reactions discussed below.

2.1.   Description of Diatomic Oxidants (E_{LUMO})

Figure 1*A* shows the molecular-orbital diagram for the isoelectronic (14 electrons) diatomic species, X_2, which include the molecular halogens, hypochlorite

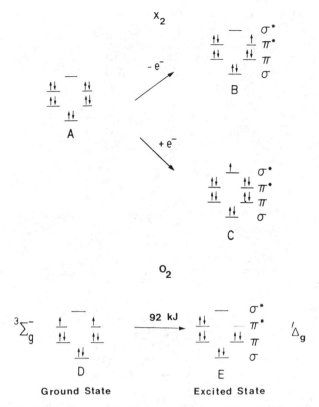

Figure 1. Idealized MO diagrams for the linear combination of $6p$ orbitals to form six molecular orbitals. (A) S_2^{2-} or O_2^{2-} (bond order 1); (B) S_2^- or O_2^- (bond order 1.5); (C) S_2^{3-} (bond order 0.5); (D) $^3\Sigma_g^-$ state of O_2; (E) $^1\Delta_g$ state of O_2.

ion, and peroxide. The diagram is essentially the same for all of these except for differences in orbital energies arising from the different effective nuclear charges of the atoms comprising these species. [Peroxide is slightly more complicated because of the two hydrogen atoms (Gimarc, 1979).] The reactivity of these oxidants stems from the nature of the σ^* orbital. These strong oxidants act as Lewis acids that accept electrons from Lewis bases (reductants). In the case of *total electron transfer* from the Lewis base to σ^*, the X_2 bond will be broken, resulting in X^- or OH^- as products.

To a first approximation, the energy of the LUMO (which is the orbital accepting electrons) for these oxidants can be judged by their electron affinities (EA) which are given in Table 1 (Lowe, 1977; Drzaic et al., 1984). Negative energies for E_{LUMO} are indicative of stable empty orbitals that can accept electrons. In principle, any difference in the rate of reaction of these oxidants with a specific Lewis base (such as pyrite or hydrogen sulfide) can be related to the differences in energy of the σ^* orbital for each oxidant. Thus, the dihalogens

TABLE 1. E_{LUMO} **Values (in Electrovolts) of Oxidants from Electron Affinity Data**

O_2	-0.44
O_2H	-1.19
O_3	-2.10
Cl_2	-2.35
Br_2	-2.53
I_2	-2.55
F_2	-3.1

TABLE 2. Electrode Potentials of Important Oxidants

Redox Couple	$E°$ (V)
$F_{2(g)} + 2e^- \rightarrow 2F^-$	2.87
$H_2O_2 + 2H^+ + 2e^- \rightarrow 2H_2O$	1.776
$O_2 + 4H^+ + 4e^- \rightarrow 2H_2O$	1.229
$Cl_{2(g)} + 2e^- \rightarrow 2Cl^-$	1.36
$Br_{2(l)} + 2e^- \rightarrow 2Br^-$	1.065
$I_{2(s)} + 2e^- \rightarrow 2I^-$	0.535

should be stronger oxidants than O_2 and H_2O_2, and this reactivity is well known. The EA of peroxide has not been reported to our knowledge, but it should be a smaller negative value than that of the halogens based on the HO_2 radical EA.

The use of EA values provides a better understanding of reactivity than do the standard redox potentials that are used in $p\varepsilon$–pH diagrams and that are a good indicator of thermodynamic stability. Table 2 shows the electrode potentials for the halogens, oxygen, and peroxide. Peroxide is expected to be reduced before all halogens other than F_2. This is not consistent with the known reactivity of these oxidants with a common reducing agent such as H_2S. The reason for this difference in chemical reactivity versus electrode reactions may be attributed to several reasons. First, the electrode material (e.g., Pt) consists of a metal with a large number of degenerate orbitals (band theory) that can easily transfer electrons to or accept electrons from a reactant over all space. Thus good overlap between the electrode's HOMO orbitals and the oxidant's LUMO orbital is generally possible. However, in chemical reactions all reactants have orbitals that are more directional (localized) in character and that affect chemical reactivity. Second, in redox experiments, electrons appear to transfer in one or many electron steps whereas in chemical reactions electron transfer occurs in one or two electron steps only, depending on the reactants. Third, in chemical reactions, steric effects and electron repulsions between reactants are more important in governing electron-transfer processes. In the case of O_2, unpairing of electrons of a reductant prior to electron transfer to a partially filled O_2 orbital requires a larger energy of activation than the transfer of an electron at an electrode. Therefore, E_{LUMO} as given by EA values is apparently a more fundamental parameter for predicting chemical reactivity of oxidants and giving insight to the kinetics of a reaction.

2.2. Speciation and Reactivity of the H_2S System (E_{HOMO})

The H_2S system has three possible species (H_2S, SH^-, S^{2-}) for reaction depending on pH. The oxidation of H_2S by O_2 and H_2O_2 involves a two-term rate law (Hoffmann, 1977; Millero et al., 1987a):

$$-d[H_2S]/dt = k_1[H_2S][\text{oxidant}] + k_2[SH^-][\text{oxidant}] \qquad (1)$$

The reaction rate increases over the pH range 4–8 and indicates that SH^- is the principal reactive species in solution. This pH dependence can be explained by a knowledge of E_{HOMO} for each species. Figure 2 shows the energies of the HOMO for H_2S (-10.47 eV), SH^- (-2.50 eV), and S^{2-} ($+6.1$ eV) with selected oxidants. These values are from the experimentally determined IP (Jolly, 1984) and EA values (Drzaic et al., 1984). More negative values indicate more stability and less electron-donation capability. The HOMO energies for these and other species can also be determined from the proton affinity (PA)–IP relationship

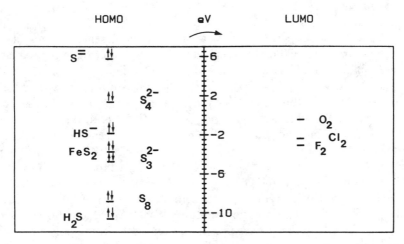

Figure 2. Relative E_{HOMO} and E_{LUMO} for selected reductants and oxidants. Pyrite has a broad ESCA valence band attributable to the HOMO of S_2^{2-} near -3.9 eV (Heide et al., 1980). The arrow indicates the direction of electron transfer in redox reactions.

given by DeKock and Barbachyn (1979),

$$PA(B) = IP(H) - IP(B) + D(B-H) \tag{2}$$

where $IP(H) = 13.598$ eV and $IP(B)$ is the first IP of the Lewis base (B) and $D(B-H)$ refers to the homolytic cleavage of the B–H bond [363 kJ mol^{-1} or 3.76 eV for the S–H bond].

The HOMO of H_2S is lower in energy than the LUMO of the common natural oxidants (O_2, H_2O_2) and *not* within 6 eV. Thus, H_2S should be resistant to oxidation in oxic natural systems of low pH. The HOMO of SH$^-$ is of similar energy to the LUMO of the oxidants and therefore should be quite reactive. The E_{HOMO} of S^{2-} is well above the E_{LUMO} of the oxidants and should be readily oxidized. The ease of oxidation of S^{2-} in $\geq$1-molar basic solutions is well known. In fact, this ease of oxidation has made it difficult to accurately determine pK_2 for the dissociation of H_2S (Myers, 1986). Thus, the relative order of the HOMO orbitals explains the kinetic observations of several workers on the oxidation of sulfide by both H_2O_2 and O_2; specifically, S^{2-} reacts faster than does SH$^-$, which reacts faster than H_2S. The relative ordering of E_{HOMO} for other protic systems will follow the same pattern as the H_2S species; namely H_2A is more stable than HA$^-$, which is more stable than A^{2-}. The reactivity of the H_2S system with oxidants is discussed below in light of overlap of the HOMO of the H_2S species with the LUMO of the various oxidants.

Figure 2 also shows the ordering of energies for S_8 [-9.04 eV, (Miller and Cusachs, 1969)], S_3^{2-} [-4.66 eV (Cotton et al., 1976)], and S_4^{2-} [$+1.36$ eV (Foti et al., 1978)]. The ability of S_8 to react as a nucleophile is low because of its stable HOMO; it reacts as an electrophile, and the reactions of SH$^-$, CN$^-$, and SO_3^{2-} exemplify the electrophilic behavior of S_8. For SH$^-$, the reaction with S_8 forms the polysulfides (S_x^{2-}), which are a source of soluble and more reactive zerovalent sulfur, $S(0)$.

The E_{HOMO} of the polysulfides shows an important difference in reactivity with the H_2S system. At pH values between 7 and 9, the S_x^{2-} ions that are predicted to be present in solution [based on pK_a values; see Hoffmann (1977), Meyer et al. (1977)] are those with $x = 3, 4, 5$; $x = 1$ and 2, the HS_x^- species are expected. Thus, at natural-water pH values, smaller polysulfides ($x = 2$ and 3) should be less reactive than SH$^-$, whereas higher polysulfides ($x = 4$ and 5) should be more reactive than SH$^-$ based on E_{HOMO}. This reactivity has been documented for the reactions of SH$^-$ and S_4^{2-} with activated olefins at pH 7–10 (Vairavamurthy and Mopper, 1989; see also LaLonde et al., 1987). However, Stahl and Jordan (1987) have observed that the H_2S system (S^{2-}) reacts faster than all S_x^{2-} ions with Hg(II) (p-hydroxymercuribenzoate solutions) at a pH of 12. Thus, through knowledge of the pH of the solution and the pK_a values of the various species, it is possible to indicate the species present for reactivity in the polyprotic sulfur systems; and through a knowledge of E_{HOMO}, the relative reactivity of these species with other reactants can be predicted.

3. HYDROGEN SULFIDE OXIDATION BY THESE STRONG OXIDANTS

In describing formally the reaction between these oxidants and sulfide, it is important to recall the structure and orbitals of hydrogen sulfide. Hydrogen sulfide has a strongly bent H–S–H bond angle of about 93° (Huheey, 1983). This geometry indicates that the mode of bonding for sulfur is with "pure" p orbitals. Thus, H_2S, SH^- and S^{2-} will donate an electron pair, which will reside in a p orbital, to a Lewis acid.

The halogens react *stoichiometrically* with sulfide and polysulfides according to reaction 3:

$$S_x^{2-} + I_2 \longrightarrow 2I^- + xS^0 \tag{3}$$

Thus, sulfide and polysulfides react as Lewis bases to transfer an electron pair from a p_z orbital to the vacant dihalogen σ^* LUMO in an inner-sphere process (Fig. 3). In Lewis structures, the transient intermediate **I** should be formed:

$$[H\text{–}\bar{S}\colon \longrightarrow \colon\bar{X}-\bar{X}\colon]^-$$

I

This intermediate is formally analogous to the well-known triiodide ion, I_3^-, which is stable relative to decomposition to I_2 and I^-. On complete electron transfer, the intermediate **I** decomposes to the expected products, elemental sulfur (S_8) and X^- (ΔG for the reaction with SH^- is $-131.63\ \text{kJ mol}^{-1}$). The formation of S–S bonds are favored because they are stronger than the X–X bonds of the oxidants. A molecular-orbital diagram for intermediate (**I**) is shown in Figure 3. The intermediate is linear because the p orbital from sulfide and the σ^* orbital from X_2 are both on the same (z) axis.

Hoffmann (1977) has explored the kinetics of the H_2S and hydrogen peroxide reaction. In his careful study over a wide range of pH, he noted the formation of elemental sulfur (S_8) as the major end product (70%) of the oxidation with

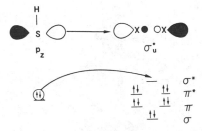

Figure 3. Molecular-orbital diagram demonstrating the intermediate (I) in which the p_z orbital of sulfide donates a pair of electrons to the σ^* orbital of an oxidant such as peroxide and halogen molecules (X_2), which are formally Lewis acids.

smaller amounts of sulfate (30%). However, he also noted the transient existence of polysulfide ions, especially at neutral and higher pH, where they are likely to be most stable. He postulated the existence of a HSOH intermediate (Eq. 4) that can react with SH^- to form polysulfides (Eq. 5):

$$SH^- + H_2O_2 \longrightarrow HSOH + OH^- \tag{4}$$

$$HSOH + SH^- \longrightarrow S_2H^- + H_2O \tag{5a}$$

$$HSOH + S_2H^- \longrightarrow S_3H^- + H_2O \tag{5b}$$

$$HSOH + S_3H^- \longrightarrow S_4H^- + H_2O \tag{5c}$$

$$HSOH + S_4H^- \longrightarrow S_5H^- + H_2O \tag{5d}$$

$$HSOH + S_5H^- \longrightarrow S_6H^- + H_2O \tag{5e}$$

$$HSOH + S_6H^- \longrightarrow S_7H^- + H_2O \tag{5f}$$

$$HSOH + S_7H^- \longrightarrow S_8H^- + H_2O \tag{5g}$$

$$HSOH + S_8H^- \longrightarrow S_9H^- + H_2O \tag{5h}$$

$$S_9H^- \longrightarrow S_8 + SH^- \tag{5i}$$

The formation of a HSOH intermediate is consistent with forming a strong S–O bond at the expense of a weak O–O bond in peroxide. However, the formation of a weak S–S bond at the expense of a strong S–O bond is not favored (Eq. 5).

From the preceding molecular-orbital theory considerations, an alternative mechanism, which is consistent with the experimental observations, can be made. This mechanism leads to homolytic bond cleavage of peroxide by sulfide as in the dihalogen–sulfide reaction above. In this alternative, an intermediate such as $[HS-O_2H_2]^-$ **(II)** can be formed that on breakdown gives water, hydroxide ion, and zerovalent S, S(0), which would react immediately with SH^- to form polysulfides (Eqs. 6–7):

$$[HS-O_2H_2]^- \longrightarrow \tfrac{1}{8}S_8 + H_2O + OH^- \tag{6}$$
$$\mathbf{II}$$

$$\tfrac{1}{8}S_8 + SH^- \longrightarrow S_2H^- \tag{7a}$$

$$\tfrac{1}{8}S_8 + S_2H^- \longrightarrow S_3H^- \tag{7b}$$

$$\tfrac{1}{8}S_8 + S_3H^- \longrightarrow S_4H^- \tag{7c}$$

$$\tfrac{1}{8}S_8 + S_4H^- \longrightarrow S_5H^- \tag{7d}$$

$$\tfrac{1}{8}S_8 + S_5H^- \longrightarrow S_6H^- \tag{7e}$$

$$\tfrac{1}{8}S_8 + S_6H^- \longrightarrow S_7H^- \tag{7f}$$

$$\tfrac{1}{8}S_8 + S_7H^- \longrightarrow S_8H^- \tag{7g}$$

$$\tfrac{1}{8}S_8 + S_8H^- \longrightarrow S_8 + SH^- \tag{7h}$$

Longer polysulfides can be built up in the reaction, which can lead to the formation of elemental sulfur.

The observed buildup of polysulfides in solution occurs because the kinetics of the reaction of sulfide species (H_2S, SH^-) with peroxide is slower than in the reaction of sulfide with the molecular halogens (that reaction is instantaneous and forms S_8). One reason is that the H atoms in H_2O_2 make the π^* orbitals less antibonding and are sterically preventing ease of attack by the sulfide. Also, the energy of the peroxide LUMO is of more positive energy than the energies of the molecular halogens. All of these provide a good indication of the slower kinetics for the peroxide–sulfide reaction as compared to dihalogen–sulfide reactions. This difference in reactivity would allow other sulfide ions to react rapidly with the zerovalent sulfur, S(0), produced. Note that there is no S–O bond formed in this mechanism. The S–O bond is quite stable and its formation should lead to the production of thiosulfate, sulfite and sulfate rather than the initial formation of a polysulfide species.

The reaction of polysulfides with peroxide depends on the polysulfide ion present (see above). Once higher polysulfides are produced, the reaction should result in a peroxide–polysulfide intermediate (similar to **II**) that transfers two electrons from the polysulfide ion to the peroxide as readily as the sulfide and peroxide reaction. At low peroxide levels, partial oxidation of the polysulfide ions should result in the direct formation of sulfate (through thiosulfate and perhaps sulfite) and S_8:

$$S_x^{2-} + 4H_2O_2 \longrightarrow \frac{x-1}{8} S_8 + SO_4^{2-} + 4H_2O \tag{8}$$

Complete oxidation of the polysulfides should lead to sulfate as the major end product. The sulfide–peroxide reaction yields different *initial* intermediates than, but similar end products as, the FeS_2 reaction with peroxide, which is discussed below.

4. HYDROGEN SULFIDE OXIDATION BY MOLECULAR OXYGEN

In the case of molecular oxygen, O_2 and Lewis bases (e.g., FeS_2 and sulfide) would *not* form a (strong) bond or Lewis acid–base adduct because of electron repulsions between the pair of electrons from a Lewis base and the partially filled LUMO (π^*) orbitals of O_2 (Figs. 1D and 4). This begins to explain the poor oxidizing ability of molecular oxygen with certain compounds and ions. (For Mn^{2+}, which is an important ion in the environment, other considerations are also important; see below.) Thus, H_2S and FeS_2 oxidation by triplet O_2 should be described more as an *outer-sphere* rather than an inner-sphere process. This is consistent with the oxygen isotopic studies of FeS_2 oxidation by Taylor et al. (1984a,b). In these studies, little or no oxygen from O_2 is found in the sulfur oxyanions formed on FeS_2 oxidation.

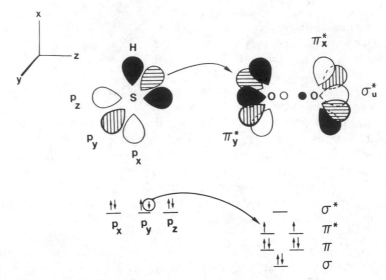

Figure 4. Molecular-orbital diagram demonstrating the attack on O_2 by SH^-. Direct p_z attack on a π^* (LUMO) of O_2 is not possible because of electron repulsions. Note that the Cartesian coordinate axes in this figure and the remaining ones are identical. The y axis is perpendicular to the plane of the paper (xz). The dark and light orbital lobes are in the xz plane; the orbital lobes with horizontal and vertical lines are in the yz plane. The dark and light shading represent different signs of the original wavefunction (Ψ); the horizontal and vertical lines also represent different signs of the original wavefunction (Ψ). Overlap or electron transfer occurs when the signs (shading or lines) match.

However, singlet oxygen (Fig. 1E) could accept a pair of electrons because it has a vacant π^* orbital (LUMO). This vacant orbital allows oxygen to be an effective Lewis acid. Thus, singlet oxygen is a powerful oxidizing agent (Greenwood and Earnshaw, 1984). For singlet O_2, accepting an electron pair would form the peroxide ion which itself is a strong oxidant. Singlet oxygen can be formed in the presence of organic sensitizers with light (Hoigné, Chapter 2, this volume).

4.1. Oxidation of the H_2S Species by Oxygen

Figure 4 shows the MO diagram and frontier orbitals that are necessary to describe the oxidation of the H_2S system by O_2. Sulfide cannot donate a pair of electrons to the σ^* of O_2 because the σ^* orbital is not the LUMO. Because the π^* orbitals of O_2 are partially filled (these are LUMO as well as HOMO orbitals), direct two-electron transfer in a σ-bonding interaction from sulfide (p_z) to O_2 $[\pi_x^*$ or $\pi_y^*]$ is not allowed, either. This electron-spin repulsion consideration requires that sulfide bind or transfer electrons to O_2 via a π-bonding interaction $[p_x (p_y)$ to $\pi_x^* (\pi_y^*)]$ (an outer-sphere process). This process is not as facile as the σ (inner-sphere) process described for the sulfide–X_2 reaction described above. Only one electron can be transferred from one p orbital on sulfide to one π^*

orbital on O_2. In addition, when more hydrogens (at lower pH) are bound to the sulfur, there is less chance for electron transfer to the O_2. (The p orbitals of the sulfur are used for bonding and cannot be used for electron transfer to O_2.) Thus, the kinetics for sulfide oxidation by O_2 (an outer-sphere process) is much slower than that for X_2 species (an inner-sphere process), and the pH dependence of sulfide oxidation is shown in bonding terms as well as energetic terms (Fig. 2).

The molecular-orbital findings discussed above for the sulfide–O_2 reaction are consistent with the mechanism proposed by Chen and Morris (1972) (Eqs. 9a–d). The rate-limiting step is Eq. 9a, which is an outer-sphere process:

$$SH^- + O_2 \xrightarrow{\text{slow}} HS\bullet + O_2^- \bullet \quad \text{(``superoxide'')} \tag{9a}$$

$$HS\bullet + O_2 \longrightarrow S + HO_2 \quad \text{(``superoxide'')} \tag{9b}$$

$$HS\bullet + O_2\bullet^- \longrightarrow S + HO_2^- \quad \text{(``peroxide'')} \tag{9c}$$

$$HS^- + (x-1)S \longrightarrow H^+ + S_x^{2-} \tag{9d}$$

4.2. Trace Metal Catalysis of H_2S Oxidation by O_2

A variety of trace metals catalyze the oxidation of H_2S by O_2 (Stumm and Morgan, 1981). This catalysis may be represented by

$$O_2 + M^{n+} + SH^- \longrightarrow [O_2\text{–}M\text{–}SH]^{(n-1)+} \tag{10a}$$
$$\text{III}$$

$$[O_2\text{–}M\text{–}SH]^{(n-1)+} \longrightarrow O_2^{2-} + M^{n+} + S^0 + H^+ \tag{10b}$$

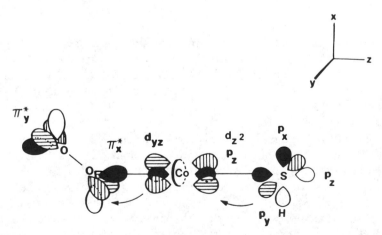

Figure 5. Diagram showing the possible π-system overlap of SH^- (p_y), Co^{2+} (d_{yz}), and O_2 (π_y^*) orbitals to effect the oxidation of SH^- by O_2. The binding of Co^{2+} and certain other metals (e.g., Mn^{2+}) to both O_2 and SH^- enhances electron transfer.

The intermediate **III** can be represented with molecular orbitals as in Figure 5. According to molecular-orbital theory, O_2 is a good π-electron acceptor from metals whereas SH^- is a good σ and π electron donor to metals (Jolly, 1984). In octahedral symmetry, the metal has e_g (σ^*) orbitals and t_{2g} (π) orbitals. Thus, the p_y orbital of SH^- is capable of electron transfer to the metal d_{yz} orbital, which, in turn, transfers electron density to the π_y^* orbital of O_2. The net effect is transfer of electrons from SH^- to O_2. During the transfer, the SH^- species adds electron density to the metal (enhances metal basicity) so that the metal can "pump" electrons for the reduction of O_2. Hydroxide and oxide ions also have the ability to enhance metal basicity (or reducing power) and are better able to stabilize high metal oxidation states (e.g., MnO_2, MnO_4^-). This concept is important to the oxidation of Mn(II) and Fe(II) by O_2 under basic conditions (see below). A major difference between oxide and sulfide ligands is that oxide ligands are not easily oxidized. Thus, Fe(II) and Mn(II) oxidation by O_2 should be important under basic and nonsulfidic conditions.

5.   SULFIDE OXIDATION BY Fe(III) MINERALS AND MnO_2

This section discusses the different reactivity of Fe(III) and Mn(IV) minerals with H_2S. The difference in reactivity is related to the differences in the symmetry of the frontier molecular orbitals for Fe(III) and Mn(IV). Whereas Fe(III) is a high-spin, labile metal cation (d^5, $t_{2g}^3 e_g^2$) that can undergo ligand exchange under appropriate conditions (e.g., acidification), Mn(IV) is an inert metal cation (d^3, t_{2g}^3) that does not undergo ligand exchange even under acidic conditions. These electron configurations of the metal cations are important to an understanding of their chemistry.

5.1.   Fe(III) Minerals

The reactions of sulfide with $Fe(OH)_3$ and $FeOOH$ have been described by Pyzik and Sommer (1981). They reported that the complete reduction of $Fe(OH)_3$ occurred within 30 min at pH 8.5. For FeOOH, the reaction is about 50 times slower. Their mechanism shows that H^+ attack facilitates exposing the Fe(III) at the mineral's surface. This allows sulfide to reduce the Fe(III) by an inner sphere process at the mineral surface.

Figure 6 shows the frontier orbitals for the reactants Fe(III) and SH^-. The LUMO for Fe(III) is one of the degenerate t_{2g} orbitals, whereas the HOMO for SH^- is one of the degenerate p orbitals not bound to hydrogen. The electron transfer for Fe(III) reduction can be described as a $\pi(p_y)$ from SH^- to a $\pi(d_{yz})$ of Fe(III). This is analogous to the O_2 oxidation of H_2S. In that process, only an outer-sphere process is possible for electron transfer because of electron repulsion effects. However, an inner-sphere process is feasible if the Fe(III) can be exposed at the mineral surface for bonding with sulfide. This is possible on acidification, which protonates OH^- and O^{2-} ions on the mineral surface and

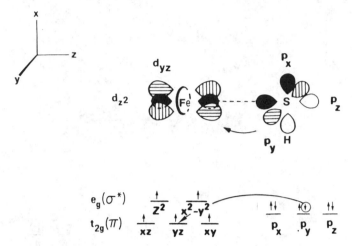

Figure 6. Molecular-orbital diagram demonstrating the reduction of Fe(III) by SH$^-$. Electron transfer occurs from SH$^-$ p_y (p_x) to Fe(III) d_{yz} (d_{xz}).

results in the formation of a Fe(III)–H$_2$O bond. This bond is labile and results in dissociation of H$_2$O. Sulfide can then bond to and reduce Fe(III). Thus, inner-sphere coordination of sulfide to Fe(III) followed by π (p_y) to π (d_{yz}) electron transfer is expected to be faster than an outer sphere process alone. The dissociation of H$_2$O is faster for Fe(OH)$_3$ than FeOOH because more protons are needed per oxygen atom to expose the FeOOH surface which contains oxide ions (e.g., Eqs. 11a,b):

$$\text{Fe(OH)}_3 + 3\text{H}^+ \longrightarrow \text{Fe}^{3+} + 3\text{H}_2\text{O} \tag{11a}$$

$$\text{FeOOH} + 3\text{H}^+ \longrightarrow \text{Fe}^{3+} + 2\text{H}_2\text{O} \tag{11b}$$

On reduction of Fe(III), the initial sulfur intermediates for these reactions will be similar to that of the O$_2$–H$_2$S reaction; specifically, S$_8$, polysulfides, and thiosulfate. Pyzik and Sommer (1981) found these intermediates in the order S$_8 > $ S$_x^{2-} > $ S$_2$O$_3^{2-}$. These intermediates are similar to those found by Chen and Morris (1972) in the O$_2$–H$_2$S reaction.

5.2. MnO$_2$

Burdige and Nealson (1986) reported that the complete reduction of MnO$_2$ by SH$^-$ occurred in 5–10 min. The observed end products were S$_8$ and sulfate. Figure 7 shows the frontier orbitals for MnO$_2$ and SH$^-$. The LUMO orbitals for Mn(IV) are the e_g (σ^*) orbitals in contrast to the Fe(III) orbitals which were t_{2g} (π). Thus, the mechanism is predicted to be different. In the MnO$_2$ reduction, direct electron pair donation (two-electron transfer) from any of the sulfide p orbitals not bound to hydrogen to the Mn(IV) ion at the surface of the crystal

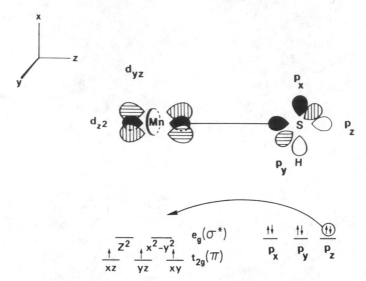

Figure 7. Molecular orbital diagram demonstrating the reduction of MnO_2 by SH^-. Electron transfer shown occurs from SH^- p_z to Mn(IV) d_{z^2}. The p_x and p_y orbitals of SH^- could also attack the d_{z^2} and $d_{x^2-y^2}$ orbitals of Mn.

would result in an electron being accepted by each of the degenerate e_g orbitals on Mn(IV). This is a σ to σ type of electron transfer, which requires an inner-sphere mechanism. This attack is necessary because the Mn in MnO_2 is inert—not susceptible to ligand lability—even under acidic conditions. This reaction is faster than the Fe(III) and the O_2 reactions with SH^- because those are one-electron transfers in the initial step. Also, there are no electron–electron repulsions between the reductant and the oxidant as in the O_2–H_2S reaction. The reaction of MnO_2 and SH^- shows a similar type of pathway as the reaction of H_2O_2 with SH^- (σ to σ inner-sphere process).

The initial sulfur intermediate in the MnO_2 reduction should again be a zero-valent sulfur atom that can form polysulfides with SH^-. The polysulfides also can react with MnO_2, leading to S_8 and sulfate. These are the major products as shown by Burdige and Nealson (1986). The products of this reaction are similar to the products of the H_2O_2–SH^- reaction discussed above.

6.   GENERAL COMMENTS ON H_2S OXIDATION

All the oxidants discussed above [X_2, H_2O_2, O_2, MnO_2, Fe(III) minerals] react with sulfide to form polysulfides as the first sulfur oxidation product. These, in turn, are able to react further depending on conditions to form S_8, thiosulfate, sulfite, and sulfate. Sulfite is typically a minor component in all sulfide oxidation reactions probably because of its high reactivity with oxidants and its reaction with S(0) to form thiosulfate. The continued reaction of polysulfides to form thiosulfate, sulfite, and sulfate are discussed below under reactions of FeS_2, which is a solid polysulfide and for which much information on reactivity is available.

7. PYRITE GEOMETRY AND MOLECULAR ORBITALS

The surface characteristics of the diamagnetic solid FeS_2 have been reviewed in Luther (1987). The Fe(II) ion is a low-spin, inert metal ion (d^6, t_{2g}^6). Briefly, the surface of a FeS_2 crystal can be viewed as a unit of FeS_2. This unit can be further divided to Fe^{2+} and S_2^{2-}. In the Lewis acid–base sense, the fragment would be

$$Fe \longleftarrow : \bar{S}_A - \bar{S}_B :$$
$$\textbf{IV}$$

where a lone pair of electrons from S_A has been donated to Fe^{2+}. Thus, there is a slight negative charge on the pyrite surface due to S_B.

In considering the S_2^{2-} ligand, it has a series of molecular orbitals that are key in understanding any reactions and electron transfer processes (oxidations or reductions) of pyrite. Figure 1A shows the arrangement and filling of molecular orbitals from the linear combinations of p orbitals for the two sulfur atoms to form S_2^{2-} (Tossell, 1983). The S_2^{2-} species formally has only one bond between the two atoms as in **IV** above. In **IV**, S_B also has a lone pair of electrons (from a π^* HOMO) available for donation to a vacant orbital of a Lewis acid. This lone pair of electrons is important for the attachment of reactants (oxidants and certain reductants) to the pyrite surface. Compounds or ions that can act as Lewis acids and which could accept a pair of electrons from pyrite are aqueous labile metal ions (Fe^{3+}, Fe^{2+}, Cr^{2+}), molecular halogens, singlet oxygen and hydrogen peroxide. These attachments to the pyrite surface are now discussed as well as the corresponding reactions with halogens and hydrogen peroxide.

7.1. Halogens and Peroxide

Pyrite can be considered a polysulfide ion, S_2^{2-}, bound to an Fe(II) ion. Polysulfide anions react readily with peroxide and halogens, and they can be quantified by iodometric titrations (Kolthoff et al., 1969) (Eq. 3). It is well known that FeS_2 reacts faster with hydrogen peroxide than dioxygen (McKibben and Barnes, 1986). Hydrogen peroxide is important in environmental chemistry (and may be important in biological oxidation of pyrite) because microorganisms through enzymes can produce it (Brock et al., 1984; Price and Morel, Chapter 8, this volume).

Bromine in excess is considered to oxidize all sulfur compounds to sulfate. To our knowledge, the reactions of F_2, Cl_2, I_2, and Br_2 with pyrite have not been studied in detail. However, Cl_2 is one of the major oxidizing reactants in aqua regia, which totally dissolves FeS_2, producing Fe(III) and sulfate. Recently, we (unpublished) have reacted molecular bromine and iodine with pyrite and have shown that there is a reaction taking place (via iodometric titrations). In our laboratory, mortars and pestles, which are used to grind FeS_2, are cleaned with aqueous bromine solutions because the reaction is so facile and produces no visual elemental sulfur precipitate. As shown below, pyrite through the S_2^{2-} ligand can bind to halogens and peroxide, and thus be oxidized in an inner sphere process.

The S_B in **IV** can donate a pair of electrons to the empty σ^* orbital of the halogens and peroxide, forming a σ bond between FeS_2 and X_2. If the donation of electrons results in a stable Lewis acid–base adduct, the intermediate without molecular orbitals would be

$$Fe\text{–}\underline{\overline{S}}_A\text{–}\underline{\overline{S}}_B: \longrightarrow :\underline{\overline{X}}\text{–}\underline{\overline{X}}:$$
$$\mathbf{V}$$

This intermediate would also be analogous to the triiodide, I_3^-, ion and **I** above. Figure 8 shows the MOT diagram for one of the HOMO orbitals (π^*) of S_2^{2-} donating (transferring) a pair of electrons to the LUMO (σ^*) orbital of X_2, forming a σ bond between FeS_2 and X_2. This intermediate (**V** and Figure 8) must

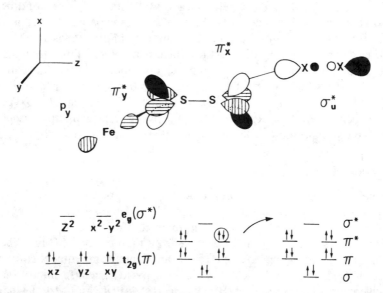

Figure 8. Molecular-orbital diagram demonstrating the intermediate (V) in which the π^* (HOMO) orbital of S_2^{2-} in FeS_2 (a Lewis base) would donate a pair of electrons to the σ^* (LUMO) orbital of an oxidant X_2. The Fe(II) and X_2 bind to the S_2^{2-} in an angular arrangement with their empty p_y and σ_u^* axis orbitals combining with the degenerate π^* (HOMO) orbitals of S_2^{2-}, which are in the yz and xz plane, respectively.

be angular because one of the degenerate π^* orbitals (HOMO) in the xz plane is combining with a σ^* orbital (LUMO) on the z axis of the reactant. For FeS_2, the molecular orbital used is analogous to the orbital used for the attachment of labile metal complexes (such as Fe^{3+}) to the pyrite surface. The *total transfer* of two electrons from pyrite to the σ^* orbital of these oxidants results in an inner sphere redox reaction with the breaking of the X–X or X–O bonds and the formation of X^- and OH^-. This process is similar to the reaction between SH^- and X_2 described above and can be represented by

$$Fe\text{–}\bar{\underline{S}}\text{–}\bar{\underline{S}}: + X_2 \longrightarrow [Fe\text{–}\bar{\underline{S}}\text{–}\bar{\underline{S}}]^{2+} + 2X^- \tag{12}$$

$$[Fe\text{–}\bar{\underline{S}}\text{–}\bar{\underline{S}}]^{2+} + H_2O \longrightarrow Fe\text{–}\bar{\underline{S}}\text{–}\bar{\underline{S}}\text{–}\bar{\underline{O}}: + 2H^+ \tag{13}$$
$$\mathbf{VI}$$

The transfer of electrons from FeS_2 to X_2 stabilizes the S–S bond (since antibonding electrons are being transferred to X_2) and produces the postulated reactive intermediate (**VI**) in Eq. 13. The bond order for the S–S bond in **VI** now approaches 2 and is formally analogous to singlet oxygen (Fig. 1E). This species should react immediately with water as in Eq. 13. The eventual formation of $S_2O_3^{2-}$ is expected by attack of water on the FeS_2 surface (see Eqs. 14–16 below).

McKibben and Barnes (1986) have studied peroxide oxidation of FeS_2 but did not propose a detailed mechanism. They report that only Fe^{3+} and sulfate are detected as products during the oxidation of pyrite by peroxide. The overall complete oxidation of FeS_2 could be written as

$$8H_2O + 7.5X_2 + FeS_2 \longrightarrow Fe^{3+} + 2SO_4^{2-} + 16H^+ + 15X^- \tag{14}$$

Equation 14 is similar to the pyrite oxidation equation of Singer and Stumm (1970). The difference is the nature of the oxidant, which for strong oxidants oxidizes Fe(II) to Fe(III). On the basis of Eqs. 12 and 13, we predict that the first surface product following the initial step of X_2 attachment to FeS_2 in the reaction of halogens and peroxide can be represented as

$$FeS_2 + X_2 + H_2O \longrightarrow Fe\text{–}\bar{\underline{S}}\text{–}\bar{\underline{S}}\text{–}\bar{\underline{O}}: + 2X^- + 2H^+ \tag{15}$$

$$Fe\text{–}\bar{\underline{S}}\text{–}\bar{\underline{S}}\text{–}\bar{\underline{O}}: + 2.5X_2 + 2H_2O \longrightarrow Fe^{3+} + S_2O_3^{2-} + 5X^- + 4H^+ \tag{16}$$

where an oxygen atom from water attaches to the surface of pyrite and acidity is generated. Continued oxidation of this Fe–S–S–O intermediate would result in thiosulfate formation [Eq. 16, as in the Fe^{3+} oxidation of FeS_2 discussed in Luther (1987)]. However, thiosulfate would immediately react with any of these oxidants and never be detected in solution. The formation and continued oxidation of thiosulfate (by Fe^{3+} and X_2) in these reactions indicates that elemental sulfur *should not* be formed as an end product. This reactivity of pyrite with peroxide is different from the sulfide–peroxide reaction discussed above. In that latter reaction, polysulfides, not thiosulfate, were formed as the first intermediates on H_2S oxidation.

Because pyrite is a polysulfide species itself, its oxidation suggests that other polysulfides (S_x^{2-}) should undergo oxidation to form thiosulfate (or perhaps sulfite) and eventually sulfate. The length of the polysulfide chain and the amount of available oxidant will dictate whether S_8 will form. The reactions of polysulfides to form thiosulfate, sulfite and S_8 can be represented by

$$3H_2O + (S)_x\text{-S-S}^{2-} + 3X_2 \longrightarrow (S)_x\text{-S-SO}_3^{2-} + 6X^- + 6H^+ \qquad (17)$$

$$(S)_x\text{-S-SO}_3^{2-} \longrightarrow (S)_x + S_2O_3^{2-} \qquad (18)$$

$$(S)_x\text{-S-SO}_3^{2-} \;\underset{\longrightarrow}{\longleftarrow}\; (S)_{x+1} + SO_3^{2-} \qquad (19)$$

$$S_8 + 8SO_3^{2-} \;\underset{\longleftarrow}{\longrightarrow}\; 8S_2O_3^{2-} \qquad (20)$$

If $x = 0$, then only $S_2O_3^{2-}$ should be formed. Equations 17–20 are consistent with the intermediate sulfur species found by Chen and Morris (1972) for the oxidation of H_2S by O_2. In their experiments, they observed a buildup of polysulfides, which, in turn, reacted with O_2 to form thiosulfate and some sulfite. Equations 19 and 20 show the relationship between $S_2O_3^{2-}$ and SO_3^{2-}. Formation of $S_2O_3^{2-}$ is favored at neutral and basic pH, whereas S_8 is favored at low pH. With strong oxidants, $S_2O_3^{2-}$ and SO_3^{2-} will react completely to form SO_4^{2-}.

8.   OXIDATION OF AQUEOUS Fe(II) AND Mn(II) BY O_2

This section discusses the different reactivity of aqueous Fe(II) and Mn(II) by O_2. Both metal ions and O_2 are formally Lewis acids. This chemical behavior suggests that reactivity may be slow if inner-sphere processes are important. As shown below, Fe(II) oxidation can occur via outer-sphere processes, whereas Mn(II) oxidation must occur via inner-sphere processes. For these oxidation reactions to be facile, it is necessary to enhance the metal ion's ability to lose an electron (become a reducing agent or a "base"). This can be accomplished with the appropriate ligating atoms. Aqueous Fe(II) and Mn(II) are both labile cations with d^6 ($t_{2g}^4 e_g^2$) and d^5 ($t_{2g}^3 e_g^2$) electron configurations, respectively. This lability allows for facile interchange of ligands.

The oxidation of Fe(II) occurs above pH 1 (Stumm and Morgan, 1981)—the reaction is slow at acid pH and faster at neutral and basic pH. However, the oxidation of Mn(II) is exceedingly slow or nonexistent until high pH (Stumm and Morgan, 1981; Diem and Stumm, 1984). A key feature of these reactions is the effect that OH^- has on enhancing metal basicity or metal reducing power and on stabilizing the higher oxidation states of the metals formed during the oxidation.

Another interesting feature of these reactions is that Mn(II) oxidation by O_2 is slow but MnO_2 reduction by SH^- is fast. However, the Fe(II) oxidation and

Fe(III) mineral reduction occur at roughly "similar" rates. This reactivity feature of the oxidation and reduction reactions for each system suggests a common pathway in the electron-transfer steps for the forward and reverse reactions in each system. However, the pathways for the iron and manganese reactions are discretely different from each other, which can be shown from the frontier-molecular-orbital approach.

8.1. Fe(II)

The reaction of Fe(II) with O_2 follows a rate law of the form given in Eq. 21 from pH 5–9 (Stumm and Morgan, 1981; Millero et al., 1987b):

$$-\frac{d[Fe(II)]}{dt} = k[Fe(II)][OH^-]^2 p_{O_2} \tag{21}$$

The reaction is 100 times slower in seawater than freshwater. This behavior can be represented by Eqs. 22–25, where Cl^- is a *weak* σ donor:

$$[Fe^{II}(H_2O)_6]^{2+} + Cl^- \underset{fast}{\longleftrightarrow} [Fe^{II}(H_2O)_5Cl]^+ \tag{22}$$

$$[Fe^{II}(H_2O)_5Cl]^+ + 2OH^- \longleftrightarrow [Fe^{II}(H_2O)_4(OH)_2] + Cl^- \tag{23}$$

$$[Fe^{II}(H_2O)_4(OH)_2] + O_2 \underset{slow}{\longrightarrow} [Fe^{III}(H_2O)_4(OH)_2]^+ + \cdot O_2^- \tag{24}$$

$$2[Fe^{II}(H_2O)_4(OH)_2] + O_2 \underset{slow}{\longrightarrow} 2[Fe^{III}(H_2O)_4(OH)_2]^+ + O_2^{2-} \tag{25}$$

In *seawater*, high Cl^- concentrations compete with OH^- for Fe(II) coordination. Reaction 22 is not applicable to *freshwater* systems. In freshwater, Cl^- is replaced by H_2O in Eq. 23. Reaction 25 is a binuclear (iron) alternative to form peroxide rather than superoxide as the initial product.

The frontier orbitals for O_2, Fe(II), and OH^- are given in Figure 9. The Fe(II) can transfer an electron to O_2 through an outer-sphere process whether OH^- is present. This results in a Fe $\pi(d_{yz})$ to O_2 π_y^*-electron transfer. This process explains the slow oxidation of Fe(II) at low pH. At basic pH, this process can be enhanced. While OH^- is a *better* σ donor than Cl^- or H_2O, it is also an excellent π donor to metals (e.g., p_y to d_{yz} electron transfer) as discussed earlier. In this process, the OH^- ligand donates electron density to the Fe(II) through both the σ and π systems, which results in metal basicity (and increased reducing power). Both of these effects now enhance Fe(II) πt_{2g} to O_2 π^*-electron transfer, which is an "ideal" orbital matchup for outer-sphere mechanisms. Unlike the metal catalysis discussed earlier, the Fe(II) actually is oxidized. The higher Fe(III) oxidation state is stabilized by the OH^- ligands. This process is still an outer-sphere process and is consistent with the linear free-energy analysis discussed by Wehrli (Chapter 11, this volume). At very high pH values (>9), an inner-sphere process may result as is discussed below for Mn(II) oxidation.

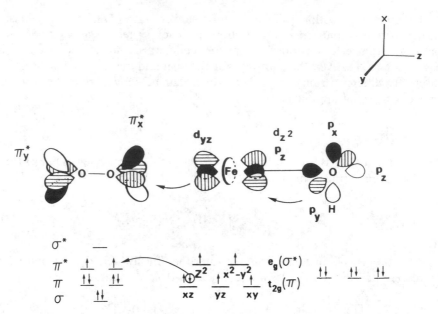

Figure 9. Molecular-orbital diagram for the oxidation of Fe(II) by O_2. An outer-sphere electron-transfer process from $\pi\,(d_{yz})$ to π_y^* [or $\pi\,(d_{xz})$ to π_x^*] is possible. OH$^-$ bound to Fe(II) can enhance the oxidation by transfer of electron density through both the σ and π systems. This stabilizes the Fe(III) formed on oxidation. Note the similarity of the π system to that in Figure 5. (The oxygen in OH$^-$ and H_2O has more "sp^3"-like orbitals than "p^3.")

8.2.   Mn(II)

In reaction with O_2, Mn(II) may undergo inner-sphere metal-centered oxidation [direct electron transfer from Mn(II) to O_2] or ligand-centered oxidation (electron transfer from the ligand to the electron acceptor, then electron transfer from Mn(II) to the ligand). Examples of metal-centered Mn(II) oxidation are provided by Hoffman et al. (1978) for a two-electron transfer process [oxidative addition of Mn(II) to O_2] and by Lever et al. (1981) for an one-electron transfer process. Richert et al. (1988) have recently provided an example of ligand-centered Mn(II) oxidation. Recent isotopic experiments by Tebo et al. (1987) show that significant quantities of oxygen from O_2 are found in the oxidized manganese minerals that are produced on Mn(II) oxidation. The following discussion will focus on metal-centered oxidation of Mn(II) as a result of the report by Tebo et al. (1987).

The oxidation of Mn(II) by O_2 follows a rate law of the form

$$\frac{-d[\text{Mn(II)}]}{dt} = k_0[\text{Mn(II)}] + k_1[\text{Mn(II)}][\text{MnO}_2] \tag{26}$$

This process requires complexation of Mn(II) to bind O_2 for initial oxidation.

The formation of manganese oxide solid phases (MnO_2) enhances the oxidation and thus the process is autocatalytic. The process is also enhanced by bacteria and organic chelates with carboxyl functional groups. The initial steps in the electron transfer process can be represented simply by

$$[Mn^{II}(H_2O)_6]^{2+} + 2OH^- \xrightarrow{\text{fast}} [Mn^{II}(H_2O)_4(OH)_2]^0 + 2H_2O \quad (27)$$

$$[Mn^{II}(H_2O)_4(OH)_2]^0 + O_2 \xrightarrow{\text{slow}} [Mn^{III}(H_2O)_4(OH)_2]^+ + \cdot O_2^- \quad (28)$$

Figure 10 shows the frontier orbitals for O_2, Mn(II), and OH^-. The electron transfer from Mn(II) to O_2 cannot occur by an outer-sphere process for two major reasons. First, the transfer of electrons from Mn $e_g(\sigma)$ to O_2 π^* is a poor symmetry mix because π to π transfers are favored for outer-sphere mechanisms. Second, the $t_{2g}(\pi)$ to π^*-electron transfer is energetically unfeasible because the t_{2g} orbital is not the HOMO orbital of Mn(II).

Thus, complexation of the Mn(II) by other ligands is necessary to remove the octahedral symmetry of the Mn(II) and rearrange the energies of the d orbitals to effect electron transfer (At least one d HOMO orbital must have more positive energy in the new geometry to effect electron transfer.) An energy-level representation similar to that for Mn–phthalocyanine complexes (Lever et al.,

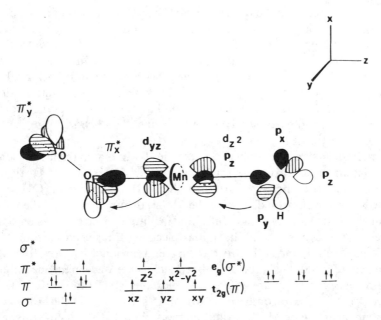

Figure 10. Molecular-orbital diagram for the oxidation of Mn(II) by O_2. An outer-sphere process is not plausible when Mn(II) is in perfect octahedral symmetry as in $Mn(H_2O)_6^{2+}$. O_2 and OH^- bound to Mn(II) will result in Mn(II) oxidation. Note the similarity of the σ and π systems to those of Figure 5.

1981) is as follows:

$$\begin{array}{ccc} & \underset{x^2\text{-}y^2}{\uparrow} & \underset{x^2\text{-}y^2}{\overline{}} \\ \underset{z^2}{\uparrow}\ \underset{x^2\text{-}y^2}{\uparrow} \end{array}$$

$$\underset{z^2}{\uparrow} \qquad\qquad \underset{z^2}{\uparrow}$$

$$\xrightarrow{} \qquad \xrightarrow{-e^-}$$

$$\underset{xy}{\uparrow} \qquad\qquad \underset{xy}{\uparrow}$$

$$\underset{xz}{\uparrow}\ \underset{yz}{\uparrow}\ \underset{xy}{\uparrow}$$

$$\underset{xz}{\uparrow}\ \underset{yz}{\uparrow} \qquad\qquad \underset{xz}{\uparrow}\ \underset{yz}{\uparrow}$$

octahedral, Mn^{2+}   **$Mn^{2+}(pc)O_2(OH)$**   **$Mn^{3+}(pc)O_2^-(OH)$**

Five- and six-coordinate geometries for Mn(II) are favored over four-coordinate (tetrahedral) in order to enhance metal basicity (and metal reducing power) and to stabilize the higher oxidation state which results on oxidation of the Mn(II). [It has been observed that Mn(II) binds to O_2 in five- (Hoffman et al., 1978) and six- (Lever et al., 1981) coordinate geometries with polydentate chelates, but not in four-coordinate geometries.] The high pH dependence for Mn(II) oxidation indicates that OH^- as a ligand is able to bind at one or more of the coordinate positions in order to effect Mn(II) electron transfer and to stabilize the higher oxidation states of Mn.

Figure 10 shows an inner-sphere process (similar to that in the metal-catalyzed H_2S oxidation shown earlier). In this example, the Mn(II) binds to O_2 through a $e_g(\sigma)$ to π_x^* bond (Lever et al., 1981). As in the Fe(II) oxidation, the OH^- ligands are able to donate electron density to the Mn(II) through the σ and the π system. This enhances the basicity of the Mn(II) and facilitates electron transfer to the O_2 via the σ or the π system. The initial electron transfer results in Mn(III) formation—Mn(III) minerals are the first end product of Mn(II) oxidation (Murray et al., 1985). This process will be much faster once manganese minerals are formed because they are able to complex the Mn(II) in solution. Organic chelates are also able to complex the Mn(II) in solution, and thus they create a similar enhancement on Mn(II) oxidation rates. Wehrli (Chapter 11, this volume) has shown through the use of linear free-energy relationships that Mn(II) oxidation should proceed through an inner-sphere process.

At very high pH, an inner-sphere process is probably available for Fe(II) oxidation also. The bonding of O_2 to Fe(II) in the precursor complex would result in faster electron transfer than that in the outer-sphere process that can occur at lower pH. The bonding between the reactants would be somewhat analogous to the bonding of Fe(II) to O_2 in hemoglobin (Huheey, 1983). This is similar to the bent Mn–O–O bond shown in Figure 10.

9. CONCLUSIONS

The series of redox reactions discussed in this chapter indicate that frontier MOT is a powerful approach and is able, without detailed quantum-mechanical computations (on which the theory is based), to predict the following:

1. The relative reactivity of oxidants and reductants through a knowledge of E_{LUMO} and E_{HOMO}.
2. The pH dependence of reactions:
    a. The sulfide (and other protic) species reactivity in solution.
    b. The importance of OH^- as a ligand to enhance reactivity of metals.
3. The importance of metals in catalysis: the role of metals as "electron pumps".
4. Whether inner- or outer-sphere mechanisms are most favored.
5. Which intermediates are most likely [for sulfide oxidation zerovalent sulfur, S(0), is the first intermediate] and what is the relative E_A of the transition state.
6. Which end products are most likely; in the case of sulfide oxidation, S_8 and sulfate predominate.

Acknowledgements

This work was supported by grants from the National Science Foundation (OCE-8696121 and OCE-8916804) and the Petroleum Research Fund administered by the American Chemical Society. The author thanks his many students (in particular, T. Ferdelman, D. Powell, and E. Tsamakis) and T. M. Church for stimulating, encouraging, and helpful discussions (and questions!).

REFERENCES

Brock, T. D., D. W. Smith, and M. T. Madigan (1984), *Biology of Microorganisms*, Prentice-Hall, Englewood Cliffs, NJ, 847 pp.

Burdige, D. J., and K. H. Nealson (1986), "Chemical and Microbiological Studies of Sulfide-Mediated Manganese Reduction," *Geomicrobio. J.* **4**, 361–387.

Chen, K. Y., and J. C. Morris (1972), "Kinetics of Oxidation of Aqueous Sulfide by O_2," *Environ. Sci. Technol.* **6**, 529–537.

Cotton, F. A., J. B. Harmon, and R. M. Hedges (1976), "Calculation of the Ground State Electronic Structures and Electronic Spectra of Di- and Trisulfide Radical Anions by the Scattered Wave–SCF–Xα Method," *J. Am. Chem. Soc.* **98**, 1417–1424.

DeKock, R. L., and M. R. Barbachyn (1979), "Proton Affinity, Ionization Energy, and the Nature of Frontier Orbital Electron Density," *J. Am. Chem. Soc.* **101**, 6516–6519.

Diem, D., and W. Stumm (1984), "Is Dissolved Mn^{2+} Being Oxidized by O_2 in Absence of Mn-Bacteria or Surface Catalysts?" *Geochim. Cosmochim. Acta* **48**, 1571–1573.

Drzaic, P. S., J. Marks, and J. I. Brauman (1984), "Electron Photodetachment from Gas Phase Molecular Anions," in *Gas Phase Ion Chemistry*, (M. T. Bowers), Ed., Vol. 3, Academic Press, New York, pp. 167–211.

Foti, A. E., V. H. Smith, and D. R. Salahub (1978), "Explanation for the Structural Differences of S_4^{2+}, S_4^0 and S_4^{2-}," *Chem. Phys. Lett.* **57**, 33–36.

Gimarc, B. M. (1979), *Molecular Structure and Bonding: The Qualitative Molecular Orbital Approach*, Academic Press, New York, 224 pp.

Greenwood, N. N., and A. Earnshaw (1984), *Chemistry of the Elements*, Pergamon Press, New York, 1542 pp.

Heide, H. van der, R. Helle, C. F. van Bruggen, and C. Haas (1980), "X-Ray Photoelectron Spectra of 3d Transition Metal Pyrites," *J. Solid State Chem.* **33**, 17–25.

Hoffman, B. M., T. Szymanski, T. G. Brown, and F. Basolo (1978), "The Dioxygen Adducts of Several Manganese(II) Porphyrins. Electron Paramagnetic Resonance Studies." *J. Am. Chem. Soc.* **100**, 7253–7259.

Hoffmann, M. R. (1977), "Kinetics and Mechanism of Oxidation of Hydrogen Sulfide by Hydrogen Peroxide in Acidic Solution," *Environ. Sci. Technol.* **11**, 61–66.

Huheey, J. E. (1983), *Inorganic Chemistry*, 3rd ed., Harper, New York, 936 pp.

Jolly, W. L. (1984), *Modern Inorganic Chemistry*, McGraw-Hill, New York, 610 pp.

Kolthoff, J. M., E. B. Sandell, E. J. Meehan, and S. Bruckenstein (1969), *Quantitative Chemical Analysis*, Macmillan, London, 1199 pp.

LaLonde, R. T., L. M. Ferrara, and M. P. Hayes (1987), "Low Temperature, Polysulfide Reactions of Conjugated Ene Carbonyls: A Reaction Model for the Geologic Origin of S-Heterocycles," *Org. Geochem.* **11**, 563–571.

Lever, A. B. P., J. P. Wilshire, and S. K. Quan (1981), "Oxidation of Manganese(II) Phthalocyanine by Molecular Oxygen," *Inorg. Chem.* **20**, 761–768.

Lowe, J. P. (1977), "Qualitative Molecular Orbital Theory of Molecular Electron Affinities," *J. Am. Chem. Soc.* **99**, 5557–5570.

Luther, G. W., III (1987), "Pyrite Oxidation and Reduction: Molecular Orbital Theory Considerations," *Geochim. Cosmochim. Acta* **51**, 3193–3199.

McKibben, M. A., and H. L. Barnes (1986), "Oxidation of Pyrite in Low Temperature Acidic Solutions: Rate Laws and Surface Textures," *Geochim. Cosmochim. Acta* **50**, 1509–1520.

Meyer, B., L. Peter, and K. Spitzer (1977), "Trends in the Charge Distribution in Sulfanes, Sulfanesulfonic Acids, Sulfanedisulfonic Acids and Sulfurous Acid," *Inorg. Chem.* **16**, 27–33.

Miller, D. J., and L. C. Cusachs (1969), "Semi-empirical Molecular Orbital Calculations on the Bonding in Sulfur Compounds I. Elemental Sulfur, S_6 and S_8," *Chem. Phys. Lett.* **3**, 501–503.

Millero, F. J., S. Hubinger, M. Fernandez, and S. Garnett (1987a), "Oxidation of H_2S in Seawater as a Function of Temperature, pH and Ionic Strength," *Environ. Sci. Technol.* **21**, 439–443.

Millero, F. J., S. Sotolongo, and M. Izaguirre (1987b), "The Oxidation Kinetics of Fe(II) in Seawater," *Geochim. Cosmochim. Acta* **51**, 793–801.

Murray, J. W., J. G. Dillard, R. Giovanoli, H. Moers, and W. Stumm (1985), "Oxidation of Mn(II): Initial Mineralogy, Oxidation State and Ageing," *Geochim. Cosmochim. Acta* **49**, 463–470.

Myers, R. J. (1986), "The New Low Value for the Second Dissociation Constant for H_2S," *J. Chem. Ed.* **63**, 687–690.

Pearson, R. G. (1976), *Symmetry Rules for Chemical Reactions*, Wiley-Interscience, New York, 548 pp.

Purcell, K. F., and J. C. Kotz (1980), *An Introduction to Inorganic Chemistry*, Saunders, Philadelphia, 637 pp.

Pyzik, A. J., and S. E. Sommer (1981), "Sedimentary Iron Monosulfides: Kinetics and Mechanism of Formation," *Geochim. Cosmochim. Acta* **45**, 687–698.

Richert, S. A., P. K. S. Tsang, and D. T. Sawyer (1988), "Ligand-Centered Oxidation of Manganese(II) Complexes," *Inorg. Chem.* **27**, 1814–1818.

Singer, P. C., and W. Stumm (1970), "Acid Mine Drainage—The Rate Limiting Step," *Science* **167**, 1121–1123.

Stahl, J. W., and J. Jordan (1987), "Thermometric Titration of Polysulfides," *Anal. Chem.* **59**, 1222–1225.

Stumm, W., and J. J. Morgan (1981), *Aquatic Chemistry*, Wiley, New York, 780 pp.

Taylor, B. E., M. C. Wheeler, and D. K. Nordstrom (1984a), "Oxygen and Sulfur Compositions of Sulfate in Acid Mine Drainage: Evidence for Oxidation Mechanisms," *Nature* **308**, 538–541.

Taylor, B. E., M. C. Wheeler, and D. K. Nordstrom (1984b), "Stable Isotope Geochemistry of Acid Mine Drainage: Experimental Oxidation of Pyrite," *Geochim. Cosmochim. Acta* **48**, 2669–2678.

Tebo, B. M., M. L. Fogel, A. Stone, and K. Mandernack (1987), "Oxygen Isotope Tracers of Manganese Oxide Precipitation," *EOS* **68**, 1702.

Tossell, J. A. (1983), "A Qualitative Molecular Orbital Study of the Stability of Polyanions in Mineral Structures," *Phys. Chem. Minerals* **9**, 115–123.

Vairavamurthy, A. and K. Mopper (1989), "Mechanistic Studies on Organosulfur (thiol) Formation in Coastal Marine Sediments," in E. S. Saltzman and W. J. Cooper, Eds. *Biogenic Sulfur in the Environment*, (ACS symposium Series No. 393), American Chemical Society, Washington DC, pp. 231–242.

APPENDIX I

Listed below are four possible linear combinations of orbitals pertinent to the understanding of the electron processes described in this chapter. Orbital overlap sketches A1, B1, and D1 can combine to form bonds or yield electron-transfer reactions, whereas C1 cannot. Overlap of two orbitals occurs when there is a matching of orbital lobes of the same sign (shaded or unshaded in the

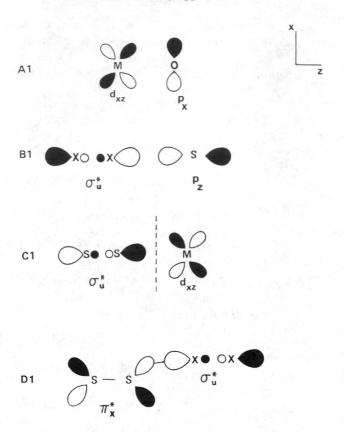

diagrams) as determined by the signs of the original wavefunctions for the orbitals involved. Sigma (σ) orbitals are located directly on the internuclear axis, whereas pi (π) orbitals are above and below the internuclear axis.

Sketch A1 demonstrates the arrangement of a d_{xz} (t_{2g}, π) metal orbital with a p_x orbital of an oxygen atom that would result in electron transfer in a $\pi \rightarrow \pi$ system. This arrangement is that for π donor ligands to transition metals (Huheey, 1983) and results in a smaller 10 Dq value for the metal ion (lower-energy UV–visible transition). The fewer the number of hydrogen atoms attached to the oxygen, the more the electron-donating ability to the metal.

Sketch B1 demonstrates the arrangement of a σ_u^* molecular orbital from X_2 and a p_z orbital. Sulfide transfers an electron pair to a molecular halogen in a $\sigma \rightarrow \sigma$ bonding system.

Sketch C1 shows the arrangement of a σ_u^* from S_2^{2-} and a d_{xz} (π) orbital from a metal in a $\sigma^* \rightarrow \pi$ system. Overlap cannot occur in this system (C1) because of symmetry and geometry considerations.

Sketch D1 shows possible orbital overlap when a π_x^* orbital on S_2^{2-} donates electrons to the σ_u^* orbital of a molecular halogen. This combination is an angular arrangement of atoms and is important in both electron-transfer processes and the formation of chemical bonds.

7

CHEMICAL TRANSFORMATIONS OF ORGANIC POLLUTANTS IN THE AQUATIC ENVIRONMENT

René P. Schwarzenbach

Swiss Federal Institute of Technology (ETHZ) and Swiss Federal Institute for Water Resources and Water Pollution Control (EAWAG), Dübendorf, Switzerland

and

Philip M. Gschwend

Ralph M. Parsons Laboratory for Water Resources and Hydrodynamics, Massachusetts Institute of Technology, Cambridge, Massachusetts

1. INTRODUCTION

In the aquatic environment, organic pollutants may be transformed by chemical, photochemical, and biological processes. "Chemical transformations" are defined as reactions that are neither induced by light nor mediated by (micro)organisms. For convenience, we may classify such "chemical" or "thermochemical" reactions according to whether there is a net electron transfer between the compound of interest and a natural reactant (i.e., redox reactions) or not. In the latter case, of most importance are reactions involving inorganic nucleophiles, above all, the solvent water (i.e., hydrolysis reactions). Among the chemical redox reactions, reductions of xenobiotic organic compounds are more interesting since only a few classes of organic pollutants seem to undergo chemical oxidation at significant rates. Furthermore, chemical reductions may lead to transformation products that may be of even greater concern than the parent compounds. A few examples illustrating the type of reactions that we will discuss in this chapter are given in Table 1.

TABLE 1. Examples of Some Important Chemical Reactions[a] of Organic Pollutants in the Environment

Reactions with Nucleophiles (Hydrolysis)

(1) $R-X (X = Cl, Br, I) + H_2O$ $\longrightarrow$ $R-OH + X^- + H^+$

(2) $RCH_2-X (X = Cl, Br, I) + HS^-$ $\longrightarrow$ $RCH_2-SH + X^-$

(3) $R_1COOR_2 + H_2O/OH^-$ $\longrightarrow$ $R_1COO(H) + HOR_2$

(4) $R_1R_2NCOOR_3 + H_2O/OH^-$ $\longrightarrow$ $R_1R_2NH + CO_2 + HOR_3$

(5) $(R_1O)_2\overset{\overset{O}{\|}}{P}-OR_2 \ + \ H_2O/OH^-$ $\longrightarrow$ $(R_1O)_2\overset{\overset{O}{\|}}{P}-O^- \ + \ HOR_2$ and

$(R_1O)(R_2O)\overset{\overset{O}{\|}}{P}-O^- \ + \ HOR_1$

Redox Reactions

(6)

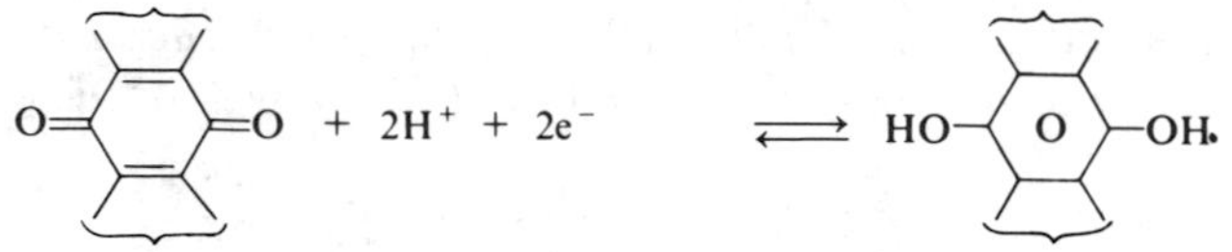

(7) $-\overset{|}{\underset{|}{C}}-X (X = Cl, Br, I) + H^+ + 2e^-$ $\longrightarrow$ $-\overset{|}{\underset{|}{C}}-H + X^-$

(8) $-\overset{\overset{\textstyle X}{|}}{\underset{\underset{\textstyle X}{|}}{C}}-\overset{|}{\underset{|}{C}}-(X = Cl, Br, I) + 2e^-$ $\longrightarrow$ $\underset{}{>}C=C\underset{}{<} + 2X^-$

(9) $-NO_2 + 6H^+ + 6e^-$ $\longrightarrow$ $-NH_2 + 2H_2O$

(10) 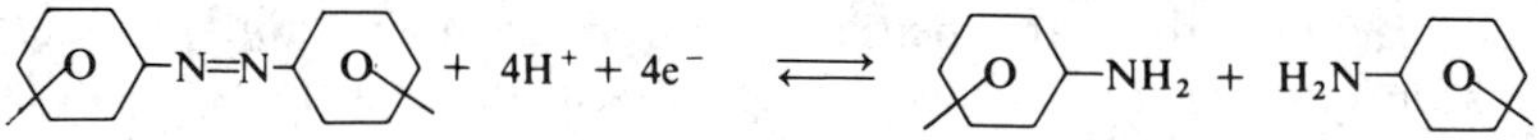 $-N{=}N-$ $+ 4H^+ + 4e^-$ $\rightleftarrows$ $-NH_2 + H_2N-$

(11) $R_1-S-S-R_2 + 2H^+ + 2e^-$ $\rightleftarrows$ $R_1-SH + HS-R_2$

(12) $R_1-\overset{\overset{O}{\|}}{S}-R_2 + 2H^+ + 2e^-$ $\longrightarrow$ $R_1-S-R_2 + H_2O$

[a] Note that some reactions are reversible (indicated by $\rightleftarrows$) while others are irreversible under environmental conditions.

In order to quantify transformation reactions of organic pollutants in the environment, several questions need to be addressed:

1. To what extent will a specific reaction occur in a given natural water? Can the reaction be regarded as more or less irreversible under any conditions, or may it be reversible under certain conditions?
2. What is the rate at which a specific transformation reaction occurs, and how is this rate influenced by various environmental variables, including temperature, pH, redox condition, ionic strength, presence of certain solutes, or concentration and type of solid surfaces?
3. Is there only one or are there several different reaction pathways by which a given compound may be transformed under given environmental conditions, and what are the reaction products formed?

The most important prerequisite for answering all these questions is knowledge of the reaction mechanism(s) by which a given compound is transformed. We recall that a reaction mechanism is defined as a set of elementary molecular changes describing the sequence in which chemical bonds are broken and new bonds are formed to convert the starting compounds to the observed product(s). Hence, when just writing down the stoichiometry of an overall reaction, as we have done in Table 1, we can answer questions only about the energetics of the overall reaction (i.e., by calculating the free-energy change of the reaction under the specific conditions), but we do not say anything about the number of elementary reaction steps, particularly, about the step(s) that is (are) rate-determining. The latter information is especially critical for derivation and application of linear free-energy relationships (LFERs; see Chapter 4, this volume) for prediction of transformation rates of organic pollutants.

The major goal of this chapter is to familiarize the reader with the most important chemical transformation reactions of organic pollutants in the aquatic environment. Using a few illustrative examples, the pertinent compound properties and environmental factors determining the kinetics of such reactions are discussed. For a more detailed treatment of this topic, the reader is referred to other texts, such as those by Mill and Mabey (1988) for hydrolysis reactions, and Schwarzenbach et al. (in press a) for a general overview of chemical transformation reactions.

2. REACTIONS INVOLVING INORGANIC NUCLEOPHILES. HYDROLYSIS

2.1. General Remarks

In an organic molecule, the atoms of a covalent polar bond (i.e., a bond between two atoms of different electronegativity) may become the site of a chemical reaction in that either a nucleophilic species is attracted by the electron-deficient

TABLE 2. Examples of Important Environmental Nucleophiles, Their Nucleophilicities Relative to Water (n Values), and Estimated Concentrations of Nucleophiles ($[nucl]_{50\%}$) Necessary to Compete with Water in an S_N2 Reaction with Primary Alkyl Bromides[a]

Nucleophile	n	$[nucl]_{50\%}$ (M)	Nucleophile	n	$[nucl]_{50\%}$ (M)
H_2O	0.0	—	HPO_4^{2-}	3.8	$\sim 9 \times 10^{-3}$
NO_3^-	1.0	~ 6	Br^-	3.9	$\sim 7 \times 10^{-3}$
SO_4^{2-}	2.5	$\sim 2 \times 10^{-1}$	OH^-	4.2	$\sim 4 \times 10^{-3}$
Cl^-	3.0	$\sim 6 \times 10^{-2}$	HS^-	5.1	$\sim 4 \times 10^{-4}$
HCO_3^-	3.8	$\sim 9 \times 10^{-3}$	CN^-	5.1	$\sim 4 \times 10^{-4}$

[a] See Eqs. 5 and 7.

atom of the bond, or an electrophilic species is attracted by the electron-rich atom. In the aquatic environment, the majority of the chemical species that may react thermochemically with organic pollutants are inorganic nucleophiles (see examples given in Table 2). Because of the large abundance of such nucleophiles in natural waters, reactive electrophiles are very short-lived, and therefore reactions of organic pollutants with such species occur usually only in light-induced (see Chapter 2, this volume) or biologically mediated processes.

As can be seen from Table 2, nucleophilic species possess a partial or full negative charge and/or have nonbonded valence electrons. As a consequence of an encounter with an organic molecule exhibiting polar bonds, the electron-rich atom of the nucleophile may form a bond with an electron-deficient atom in the organic molecule. Since a new bond is formed by this process, another bond has to be broken, which usually (but not always) means that a group (or atom) is split off from the organic compound. Such a group (or atom) is commonly referred to as a "leaving group."

Because of its great abundance, water plays a pivotal role among the nucleophiles present in the aquatic environment. As is illustrated by Table 2 and discussed in Section 2.2, in uncontaminated freshwaters, the reaction with water, that is, neutral hydrolysis, is the dominant process with respect to trans-formation of alkyl halides. Neutral hydrolysis and/or acid- or base-catalyzed hydrolysis are also the most important reactions occurring at a variety of other functional groups present in organic chemicals (see Fig. 1), including carboxylic acid esters and amides, carbamates, and phosphoric and thiophosphoric acid esters and thioesters. Note that when undergoing hydrolysis, a compound is transformed into more polar products that have quite different properties, and, therefore, a different environmental behavior than the starting chemical. It should also be noted that hydrolysis products are often of less environmental concern as compared to the parent compound, which is not necessarily true for reactions with other nucleophiles (e.g., HS^-, CN^-). Finally, for typical ambient conditions and concentrations, hydrolysis reactions usually exhibit large nega-tive ΔG values. Consequently, as a practical matter, for the following discussion, we shall consider hydrolysis and other reactions involving nucleophiles to proceed in only one direction; that is, we consider them to be irreversible.

2.2 Nucleophilic Substitution Reactions

The most common case of nucleophilic substitution reactions involving organic pollutants is nucleophilic substitution at a saturated carbon atom. To describe such a process, it is useful to consider two different reaction mechanisms representing two extreme cases. In the first case, the reaction occurs because a nucleophile Nu^{v-} attacks the carbon atom from the side opposite to the leaving group $-X$, thus forming an activated complex in which both the nucleophile as well as the leaving group are partly bound to the carbon atom:

$$Nu^{v-} \; + \; \overset{|}{\underset{|}{C}}{}^{\delta+}\!-\!X^{\delta-} \quad \xrightarrow{\text{slow}} \quad [Nu^{v-}\cdots\overset{|}{C}{}^{\delta+}\cdots X^{\delta-}]^{\neq}$$

activated complex

$$\longrightarrow \quad Nu^{(v-1)-}\!-\!\overset{|}{\underset{|}{C}} \; + \; X^- \qquad (1)$$

If a reaction occurs by this mechanism, it is commonly referred to as an "S_N2" (substitution, nucleophilic, bimolecular) reaction. It represents an example of a simple elementary bimolecular reaction and, therefore, obeys a second-order kinetic rate law:

$$\text{Rate} = -k_{Nu}\,[Nu^{v-}]\,[\overset{|}{\underset{|}{-C}}\!-\!X] \qquad (2)$$

where k_{Nu} is a second-order rate constant ($M^{-1}\,s^{-1}$).

The second mechanism that differs substantially from the first one, is one in which one postulates the substitution reaction to occur in two steps: (1) complete dissociation of the leaving group and (2) reaction of the carbocation with a nucleophile:

$$\overset{|}{\underset{|}{C}}{}^{\delta+}\!-\!X^{\delta-} \;\xrightarrow{\text{slow}}\; \left[\overset{|}{C}{}^{+} \quad X^{-}\right] \;\xrightarrow{+\,Nu^{v-}}\; \overset{|}{\underset{|}{-C}}\!-\!Nu^{(v-1)} \qquad (3)$$

intermediate

If a reaction occurs exclusively by this second mechanism, the reaction rate law will be first-order:

$$\text{Rate} = -k\,[\overset{|}{\underset{|}{-C}}\!-\!X] \qquad (4)$$

where k is a first-order rate constant (s^{-1}), and the reaction is called an 'S_N1" (substitution, nucleophilic, monomolecular) reaction. Note that in aqueous solution, the S_N1 mechanism will, in general, strongly favor the formation of the hydrolysis product (i.e., substitution of $-X$ by $-OH$) because the nucleophile is not involved in the rate-limiting step.

Let us first look at some reactions that occur predominantly by an S_N2 mechanism. We consider the reactions of primary alkyl halides (i.e., $R-CH_2-X$, where $X = F$, Cl, Br, I) with a series of inorganic nucleophiles. For a given primary alkyl halide (e.g., CH_3-Br, $R-CH_2-Cl$), the relative reactivity toward various nucleophiles may be related by a LFER, the Swain–Scott relationship (e.g., Hine, 1962):

$$\log \frac{k_{Nu}}{k_{H_2O}} = s\,n \tag{5}$$

where k_{H_2O} is the second-order rate constant for the reaction with water (hence, $Nu^{v-} = H_2O$ in Eq. 2), the reference nucleophile; n is a measure of the attacking aptitude or nucleophilicity of the nucleophile of interest (note that $n_{H_2O} = O$), and s reflects the sensitivity of the organic molecule to nucleophilic attack. As a standard for the sensitivity values, s is set equal to 1.0 for S_N2 reactions of methyl bromide:

$$CH_3-Br + Nu^{v-} \longrightarrow CH_3-Nu^{(v-1)-} + Br^-; \quad (s = 1) \tag{6}$$

Table 2 gives the n values of some important environmental nucleophiles together with the calculated concentrations, $[nucl]_{50\%}$, at which the reaction of the nucleophile (with CH_3-Br and other alkyl halides that exhibit s values close to 1.0) is equally important as the reaction with H_2O, that is, $k_{Nu}[nucl]_{50\%} = k_{H_2O}[H_2O]$:

$$[nucl]_{50\%} = 55.5 \times 10^{-n}\ M \tag{7}$$

Hence the larger the n value, the stronger the nucleophile, and the smaller the $[nucl]_{50\%}$. As already pointed out earlier, the $[nucl]_{50\%}$ values given in Table 2 show that in uncontaminated freshwaters, hydrolysis is by far the most important nucleophilic substitution reaction. Furthermore, since the hydrolysis of a carbon–halogen bond is generally not catalyzed by acids, one can assume that the hydrolysis rate of aliphatic halides will be independent of pH at typical ambient conditions (i.e., pH $\leqslant 10$). In this context it is also important to note that no catalysis of the hydrolysis of alkyl halides by solid surfaces has been observed (El-Amamy and Mill, 1984; Haag and Mill, 1988). In salty or contaminated waters, reactions of organic chemicals with nucleophiles other than water or hydroxide ion may be important. Zafiriou (1975), for example, has demonstrated that in seawater ($[Cl^-] \cong 0.5\ M$), a major sink for naturally produced methyl iodide is transformation to methyl chloride:

$$CH_3-I + Cl^- \longrightarrow CH_3-Cl + I^- \tag{8}$$

The half-life with respect to chemical transformation of CH_3-I in seawater at 20°C was determined to be 20 days, as compared to about 200 days in freshwater (reaction with H_2O). In a case of a groundwater contamination with several alkyl bromides and alkyl chlorides, Schwarzenbach et al. (1985) reported the formation of dialkyl sulfides under sulfate reducing conditions in an aquifer. They postulated that in an initial reaction, primary alkyl bromides reacted with the strong nucleophile HS^- by an S_N2 mechanism to yield the corresponding mercaptans:

$$R-CH_2-Br + HS^- \longrightarrow R-CH_2-SH + Br^- \qquad (9)$$

So far, we have focused our discussion on the relative nucleophilicities of natural nucleophiles. Let us now consider how certain structural features of the organic molecule (i.e., type of leaving group, type of carbon skeleton) determine the kinetics of nucleophilic substitution. In Table 3, the (neutral) hydrolysis half-lives are given for various monohalogenated compounds at 25°C. Also indicated are the postulated reaction mechanisms with which these compounds undergo S_N reactions. As can be seen from Table 3, the carbon–bromide and carbon–iodine bonds hydrolyze fastest, about one to two orders of magnitude

TABLE 3. Postulated Reaction Mechanisms and Hydrolysis Half-Lives at 25°C of Some Monohalogenated Hydrocarbons at Neutral pH

Compound	$t_{1/2}$ (Hydrolysis)				Dominant Mechanism in Nucleophilic Substitution Reactions
	X = F	Cl	Br	I	
$R-CH_2-X$	≈ 30 yrs[a]	340 days[a]	20–40 days[b]	50–110 days[c]	S_N2
$\begin{array}{c}CH_3 \\ >CH-X \\ CH_3\end{array}$		38 days	2 days	3 days	$S_N2 \cdots S_N1$
$\begin{array}{c}CH_3 \\ | \\ CH_3-C-X \\ | \\ CH_3\end{array}$	50 days	23 s			S_N1
$CH_2=CH-CH_2-X$		69 days	0.5 days	2 days	$(S_N2) \cdots S_N1$
$⟨O⟩-CH_2-X$		15 h	0.4 h		S_N1

Data taken from Mabey and Mill, 1978.
[a] $R = H$.
[b] $R = H$, C_1 to C_5-n-alkyl.
[c] $R = H$, CH_3.

faster than the carbon–chlorine bond. Furthermore, in most cases the hydrolysis of carbon–fluorine bonds is likely to be too slow to be of great environmental significance. The major reason for these findings is the increased strength of the carbon–halogen bond (that has to be broken) in the sequence C–F > C–Cl > C–Br > C–I.

When comparing the hydrolysis half-lives of the various types of monohalogenated compounds in Table 3, we notice that the reaction rates increase dramatically when going from primary ($-CH_2-X$) to secondary ($>CH-X$) to tertiary ($-\overset{|}{\underset{|}{C}}-X$) carbon–halogen bonds. In this series, increased stabilization of the carbocation by hyperconjugation decreases the activation energy needed to form this intermediate, thereby shifting the reaction to an increasingly S_N1-like mechanism. Similarly, faster hydrolysis rates and increasing S_N1 character can be expected if stabilization is possible by resonance with a double bond or an aromatic ring.

If a carbon atom is bound to more than one halogen, as, for example, is the case in polyhalogenated methanes (e.g., CH_2Cl_2, $CHCl_3$, $CHBr_3$), nucleophilic substitution at this carbon atom is very slow, primarily because of steric hindrance by the relatively bulky halogens (Hughes, 1971). For example, the hydrolysis half-lives of polyhalogenated methanes are estimated to be hundreds to thousands of years under ambient conditions (Mabey and Mill, 1978). However, as we will see in Section 3, the polyhalogenated methanes as well as other polyhalogenated compounds may, under certain conditions, react by another reaction pathway, namely, reductive dehalogenation.

In contrast to polyhalogenated methanes, polyhalogenated ethanes react in water at significantly faster rates. For example, the half-lives at 25°C of 1,1,1-trichloroethane (CCl_3-CH_3) and 1,1,2,2-tetrachloroethane ($CHCl_2-CHCl_2$) are 400 days and 40 days, respectively. One of the reasons for these findings is that such halogenated compounds may also undergo another type of reaction in aqueous solution, a so-called β-elimination:

$$\underset{H}{\overset{X}{\underset{|}{\overset{|}{C}}-\underset{|}{\overset{|}{C}}}} \xrightarrow[\beta\text{-elimination}]{-HX} \quad >C=C< \qquad (10)$$

This reaction occurs (in competition with hydrolysis) preferentially in compounds in which a relatively acidic proton is located at a carbon atom adjacent to the carbon atom carrying the leaving group (i.e., the halogen). These criteria are optimally met in 1,1,2,2-tetrachloroethane, where the four electron-withdrawing chlorine atoms render the hydrogens more acidic and, simultaneously, hinder nucleophilic attack. In aqueous solution, 1,1,2,2-tetrachloroethane is converted more or less quantitatively to trichloroethylene (Haag and Mill, 1987) by a so-called E2 (elimination, bimolecular) mechanism; that is, the elimination takes place in a concerted reaction with OH:

$$
\begin{array}{c}
\underset{\substack{|\\H}}{\overset{\substack{Cl\ \ Cl\\|}}{Cl\text{''''''}C}}\text{—}\underset{\substack{|\\Cl}}{\overset{\substack{\\|}}{C}}\text{''''}H \ + \ OH^- \ \longrightarrow \ \left[\ \underset{\substack{|\\H}}{\overset{\substack{Cl\ \ Cl\\|}}{Cl\text{''''}C}}\text{—}\underset{\substack{:\\Cl}}{\overset{\substack{\\:}}{C}}\text{''''}H\ \right] \\
HO^-
\end{array}
\tag{11}
$$

$$
\longrightarrow \ Cl_2C\text{=}CHCl \ + \ H_2O \ + \ Cl^-
$$

Hence, reaction 11 follows a second-order rate law, that is, it is dependent on both 1,1,2,2-tetrachlorethane and hydroxide ion concentration.

Let us now turn to an example of nucleophilic substitution involving a group of pollutants other than alkyl halides. We consider the hydrolysis of thiometon and disulfoton, two insecticides that were among the major contaminants that entered the Rhine River after the famous accident at Schweizerhalle in Switzerland in 1986 (Capel et al., 1988). This example is representative for the hydrolysis of a variety of phosphoric and thiophosphoric acid derivatives (e.g., esters, thioesters, see Fig. 1), and it illustrates that hydrolysis of a more complex molecule may be somewhat more complicated. The kinetic data, as well as the proposed mechanisms of hydrolysis of thiometon and disulfoton, are presented in Table 4 and Figure 2, respectively. In these cases, the base catalyzed reaction

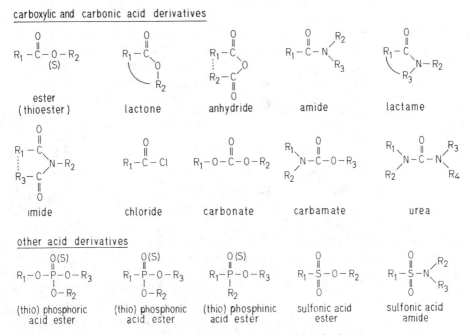

Figure 1. Examples of hydrolyzable acid derivatives.

TABLE 4. Kinetic Data for the Abiotic Hydrolysis of Disulfoton and Thiometon

	Neutral Hydrolysis[b]		Base-Catalyzed Hydrolysis[c]		
Compound Name[a]	k_N^d (25°C) (s^{-1})	E_a^e $(kJ\ mol^{-1})$	k_B (25°C) $(M^{-1}\ s^{-1})$	E_a^f $(kJ\ mol^{-1})$	I_{NB}^g (25°C)
Disulfoton	$(2.2 \pm 0.1) \times 10^{-7}$	76 ± 2	$(3.0 \pm 0.1) \times 10^{-3}$	≈ 61	9.9
Thiometon	$(2.0 \pm 0.1) \times 10^{-7}$	89 ± 2	$(9.9 \pm 0.3) \times 10^{-3}$	≈ 61	9.3

Data from Wanner et al., 1989
[a] For the structural formula, see Figure 2.
[b] In 10 mM phosphate buffer of pH 6.
[c] In 10^{-2} M NaOH.
[d] Extrapolated from measurements at higher temperatures (30–50°C).
[e] Determined for the temperature range between 30 and 50°C.
[f] Only two temperature points (20 and 30°C).
[g] pH value at which neutral and base-catalyzed reactions are equally important, i.e. $k_N = k_B$ [OH$^-$], see also Eq. 13.

occurs predominantly as S_N2 reaction at the phosphorus atom (P–S cleavage, mechanism II) yielding different products than the neutral reaction that takes place at both the ester (C–O cleavage, mechanism I) and the thioester (C–S cleavage, mechanism III) moiety. In the latter case, the reaction may proceed by either an S_N2 (mechanism IIIa) or a so-called S_Ni (intramolecular nucleophilic substitution, mechanism IIIb) mechanism. Consequently, the rate constants and activation energies given in Table 4 for neutral hydrolysis (which is the dominant process at ambient pH) are only apparent values, each representing a composite value of the contribution of the various different reaction pathways. Note that the relative importance of the different pathways, and thus the product distribution, is dependent on temperature.

2.3   Hydrolysis of Carboxylic and Carbonic Acid Derivatives

As is indicated in Figure 1, there are a variety of carboxylic and carbonic acid derivatives that may undergo hydrolysis. To illustrate some important features of the hydrolysis of such functional groups, we consider the hydrolytic transformation of some carboxylic acid esters and of some carbamates.

Figure 3 shows the hydrolysis half-lives

$$t_{1/2} = \ln 2/k_h \qquad (12)$$

at 25°C in homogeneous aqueous solution for some simple carboxylic acid esters as a function of pH, where k_h is the pseudo-first-order hydrolysis rate constant at

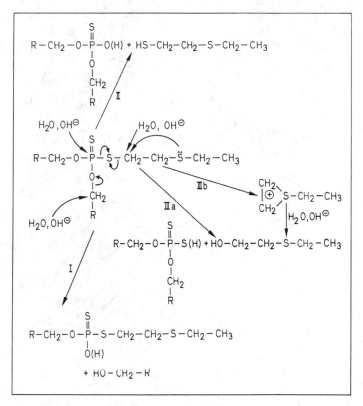

Figure 2. Possible hydrolysis mechanisms for disulfoton ($R = CH_3$) and thiometon ($R = H$). I: Nucleophilic displacement (S_N2) at the ethyl or methyl group (C–O cleavage). II: Nucleophilic displacement (S_N2) at the phosphorus atom (P–S cleavage). III: Nucleophilic displacement (S_N2, IIIa; S_Ni, IIIb) at the 2-(ethylthio)–ethyl group (C–S cleavage). For details, see Wanner et al. (1989).

a given pH. As can be seen, in general, the hydrolysis rate of carboxylic acid esters (as well as of other carboxylic and carbonic acid derivatives) is pH-dependent over the ambient pH range. Recognizing that the curve sections which decrease with a slope of -1 as a function of pH reflect reactions mediated by OH⁻, we notice that for all compounds, reaction with OH⁻ ("base catalysis") is important even at pH values below pH 7, and that acid catalysis (curve portions with slope of $+1$) is relevant only at relatively low pHs and only for compounds showing rather slow hydrolysis kinetics. By taking into account the acid-catalyzed (k_A, $M^{-1} s^{-1}$), neutral ($k_N = k_{H_2O}[H_2O]$, s^{-1}), and base-catalyzed (k_B, $M^{-1} s^{-1}$) reactions, we can express the observed (pseudo-first-order) hydrolysis rate constant, k_h (s^{-1}), as

$$k_h = k_A[H^+] + k_N + k_B[OH^-] \tag{13}$$

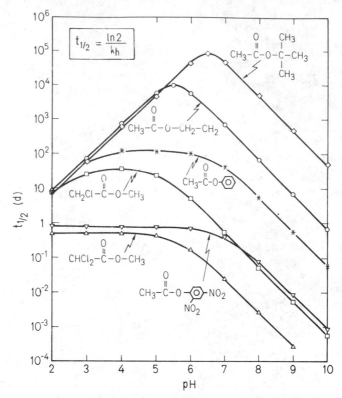

Figure 3. Variation of hydrolysis half-life for several carboxylic acid esters as a function of solution pH. [Source of data: Mabey and Mill (1978).]

In some special cases, hydrolysis of carboxylic and carbonic acid derivatives in homogeneous solution may also be catalyzed by metal ions (see Chapter 3, this volume). The effects of dissolved organic matter on hydrolysis rates seems, however, to be of secondary importance (Macalady et al., 1989).

Regardless of whether the hydrolysis of a carboxylic acid ester occurs via the acid-catalyzed, neutral, or base-catalyzed mechanism, the products are the same; that is, the compound is transformed to the corresponding acid and alcohol moiety:

$$R_1-\overset{O}{\overset{\|}{C}}-O-R_2 \xrightarrow{\quad H_3O^+|H_2O|OH^- \quad} R_1-\overset{O}{\overset{\|}{C}}-O(H)+HO-R_2 \qquad (14)$$

In all cases, addition of H_2O or OH^- to form a tetrahedral intermediate (as is illustrated by reaction (1) in Figure 4 for the base-catalyzed reaction) is a rate-determining step.

As is indicated by Figure 3, at very low pH values where acid catalysis is important, there are no large differences in the hydrolysis half-lives of esters

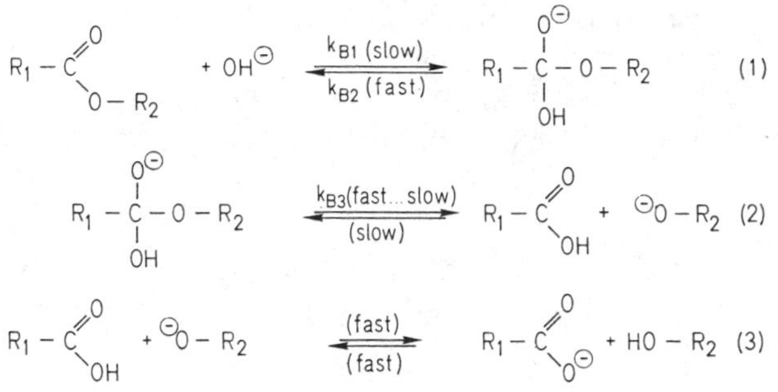

$$R_1 - C\overset{\displaystyle O}{\underset{\displaystyle O - R_2}{\Big\langle}} \; + \; OH^{\ominus} \quad \underset{k_{B2} \text{ (fast)}}{\overset{k_{B1} \text{ (slow)}}{\rightleftharpoons}} \quad R_1 - \overset{\displaystyle O^{\ominus}}{\underset{\displaystyle OH}{\overset{\displaystyle |}{\underset{\displaystyle |}{C}}}} - O - R_2 \qquad (1)$$

$$R_1 - \overset{\displaystyle O^{\ominus}}{\underset{\displaystyle OH}{\overset{\displaystyle |}{\underset{\displaystyle |}{C}}}} - O - R_2 \quad \underset{\text{(slow)}}{\overset{k_{B3}\text{(fast...slow)}}{\rightleftharpoons}} \quad R_1 - C\overset{\displaystyle O}{\underset{\displaystyle OH}{\Big\langle}} \; + \; {}^{\ominus}O - R_2 \qquad (2)$$

$$R_1 - C\overset{\displaystyle O}{\underset{\displaystyle OH}{\Big\langle}} \; + \; {}^{\ominus}O - R_2 \quad \underset{\text{(fast)}}{\overset{\text{(fast)}}{\rightleftharpoons}} \quad R_1 - C\overset{\displaystyle O}{\underset{\displaystyle O^{\ominus}}{\Big\langle}} \; + \; HO - R_2 \qquad (3)$$

Figure 4. Reaction scheme for the base-catalyzed hydrolysis of carboxylic acid esters.

exhibiting very different structures. The reason for these findings is that the effect of substituents on the free energy of activation for formation of the tetrahedral intermediate is more or less compensated by the effect of the substituent on the affinity of the ester group to become protonated. For example, an electron-withdrawing substituent such as a chlorine atom bound to the carbon atom adjacent to the ester moiety will favor the formation of the tetrahedral intermediate, but will at the same time render the ester group less basic, thus decreasing the fraction of molecules present as the reactive protonated species. We also note that in the acid-catalyzed hydrolysis the dissociation of the leaving group (i.e., the alcohol moiety) is not determining the rate of hydrolysis, since this step is fast because the alcohol is leaving as a neutral species and not as an anion.

In contrast to acid-catalyzed hydrolysis, the rate of neutral and base-catalyzed hydrolysis is strongly dependent on both the structure of the acid and the alcohol moiety (see Fig. 3). As is illustrated in Figure 4, for the base-catalyzed reaction, the dissociation of the alcohol moiety [reaction (2)] may or may not be rate-determining, depending on how good a leaving group the alcohol moiety (actually the alcoholate species, i.e., RO^-) is. As a rule of thumb, we can relate the ease with which the RO^- group dissociates with the ease with which the corresponding alcohol dissociates in aqueous solution, which is expressed by its pK_a value. Consequently, a phenolic group, particularly if it is substituted with electron-withdrawing substituents (thus exhibiting a low pK_a value), will generally be a much better leaving group than will an aliphatic alcohol.

While there are abundant rate data available on ester hydrolysis in homogeneous aqueous solution (e.g., Mabey and Mill, 1978), quantitative data on the effect of surfaces on reaction rates are rather scarce. Hoffmann (Chapter 3, this volume) and Stone (1989) have investigated the catalytic effect of oxide surfaces on the hydrolysis of a few carboxylic acid esters, and have found a rate enhancement for compounds for which base catalysis is important at neutral pH. The observed acceleration of hydrolysis may be attributed to either reaction with surface hydroxyl groups (Hoffmann, Chapter 3, this volume), or to reaction with hydroxide ions postulated to be present at higher concentrations in the diffuse layer near the surface as compared to the bulk solution (Stone, 1989).

In the case of hydrolysis of carbamates, the second group of compounds that we want to look at more closely, the base-catalyzed reaction plays the most important role. Regardless of the reaction mechanism, the hydrolysis of carbamates yields the alcohol (R_3–OH), the amine (R_1R_2NH), and CO_2:

$$\begin{array}{c} R_1 \\ {\large >} N-\overset{\overset{\displaystyle O}{\|}}{C}-O-R_3 \\ R_2 \end{array} \xrightarrow{\ H_2O|OH^-\ } \begin{array}{c} R_1 \\ {\large >} NH+CO_2+HO-R_3 \\ R_2 \end{array} \tag{15}$$

where R_1 and R_2 are hydrogen or carbon-centered substituents, and R_3 is a carbon-centered substituent. Since, in most cases, the alcohol moiety will be the better leaving group, the initial hydrolysis reaction occurs commonly by cleavage of the ester-like bond.

In the case of base-catalyzed hydrolysis of carbamates, the critical question is whether one of the groups bound to the nitrogen (R_1,R_2) is a hydrogen atom. This becomes obvious when comparing the k_B values at 25°C of the two carbamates I and II (Williams, 1972):

I: $k_B = 8 \times 10^{-4}\ M^{-1}\ s^{-1}$ **II**: $k_B = 2.7 \times 10^{+5}\ M^{-1}\ s^{-1}$

First, we realize that although p-nitrophenolate is a good leaving group, the base-catalyzed hydrolysis of 4-nitrophenyl N-methyl-N-phenylcarbamate (**I**) is extremely slow. In this case, by analogy to what we have postulated above for carboxylic acid esters, the rate-determining step is the formation of a tetrahedral intermediate, which is not very favorable in the case of the carbamates. Note that the hydrolysis of the ester bond is generally followed by a fast decarboxylation reaction yielding the corresponding amine.

Considering the large difference between the k_B values of compounds **I** and **II**, the base catalyzed transformation of 4-nitrophenyl-N-phenylcarbamate (**II**) must proceed by a different reaction mechanism. In this case, the first step of the reaction is an elimination that is similar to the β-elimination discussed earlier (see Eq. 10):

$$\tag{16}$$

The resulting isocyanate is then converted in fast reactions to the amine and CO_2 by addition of water and subsequent decarboxylation. In reaction 16, the dissociation of the leaving group is rate-determining. Thus, it is not surprising that an excellent correlation (LFER) is found between the log k_B values of a series of N-phenylcarbamates and the pK_a values of the corresponding alcohol moieties (see Fig. 5).

We conclude this section on chemical transformations of organic pollutants involving inorganic nucleophiles by a few remarks on the temperature dependence of such reactions, particularly hydrolysis. As can be easily deduced from the Arrhenius equation (see Chapter 1, this volume), for a given chemical reaction, the ratio of the rate constants (and thus of the reaction rates) at two different temperatures T_1 [in kelvins (K)] and T_2 (K) is given by

$$\frac{k(T_1)}{k(T_2)} = \exp\left[\frac{E_a}{R}\left(\frac{1}{T_2} - \frac{1}{T_1}\right)\right] \tag{17}$$

where E_a is the activation energy in kilojoules per mole and R is the gas constant

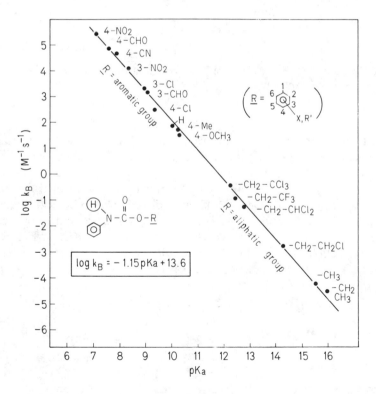

Figure 5. Base-catalyzed hydrolysis rate (expressed by log k_B) as a function of the pK_a of the alcohol moiety (i.e., the leaving group) for a series of N-phenylcarbamates. Data collected by Wolfe et al. (1978).

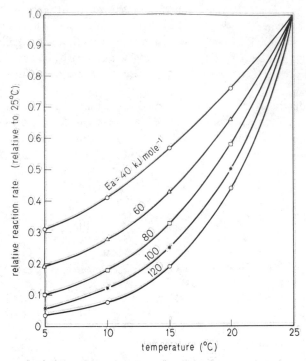

Figure 6. Change of relative reaction rate as a function of temperature for various activation energies. Rates are normalized to 25°C.

$(R = 8.31 \text{ J K}^{-1} \text{ mol}^{-1})$. For various values of E_a, Eq. 17 is graphically presented in Figure 6, where rate constants are plotted as a function of temperature relative to k (25°C). For all hydrolysis reactions discussed in this section, E_a values may vary considerably between compounds exhibiting the same hydrolyzable functional groups (i.e., alkyl halides, 80–120 kJ mol^{-1}; carboxylic acid derivatives, 50–90 kJ mol^{-1}; carbamates, 50–100 kJ mol^{-1}; phosphoric and thiophosphoric acid derivatives, 50–100 kJ mol^{-1}). Thus, as can be derived from Figure 6, for hydrolysis reactions of a variety of organic pollutants, a 10° change in temperature may result in a change in reaction rate between a factor of 2 $(E_a = 50 \text{ kJ mol}^{-1})$ and 5 $(E_a = 120 \text{ kJ mol}^{-1})$.

3.   REDOX REACTIONS

3.1. General Remarks

So far we have confined our discussion to transformation reactions in which no net electron transfer from the organic compound (i.e., oxidation) or to the organic compound (i.e., reduction) of interest occurred. The major pathways by which organic pollutants are transformed in the environment include, however, oxidative and reductive steps, especially when we consider photochemical and biologically mediated transformation processes. There is, however, good evidence that some of these reactions also occur abiotically in the dark, although it is not always easy to distinguish whether, in a given environmental system, a redox reaction occurs strictly abiotically, whether it is mediated by microorganisms, or whether both types of processes play a role (e.g., Macalady et al., 1986).

At this point, it is useful to recall how one can recognize whether an organic compound has been oxidized or reduced during a reaction. The easiest way is to check whether there has been a net change in the oxidation state(s) of the atom(s) (commonly C, N, or S; see Table 1) involved in the reaction. For example, if a chlorine atom in an organic molecule is substituted by a hydrogen atom, as is observed in the transformation of DDT to DDD,

$$+ \ Cl^- \qquad (18)$$

the oxidation state of the carbon atom at which the reaction occurs changes from $+III$ to $+I$, while the oxidation states of all other atoms remain the same. Hence, transformation of DDT to DDD requires a total of two electrons to be transferred from an electron donor to DDT. This type of reaction is termed a *reductive dechlorination*. Other important types of redox reactions include (1) vicinal dehalogenation (reaction 8 in Table 1), in which the oxidation states of both carbon atoms involved in the reaction are altered by $-I$; (2) the reduction of aromatic nitro groups to the corresponding anilines (reaction 9, change of the oxidation state of N from $+III$ to $-III$); (3) the reduction of aromatic diazo compounds to the corresponding anilines (reaction 10, change of the oxidation states of the two N atoms from $-I$ to $-III$); (4) the reduction of disulfides to the corresponding mercaptans (reaction 11, change of the oxidation states of the two S atoms from $-I$ to $-II$); and (5) the reduction of sulfoxide groups (reaction 12, change of the oxidation state of the S atom from O to $-II$). Note that some reactions may also occur in the opposite direction; that is, the reduced species may be oxidized under certain conditions.

In Table 1 only half-reactions involving the organic pollutants are indicated, and the species that act as a sink or source of electrons (i.e., the oxidants or reductants, respectively) are not specified. Unfortunately, in environmental systems, in contrast to the reactions involving nucleophiles, to date, it has not been possible to identify which species react with an organic pollutant in a redox reaction. Therefore, it is often not possible to assess exact reaction pathways and to derive kinetic data that can be generalized. Consequently, with our present knowledge of redox reactions of organic pollutants in the environment, we frequently have to content ourselves with a rather qualitative description of such processes that may include an assessment of the environmental (redox) conditions that must prevail to allow a reaction to occur spontaneously, and an assessment of the relative reactivities for a series of related compounds in a given system.

3.2. Energetic Considerations of Redox Processes

In our earlier discussion of reactions such as hydrolysis, we have (correctly) assumed that under typical environmental conditions, the free-energy change, ΔG, of the reaction considered is negative, that is, that the reaction occurs spontaneously in one direction. We did not, therefore, bother evaluating the thermodynamics of such reactions. When looking at redox reactions of organic pollutants, however, the situation is quite different. Depending on the redox conditions (that are determined largely by microbially mediated processes), naturally occurring oxidants or reductants that may react chemically in a thermodynamically favorable reaction with an organic pollutant may or may not be present in sufficient abundance. Of course, as we have seen when discussing reactions involving nucleophiles, a thermodynamically favorable reaction may still not occur at a significant rate for kinetic reasons. Nevertheless, thermodynamic considerations are very helpful as a first step in evaluating the redox conditions under which a given organic pollutant might undergo an oxidation or reduction reaction.

The most convenient way of assessing the free-energy change, ΔG, of a redox reaction is the use of the reduction potentials of the half-reactions of the redox couples involved in the reaction. [For definitions and for a more detailed discussion of this topic, see Stumm and Morgan (1981) or Schwarzenbach et al. (in press a).] Hence, one needs to know the (standard) reduction potentials of the half-reactions of the pollutant of interest and its oxidized and reduced transformation products, and of the natural oxidant–reductant couples present in a given system. Unfortunately, such information is commonly not readily available. In some cases, reduction potentials of organic pollutants exhibiting oxidizable or reducible structural moieties may be estimated from thermodynamic data or from polarographic data (e.g., Vogel et al., 1987; Schwarzenbach et al., in press a). Furthermore, because of the difficulties in assigning meaningful reduction potentials to a given natural system (Stumm and Morgan, 1981), it is helpful to use the (standard) reduction potentials of the most important biogeochemical redox processes as a framework for evaluating under which general redox conditions a given organic compound might undergo a certain redox reaction.

The (standard) reduction potentials at pH 7 of some important biogeochemical redox couples are given in Table 5 together with the reduction potentials of some half-reactions involving xenobiotic organic species. From the data in Table 5 we can, for example, conclude that, from a thermodynamic point of view, the

TABLE 5. (Standard) Reduction Potentials at 25°C and pH 7 of Some Redox Couples Important in Natural Redox Processes[a] and of Some Half-Reactions Involving Organic Compounds[b]. Half-Reactions Ordered in Decreasing $E_H^0(W)$ Values[c]

	Half-reaction		$E_H^0(W)$ (V)	$\Delta G_H^0(W)$ per e^- Transferred[d] (kJ mol^{-1})
	Oxidized Species	Reduced Species		
(1)	$Cl_3C-CCl_3 + 2e^-$	$= Cl_2C=CCl_2 + 2Cl^-$	$+1.13$	-109.0
(2)	$O_2(g) + 4H^+ + 4e^-$	$= 2H_2O$	$+0.82$	-78.3
(3)	$2NO_3^- + 12H^+ + 10e^-$	$= N_2(g) + 6H_2O$	$+0.74$	-71.4
(4)	$CCl_4 + H^+ + 2e^-$	$= CHCl_3 + Cl^-$	$+0.67$	-64.6
(5)	$CHCl_3 + H^+ + 2e^-$	$= CH_2Cl_2 + Cl^-$	$+0.56$	-54.0
(6)	$MnO_2(s) + HCO_3^- + 3H^+ + 2e^-$	$= MnCO_3(s) + 2H_2O$	$+0.52$	-50.2

(7) $\langle O \rangle-NO_2 + 6H^+ + 6e^- = \langle O \rangle-NH_2 + 2H_2O$ $+0.42$ -40.5

(8) $\langle O \rangle-NO + 2H^+ + 2e^- = \langle O \rangle-NHOH$ $+0.16$ -15.4

(9) $CH_3-\overset{O}{\overset{\|}{S}}-CH_3 + 2H^+ + 2e^- = CH_3-S-CH_3 + H_2O$ $+0.16$ -15.4

(10) [bicyclic OH, O quinone structure] $+ 2H^+ + 2e^- =$ [bicyclic OH, OH, OH reduced structure] $+0.03$ -2.9

| (11) | $FeOOH(s) + HCO_3^- + 2H^+ + e^-$ | $= FeCO_3(s) + 2H_2O$ | -0.05 | $+4.8$ |

(12) $\langle O \rangle-N=N-\langle O \rangle = \langle O \rangle-NH-NH-\langle O \rangle$

   $+ 2H^+ + 2e^-$ -0.11 $+10.6$

(13) [bicyclic O, OH quinone structure] $+ 2H^+ + 2e^- =$ [bicyclic OH, OH, OH reduced structure] -0.15 $+14.5$

(14) [anthraquinone sulfonate structure, ^-O_3S, SO_3^-] $=$ [reduced anthraquinone structure, ^-O_3S, SO_3^-, OH, OH]

   $+ 2H^+ + 2e^-$ -0.18 $+17.4$

TABLE 5. (Continued)

	Half-reaction		$E_H^0(W)$	$\Delta G_H^0(W)$ per e$^-$
	Oxidized Species	Reduced Species	(V)	Transferred[d] (kJ mol^{-1})
(15)	$SO_4^{2-} + 9H^+ + 8e^-$	$= HS^- + 4H_2O$	-0.22	$+21.2$
(16)	$S(s) + 2H^+ + 2e^-$	$= H_2S(g)$	-0.24	$+23.2$
(17)	$CO_2(g) + 8H^+ + 8e^-$	$= CH_4(g) + 2H_2O$	-0.24	$+23.5$
(18)	$2H^+ + 2e^-$	$= H_2(g)$	-0.41	$+39.6$
(19)	$6CO_2(g) + 24H^+ + 24e^-$	$= C_6H_{12}O_6 + 6H_2O$ (glucose)	-0.43	$+41.5$

[a] Data from Clark (1960) and Stumm and Morgan (1981).
[b] Estimated from thermodynamic and polarographic data.
[c] Values for aqueous solution relative to the hydrogen electrode, pH = 7, $[HCO_3^-] = 10^{-3}\,M$, $[Cl^-] = 10^{-3}\,M$.
[d] Note that $\Delta G_H^0(W) = -nFE_H^0(W)$.

reduction of hexachloroethane to tetrachloroethene (redox couple 1) may occur under any environmental condition including aerobic conditions. Indeed, the aerobic transformation of hexachloroethane to tetrachloroethane has been observed in the field (Criddle et al., 1986). In this case, it was, however, not clear whether the reaction occurred abiotically or was biologically mediated (or both).

At this point, one might wonder how reasonable it is to use $E_H^\circ(W)$ values as given in Table 5 for characterization of the redox conditions in a natural system, since the species that define these conditions will, of course, seldom be present at standard concentrations (i.e., at 1 M or 1 atm). To evaluate this problem and to illustrate the pH dependence of $E_H(W)$ values, we compare the E_H value of a $10^{-4}\,M$ aqueous hydrogen sulfide solution at pH 8 with the $E_H^\circ(W)$ value of reaction 16 in Table 5, which is defined for gaseous H_2S at 1 atm. We assume that any oxidation of hydrogen sulfide in our solution will yield elementary sulfur to which, by convention, we assign an activity of one. However, in contrast to reaction 16 in Table 5, we do not consider hydrogen sulfide in the gaseous form but dissolved in water, which means that we also have to take into account that H_2S dissociates in aqueous solution (the pK_a of H_2S is 7.0 at 25°C). The Nernst equation of our system is then given by

$$E_H = E_H^\circ(H_2S_{aq}) + 0.03 \log \frac{[H^+]^2}{[H_2S_{aq}]} \tag{19}$$

with $E_H^\circ(H_2S_{aq}) = +0.14$ V (CRC, 1985). The fraction of the total hydrogen sulfide present as H_2S_{aq} is given by

$$[H_2S_{aq}] = \frac{1}{1 + K_a/[H^+]}[H_2S]_T \tag{20}$$

Substitution of $E_H^0(H_2S_{aq})$ and Eq. 20 into Eq. 19 yields the Nernst equation for calculating the E_H value at 25°C at a given pH for a total hydrogen sulfide concentration $[H_2S]_T$:

$$E_H(pH) = +0.14 + 0.03 \log \frac{[H^+]\{[H^+] + K_a\}}{[H_2S]_T} \tag{21}$$

where E_H (pH) is expressed in volts. In our example, we set $[H^+] = 10^{-8} M$, $K_a = 10^{-7} M$, and $[H_2S]_T = 10^{-4} M$, and we obtain an E_H value of -0.19 V, which is not too much different from the $E_H^0(W)$ value of reaction 16 in Table 5.

To illustrate how we may now calculate more precisely to what extent a given organic compound may be reduced or oxidized in a given system, we consider a dilute solution (e.g., 0.1 mM) of 2-hydroxy naphthoquinone [trivial name: "lawson" (LAW), reaction 13 in Table 5] in 5 mM aqueous hydrogen sulfide at various pH values. As we will see later, since such quinoid compounds undergo reversible redox reactions in natural systems, they may play a pivotal role in the transformation of organic pollutants. Hence, we consider the reversible reaction

$$\tag{22}$$

LAW   H$_2$LAW

Since LAW and H$_2$LAW are present at much lower concentrations than hydrogen sulfide, the redox potential, $E_H(pH)$ of the system is essentially determined by Eq. 21 with $[H_2S]_T = 5 \times 10^{-3} M$. Hence, in analogy to a pH buffer for proton-transfer reactions, the H$_2$S/S(s) couple is used as a redox buffer for electron transfer.

Since both LAW and H$_2$LAW are weak acids with pK_a values of 3.98 (pK_a^o, LAW), 8.68 (pK_{a1}^r, H$_2$LAW), and 10.71 (pK_{a2}^r, H$_2$LAW), the pH dependence of the E_H value of this redox couple is somewhat more complicated (Clark, 1960):

$$E_{H,LAW}(pH) = 0.35 + 0.03 \log \frac{\{[H^+]^3 + K_{a1}^r[H^+]^2 + K_{a1}^r K_{a2}^r[H^+]\}[LAW]}{\{[H^+] + K_a^o\}[H_2LAW]} \tag{23}$$

Note that we have introduced the superscripts "o" and "r" to distinguish among the oxidized (o) and reduced (r) species.

By setting $E_{H,LAW}(pH) = E_H(pH)$ (Eq. 21), we may now calculate the ratio of the reduced and oxidized form of "lawson" in our solution at equilibrium at a given pH by

$$\log \frac{[H_2LAW]}{[LAW]} = 0.21 + 0.03 \log \frac{\{[H^+]^3 + K^r_{a1}[H^+]^2 + K^r_{a1}K^r_{a2}[H^+]\}[H_2S]_T}{[H^+]\{[H^+] + K^o_a\}} \tag{24}$$

Using the acidity constants given above and a total H_2S concentration of 5×10^{-3} M, calculation of the $[H_2LAW]/[LAW]$ ratio shows that at pH 7, more than 95% of the lawson will be present as H_2LAW, while at pH 8, only about 35% will be present in the reduced form. This example illustrates that the abundance of potential reactive reductants or oxidants in a given system may be strongly pH-dependent.

3.3. Kinetics of Redox Reactions

General Remarks.   As is the case for any chemical reaction, the partners in a bimolecular electron-transfer reaction must encounter one another (i.e., to colloide) in order to allow the reaction to proceed. Since the majority of chemical redox reactions of organic pollutants proceed in sequential one-electron steps, we consider a reaction in which one electron is transferred between two reactants P and R, where P is, for example, an organic pollutant and R is a reductant. In a general way, one may express such a one-electron transfer reaction schematically as (Eberson, 1987):

$$P + R \ \rightleftharpoons \ (PR) \ \longrightarrow \ [PR \leftrightarrow P^{\bullet -}R^{\bullet +}]^{\neq} \longrightarrow \ (P^{\bullet -}R^{\bullet +})$$

educts   precursor   activated   successor
complex   complex   complex

$$P^{\bullet -} + R^{\bullet +} \tag{25}$$

products

If there is a strong electronic coupling between P and R in the transition state, one commonly speaks of an "inner-sphere mechanism," and, conversely, if the interaction is weak, one uses the term "outer-sphere mechanism." There are various theoretical approaches for quantifying the rates of redox reactions, including the so-called Marcus theory. For a description of these approaches, we refer to the literature (e.g., Eberson, 1987). For our discussion here, we content ourselves with trying to identify the factors that determine the rate at which a given organic pollutant is reduced or oxidized in the environment.

From Eq. 25 we see that a one-electron transfer may (perhaps sometimes somewhat artificially) be divided into various steps. First, a precursor complex has to be formed; that is, the reactants have to meet and interact. Hence, electronic as well as steric effects will influence the rate and extent at which precursor complex formation occurs. Consequently, in heterogeneous systems, where redox reactions often occur at surfaces, the sorption behavior of the compound may be pivotal for determining the rate of transformation. In the next step, the actual electron transfer between P and R occurs. The activation energy required for this step depends primarily on the "willingness" of a given chemical species to lose or gain, respectively, an electron. A measure of the "willingness" of a given chemical species to lose or gain an electron (and thus the reactivity of the compound) is given by the reduction potential of the respective one-electron half-reaction. Thus, the activation energy for the formation of the transition state should be proportional to the difference in the reduction potentials of the corresponding one-electron half-reactions. Hence, in most cases, the difference in the one-electron reduction potentials will be reflected in the reaction rate. Finally, in the last steps of the reaction mechanism in Eq. 25, a successor complex is postulated that decays into the products.

As can be seen from Tables 1 and 5, in most cases, oxidative or reductive transformation of an organic pollutant requires at least two electrons to be transferred from or to the compound to yield a stable product. With the first electron, a radical species is formed that is, in general, much more reactive than the parent compound. Hence, generally transfer of the first electron from or to the pollutant will be rate-determining. Thus, it is not surprising that reduction potentials as given in Table 5 for multielectron transfers do not correlate well with observed reaction rates (e.g., Weber and Wolfe, 1987). Unfortunately, for aqueous solution, data on one-electron reduction (or oxidation) potentials of organic chemicals are very scarce. One-electron reduction (or oxidation) potentials that are available in the literature have been determined either by pulse radiolysis techniques or by electrochemical methods including polarographic techniques (e.g., Meisel and Czapski, 1975; Wardman, 1977; Suatoni et al., 1961; Bard and Lund, 1978–1984).

Oxidation Reactions.   When thinking about the oxidation of organic pollutants in the environment, one immediately wonders about the importance of molecular oxygen (triplet oxygen, $^3O_2$) in such reactions. From our daily experience, we know that (fortunately) most organic compounds do not react spontaneously at significant rates with molecular oxygen, although the overall reaction would, in general, be exergonic. Hence, the reason for the inertness of organic pollutants with respect to molecular oxygen (if not activated by photolytic or biological processes) must be a kinetic one. Indeed, as is indicated by the (standard) reduction potential for transfer of one electron to molecular oxygen yielding superoxide [pK_a of $O_2H^{\cdot} = 4.88$ (Ilan et al., 1987)]

$$^3O_2 \, (1 \, M) + e^- = {}^3O_2^{\cdot -} (1 \, M), \quad E_H^{\circ}(W) = -0.16 \text{ V} \tag{26}$$

in aqueous solution at pH 7, molecular oxygen is only a very weak oxidant. (Note that we are using a thermodynamic entity for a kinetic argument.) Consequently, only compounds that are very easily oxidized (e.g., mercaptans, or phenols and anilines that are substituted with electron donating substituents such as alkyl or alhoxy groups) will react with $^3O_2$ at significant rates. Note that, as is discussed in Chapter 2, photolytically activated oxygen, so-called singlet oxygen ($^1O_2$), is much more reactive than $^3O_2$ and it has a significantly higher one-electron reduction potential (Eberson, 1987):

$$^1O_2(1\ M) + e^- = {}^1O_2^{\cdot-}(1\ M), \qquad E_H^\circ(W) = +0.79\ V \qquad (27)$$

Besides molecular oxygen, iron(III) and manganese(III/IV) oxides are the most abundant natural oxidants that may undergo chemical reactions with organic pollutants but also only with very easily oxidizable compounds. For example, Stone (1987) and Ulrich and Stone (1989) have investigated the oxidation of a series of substituted phenols by well-characterized manganese oxide particles. The results of their study indicate that reaction rates are influenced by the tendency of a given compound to bind both to the surface (i.e., form a precursor complex), and lose an electron (expressed by the polarographically determined one-electron oxidation potential).

Reduction Reactions.   As we have just noted, most organic pollutants are quite inert to chemical oxidation in the dark. This is actually not surprising since most xenobiotic compounds are designed to survive in an aerobic world. When introduced into an anaerobic environment, however, quite a few classes of compounds may undergo reduction (see examples given in Table 1).

We have already pointed out that one major difficulty encountered when working with natural samples is to distinguish between abiotic and biologically mediated reactions. In cases such as hydrolysis, in which the reactants and their abundances are well known, it is possible to apply rate constants determined in sterile systems in the laboratory to natural systems. Unfortunately, for most reduction processes, one neither knows the identities nor the abundances of the natural reductants responsible for chemical transformation of a given organic pollutant. Furthermore, autoclaving or addition of poisons such as formaldehyde or sodium azide to inhibit biological activity may also effect such "abiotic reductants" (e.g., Jafvert and Wolfe, 1987; Weber and Wolfe, 1987). Consequently, from the data available, it has not been possible to derive rate data for predicting absolute reaction rates, but only relative rates within a series of related compounds.

As an example, in Table 6, the pseudo-first-order rate constants determined by Jafvert and Wolfe (1987) are given for the disappearance of some polyhalogenated ethanes in an anaerobic sediment–water mixture. For all compounds, the most important reaction mechanism was found to be vicinal dehalogenation (reaction 8 in Table 1). As can be seen, in this system, an appreciable fraction of all compounds (particularly, the hydrophobic compound hexachloroethane) was present in the sorbed form. We should note, however, that unlike the case of surface complexation of hydrophilic compounds at oxide surfaces, (e.g., as encountered in the oxidation of phenols by manganese oxides, see Stone, 1987),

TABLE 6. Rates of Disappearance of Some Halogenated Ethanes in an Anaerobic Sediment–Water Slurry[a]

Compound Name	Structure	k_{obs} (s^{-1})	$t_{1/2}$ (h)	Sediment–Water Coefficient K_d (L kg$_s^{-1}$)	% sorbed	Initial Concentration (mol L^{-1} of Suspension)
1,2-Dichloroethane	CH_2Cl-CH_2Cl	$\ll 2 \times 10^{-7}$	$\geqslant 950$	1.3	9	1.0×10^{-5}
1,2-Dibromoethane	CH_2Br-CH_2Br	3.5×10^{-6}	55	2.0	13	3.0×10^{-7}
1,2-Diiodoethane	CH_2I-CH_2I	4.8×10^{-4}	0.4	3.5	21	2.5×10^{-7}
1,1,2,2-Tetrachloroethane	$CHCl_2-CHCl_2$	1.2×10^{-6}	160	3.4	20	3.5×10^{-7}
Hexachloroethane	CCl_3-CCl_3	3.2×10^{-4}	0.6	29	69	1.0×10^{-7}

Data taken from Jafvert and Wolfe (1987).
[a] Sediment: water ratio = 0.075 (w:w), pH 6.5, apparent E_H value = −0.140 V.

strong sorption of a hydrophobic compound does not necessarily mean en-hanced reactivity. In fact, Weber and Wolfe (1987) concluded that sorption to non-reactive sites (e.g., into nonreactive parts of organic material present in a natural sorbent) was responsible for a decrease in the reduction rates observed in an anoxic sediment–water slurry for a series of substituted azobenzenes exhibiting different hydrophobicities. Although sorptive processes may have had an effect on the transformation rates observed for the compounds in listed Table 6, their relative reactivities also reflect the differences that one would expect in the reduction potentials for transfer of the first electron to the various compounds. Since no one-electron reduction potentials are available in aqueous solution for the compounds listed, we must confine our discussion to a qualitative evaluation of the factors that determine the free energy of activation for adding an electron to a halogenated compound.

If an electron is added to a carbon–halogen (C–X) bond, the electron causes a partial dissociation of the C–X bond that is subsequently cleaved to yield a carbon radical and a halide anion. Hence, the factors that will determine the free energy of the activated complex include the tendency of the C–X bond to accept an electron (which is reflected in the electronegativity of the halogen), the ease with which the C–X bond is cleaved (i.e., the strength of the C–X bond), and the stability of the carbon radical formed. From the relative reactivities of dichloro-ethane, dibromoethane, and diiodoethane, we may conclude that the latter two factors are dominating, since the electronegativities of the halogens decrease from chlorine to bromine to iodine. However, as we have already noted when discussing halogens as leaving groups in S_N2 reactions, the bond strengths also decrease significantly in the same sequence. Finally, we may also speculate that stabilization of the radical by bridging of the second halogen (March, 1985) increases from chlorine to bromine to iodine:

$$
\underset{\substack{|\\ \text{—}\overset{\displaystyle\cdot}{\text{C}}\text{—}\overset{\displaystyle\mid}{\underset{\mid}{\text{C}}}\text{—}}}{\overset{\displaystyle\overset{\text{X}}{\mid}}{}} \quad \longleftrightarrow \quad \underset{\substack{|\qquad|\\ \text{—}\text{C}\text{——}\text{C}\text{—}}}{\overset{\displaystyle\overset{\text{X}}{\cdot}}{\triangle}} \tag{28}
$$

The increasing reactivity of the chlorinated ethanes with increasing number of chlorines may be rationalized by the decreasing C–Cl bond dissociation energy with increasing chlorination (Goldfinger and Martens, 1961), and by the availability of an increasing number of chlorines to stabilize the radical.

We conclude this chapter by hypothesizing which water or solid constituents might cause reductions of organic pollutants in anaerobic environments. The most abundant natural reductants in anaerobic soils and sediments include reduced inorganic forms of iron and sulfur, such as iron(II) sulfides, iron(II) carbonates, and hydrogen sulfide. Although some of these reductants have been found to react with reducible organic pollutants, the reaction rates are, in general, much too slow to account for the extremely rapid transformation rates often observed in natural systems. For example, half-lives of seconds to minutes have been determined for the reduction of the nitro group of parathion (Wahid et al., 1980) and methyl parathion (Wolfe et al., 1986) in anaerobic soils and sediments. Consequently, there must be more reactive reductants present. Such reactive species may play the role of electron-transfer mediators, that is, after reducing a pollutant, they may themselves get reduced again by the bulk of inorganic reductants present:

$$(\text{bulk})_{\text{ox}} \xleftarrow{} \quad (\text{``mediator''})_{\text{ox}} \xleftarrow{} \quad (\text{pollutant})_{\text{ox}}$$

$$-\,\text{ne}^- \quad +\,\text{ne}^- \qquad\qquad -\,\text{ne}^- \quad +\,\text{ne}^- \qquad (29)$$

$$(\text{bulk})_{\text{red}} \xrightarrow{} (\text{``mediator''})_{\text{red}} \xrightarrow{} (\text{pollutant})_{\text{red}}$$

Candidates for such "mediators" of chemical reductions of organic pollutants include iron porphyrins or other iron complexes (e.g., Zoro et al., 1974; Holmstead, 1976; Klecka and Gonsior, 1984), as well as quinoid type compounds (Tratnyek and Macalady, 1989; Schwarzenbach et al., in press b). Both types of compounds are involved in electron-transfer reactions in biological systems, and both iron complexes and quinoid-type structures are postulated to occur in natural organic matter including humic and fulvic materials (Thurman, 1985; Buffle and Altmann, 1987). However, in a natural system, such species will often not be present in large abundances. Hence, in order to be effective "mediators" as depicted by Eq. 29, the oxidized forms of these electron-transfer molecules have to be rapidly reduced again by the bulk of reductants present. To demonstrate that such fast reduction of "mediators" may indeed take place, and to illustrate

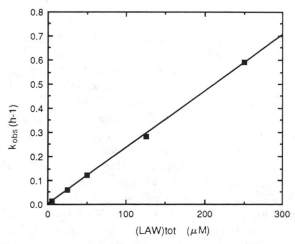

Figure 7. Observed pseudo-first-order rate constants for the disappearance of 3-chloro-nitrobenzene (initial concentration 100 μM) in 5 mM aqueous H_2S solution at pH 7.24 ± 0.02 (50 mM phosphate buffer) as a function of total lawson concentration.

some of the already mentioned problems encountered when trying to predict the rates of such reactions in natural system, we consider the reduction of some monosubstituted benzenes in an aqueous solution containing 5 mM hydrogen sulfide and various concentrations of a given quinone. We recall that we have already discussed some thermodynamic aspects of such a system in Section 3.2 (Eqs. 19–24). The following data stem from a recent study by Schwarzenbach et al. (in press b), to which the reader is referred to for further detail.

In Figure 7, the observed (pseudo-) first-order rate constants for the disappearance of 3-chloronitrobenzene (3-Cl) in 5 mM H$_2$S at pH 7.24 (± 0.02) are plotted as a function of the total concentration of the quinone lawson (LAW). As discussed earlier, at pH 7.24, more than 90% of the lawson is present in the reduced form (H$_2$LAW; see Eq. 24). The initial concentration of 3-Cl was 100 μM in all experiments. For all concentrations of lawson used, the rate of disappearance of 3-Cl followed first-order kinetics over several half-lives indicating that rereduction of lawson by H$_2$S occurred fast as compared to its oxidation by 3-Cl (**III**) and its reduced intermediates (**IV, V**):

$$\underset{\text{III}}{\text{Cl}-\bigcirc-\text{NO}_2} \xrightarrow{+2e^- + 2H^+} \underset{\text{IV}}{\text{Cl}-\bigcirc-\text{NO}} \xrightarrow{+2e^- + 2H^+}$$

$$\underset{\text{V}}{\text{Cl}-\bigcirc-\text{NHOH}} \xrightarrow{+2e^- + 2H^+} \underset{\text{VI}}{\text{Cl}-\bigcirc-\text{NH}_2} \qquad (30)$$

Furthermore, Figure 7 also indicates that in the absence of a mediator, the reaction of 3-Cl with H$_2$S is extremely slow. Finally, it should be noted that, in general, the reduction of a nitroaromatic compound with lawson as the mediator yielded the hydroxylamine **V** as the primary product, which was subsequently converted very slowly to the corresponding amine **VI**.

As could have been expected from other studies dealing with the oxidation of phenolic compounds (e.g., oxidation with ^{1}O$_2$; see Chapter 2, this volume), the monophenolate species (HQ$^-$) of a given hydroquinone is a by far more reactive reductant as compared to the nondissociated hydroquinone (H$_2$Q). This is illustrated by Figure 8, which shows the pH dependence of the rate of transformation of 4-chloro-nitrobenzene (4-Cl) in 5 mM H$_2$S at a fixed total lawson concentration [[LAW]$_T$ = 250 μM, pK_{a1}^r (H$_2$LAW) = 8.68].

From the data presented in Figure 8 (and from other data not presented here), it can be shown (Schwarzenbach et al., in press b) that the rate of reduction of 4-Cl and of other substituted nitroaromatic compounds (ArNO$_2$) with lawson as

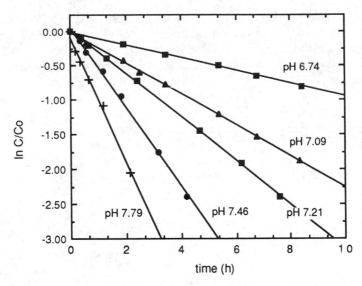

Figure 8. Rate of disappearance of 4-chloro nitrobenzene (initial concentration 100 μM) in 5 mM aqueous H_2S solution containing 250 μM total lawson at different pH values (50 mM phosphate buffers).

mediator may be described by a second-order rate law:

$$\text{Rate} = -\frac{d[\text{ArNO}_2]}{dt} = k[\text{ArNO}_2][\text{HLAW}^-] \tag{31}$$

Note that in the case of lawson, in the system considered, the total fraction of reduced quinone is also pH-dependent (see Eq. 24), which has to be taken into account when calculating $[\text{HLAW}^-]$. Furthermore, at pH values $> \sim 7.2$, the reaction with the biphenolate species (LAW^{2-}) also becomes important.

From the rate law found for the reduction of nitroaromatic compounds with lawson, one may postulate that the transfer of the first electron from HLAW^- to ArNO_2 is the rate-limiting step for transforming ArNO_2 to ArNO (which then reacts in a fast reaction to yield ArNHOH):

$$\text{ArNO}_2 + \text{(HLAW-)} \longrightarrow \text{ArNO}_2^{\cdot-} + \text{(HLAW-)} \tag{32}$$

It should be pointed out that when expressing the rate-determining step by Eq. 32, we have assumed that the pK_a values of the various species involved

$(ArNO_2^{\cdot-}, HLAW^-, HLAW^{\cdot})$ are either sufficiently low $(ArNO_2^{\cdot-})$ or sufficiently high $(HLAW^-, HLAW^{\cdot})$, so that no other species are present in significant concentration in the pH range considered. For some $ArNO_2^{\cdot-}$ species, pK_a values of below 5 have been reported (Neta and Meisel, 1976) and, as already indicated earlier, the pK_a of $HLAW^-$ is 10.7. For $HLAW^{\cdot}$, no pK_a value is available, and it cannot be *a priori* excluded that at higher pH values (i.e., pH > 7.5), the dissociated form of the semiquinone might play a role in determining the observed overall transformation rate. For the following discussion, we assume, however, that Eq. 32 properly describes the rate-limiting step.

The standard free-energy change, $\Delta G°(W)$, of reaction 32 may be expressed by the difference of the (pH-independent) reduction potentials of the two half-reactions:

$$ArNO_2 + e^- = ArNO_2^{\cdot-}; \qquad E_H^1(W)(ArNO_2) \qquad (33)$$

$$HLAW^{\cdot} + e^- = HLAW^-; \qquad E_H^1(W)(HLAW^{\cdot})$$

$$ArNO_2 + HLAW^{\cdot} = ArNO_2^{\cdot-} + HLAW^{\cdot} \qquad \Delta G°(W) = -F[E_H^1(W)(ArNO_2) \\ - E_H^1(W)(HLAW^{\cdot})]$$

where $F = 96.5 \text{ kJ V}^{-1}$, and where we have introduced the superscript 1 to denote that we are dealing with a one-electron step. Unfortunately, as is the case for many quinoid compounds, $E_H^1(W)(HLAW^{\cdot})$ is not available. However, for a series of substituted nitrobenzenes, $E_H^1(W)(ArNO_2)$ values are known and are given in Table 7 together with the second-order rate constants, k, for reaction with $HLAW^-$ (Eq. 31).

TABLE 7. One-Electron Reduction Potentials E_H^1 of a Series of Nitrobenzenes, and Second-Order Rate Constants k for Reaction with $HLAW^-$

Compound Name	Abbreviation	E_H^1 (W)[b] (V)	k^c (M^{-1} s^{-1})
Nitrobenzene	H	−0.485	2.5
2-Methylnitrobenzene	2-CH₃	−0.590	0.04
3-Methylnitrobenzene	3-CH₃	−0.475	2.9
4-Methylnitrobenzene	4-CH₃	−0.500	1.1
3-Chloronitrobenzene	3-Cl	−0.405	31
4-Chloronitrobenzene	4-Cl	−0.450	11
4-Acetylnitrobenzene	4-COCH₃	−0.360	340

[a] The k values were determined at pH 6.98 ± 0.02 at 25°C.
[b] Data from Meisel and Neta (1975); Wardman (1977); Kemula and Krygowski (1979).
[c] See Eq. 31; data from Schwarzenbach et al. (in press b).

From the data given in Table 7, it can be seen that the presence of an electron-withdrawing substituents (i.e., Cl, $COCH_3$) leads to a decrease in the one-electron reduction potential and to an increase in the reaction rate. This is due to the stabilizing effect of these substituents on the radial anion $ArNO_2^{\cdot-}$. Electron-donating substituents (e.g., CH_3), on the other hand, have the opposite effect. In addition, as is illustrated by 2-methylnitrobenzene, as a result of steric interactions, substituents in a position ortho to that of the nitro group may hinder proper resonance of the nitro group with the aromatic ring, thus increasing the reduction potential as compared to the isomer exhibiting the same substituent in the para position. Let us now try to quantity the relationship between single-electron reduction potentials and reaction rates.

According to the transition-state theory (see Chapter 1, this volume), the rate constant for a bimolecular reaction may be expressed by

$$k = \frac{\hbar T}{h} \exp\left[-\frac{\Delta G^{\neq}(W)}{RT}\right] \tag{34}$$

where, in our case, $\Delta G^{\neq}(W)$ is the free energy of activation for the reaction of $ArNO_2$ with $HLAW^-$. When considering, for example, the rate of reduction of a substituted nitrobenzene (SNB) relative to the nonsubstituted compound (NB), we may write

$$\frac{k(SNB)}{k(NB)} = \exp\left[-\frac{\Delta G^{\neq}(W)(SNB) - \Delta G^{\neq}(W)(NB)}{RT}\right]$$

or

$$\log\frac{k(SNB)}{k(NB)} = \log k_{rel} = \frac{\Delta G^{\neq}(W)(SNB) - \Delta G^{\neq}(W)(NB)}{2.303RT} \tag{35}$$

If we postulate that the electron transfer between $HLAW^-$ and a given nitrobenzene occurs by an outer-sphere mechanism, and, as a special case, that the difference in $\Delta G^{\neq}(W)$ for the reduction of two different nitrobenzenes is equal to the difference between the respective $\Delta G^{\circ}(W)$ values, we obtain

$$\Delta G^{\neq}(W)(SNB) - \Delta G^{\neq}(W)(NB) = \Delta G^{\circ}(W)(SNB) - \Delta G^{\circ}(W)(NB) \tag{36}$$

Substitution of Eqs. 33 and 36 into Eq 35 yields

$$\log\frac{k(SNB)}{k(SN)} = \log k_{rel} = \frac{F}{RT}[E_H^1(W)(SNB) - E_H^1(W)(NB)] \tag{37}$$

At 25°C:

$$\log k_{rel} = \frac{1}{0.059\ V}[E_H^1(W)(SNB) - E_H^1(W)(NB)] \tag{38}$$

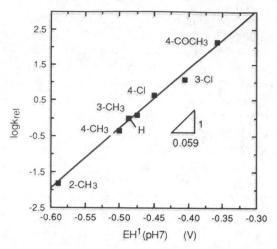

Figure 9. Relationship between relative reaction rates and one-electron reduction potentials for the reaction of a series of substituted nitrobenzenes with HLAW⁻ (see Table 7).

In this case, a plot of log k_{rel} versus $E_H^1(W)$ for a series of substituted nitrobenzenes should yield a straight line with a slope of (1/0.059), which as is shown in Figure 9, is actually found for the compounds listed in Table 7. Hence, for the reduction of neutral substituted nitrobenzenes in homogeneous aqueous hydrogen sulfide solution with lawson as electron-transfer "mediator," the relative reaction rate of a given compound may be predicted from its one-electron reduction potential. Similar results have been obtained for the same set of compounds with another quinone (i.e., 5-hydroxynaphthoquinone (reaction 10 in Table 5); see Schwarzenbach et al. (in press b).

With this example, we have illustrated that quinoid-type compounds may mediate reductions of organic pollutants in aqueous solution. If such reactive species, as well as other possible mediator compounds (e.g., iron complexes), play an important role in environmental redox reactions involving organic pollutants, it is easy to see that it will be very difficult (if not impossible) to predict the absolute rate at which a given pollutant will be reduced or oxidized in the environment, since, in a given natural system, there will be many potential redox mediators present that exhibit very different (pH-dependent) reactivities. Furthermore, many redox reactions occur at solid–water interfaces, and they may also be catalyzed by microorganisms. Thus, in contrast to the reactions involving inorganic nucleophiles, for redox reactions it may only be possible to establish relationships between compound properties and relative reactivities. For assessing absolute reaction rates in a given system, it is then necessary to take a calibration approach using an appropriate model compound (see also Chapter 2, this volume).

Acknowledgements

We thank P. Tratnyek for reviewing the manuscript, and M. Hoffmann and J. Hoigné for valuable comments.

REFERENCES

Bard, A. J. and H. Lund, Eds. (1978–1984), *Encyclopedia of Electrochemistry of the Elements*, Vols. XI–XV, Dekker, New York.

Buffle, J., and R. S. Altmann (1987), "Interpretation of Metal Complexation by Heterogeneous Complexants," in W. Stumm, Ed., *Aquatic Surface Chemistry*, Wiley, New York, 1987.

Capel, P. D., W. Giger, P. Reichert, and O. Wanner (1988), "Accidental Input of Pesticides into the Rhine River," *Environ. Sci. Technol.* **22**, 992–997.

Clark, W. M. (1960), *Oxidation–Reduction Potentials of Organic Systems*, Williams and Wilkins, Baltimore.

CRC Handbook of Chemistry and Physics (1985), R. C. Weast and M. J. Astle, Eds., CRC Press, Boca Raton, FL.

Criddle, C. S., P. L. McCarty, M. C. Elliot, and J. F. Barker (1986), "Reductions of Hexachloroethane to Tetrachloroethylene in Groundwater," *J. Contam. Hydrol.* **1**, 133–142.

Eberson, L. (1987), *Electron Transfer Reactions in Organic Chemistry*, Springer, Berlin.

El-Amamy, M. M., and T. Mill (1984), "Hydrolysis Kinetics of Organic Chemicals on Montmorillonite and Kaolinite Surfaces as Related to Moisture Content," *Clays Clay Minerals* **32**, 67–73.

Goldfinger, P., and G. Martens (1961), "Elementary Rate Constants in Atomic Chlorination Reactions. Part 3: Bond Dissociation Energies and Entropies of the Activated State," *Faraday Soc. Trans.* **57**, 2220–2225.

Haag, W. R., and T. Mill (1988), "Effects of a Subsurface Sediment on Hydrolysis of Haloalkanes and Epoxides," *Environ. Sci. Technol.* **22**, 658–663.

Hine, J. (1962), *Physical Organic Chemistry*, McGraw-Hill, New York.

Holmstead, R. L. (1976), "Studies of the Degradation of Mirex with an Iron (II) Porphyrin Model System," *J. Agric. Food Chem.* **24**, 620–624.

Hughes, E. A. M. (1971), *The Chemical Statics and Kinetics of Solutions*, Academic Press, London.

Ilan, Y. A., G. Czapski, and D. Meisel (1987), "The One-Electron Transfer Redox Potentials of Free Radicals: 1. The Oxygen Superoxide System," *Biochim. Biophys. Acta*, **430**, 209–244.

Jafvert, C. T., and N. L. Wolfe (1987), "Degradation of Selected Halogenated Ethanes in Anoxic Sediment–Water Systems," *Environ. Toxicol. Chem.* **6**, 827–837.

Kemula, W., and T. M. Krygowski (1979), "Nitro Compounds," in A. J. Bard and H. Lund, Eds., *Encyclopedia of Electrochemistry of the Elements*, Vol. XIII, Dekker, New York.

Klecka, G. M., and S. J. Gonsior (1984), "Reductive Dechlorination of Chlorinated Methanes and Ethanes by Reduced Iron (II) Porphyrins," *Chemosphere* **13**, 391–402.

Mabey, W., and T. Mill (1978), "Critical Review of Hydrolysis of Organic Compounds in Water under Environmental Conditions," *J. Phys. Ref. Data* **7** (2), 383–415.

Macalady, D. L., P. G. Tratnyek, and T. J. Grundl (1986), "Abiotic Reduction Reactions of Anthropogenic Organic Chemicals in Anaerobic Systems: A Critical Review," *J. Contam. Hydrol.* **1**, 1–28.

Macalady, D. L., P. G. Tratnyek, and N. L. Wolfe (1989), "Influences of Natural Organic Matter on the Abiotic Hydrolysis of Organic Contaminants in Aqueous Systems," in *Aquatic Humic Substances. Influence on Fate and Treatment of Pollutants*, I. H. Suffet and P. MacCarthy, Eds. (Advances in Chemistry Series No. 219), American Chemistry Society, Washington, DC.

March, J. (1985), *Advanced Organic Chemistry*, 3rd Ed., Wiley, New York.

Meisel, D., and G. Czapski (1975), "One-Electron Transfer Equilibria and Redox Potentials of Radicals Studied by Pulse Radiolysis," *J. Phys. Chem.* **79**, 1503–1509.

Meisel, D., and P. Neta (1975), "One-Electron Redox Potentials of Nitro Compounds and Radiosensitizers. Correlation with Spin Densities of Their Radical Anions," *J. Am. Chem. Soc.* **97**, 5198–5203.

Mill, T., and W. Mabey (1988), "Hydrolysis of Organic Chemicals," in O. Hutzinger, Ed., *The Handbook of Environmental Chemistry*, Vol. 2, Part D. Springer, New York.

Neta, P., and D. Meisel (1976), "Substituent Effects of Nitroaromatic Radical Anions in Aqueous Solution", *J. Phys. Chem.*, **80**, 519–524.

Schwarzenbach, R. P., P. M. Gschwend, and D. M. Imboden (in press a), *Environmental Organic Chemistry of Aquatic Systems*, Wiley, New York.

Schwarzenbach, R. P., R. Stierli, K. Lanz, and J. Zeyer (in press b), "Quinone and Iron Porphyrin Mediated Reduction of Nitroaromatic Compounds in Homogeneous Aqueous Solution", *Environ. Sci. Technol.*

Schwarzenbach, R. P., W. Giger, C. Schaffner, and O. Wanner (1985), "Groundwater Contamination by Volatile Halogenated Alkanes: Abiotic Formation of Volatile Sulfur Compounds under Anaerobic Conditions," *Environ. Sci. Technol.* **19**, 322–327.

Stone, A. T. (1987), "Reductive Dissolution of Manganese (III/IV) Oxides by Substituted Phenols," *Environ. Sci. Technol.* **21**, 979–988.

Stone, A. T. (1989), "Enhanced Rates of Monophenyl Terephthalate Hydrolysis in Aluminium Oxide Suspensions," *J. Colloid Interface Sci.* **127**, 429–441.

Stumm, W., and J. J. Morgan (1981), *Aquatic Chemistry*, 2nd ed., Wiley, New York.

Suatoni, J. C., R. E. Snyder, and R. O. Clark (1961), "Voltametric Studies of Phenol and Aniline Using Ring Substitution," *Anal. Chem.* **33**, 1894–1897.

Thurman, E. M. (1985), *Organic Geochemistry of Natural Waters*, Nijhoff, Boston.

Tratnyek, P. G., and D. L. Macalady (1989), "Abiotic Reduction of Nitro Aromatic Pesticides in Anaerobic Laboratory Systems," *J. Agric. Food Chem.* **37**, 248–254.

Ulrich, M. J., and A. T. Stone (1989), "The Oxidation of Chlorophenols Adsorbed to Manganese Oxide Surfaces," *Environ. Sci. Technol.* **23**, 421–428.

Vogel, T. M., C. S. Criddle, and P. L. McCarty (1987), "Transformations of Halogenated Aliphatic Compounds," *Environ. Sci. Technol.* **21**, 722–736.

Wahid, P. A., C. Ramakrishna, and N. Sethunathan (1980), "Instantaneous Degradation of Parathion in Anaerobic Soils," *J. Environ. Qual.* **9**, 127–130.

Wanner, O., T. Egli, T. Fleischmann, K. Lanz, P. Reichert, and R. P. Schwarzenbach (1989), "The Behavior of the Insecticides Disulfoton and Thiometon in the Rhine River—A Chemodynamic Study," *Environ. Sci. Technol.*, **23**, 1232–1242.

Wardman, P. (1977), "The Use of Nitroaromatic Compounds as Hypoxic Cell Radiosensitizers," *Current Top. Radiat. Res. Q.* **11**, 347–398.

Weber, E. J., and N. L. Wolfe (1987), "Kinetic Studies of the Reduction of Aromatic Azo Compounds in Anaerobic Sediment/Water Systems," *Environ. Toxicol. Chem.* **6**, 911–919.

Williams, A. (1972), "Alkaline Hydrolysis of Substituted Phenylcarbamates. Structure–Reactivity Relationships Consistent with an E1cB Mechanism," *J. Chem. Soc. Perkins II*, 808–812.

Wolfe, N. L., B. E. Kitchens, D. L. Macalady, and T. J. Grundl (1986), "Physical and Chemical Factors that Influence the Anaerobic Degradation of Methyl Parathion in Sediment Systems," *Environ. Toxicol. Chem.* **5**, 1019–1026.

Wolfe, N. L., R. G. Zepp, and D. F. Paris (1978), "Use of Structure–Reactivity Relationships to Estimate Hydrolytic Persistence of Carbamate Pesticides," *Water Res.* **12**, 561–563.

Zafiriou, O. (1975), "Reactions of Methyl Halides with Seawater and Marine Aerosols," *J. Marine Res.* **33**, 75–81.

Zoro, J. A., J. M. Hunter, G. Eglington, and G. C. Ware (1974), "Degradation of *p,p'*-DDT in Reducing Environments," *Nature* **247**, 235–237.

8

ROLE OF EXTRACELLULAR ENZYMATIC REACTIONS IN NATURAL WATERS

Neil M. Price and François M. M. Morel

R. M. Parsons Laboratory for Water Resources and Hydrodynamics, Massachusetts Institute of Technology, Cambridge, Massachusetts

1. INTRODUCTION

Aquatic microorganisms regulate and modify the chemical composition of their environment through their production and consumption of organic matter, their involvement in nutrient cycling, and their scavenging of trace metals and pollutants. In addition to the general biochemical processing of carbon and nutrients, microorganisms also mediate specific reactions through extracellular, chiefly enzymatic, reactions. Extracellular enzymes either on the cell surface or free in solution hydrolyze, oxidize, or reduce a variety of solutes. These enzymatically catalyzed reactions enable microbes to obtain what they need for growth: major nutrients from macromolecules that are not readily utilizable and trace elements from unavailable redox species or complexes. They alter the reactivity and redox state of existing chemical species, change the nature of the dissolved organic matter, and, in some cases, produce reactive intermediates that can interact with other solutes such as trace metals. As a result of these extracellular transformations, aquatic microbes both catalyze thermodynamically favorable reactions and maintain natural waters in a state of chemical disequilibrium.

The study of extracellular enzymatic reactions in aquatic organisms is a fertile area of research, and much remains to be discovered. The importance of such reactions in the ecology and physiology of microbes is beginning to be recognized, but its role in the chemistry of natural waters has been heretofore practically ignored. In localized aquatic environments, such as biofilms, marine snow, and in the midst of algal blooms, we expect that these extracellular reactions markedly influence water chemistry. However, their general importance in lakes and oceans is presently unknown.

While we may not yet be able to assess the quantitative role of extracellular enzymatic processes in aquatic chemistry, some qualitative features are already apparent. Through the activity of transmembrane reductases, the microbiota can serve as a source of electrons for aquatic solutes apart from the well-known fixation of inorganic carbon and nitrogen. In addition, many redox and hydrolysis reactions may be catalyzed selectively and effectively (to the limit of diffusion rate) for solutes with active biological roles. Because trace elements are often involved in the activation and inhibition of enzymes, extracellular enzymatic reactions render the cycles of major nutrients and trace elements interdependent on each other. Finally, we must note that our ability to predict where and when a particular enzymatic reaction occurs depends on our understanding of the physiological function and regulation of the enzyme as well as our knowledge of the chemistry of the medium.

In this chapter, we discuss the different types of extracellular enzymes of aquatic microbes, including their reactions and their influence on the chemistry of the medium. Our objective is to point out the possible chemical significance of these processes in natural waters using primarily, but not exclusively, examples of cell surface enzymes found in phytoplankton. While we are now beginning to document the nature and activity of some of these extracellular enzymes in aquatic microbes, the vast majority of our information on such enzymes comes from mammals, bacteria, yeast, and higher plants. Out of necessity, we draw on results obtained from these other biological systems as examples relevant to our discussion. We do not suggest that mammalian tissues typify aquatic bacteria and phytoplankton, but merely that the universality of many biochemical processes is a sufficient criterion to warrant discussions of cellular reactions of higher organisms in a chapter dealing with aquatic microbes.

2.   CELL SURFACES AND THEIR ENZYMES

In most aquatic microorganisms the cell membrane or plasmamembrane is surrounded by a porous cell wall composed of polysaccharides and proteins, and, in some cases, inorganic matrices (SiO_2 and $CaCO_3$) (Fig. 1). Reactive functional groups in both the cell membrane and cell wall constituents, such as $R-OH$, $R-COOH$, $R-NH_2$, and $R-SH$, act as coordinating sites on the cell surface with which dissolved metals may interact. The plasmamembrane, a lipid bilayer of hydrophobic interior and hydrophilic exterior with embedded proteins and carbohydrates (Singer and Nicholson, 1972), maintains the integrity of the cell. It also plays a vital role in regulating the flow of materials and energy into and out of the cell.

Proteins account for a large portion of the plasmamembrane weight (30–40%) (Leonard and Hodges, 1980), and are necessary for structure, cell–cell recognition, and nutrient transport. Some of these exoproteins in the plasmamembrane and cell wall are externally oriented enzymes (called *ectoenzymes*) that catalyze reactions with external substrates. The first such ectoenzyme was described in

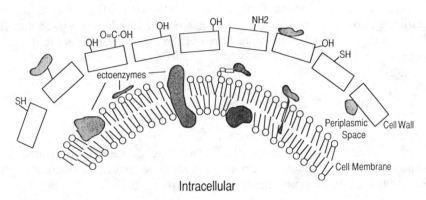

Figure 1. A conceptualized cross section through a portion of the cell wall (rectangles), periplasmic space, and cell membrane (lipid bilayer with polar head groups in contact with cytoplasm and external medium, and hydrophobic hydrocarbon chains) of an aquatic microbe. Reactive functional groups ($-SH$, $-COOH$, $-OH$, $-NH_2$) present on the wall constituents and extracellular enzymes (depicted as shaded objects) attached by various means promote and catalyze chemical reactions extracellularly. Specific examples of these enzymes and their mode of attachment to the cell are outlined in the text.

the 1950s by Rothstein et al. (1953) and at least 30 others have been identified since (Kenny and Turner, 1987). While some catalytic properties of these enzymes are known, in many cases, their actual physiological role is not.

Extracellular enzymes are attached to the cell surface by hydrophobic interaction with membrane lipids, and sometimes, as sketched in Figure 1, by unique, specific anchoring segments. A few ectoenzymes contain a hydrophobic amino acid segment that is embedded in the lipid bilayer; the hydrophilic region containing the catalytic site is then externally oriented in its active conformation. Other enzymes are inserted through the plasmamembrane and are exposed on both sides. Alternatively, some proteins are covalently linked to lipids through amide, thioester or O-acyl bonds and are attached to the cell membrane in this manner. Alkaline phosphatase of mammalian tissues, for example, is an ectoenzyme linked through a sugar phospholipid linkage (Low, 1987). For some enzymes, however, such as diamine oxidase in plant cell walls, their specific mode of attachment is unknown (Kaur-Sawhney et al., 1981).

Ectoenzymes are synthesized intracellularly and then transported to the cell surface. For example, the cell surface enzyme 5'-nucleotidase, is synthesized intracellularly, and then shuttled through the Golgi apparatus (intracellular membrane network) to the external surface of the plasmamembrane (Luzio et al., 1988). After its synthesis and during this pathway, much modification to the enzyme occurs intracellularly, including glycosylation (attachment of sugars) and proteolysis (removal of amino acid fragments). Certainly, the mechanisms of activation and the extracellular insertion of ectoenzymes must be as diverse as the enzymes themselves. For example, in the case of the cell-wall-bound alkaline phosphatase of the diatom *Phaeodactylum tricornutum* (Flynn et al., 1986) additional steps in the processing of this enzyme are required to attach it to the cell wall, and these are likely to take place outside the cell. We expect that similar complexity is involved in the processing of all surface enzymes.

3.   REACTIONS AT CELL SURFACES

Chemical reactions in aquatic media that are promoted and catalyzed by microbes occur through a number of pathways (Fig. 2). Reductants produced by photosynthetic light reactions drive the metabolic processes of autotrophic organisms, including carbon fixation and nitrate reduction:

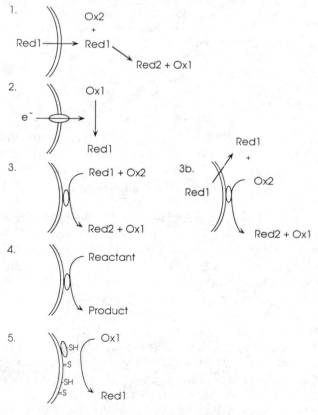

Figure 2. Examples of the types of chemical reactions promoted by aquatic microbes, including reductant excretion (1), transplasma membrane electron transport (2), redox reactions involving extracellular solutes (3) and reductants excreted by the cell (3*b*), hydrolysis reactions (4), and nonenzymatically mediated redox reactions (5). The cell wall and membrane are represented by the curved parallel lines and extracellular enzymes by ovals.

Light Reaction

$$2H_2O + hv = O_2 + 4e^- + 4H^+$$

"Dark" Reactions

$$CO_2 + 4e^- + 4H^+ = CH_2O + H_2O$$

$$NO_3^- + 8e^- + 10H^+ = NH_4^+ + 3H_2O$$

During cell growth, dissolved organic carbon is excreted to the medium and is a source of reductant for chemically and biologically mediated reactions. The magnitude of this process is open to debate (Fogg, 1983; Sharp, 1984), but is likely to be on the order of 5–10% of the total phytoplankton carbon.

Release of reduced carbon by photosynthetic phytoplankton and its oxidative degradation by heterotrophic bacteria represents the most general case of biologically controlled chemical reactions (Fig. 2.1). Included in this category are the production of exudates that participate in specific and general complexation reactions with metals, as well as those that promote redox reactions in solution. For example, some aquatic microbes are capable of producing siderophores to complex iron (Kerry et al., 1988). Other bacteria oxidize organic matter with sulfate and release sulfide, which can then reduce a number of metals (Burdige and Nealson, 1986).

The electrons provided in the light reaction, however, may also be directly exported from the cells and used to reduce a variety of extracellular substrates. This electron export is effected by surface enzymes (called *transplasmamembrane reductases*) spanning the plasmamembrane from the inside surface to the outside. They transfer electrons from an internal electron donor [chiefly NADH and NADPH; see Crane et al. (1985)] to an external electron acceptor. Direct reduction of extracellular compounds by transplasmamembrane electron transport proteins is prevalent in all cells thus far examined (Fig. 2.2). Although the function of this redox system is still subject to speculation, in phytoplankton it shows considerable activity, relative to other biochemical processes. A host of membrane-impermeable substrates, including ferricyanide, cytochrome c, and copper complexes, are reduced directly at the cells surface by electrons originating from within the cell. In phytoplankton, where the source of electrons is the light reactions of photosynthesis, the other half-redox reaction is the evolution of O_2 from H_2O. In heterotrophs, the electrons originate in the respiration of reduced substances.

Apart from releasing photosynthetically reduced carbon and exporting electrons, microbes also catalyze extracellular redox and hydrolysis reactions on their surfaces using external substrates. Of particular interest in our discussion is a class of surface enzymes in phytoplankton, deaminases, that catalyze the release of NH_4^+ by oxidation of nitrogen containing compounds (Palenik, 1989)

(Fig. 2.3). The balancing half-reaction for this oxidation is the reduction of O_2 to H_2O_2. These ectoenzymes appear to have evolved to permit some phytoplankton to exploit primary amines and amino acids as nitrogen sources for growth.

In some cases, mechanisms 2.1 and 2.3 are combined to yield a sequence of reactions where reductants excreted by the cell participate in redox reactions catalyzed by ectoenzymes (Fig. 2.3). Although no such reaction to our knowledge is yet characterized for aquatic microbes, in higher plants a cell surface enzyme that oxidizes malate to reduce extracellular NAD^+ operates in this manner.

Hydrolysis reactions of great variety commonly occur on microbial cell surfaces (Fig. 2.4). Phosphatases that degrade phosphomonoesters, urease that degrades urea, carbonic anhydrase that catalyzes the interconversion between bicarbonate and carbon dioxide, and proteases that attack amide bonds in proteins, to name a few, are present. Some of these enzymes are found inside as well as on the surfaces of cells.

Reduction of extracellular substrates can also occur at the expense of reduced moieties on the cell surface (Fig. 2.5). These reactions may occur in the dark or require promotion by light. They are nonenzymatic, and the reductant must be regenerated while it is consumed in order to continue catalyzing the reaction. Examples of this reaction are Cu(II) and Fe(III) reduction by algal cell walls (Anderson and Morel, 1980; Jones et al., 1987).

In the remainder of this chapter after discussing the issues of the chemistry of the external milieu and enzyme kinetics, we provide some examples of extracellular reactions, with emphasis on those that are catalyzed by enzymes and discuss their implications for water chemistry. We focus on transplasmamembrane redox proteins (Fig. 2.2), deaminases (Fig. 2.3), and phosphatases (Fig. 2.4), which are the objects of active research among biologists.

4.   THE EXTERNAL ENVIRONMENT

Intracellular enzymes function in an internal milieu of well-controlled composition. By contrast, enzymatic reactions outside the cell must occur in an external medium of variable chemistry. Many enzymes, however, are known or thought to be located between the cell wall and the cell membrane in the periplasmic space. This extracellular compartment may afford some degree of regulation of the composition of the medium for ectoenzymes. For example, both the activity of the ectoenzymes themselves and the diffusion barrier created by the cell wall may conspire to maintain a periplasmic composition different from that of the bulk medium.

Ionic strength, pH, and Ca^{2+} and Mg^{2+} concentrations affect the activity of intracellular enzymes, and are expected to influence ectoenzyme function in aquatic microorganisms. Such effects should be most pronounced in estuarine systems where these parameters are most variable. Reactive trace metals deactivate enzymes by exchanging with essential metals at the active center of metalloenzymes (nonreversible competitive binding) or by reacting with functional groups on enzymes such as sulfhydryls and changing their conformation (noncompetitive inhibition). High concentrations of dissolved metals in some aquatic systems may thus be detrimental to ectoenzyme activity. For example, alkaline phosphatase in phytoplankton culture and in natural samples is inhibited by copper at quite low cupric ion activities (Reuter, 1983). Because phytoplankton have the capacity to exclude some toxic metals and regulate their internal composition, ectoenzymes are obviously more vulnerable to increases in dissolved metal concentration than are intracellular enzymes. Consequently, analysis of ectoenzyme activity may be a more discriminating way of assessing the subtle effects of trace metal toxicity to microorganisms than measuring metabolic reactions, such as photosynthesis and respiration, which integrate a large number of biochemical reactions.

Two immediate consequences of the inhibition of an ectoenzyme, such as alkaline phosphatase, are the cessation of algal growth by nutrient limitation and the disruption of phosphorus cycling. As emphasized later, ectoenzymes promote an interdependence among the cycles of major and minor elements in aquatic systems.

5. CELL SURFACE ENZYME KINETICS

The kinetics of transformation of substrates by enzymes follows a well-known rate law, described by the Michaelis–Menten equation:

$$-\frac{dS}{dt} = V = \frac{V_{\max} S}{(K_s + S)} \tag{1}$$

where $V_{\max}$ is the maximum catalytic rate, S the substrate concentration, and K_s the half-saturation constant—that is, the concentration of S at which $V = \frac{1}{2} V_{\max}$. If S is small relative to K_s, then enzyme kinetics are first-order in substrate concentration:

$$V = \left(\frac{V_{\max}}{K_s}\right) S \tag{2}$$

At high substrate concentrations, when $S \gg K_s$, the kinetics reach the enzyme saturated rate and become zero-order:

$$V = V_{\max} \tag{3}$$

The Michaelis–Menten rate equation can be derived by considering a reversible binding of the substrate to the enzyme

$$S + E \underset{k_b}{\overset{k_f}{\rightleftharpoons}} SE \qquad (4)$$

and an irreversible catalytic transformation (or uptake) step

$$SE \overset{k_t}{\rightleftharpoons} S + P \qquad (5)$$

At steady state, when $d(SE)/dt = 0$, the rate of transformation of the substrate is given by

$$-\frac{dS}{dt} = V = \frac{E_T k_t \, k_f/(k_b + k_t) \, S}{(1 + k_f/(k_b + k_t) \, S)} \qquad (6)$$

where E_T is the total enzyme concentration. Thus the maximum rate and the half saturation constants can be expressed as a function of the chemical kinetic constants:

$$V_{max} = E_T k_t \qquad (7)$$

$$K_s = \frac{k_b + k_t}{k_f} \qquad (8)$$

The preceding derivation allows us to highlight a few important points regarding enzyme kinetics. The most fundamental of these is that Michaelis–Menten kinetics are applicable only at steady state. On introduction of a substrate into a cell culture or a natural sample, for example, the initial rate of transformation is zero and increases to its steady-state value over a time whose duration is given by the characteristic $t = (k_f E_T + k_b + k_t)^{-1}$. This transient time may not be very short and may confound the results of short-term tracer experiments.

It is often assumed that the transformation rate of an enzyme is slow compared to the dissociation of the enzyme–substrate complex $(k_t \ll k_b)$. In that case, the saturation constant is simply the inverse of the stability constant for the formation of the enzyme–substrate complex $(K = k_f/k_b)$, that is, of the affinity of the enzyme for the complex. Under these particular conditions the steady-state enzymatic rate corresponds to a pseudoequilibrium between the substrate and the enzyme. That is, the substrate and the enzyme remain at equilibrium with each other while the substrate is being transformed and depleted.

In many cases of interest in natural waters, however, enzymatic systems have evolved to use a rare, sometimes limiting substrate, and there is little or no advantage in a slow transformation rate compared to the back reaction. Characteristically, in this case, $k_t \gg k_b$. The system then does not reach a true pseudoequilibrium, but a steady state where the rate of binding of the substrate to the enzyme equals the rate of transformation by the enzyme (Hudson, 1989). Under these conditions, the half-saturation constant is the ratio of binding and transformation rate constants, $K_s = k_t/k_f$, and is not simply related to the affinity of the enzyme for the substrate.

Often extracellular enzymes in natural waters are induced by a low concentration of a necessary substrate. For example, some deaminases and phosphatases are induced in phytoplankton when inorganic nitrogen and phosphorus are scarce, so as to use organic nutrient sources. Further, the enzymatic system may be made more efficient if the substrate itself (e.g., the organic nutrient) becomes low.

As seen in the preceding equations, the rate of catalysis may be accelerated by increasing V_{max} and, to a point, by increasing k_t. Above a certain value for k_t ($> k_b$), the enzyme transformation rate becomes

$$V = E_T k_f \tag{9}$$

Then a further increase in the turnover rate of the enzyme (k_t) has no net effect on the enzymatic rate. In this situation, the enzyme kinetics become limited by the rate of coordination of the enzyme with the substrate and by the maximum density of enzyme that can be accommodated by the cell surface (Hudson, 1989).

Finally, we note that in a biochemically versatile world, enzyme kinetics may often reach the limit of the possible as defined by the rate of diffusion of the substrate. In other words, it is the rule rather than the exception that enzyme kinetics for a limiting substrate become so effective in nature that the rate of transformation of the substrate is limited by the rate of diffusion to the cell surface as much as by the enzyme reactions themselves. The kinetics of such a situation have been worked out by Pasciak and Gavis (1974). The net result is a rate equation of the form

$$V = \frac{V_{max}\, PS}{(K_s + PS)} \tag{10}$$

where $P = 4\pi RDK_s/V_{max}$ is simply the ratio of the rate of supply of the nutrient by diffusion to the maximum rate of enzyme catalysis, R (in centimeters) the radius of the cell, and D (in square centimeters per second) the diffusivity of the substrate. As seen in this expression, for the same density of cell surface enzymes, large cells are more likely than small ones to operate at a diffusion-limited rate.

The absolute diffusion-limited rate (considering spherical cells that are perfect sinks for the substrate) is given by

$$V_D = 4\pi RDC \tag{11}$$

Thus, with a typical diffusivity of 10^{-5} cm^2 s^{-1}, a 3-μm cell obtains at the most 4×10^{-20} mol cell^{-1} s^{-1} from a substrate at 10^{-9} M. This rate increases proportionally with C and R, but the need for substrate increases with R^3, of course. (*Note*: This discussion of enzyme kinetics is applicable *in toto* to uptake kinetics, another mechanism by which cells influence the chemistry of their surroundings.)

6. TRACE ELEMENT REDUCTION BY ALGAE

A diverse assemblage of phytoplankton and macroalgae, including diatoms, coccolithophorids (Palenik and Morel, 1988), charophytes (Ivankina and Novak, 1988), and cyanobacteria (Peschek et al., 1988) reduce extracellular electron acceptors. For example, the coastal diatom, *Thalassiosira weissflogii*, reduces Cu(II) and Fe(III) complexes via three independent reductants as shown in Fig. 3 (Jones et al., 1987): (1) Reductants released by the cells and free in solution that are quickly consumed and are not regenerated in the absence of cells, (2) cell surface moieties that are also rapidly exhausted and can be inactivated by pretreatment with oxidants, and (3) a cell surface reductase whose activity is light-independent.

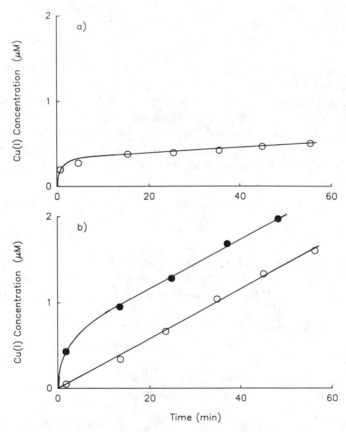

Figure 3. (*a*) Cu(BPDS)$_2$ reduction by *Thalassiosira weissflogii* culture medium after cells were removed. Reduction of Cu(II) by BPDS has been substracted. (*b*) Cu(BPDS)$_2$ reduction by *Thalassiosira weissflogii*. Cellular component of reduction: control (●), after copper treatment (○). Reduction of Cu by extracellular solutes and the blank have been subtrated. [After Jones et al. (1987).]

6.1. Soluble Reductants

In a culture medium in which *T. weissflogii* cells have grown and are then removed, dissolved solutes are present that reduce Cu complexes (Jones et al., 1987) (Fig. 3*a*), These same or similar solutes also reduce Fe(III) (Anderson and Morel, 1980). It is premature at this time to rule out the possibility that disruption of cells during their removal releases reductants into the medium, since the nature of this reducing material is unknown. However, Jones et al. (1987) have shown that polygalacturonic acid, a common cell wall constituent, is able to reduce $Cu(BPDS)_2$ in the absence of cells. Conceivably, such cell wall components may be normally sloughed off, particularly during filtration or centrifugation of the cells from the medium, and be responsible for this activity in cell-free medium. Other cell surface constituents that could promote these redox reactions, in addition to the cell wall polysaccharides, are the sulfhydryl-containing proteins and amino acids.

6.2. Cell Surface Reductants

Direct, nonenzymatic reduction of metal complexes by the cell wall or cell membrane of *T. weissflogii* is evidenced in two ways. First, a nonlinear rate of $Cu(BPDS)_2$ reduction occurs initially after addition of the substrate. Second, this reductive capacity of the cells is sensitive to pretreatment with an oxidant such as Cu(II) (Fig. 3*b*). These observations demonstrate that the reducing sites are irreversibly consumed and are not replaced or regenerated quickly. The copper pretreatment does not interfere with the cell's ability to reduce metal complexes through an enzymatic pathway.

Similar cell surface redox reactions catalyzed by reactive surface groups such as sulfhydryl groups are believed to effect the reduction of Cu(II) in *Dunaliella tertiolecta* (Jones et al., 1985). In this organism, reduction of Cu(II) is observed only in the light (>425 nm). Reduction of molecular oxygen by Cu(I) to superoxide (O_2^-) regenerates the metal ion to its original valence. Maintenance of these reduced groups on the cell surface may depend on extracellular photochemistry or on photosynthetic replacement as the cell grows rather than on transmembrane electron export.

6.3. Cell Surface Reductases

Reduction of membrane-impermeable oxidants by intact cells of *T. weissflogii* occurs without the addition of extracellular reductants, by a process mediated by a transplasmamembrane redox system (Fig. 3*b*). In the light and in the dark, enzymatic processes actively reduce Cu and Fe complexes with reduction potentials between -100 and $+100$ mV. The rate of reductant consumption in this process [a few femtomoles of Cu(II) reduced per cell per hour] can be as high as 10% of the rates of carbon fixation and nitrate reduction. In other words, reduction of extracellular solutes by cells potentially diverts a significant portion of photochemically generated electrons out of the cell.

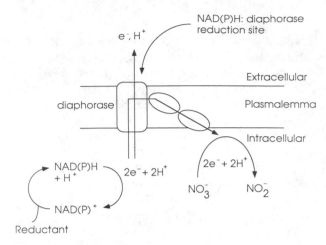

Figure 4. Model of membrane reductase in *T. weissflogii*. [After Jones and Morel (1988).]

Reduction of Cu and Fe organic complexes is saturable, and is inhibited by cell-impermeable probes, including polyclonal antibodies raised to *Chlorella* nitrate reductase (Jones and Morel, 1987). The model proposed suggests that a fraction of a membrane-bound nitrate reductase (the so-called diaphorase component that catalyzes NADH reduction), spans the cell membrane (Fig. 4). Organic metal complexes intercept electrons (destined for nitrate reduction intracellularly) at the outer cell surface.

Similar transplasmamembrane reductases are present in other algae and aquatic macrophytes, and probably aquatic bacteria. As in the preceding example, evidence for their existence comes from the observation that a variety of membrane-impermeable solutes are oxidized and reduced by intact cells. Distinct from these enzymes are other redox enzymes present on the inside and outside surfaces of the plasmamembrane that may somehow be linked to the transplasmamembrane redox system.

The true physiological role of these reductases in phytoplankton is not known and it is unclear whether electron transport out of the cell occurs in nature. Although the oxidation and reduction of the extracellular solutes may just be an adventitious reaction of these enzymes with no significance to the microorganism, under certain conditions, such as the presence of favorable redox couples, such electron export may occur. Because of the high half-saturation constants measured for metal reduction by Jones et al. (1987), it seems unlikely that these or similar trace metal complexes are the major electron acceptors in nature. We cannot rule out, however, the possibility of reduction of other (major) solutes such as sulfate or intermediate redox sulfur species.

For other biological systems, the role of plasmamembrane reductases is more obvious. In plant cells, an electrochemical gradient across the plasmamembrane is the driving force behind ion and solute transport (Spanswick, 1981). This gradient is established by the transport of protons out of the cell, thereby rendering the cell interior negative with respect to the outside and establishing a proton concentration gradient across the cell membrane. Transplasma-membrane redox enzymes are thought to be responsible in part for the establishment of this electrochemical gradient, although the degree to which redox processes and the proton pumping ATPases (Sze, 1985) contribute to the membrane potential is not yet established. In several instances, acidificaion of the medium (interpreted as proton export) occurs concomitantly with the reduction of model substrates such as ferricyanide (Neufeld and Bown, 1987; Peschek et al., 1988). Thus, cell surface redox activity is thought to be responsible for the maintenance of a transmembrane electrochemical gradient that is used to drive, either directly or indirectly, ion and solute transport into the cell (Crane et al., 1985a). Interestingly, in *T. weissflogii* nitrate uptake is inhibited by 30% in the present of 10 μM Cu(BPDS)$_2$, but carbon fixation rates are unaffected even at 100 μM Cu(BPDS)$_2$. These results present the possibility that nitrate uptake and plasmamembrane redox activity may be related in this organism.

In higher plants and mammalian cells, the plasmamembrane redox system also appears to be involved in iron reduction (Crane et al., 1985b). An inducible redox activity in plant cells is related to iron deficiency and is distinct from normal redox activity in its ability to reduce a variety of iron chelates, including FeEDTA (Bienfait, 1985). Two pathways may be operative; whose relative contributions are dependent on extracellular pH. In one, excreted malate provides reducing power for NAD$^+$ reduction via a cell-wall-bound malate dehydrogenase (as in the mechanism of Fig. 2.3b). The subsequently formed NADH reduces ferric chelates through a ferric NADH oxidoreductase in the cell wall (Tipton and Thowsen, 1985). In the other, superoxide, formed through plasmamembrane redox enzyme activity, promotes Fe(III) reduction (Cakmak et al., 1987).

Recent evidence has been gathered that indicates that extracellular iron reduction and iron uptake in some phytoplankton may also be linked. Anderson and Morel (1981) showed that the Fe(II) complexing agent BPDS inhibited Fe(III) uptake by *T. weissflogii* in medium spiked with Fe(III)EDTA, suggesting a role for Fe reduction in making Fe more available to the cells. In *Chlorella vulgaris*, the Fe in the hydroxamate complexes ferrioxamine B and rhodotorulic acid is taken up only by iron-starved cells, and Fe reduction also appears to be involved in this process (Allnutt and Bonner, 1987).

7. DEAMINASES

Cell surfaces may also catalyze redox reactions at their surfaces without supplying the electrons from internal reductants. The net formula is that of a "normal" redox reaction

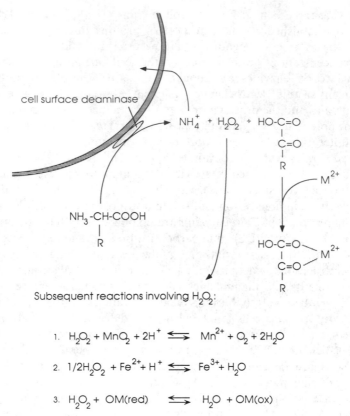

cell surface deaminase

NH_4^+ + H_2O_2 + HO-C=O
                              |
                              C=O
                              |
                              R
                                    M^{2+}

NH_3-CH-COOH
        |
        R

HO-C=O
    | M^{2+}
    C=O
    |
    R

Subsequent reactions involving H_2O_2:

1. H_2O_2 + MnO_2 + $2H^+$ $\rightleftharpoons$ Mn^{2+} + O_2 + $2H_2O$

2. $1/2H_2O_2$ + Fe^{2+} + H^+ $\rightleftharpoons$ Fe^{3+} + H_2O

3. H_2O_2 + OM(red) $\rightleftharpoons$ H_2O + OM(ox)

Figure 5. L-Amino acid oxidase of *Pleurochrysis carterae*. L-Amino acids are oxidized by a cell surface enzyme producing extracellularly the corresponding α-keto acid, H_2O_2, and NH_4^+. The NH_4^+ is taken up by the alga and used for growth, the α-keto acid in this is diagram hypothesized to complex metals; and the H_2O_2, to oxidize and reduce a variety of solutes (Palenik, 1989).

$$Red_1 + Ox_2 = Ox_1 + Red_2$$

but the electron transfer is effected by surface enzymes. The best example we have of this reaction to date is the extracellular L-amino acid oxidase of *Pleurochrysis carterae* (Fig. 5) (Palenik and Morel, in press). This enzyme oxidizes some amino acids found in natural waters, producing NH_4^+, H_2O_2, and the corresponding α-keto acid

$$R-CHNH_3COO + H_2O + O_2 = H_2O_2 + NH_4^+ + R-CHOCOO^-$$

This enzymatic reaction has been demonstrated to occur extracellularly through the use of cell-impermeable protein modifying reagents that inhibited H_2O_2 production. Moreover, the products of the reaction (H_2O_2 and pyruvate—an α-keto acid) accumulate in the external medium in equimolar concentrations when cells are exposed to the amino acid L-alanine. The ammonium liberated by the reaction is taken up by the cells and used as the nitrogen source for growth.

Not only does this deaminase reaction transform dissolved organic nitrogen, but the reactive H_2O_2 produced may also oxidize or reduce some aquatic substrates such as metals (Cu, Fe, Mn) or hydrolyze organic compounds. The quantitative role that these and similar enzymes may play in the production of H_2O_2 in natural waters is not yet known. Initial results suggest that while abiotic photochemical formation rates of H_2O_2 (Cooper et al., 1988) are faster than biological production rates in surface waters, the biota may be significant producers at least at some locales, at certain depths in the water column. Palenik et al. (1987) estimated that H_2O_2 production by *Pleurochrysis carterae* alone could be $1-2 \, nM \, h^{-1}$. Given that such amino acid oxidases are widespread among phytoplankton species (Palenik, 1989), and that other deaminase reactions also lead to the production of H_2O_2, biological production of H_2O_2 may be more important than is currently realized.

The activity of cell surface deaminases is typically greatest when phytoplankton are nitrogen-limited, and least, or nonexistent, when ammonium is present. For this reason, in aquatic habitats, such as the open ocean, where the concentrations of ammonium and other dissolved inorganic nitrogen species are exceedingly low these extracellular enzymes will be most active.

Other cell surface deaminases may be important in nitrogen transformations in seawater. An amine oxidase which degrades primary amines, including ethanolamine and putrescine, is present in some phytoplankton (Palenik, 1989). This enzyme is distinct from the L-amino acid oxidase, and in *Pleurochrysis carterae* both enzymes may be present at the same time. A cell-wall-localized polyamine oxidase in oat seedlings that uses spermine and spermidine as substrates (Kaur-Sawhney et al., 1981), produces H_2O_2 as a by-product of this oxidation. We have yet to assess whether similar enzymes are present on the surfaces of aquatic microbes, but the marine diatom *T. weissflogii* can use spermine and spermidine as nitrogen sources for growth.

The picture that is now emerging is one where phytoplankton oxidize a variety of organic nitrogen substrates via ectoenzymes and use the liberated ammonium for growth. As exemplified by Palenik's work, it appears that some of these organisms (and perhaps all) have become specialized for using certain forms of nitrogen present in natural waters. These results also suggest that phytoplankton may be instrumental in the transformation of dissolved organic nitrogen, a role previously attributed solely to bacteria.

8. NONREDOX ECTOENZYMES

A variety of extracellular enzymes present in bacterioplankton and phytoplankton are not involved in redox reactions (Table 1). Alkaline phosphatase,

TABLE 1. Some Known Ectoenzymes

Enzyme	Function	Remarks
Deaminases	N aquisition	O_2 reduced to H_2O_2
Phosphatases	P acquisition	Present in algae and bacteria
Plasmamembrane Reductases	Metal reduction (Fe uptake?)	Physiological role uncertain
Proteases	Degradation of proteins: C and N aquisition	Present in bacteria
Urease	N aquisition	Ectoenzyme in some bacteria, and a diatom

carbonic anhydrase, urease, and proteases are examples which catalyze the following reactions:

$$R-OPO_3^{2-} + H_2O = R-OH + HPO_4^{2-}$$
$$HCO_3^- + H^+ = CO_2 + H_2O$$
$$CO(NH_2)_2 + H_2O = 2NH_3 + CO_2$$
$$R_1-CONH-R_2 + H_2O = R_1-COOH + R_2-NH_2$$

The first two enzymes are well documented to occur extracellularly. We have studied urease and have also found evidence that this enzyme is on the cell surface of at least one phytoplankton species. Possibly with the exception of the proteases, these enzymes contain essential metal cofactors which are necessary for their activity. There are reports, however, which document the inactivation of microbial proteases by chelating agents, suggesting that they too are metal-loenzymes (Matsubara and Feder, 1971).

8.1. Urease

Urease, a nickel metalloenzyme, is required by marine diatoms for growth on urea, but is not required when other nitrogen substrates are used for growth. Urease activity is regulated by the chemistry of the medium, by both the availability of nickel and the type of dissolved nitrogenous substrates present. Similarly, the uptake of nickel is also influenced by the nitrogenous substrates that are used for growth. Cells regulate accordingly the transport of this essential cofactor depending on their need for urease to degrade urea. In *T weissflogii*, uptake rates of nickel are 5–10 times faster in urea-grown cells than in cells using ammonium as a nitrogen source. These results demonstrate the interaction between micronutrients and macronutrients. In other words, the cell's requirement for nickel is affected by the nitrogenous compounds used for growth, while the cell's nitrogen requirements cannot be satisfied by urea without nickel. This nutrient interaction should be generally applicable for other metal-containing enzymes (such as phosphatases) that catalyze the degradation of macronutrients. It is through this type of mechanism that the cycles of the major and minor nutrients are inseparably related.

8.2. Proteases

Extracellular proteases, a poorly characterized, heterogeneous group of enzymes, have been detected in natural waters using model substrates and are abundant (Hoppe et al., 1988). Their identification and study has been limited because of a lack of knowledge of their natural substrates. In natural waters, a major fraction of dissolved nitrogen is in organic compounds, and according to some estimates (Tuschall and Brezonik, 1980) a large portion of this is protein. These enzymes then may play an important role in the cycling of dissolved organic nitrogen, by degrading proteins into readily assimilable amino acids. As we have discussed, phytoplankton cell surfaces deaminases can then function to recover the ammonium from these compounds.

8.3. Phosphatases

Utilization of phosphate monoesters by microalgae and bacteria is effected by phosphomonoesterases (phosphatases) of broad specificity present at the cell surface. Hydrolytic release of PO_4^{3-} from sugar phosphates, nucleotide phosphates, phospholipids, and phenyl phosphates, to name a few, enables a wide variety of phosphorus containing compounds to be utilized as phosphorus sources for growth of microbes. Ultrastructural observations and results from biochemical experiments indicate that extracellular phosphatases cleave the phosphate moiety from dissolved organic phosphorus compounds, which is then internalized, leaving the carbon skeleton outside the cell (Kuenzler and Perras, 1965; Doonan and Jensen, 1977).

Extracellular, acidic, and alkaline phosphatases differ not only in their pH optima, but also in their requirements for Mg^{2+} and Zn^{2+}, in their inhibition by chelators such as EDTA, and in their sensitivity to fluoride (Cembella et al., 1984). These enzymes, known to occur simultaneously in certain algal species, appear to be common among algae and bacteria, although in some, such as *Dunaliella tertiolecta*, they are lacking. Some phosphatases, particularly the intracellular ones are always present and active in cells, but the activity of their extracellular counterparts is dependent on phosphorus availability. Like the extracellular deaminases, the synthesis and activity of phosphatases is regulated by the nutritional state of the organisms.

Extracellular phosphatase activity has been detected in the oligotrophic Pacific Ocean (Perry, 1972), in the Sargasso Sea (Rivkin and Swift, 1979), in coastal waters (Taft et al., 1977) and in lakes (Berman, 1970). Under conditions where ample dissolved phosphate is present for growth or intracellular concentration of the storage polyphosphate is high, synthesis of these enzymes is impaired. Because enzyme synthesis and activity is induced in response to phosphorus limitation, the presence of alkaline phosphatase activity in natural waters is interpreted as an indication of phosphate limitation.

These ectoenzymes play a significant part in phosphorus cycling in natural waters. In lakes and oceans, phosphorus is partitioned among particulate and dissolved inorganic and organic fractions and is rapidly transformed from one fraction to another. Estimates of the size of the labile dissolved organic phosphorus pool in waters off the coast of Hawaii (0.01–0.2 μg at P L^{-1}), and its rapid turnover (0.008–0.04 h^{-1}), presumably facilitated by extracellular phosphatases, are comparable with those of PO$_4^{3-}$ (Smith et al., 1985) and indicate the importance of ectoenzymes in the major nutrient cycles.

Some of the enzymes discussed above have been detected in filtered water samples free of particulate matter. These include the phosphatases and the starch-degrading amylases. Generally, free enzymes appear to be more abundant in sediments than in the water column (Kim and Zobell, 1972), but at times some of them, such as alkaline phosphatase, can be as active as their particle-bound forms (Berman, 1970). Dissolved extracellular enzymes are presumably excreted by various microorganisms, or perhaps are released during cell degradation or lysis.

9.   ROLE OF ECTOENZYMES IN TRACE ELEMENT CYCLING

The reactions of ectoenzymes transform dissolved organic nitrogen and phosphorus compounds and thus are intimately involved in the biogeochemical cycling of these elements. Ectoenzymes also influence the cycling of trace metals by promoting and catalyzing redox reactions through either direct or indirect mechanisms. Reduction of Cu and Fe complexes by the plasmamembrane reductase in *T. weissflogii* exhibits characteristics of an enzyme–substrate complex, implying that there is direct participation of the ectoenzyme in this reaction. While we recognize that the physiological role of this enzyme may not be for metal reduction, it will, under appropriate conditions, mediate this reaction and it may be active in natural waters. Oxidation and reduction of metals by H$_2$O$_2$, produced by deaminases, represents an indirect effect of an ectoenzyme on trace element cycling.

In laboratory cultures of marine phytoplankton, a variety of Cu(II) complexes are reduced by plasmamembrane reductases. While we have no evidence for their activity in natural waters, the maximum contribution of these enzymes to Cu(II) reduction can be calculated. If we assume that all Cu(II) is reducible by a plasmamembrane reductase similar to that present in *T. weissflogii*, and that cell densities are 10^5 to 10^7 cells per liter, then the reduction rate in natural waters would be between 2.5–250 pM day^{-1}. This reduction rate would increase twofold if the activity of the cell wall and soluble reductants were considered. In surface coastal waters, Moffett and Zika (1987) found a Cu(I) production rate of $\sim$200 pM h^{-1}. Supposing that this rate is maintained throughout the light portion of the day (10 h), then the daily production rate is 2000 pM day^{-1}, 10 times the maximum potential biological production rate. Although the abiotic, photochemical production of Cu(I) exceeds potential biological production in surface waters, at greater depths or in highly colored water where light penetration is reduced, biological Cu(II) reduction may be more significant. The relative contributions of these processes to Cu(II) reduction will vary not only spatially, but also temporally. To assess the importance of biological reduction in natural waters, we need to determine the physiological role of the plasmamembrane reductases of aquatic microbes and to characterize potential electron acceptors.

In the case of Mn, there is evidence for a role of ectoenzymes in both oxidation and reduction reactions (Fig. 6). Manganese cycling in the photic zone of natural waters involves a redox transition between dissolved Mn(II) and particulate Mn(IV). In freshwater, Mn(II) oxidation by algal photosynthetic activity has been reported (Richardson et al., 1988) and may indirectly involve the enzyme carbonic anhydrase, which catalyzes the conversion of bicarbonate to carbon dioxide extracellularly. For bicarbonate utilizing species of algae, a more rapid

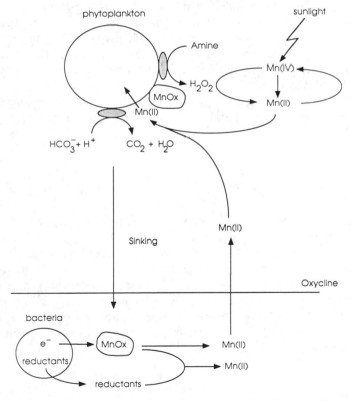

Figure 6. Manganese cycling and the role of cell surface enzymes. Consult the text for an explanation of this cycle.

HCO_3^- influx in the presence of extracellular carbonic anhydrase could lead to more OH^- production by the cells, resulting in an extracellular pH rise. Such pH conditions are known to favor Mn(II) oxidation (Stumm and Morgan, 1981). Aquatic microbes in oxygenated (Sunda and Huntsman, 1987) and anaerobic (Emerson et al., 1982) seawater oxidize Mn(II) by another mechanism that is not yet understood. This catalysis is not an indirect result of microbial reactions as described in the preceding example. The measured rates of Mn(II) oxidation exceed by 100 times the rates of Mn(II) uptake by phytoplankton (Sunda and Huntsman, 1987) and are greater than abiotic oxidation rates in sterile medium. Thus, these processes cannot explain the high rates of Mn(II) oxidation. It seems reasonable then to assume that Mn(II) oxidation involves biota and that it must occur extracellularly. This may be one example of an oxidation reaction involving cell surface enzymes or functionalities of aquatic microbes.

Manganese oxide (MnO_x) photoreduction is an important mechanism maintaining manganese in a reduced state in surface waters (Sunda et al., 1983), but reductive dissolution of MnO_x by H_2O_2 (Sunda and Huntsman, 1983), produced in part by phytoplankton amine oxidases, may be another pathway by which manganese is kept in a thermodynamically unfavorable reduced state.

The role of ectoenzymes in Fe cycling is not known presently, but we have some evidence of Fe(III) reduction in phytoplankton cultures, and Fe(III) complexes are reduced by plasmamembrane reductases. Hydrogen peroxide can oxidize Fe(II) (Waite and Morel, 1984) and probably reduce Fe(III) in light, thus implying that the biota may promote this reaction through the activity of their extracellular enzymes.

These observations present a strong case for a role of ectoenzymes in manganese and perhaps iron and copper cycling in natural waters.

10. CONCLUSIONS

Aquatic microorganisms supply electrons through transplasmamembrane reductases to external solutes, enzymatically catalyze a variety of redox and other reactions on the cell surface, and are a source of dissolved extracellular enzymes. Both bound and dissolved extracellular enzymes are probably significant in maintaining a state of disequilibrium for some redox processes in natural waters and in accelerating some thermodynamically favorable reactions. In addition, as described for nickel and nitrogen in the urease example, these enzymes may also render the chemistry of the various components of aquatic systems highly interdependent.

Acknowledgments

Support for this work was provided by ONR Contract N-00014-80-C-0273, NSF Grant OCE 3-17532, and EPA and by NSERC and Killam postdoctoral fellowships to NMP. We thank J. Bowen, J. Hering, R. Hudson, B. Palenik, and J. A. Raven for reviewing the manuscript.

REFERENCES

Allnutt, F. C. T., and W. D. Bonner, Jr. (1987), "Characterization of Iron Uptake from Ferrioxamine B by *Chlorella vulgaris*," *Plant Physiol.* **85**, 746–750, 751–756.

Anderson, M. A., and F. M. M. Morel (1980), "Uptake of Fe(II) by a Diatom in Oxic Culture Medium," *Mar. Biol. Lett.* **1**, 263–268.

Anderson, M. A., and F. M. M. Morel, 1981. "The Influence of Aqueous Iron Chemistry on the Uptake of Iron by the Coastal Diatom *Thalassiosira weissflogii*," *Limnol. Oceanogr.* **27**, 798–813.

Berman, T. (1970), "Alkaline Phosphatases and Phosphorus Availability in Lake Kinneret," *Limnol. Oceanogr.* **15**, 663–674.

Bienfait, H. F. (1985), "Regulated Redox Processes at the Plasmalemma of Plant Root Cells and Their Function in Iron Uptake," *J. Bioenerg. Biomem.* **17**, 73–83.

Burdige, D. J., and K. H. Nealson (1986), "Chemical and Microbiological Studies of Sulfide-Mediated Manganese Reduction," *Geomicrobiol. J.* **4**, 361–387.

Cakmak, I., D. A. M. Van De Wetering, H. Marschner, and H. F. Bienfait (1987), "Involvement of Superoxide Radical in Extracellular Ferric Reduction by Iron-Deficient Bean Roots," *Plant Physiol.* **85**, 310–314.

Cembella, A. D., N. J. Antia, and P. J. Harrison (1984), "The Utilization of Inorganic and Organic Phosphorus Compounds as Nutrients by Eukaryotic Microalgae: A Multidisciplinary Perspective: Part 1," *CRC Crit. Revi. Microbiol.* **10**, 317–391.

Cooper, W. J., R. G. Zika, R. G. Petasne, and J. M. C. Plane (1988), "Photochemical Formation of H_2O_2 in Natural Waters Exposed to Sunlight," *Environ. Sci. Technol.* **22**, 1156–1160.

Crane, F. L., H. Low, and M. G. Clark (1985a), "Plasma Membrane Redox Enzymes," in A. N. Maronosi, Ed., *The Enzymes of Biological Membranes*, Vol. 4, pp. 465–510.

Crane, F. L., I. L. Sun, M. G. Clark, C. Grebing, and H. Low (1985b), "Transplasma-Membrane Redox Systems in Growth and Development," *Biochim. Biophys. Acta* **861**, 233–264.

Doonan, B. B., and T. E. Jensen (1977), "Ultrastructural Localization of Alkaline Phosphatase, in the Blue–Green Bacterium *Plectonema boryanum*," *J. Bacteriol.* **132**, 967–973.

Emerson, S., S. Kalhorn, L. Jacobs, B. M. Tebo, K. H. Nealson, and R. A. Rosson (1982), "Environmental Oxidation Rate of Manganese(II): Bacterial Catalysis," *Geochim. Cosmochim. Acta* **46**, 1073–1079.

Flynn, K. J., H. Opik, and P. J. Syrett (1986), "Localization of the Alkaline Phosphatase and 5′-Nucleotidase Activities of the Diatom *Phaeodactylum tricornutum*," *J. Gen. Microbiol.* **132**, 289–298.

Fogg, G. E. (1983), "The Ecological Significance of Extracellular Products of Phytoplankton Photosynthesis," *Bot. Mar.* **26**, 3–14.

Hoppe, H.-G., S.-J. Kim, and K. Gocke (1988), "Microbial Decomposition in Aquatic Environments: Combined Process of Extracellular Enzyme Activity and Substrate Uptake," *Appl. Environ. Microbiol.* **54**, 784–790.

Hudson, R. J. M. (1989), "The Chemical Kinetics of Iron Uptake by Marine Phytoplankton," Ph.D. thesis, M.I.T., Cambridge, MA 199p.

Ivankina, N. G., and V. A. Novak (1988), "Transplasmalemma Redox Reactions and Ion Transport in Photosynthetic and Heterotrophic Plant Cells," *Physiol. Plant.* **73**, 161–164

Jones, G. J., B. P. Palenik, and F. M. M. Morel (1987), "Trace Metal Reduction by Phytoplankton: The Role of Plasmalemma Redox Enzymes," *J. Phycol.* **23**, 237–244.

Jones, G. J., and F. M. M. Morel (1988), "Plasmalemma Redox Activity in the Diatom *Thalassiosira*. A Possible Role for Nitrate Reductase," *Plant Physiol.* **87**, 143–147.

Jones, G. J., T. D. Waite, and J. D. Smith (1985), "Light-Dependent Reduction of Cu(II) and Its Effect on Cell-Mediated, Thiol-Dependent Superoxide Production," *Biochem. Biophys. Res. Commun.* **128**, 1031–1036.

Kaur-Sawhney, R., H. E. Flores, A. W. Galston (1981), "Polyamine Oxidase in Oat Leaves: A Cell Wall Localized Enzyme," *Plant Physiol.* **68**, 494–498.

Kenny, A. J., and A. J. Turner (1987), "What Are Ectoenzymes?," in A. J. Kenny and A. J. Turner, Eds., *Mammalian Ectoenzymes* (research monographs in cell and tissue physiology), Vol. 14, Elsevier, Amsterdam, pp. 1–13.

Kerry, A., D. E. Laudenbach, and C. G. Trick (1988), "Influence of Iron Limitation and Nitrogen Source on Growth and Siderophore Production by Cyanobacteria," *J. Phycol.* **24**, 566–571.

Kim, J., and C. E. Zobell (1972), "Occurrence and Activities of Cell-Free Enzymes in Oceanic Environments," in R. R. Cowell and R. Y. Morita, Eds., *Effect of the Ocean Environment on Microbial Activities*, University Park Press, University Park, TX, pp. 368–385.

Kuenzler, E. J., and J. P. Perras (1965), "Phosphatases of Marine Algae," *Biol. Bull.* **128**, 271–284.

Leonard, R. T., and T. K. Hodges (1980), "The Plasma Membrane," in *The Biochemistry of Plants*, Vol. 1, Academic Press, Orlando, FL, pp. 163–182.

Low, M. G. (1987), "Biochemistry of the Glycosyl–Phosphatidylinositol Membrane Protein Anchors." *Biochem. J.* **244**, 1–13.

Luzio. P. J., M. D. Barron, and E. M. Bailyer (1988), "Cell Biology," in A. J. Kenny and A. J. Turner, Eds., *Mammalian Ectoenzymes* (research monographs in cell and tissue physiology), Vol. 14, Elsevier, Amsterdam, pp. 111–138.

Matsubara, H., and J. Feder (1971), "Other Bacterial, Mold, and Yeast Proteases," in P. D. Boyer, Ed., *The Enzymes*, Vol. 3, Hydrolysis: Peptide Bonds, 3rd ed., Academic Press, New York, pp. 721–795.

Moffett, J. W., and R. G. Zika (1987), "Photochemistry of Copper Complexes in Seawater," in R. G. Zika and W. J. Cooper, Eds., *Photochemistry of Environmental Aquatic Systems*, American Chemical Society, Washington DC, pp. 116–130.

Neufeld, E., and A. W. Bown (1987), "A Plasmamembrane Redox System and Proton Transport in Isolated Mesophyll Cells," *Plant Physiol.* **83**, 895–899.

Palenik, B. (1989), "Organic Nitrogen Utilization by Phytoplankton: The Role of Cell Surface Deaminases," Ph.D. thesis, MIT/WHOI 134p.

Palenik, B., and F. M. M. Morel (1988), "Dark Production of H_2O_2 in the Sargasso Sea," *Limnol. Oceanogr.* **33**, 1606–1611.

Palenik, B., and F. M. M. Morel (in press), "Amino Acid Utilization by Marine Phytoplankton: A Novel Mechanism," *Limnol. Oceanogr.*

Palenik, B., O. C. Zafiriou, and F. M. M. Morel (1987), "Hydrogen Peroxide Production by a Marine Phytoplankter," *Limnol. Oceanogr.* **32**, 1365–1369.

Pasciak, W. J., and J. Gavis (1974), "Transport Limitation of Nutrient Uptake in Phytoplankton," *Limnol. Oceanogr.* **19**, 881–888.

Perry, M. J. (1972), "Alkaline Phosphatase Activity in Subtropical Central North Pacific

Waters Using a Sensitive Fluorometric Method," *Mar. Biol.* **15**, 113–119.

Peschek, G. A., M. A. Kury, and W. W. A. Erber (1988), "Impermeant Electron Acceptors and Donors to the Plasma Membrane-Bound Respiratory Chain of Intact Cyanobacterium *Anacystis nidulans*," *Physiol. Plant.* **73**, 175–181.

Reuter, Jr., J. G. (1983), "Alkaline Phosphatase Inhibition by Copper: Implications to Phosphorus Nutrition and Use as a Biochemical Marker of Toxicity," *Limnol. Oceanogr.* **28**, 743–748.

Richardson, L. L., C. Aguilar, and K. H. Nealson (1988), "Manganese Oxidation in pH and O_2 Microenvironments Produced by Phytoplankton," *Limnol. Oceanogr.* **33**, 352–363.

Rivkin, R. B., and E. Swift (1979), "Diel and Vertical Patterns of Alkaline Phosphatase Activity in the Oceanic Dinoflagellate *Pyrocystis noctiluca*," *Limnol. Oceanogr.* **27**, 107–116.

Rothstein, A., R. C. Meier, and T. G. Scharff (1953), "Relation of Cell Surface to Metabolism. IX. Digestion of Phosphorylated Compounds by Enzymes Located on Surface of Intestinal Cells," *Am. J. Physiol.* **137**, 41–46.

Sharp, J. H. (1984), "Inputs into Microbial Food Chains," in J. E. Hobbie and P. J. leB. Williams, Eds., *Heterotrophic Activity in the Sea*, Plenum Press, New York, pp. 101–120.

Singer, S. J., and G. L. Nicholson (1972), "The Fluid Mosaic Model of the Structure of Membranes," *Science* **175**, 720–731.

Smith, R. E. H., W. G. Harrison, and L. Harris (1985), "Phosphorus Exchange in Marine Microplanktonic Communities Near Hawaii," *Mar. Biol.* **86**, 75–84.

Spanswick, R. M. (1981), "Electrogenic Ion Pumps," *Ann. Rev. Plant Physiol.* **32**, 267–289.

Stumm, W. G., and J. J. Morgan (1981), *Aquatic Chemistry*, Wiley, New York.

Sunda, W. G., and S. A. Huntsman (1983), "Reduction of Manganese Oxides in Near-Surface Waters by H_2O_2," *EOS* **64**, 1029.

Sunda, W. G., and S. A. Huntsman (1987), "Microbial Oxidation of Manganese in a North Carolina Estuary," *Limnol. Oceanogr.* **32**, 552–564.

Sunda, W. G., S. A. Huntsman, and G. R. Harvey (1983), "Photoreduction of Manganese Oxides in Seawater and Its Geochemical and Biological Implications," *Nature* **301**, 234–236.

Sze, H. (1985), "H^+-Translocating ATPases: Advances Using Membrane Vesicles," *Ann. Rev. Plant Physiol.* **36**, 175–208.

Taft, J. L., M. E. Loftus, and W. R. Taylor (1977), "Phosphate Uptake from Phosphomonesters by Phytoplankton in the Chesapeake Bay," *Limnol. Oceanogr.* **22**, 1012–1021.

Tipton, C. L., and J. Thowsen (1985), "Fe(III) Reduction in Cell Walls of Soybean Roots," *Plant Physiol.* **79**, 432–435.

Tuschall, Jr., J. R., and P. L. Brezonik (1980), "Characterization of Organic Nitrogen in Natural Waters: Its Molecular Size, Protein Content, and Interactions with Heavy Metals," *Limnol. Oceanogr.* **25**, 495–504.

Waite, T. D., and F. M. M. Morel (1984), "Photoreductive Dissolution of Colloidal Iron Oxides in Natural Waters," *Environ. Sci. Technol.* **18**, 860–868.

Ward, M. R., R. Tischner, and R. C. Huffaker (1988), "Inhibition of Nitrate Transport by Anti-nitrate Reductase IgG Fragments and the Identification of Plasma Membrane Associated Nitrate Reductase in Roots of Barley Seedlings," *Plant Physiol.* **88**, 1141–1145.

9

AB INITIO QUANTUM-MECHANICAL CALCULATIONS OF SURFACE REACTIONS— A NEW ERA?

Antonio C. Lasaga

Kline Geology Laboratory, Yale University, New Haven, Connecticut

and

Gerald V. Gibbs

Department of Geological Sciences, Virginia Polytechnic Institute, Blacksburg, Virginia

1. INTRODUCTION

It has become increasingly clear that any quantitative study of heterogeneous kinetics must be based on a thorough understanding of the structure and dynamics of the mineral surfaces. Great advances in our understanding of surface processes have resulted from application of new high technologies, in particular spectroscopic methods. These advances, coupled with the more traditional approaches to surface chemistry, have forced us to take a serious look at the atomic nature of mineral surfaces and to try to model the atomic processes observed by the various experimental probes. The roles of adsorption and desorption in surface chemical kinetics have been discussed in other chapters of this book as well as numerous previous papers and texts cited in those chapters. Kinetic mechanisms proposed to understand the effect of pH on dissolution rates, the effect of ionic strength on surface rates, the dependence of rates on supersaturation or undersaturation of solutions, and the catalytic or inhibitory effects of dissolved constituents require a good knowledge of the main atomic species involved and of the chemical and physical nature of the mineral surface. A

259

rigorous attack on the atomic nature of surface kinetics including the types of activated complexes relevant to adsorption, dissolution, and precipitation will clarify greatly the nature of these processes.

A first-principles investigation of the bonding and atomic dynamics of minerals and their interaction with both hydrated ions and water can answer significant questions such as the link between local defects or impurities and the bulk kinetic properties or the nature of the bonds and forces involved in surface reactions or the association of a local structure with a particular catalytic or inhibitory effect. The task at hand is one of evaluating the interatomic forces from the fundamental laws of physics and a handful of universal constants (such as Planck's constant, the electron charge, the speed of light, the mass of nuclei and electrons)—hence the term "ab initio." No empirical or semi empirical constants are introduced into the calculations. Ab initio methods have become as good as experiments for the accurate prediction of structures and vibrational frequencies of new gas-phase molecules. Fortunately, current computational levels enable us to extend the applications to mineral surface reactions. As will be emphasized again later, an important component of successfully carrying out ab initio calculations on silicates and their surface reactions is the local nature of the chemical forces shaping the dynamics of silicates. Our earlier work (Gibbs, 1982; Lasaga and Gibbs, 1987, 1988) has shown that ab initio calculations on molecular clusters are capable of predicting the crystal structure and the equations of state of many silicates quite successfully.

In this chapter we will first discuss briefly the basic theory behind ab nitio methods. Then a discussion of transition state theory, in light of the ab initio capabilities, will be taken up. This discussion will be followed by an analysis of the molecular mechanisms in water–silicate reactions and the ab initio eluci-dation of the adsorption and kinetic barriers involved in the bulk chemical reactions occurring at mineral–water interfaces.

2. AB INITIO THEORY

The basic problem in the quantum treatment of chemical bonds is obtaining solution to Schrödinger's nonrelativistic time-independent equation:

$$\hat{H}\Psi = E\Psi \qquad (1)$$

The driving force for all the surface chemical reactions is the chemical interaction between the atoms in the system. If the variation in energy of the whole system is known as a function of the positions of the atoms in a cluster, the forces and thereby the motion of atoms can be readily computed. Such potential surfaces are the topic of this chapter.

The evaluation of potential surfaces is feasible nowadays at the level of the Born–Oppenheimer approximation. In the Born–Oppenheimer approximation the positions of the nuclei are fixed and equation 1 is solved for the wavefunction of the electrons. This separation of electron and nuclear motion is allowed because the nuclear masses are much greater than the mass of the electrons and, as a result, nuclei move much more slowly. In the Born–Oppenheimer approximation the eigenvalue, E, will then be a function of the atomic positions i.e. $E(\mathbf{R})$. Equation 1 now becomes

$$\hat{H}^{\text{elec}} \Psi^{\text{elec}} (\mathbf{r}, \mathbf{R}) = E(\mathbf{R}) \Psi^{\text{elec}} (\mathbf{r}, \mathbf{R}) \tag{2}$$

where the electronic Hamiltonian is given by the sum of the kinetic and potential energies of the electrons and the nuclear–nuclear electrostatic repulsion is added as well:

$$\hat{H}^{\text{elec}} = \hat{T}^{\text{elec}} + \hat{V} \tag{3}$$

$$\hat{T}^{\text{elec}} = -\frac{h^2}{8\pi^2 m} \sum_i^{\text{elec}} \left(\frac{\partial^2}{\partial x_i^2} + \frac{\partial^2}{\partial y_i^2} + \frac{\partial^2}{\partial z_i^2} \right) \tag{4}$$

$$\hat{V} = -\sum_i^{\text{elec}} \sum_s^{\text{nucl}} \frac{Z_s e^2}{r_{is}} + \sum_{i<j}^{\text{elec}} \frac{e^2}{r_{ij}} + \sum_{s<t}^{\text{nucl}} \frac{Z_s Z_t e^2}{R_{st}} \tag{5}$$

The eigenvalue $E(\mathbf{R})$ in equation (2) yields the Born–Oppenheimer potential surface if the nuclear positions, $\mathbf{R}$, are all varied. In particular, because the energy obtained is that of the lowest energy state, that is, the ground electronic state, the surface is the ground-state potential-energy surface. If we know $E(\mathbf{R})$ accurately, then we could predict the detailed atomic forces and the chemical behavior of the entire system.

For a closed shell system, there will be an even number of electrons, $2n$. Equation 2 is a partial differential equation for the wavefunction Ψ^{elec}, which depends on the positions of all $2n$ electrons. Most people familiar with the technique of separation of variables will recognize that, except for the electron–electron repulsion term, the solution to equation 2 could be written using separation of variables. Thus we could write Ψ^{elec} as a product of functions of only one-electron coordinates, $\psi_i(x, y, z)$. In the molecular-orbital parlance, this next move is called the *Hartree–Fock approximation*. This separation of variables allows us to write

$$\Psi = \sum_{\mathbf{P}} (-1)^{\mathbf{P}} \mathbf{P} \left[\psi_1(1)\alpha(1)\psi_1(2)\beta(2) \cdots \psi_n(2n)\beta(2n) \right] \tag{6}$$

where the summation in front of the usual separation of variables product is needed to satisfy the requirement that the electronic wavefunction, Ψ, be antisymmetric with respect to interchange of any two electrons. Therefore, we sum over all the possible permutations, $\mathbf{P}$, of the $2n$ individual functions, except that $+1$ or -1 is inserted in front of each product, depending on whether the permutation is even or odd. The product in equation 6 assumes that each spatial function can accommodate two electrons, one with spin up (α) and one with spin down (β). Thus Eq. 6 is correct for closed-shell systems, which form the majority of systems relevant to our discussion.

The one-electron functions, ψ_i, in the separation of variables schemes are called *molecular orbitals*. These molecular orbitals form the basis for the conceptual treatment of bonding in molecules (see Luther, Chapter 6, this volume). If the approximate solution (Eq. 6) is inserted into equation 3, one obtains a set of differential equations for the one-electron molecular orbitals and for the electronic energy eigenvalue, ε_i, often termed the *molecular-orbital energy*. The ε_i are, in fact, closely related to the ionization potentials of the molecule. In principle, we could solve these equations numerically for ψ_i; however, in practice, it is much more efficient to expand the molecular orbitals as a sum over some set of prescribed atomic orbitals, ϕ_μ

$$\psi_i = \sum_{\mu=1}^{N} c_{\mu i} \phi_\mu \tag{7}$$

and then find the set of coefficients, $c_{\mu i}$, which most closely make ψ_i satisfy the original one-electron differential equation. Because Schrödinger's equation is a linear equation, the equations for the coefficients become simple eigenvalue matrix equations. One problem with this scheme is that the unknown coefficients, $c_{\mu i}$, appear also in the potential-energy term of the one-electron differential equation. Therefore, the $c_{\mu i}$ are also needed to compute the matrix elements in the eigenvalue equation. This need to include the coefficients arises because in the molecular Hartree–Fock scheme each electron is moving in the average electrostatic potential of all other electrons. Hence, after a set of coefficients is solved for, this set is input into the potential-energy term to update the differential equation and obtain a new matrix. Then this new equation is used to obtain a new set of coefficients. This process is repeated until the set of coefficients does not change and we reach what is termed a "self-consistent" solution.

The set of atomic orbitals, $\{\phi_\mu\}$, used to obtain the molecular orbitals (Eq. 7) is termed the *basis set*. The size of the atomic orbital set, N, varies with the accuracy demanded of the calculation. A *minimal basis set* is one that merely uses the minimum atomic orbitals needed to accommodate all the electrons in the system up to the valence electrons. For example, a minimal basis set on oxygen would have one $1s$ function, one $2s$ function, and three $2p$ functions (see Table 1). Larger basis sets are labeled, extended basis sets (see Table 1). Mathematically, if the number of "different" atomic orbitals increases to infinity (i.e., this set forms what is termed a "complete" set), then the orbitals, ψ_i, obtained by the coefficients $c_{\mu i}$ will be the exact solutions to the one-electron differential equation. In turn, this limit would yield the best possible solution to the full Born–Oppenheimer Schrödinger equation *within* the "separation of variables" scheme represented by Eq. 6. Such a solution is termed the *Hartree–Fock limit*.

To go beyond the Hartree–Fock limit and obtain the full solution to the Schrödinger equation (in the nonrelativistic and Born–Oppenheimer limit), one would have to combine various solutions of the type in Eq. 6. In general, the n

TABLE 1. Sample Basis Sets

	Minimal Basis Set
H, He:	$1s$
Li to Ne:	$1s$
	$2s, 2p_x, 2p_y, 2p_z$
Na to Ar:	$1s$
	$2s, 2p_x, 2p_y, 2p_z$
	$3s, 3p_x, 3p_y, 3p_z$

	Split-Valence Basis Set
H, He:	$1s'$
	$1s''$
Li to Ne:	$1s$
	$2s', 2p'_x, 2p'_y, 2p'_z$
	$2s'', 2p''_x, 2p''_y, 2p''_z$
Na to Ar:	$1s$
	$2s, 2p_x, 2p_y, 2p_z$
	$3s', 3p'_x, 3p'_y, 3p'_z$
	$3s'', 3p''_x, 3p''_y, 3p''_z$

molecular orbitals with the *lowest* molecular-orbital energies are used in the Hartree–Fock solution, Eq. 6, for the ground state of a $2n$ electron system. The rest of the molecular orbitals obtained will be *excited state molecular orbitals*. Other possible wavefunctions of the type given in Eq. 6 can be formed by using excited state molecular orbitals in the product. The set of all possible products can now be used as a basis set to solve the full Schrödinger equation:

$$\Psi = \sum_i C_i \sum_P (-1)^P \mathbf{P} \left[\psi_{i_1}(1)\alpha(1) \cdots \psi_{i_n}(2n)\beta(2n) \right] \qquad (8)$$

where $i = \{i_1, i_2, \ldots, i_n\}$ stands for the set of n molecular orbitals used in the particular ith product and the n orbitals are picked from the (in principle) infinite set of molecular orbitals allowable in the system. Such a calculation corrects the energy for what is termed *electron correlation* (i.e., it corrects for the assumption of average motion used in the one-electron differential equation for the ψ_i). The method of Eq. 8 is termed *configuration interaction*, because the sum in Eq. 8 is one over many electronic configurations involving excited electronic states (i.e., the new molecular orbitals added to the list). For most chemical systems, the solutions obtained by extensive application of Eq. 8 (e.g., high N) become equivalent to the exact solution of the potential-energy surface discussed earlier.

The form of the one-electron functions used in the expansion, Eq. 7, of the molecular orbital is obviously important for an accurate solution. We expect that the better the one-electron functions mimic the atomic orbitals of the atoms involved, the better our solution will be for ψ_i even with a small number, N, of atomic orbitals. We have a fairly good idea of what the atomic orbitals look like from the exact solution of the hydrogen atom. The most salient feature of the hydrogenic atomic orbitals is their exponential dependence on the distance from the nucleus. Slater extended these orbitals to higher atoms. The so-called Slater-type orbitals have the form

$$\phi_\mu = A \, Y_{lm}(\theta, \phi) r^{n-1} \, e^{-\alpha r} \tag{9}$$

where the spherical harmonics, Y_{lm}, are a function of the spherical coordinate angles of the electron position and where r is the distance of the electron from the nucleus (each ϕ_μ is anchored at some particular nucleus). The integers n, l, m in Eq. 9 refer to the usual notation in atomic orbitals, that is, the principal quantum number, the angular momentum quantum number, and the azimuthal quantum number. These atomic orbitals form a very good set in which to expand the molecular orbitals in Eq. 7. However, in computing the average electrostatic potential a huge number of electron repulsion integrals involving the atomic orbitals on different atoms are needed. In the calculations to be discussed later, we typically carry out over one to ten million electron integrals per energy point. Therefore, the method requires that the integrals be calculated very efficiently. Efficient integral evaluation is not possible with Slater-type orbitals (STO for short). Instead, the next new trick [introduced by Boys (1950)] is to write the STO themselves as a sum of Gaussian functions. Gaussian functions are very convenient because any multielectron and multiatom integral can be computed immediately with analytic formulas. Hence the huge savings in time more than offsets the need to expand each atomic orbital, ϕ_μ, in terms of a number M of Gaussian functions:

$$\phi_\mu = \sum_{s=1}^{M} d_{\mu s} g_s \tag{10}$$

(The reader may have noticed that at this point we have sums within sums within sums!) The Gaussian functions themselves would have the appropriate behavior as the particular atomic orbital being approximated. For example, for s orbitals one would use a general Gaussian of the form:

$$g_s = \left(\frac{2\alpha}{\pi}\right)^{3/4} \exp\left(-\alpha r^2\right)$$

while for p_x, p_y, p_z orbitals, the general Gaussians would all be

$$g_{p_x} = \left(\frac{128\alpha^5}{\pi^3}\right)^{1/4} x \exp\left(-\alpha r^2\right)$$

$$g_{p_y} = \left(\frac{128\alpha^5}{\pi^3}\right)^{1/2} y \exp\left(-\alpha r^2\right)$$

$$g_{p_z} = \left(\frac{128\alpha^5}{\pi^3}\right)^{1/4} z \exp\left(-\alpha r^2\right)$$

Note that in each case the *only* difference between the Gaussians of a given type is in the exponential, α. The coefficient in front simply normalizes the function so that the integral over all space is unity.

The coefficients, $d_{\mu s}$, in Eq. 10 are chosen so as to minimize the difference between the atomic orbitals given by Eqs. 9 and 10. Once chosen, these coefficients are *fixed* in all subsequent ab initio calculations; that is, ϕ_μ is completely specified by Eq. 10. The size of the exponent α in the Gaussian determines how close to the nucleus the electron charge is or conversely how "diffuse" the electron charge is. For higher-level basis sets, two sets of valence atomic orbitals are used (these are called "split-valence" basis sets). One set lies close to the nucleus and mimics closely the true-valence atomic orbitals. The other set is more diffuse (smaller α) and enables the molecular orbital to respond to electron cloud deformation away from the nucleus due to chemical bond formation (see Table 1).

For example, the particular basis set 3-21G would have the following expansion for the inner silicon atom orbitals:

$$1s_{si} = 0.066 N_\alpha e^{-910.655r^2} + 0.386 N_\alpha e^{-137.336r^2} + 0.672 N_\alpha e^{-29.760r^2}$$

$$2p_{y,si} = 0.113 N_\alpha y e^{-36.672r^2} + 0.458 N_\alpha y e^{-8.317r^2} + 0.607 N_\alpha y e^{-2.216r^2}$$

Or for the oxygen *valence* orbitals, it would have the following split functions:

$$2p_{x,O} = 0.245 N_\alpha x e^{-7.403r^2} + 0.854 N_\alpha x e^{-1.576r^2} \quad \text{(inner)}$$

and

$$2p'_{x,O} = N_\alpha x e^{-0.374r^2} \quad \text{(outer)}$$

The N_α in front of each Gaussian refers to the normalization factor (which depends on the size of α) given earlier. Note the much higher values of α for the 1s orbital of Si than the $2p_y$, which keeps the 1s orbital closer to the nucleus. Likewise, the $2p'_{x,O}$ orbital is much more diffuse than the $2p_{x,O}$.

The size of the M used in expanding each atomic orbital in terms of Gaussians (Eq. 10) is usually included in the basis-set description. Thus a STO-3G set is a minimal basis set with each atomic orbital (i.e., each STO) expanded by three Gaussian functions in Eq. 10 (i.e., $M = 3$). For extended sets, one normally uses more Gaussians to describe the inner (core) atomic orbitals; therefore, more numbers are given in the label. For example, a 3-21G basis set is an extended basis set with three Gaussians used to expand the core atomic orbitals, two Gaussians used to expand one set of valence atomic orbitals, and one Gaussian used to expand a more "diffuse" set of atomic orbitals. (This is a split-valence basis set.) If, in addition, orbitals of higher angular momentum than required by the electrons in a given atom are used (termed *polarization functions*), an asterisk is added. For example, the 3-21G* basis would add 3*d* orbitals on all second row atoms.

As an illustration, a Gaussian 3-21G* basis set for silicon will consist of the following atomic orbitals:

$1s$, $2s$, $2p_x$, $2p_y$, $2p_z$—all described with 3 Gaussians in Eq. 10
$3s$, $3p_x$, $3p_y$, $3p_z$ inner orbitals—2 Gaussians
$3s$, $3p_x$, $3p_y$, $3p_z$ outer orbitals—1 Gaussian
$3d_{x^2}$, $3d_{y^2}$, $3d_{z^2}$, $3d_{xy}$, $3d_{xz}$, $3d_{yz}$—1 Gaussian (polarization)

Altogether, 19 atomic orbitals will be inputed into the ab initio calculations per Si atom, requiring 33 Gaussian functions to describe them. For a calculation on H_3SiOH, there would be 36 atomic orbitals (e.g., 19 on Si, 9 on O, and 2 on each H) input to obtain the molecular orbitals. The 36 atomic orbitals would be expanded by 60 Gaussians.

Once an accurate wavefunction has been obtained in ab initio calculations, the forces on *all* the atoms in a cluster can be computed exactly and analytically using well-developed quantum-mechanical techniques. This ability enables us to carry out a *full* ab initio minimization of the cluster geometry and extract the optimal equilibrium geometry. In general, the optimization algorithm searches for a "stationary point", that is, a molecular structure such that for all atomic coordinates the force is zero. Mathematically, this means that

$$\frac{\partial E}{\partial x_i} = 0, \quad i = 1, \ldots, 3N \tag{11}$$

for a cluster of N atoms, where E is the potential energy as a function of the atomic nuclear positions, that is, the Born–Oppenheimer energy, $E(\mathbf{R})$, in Eq. 2. Having reached a stationary point, it is important to ascertain whether the structure is a *true minimum*. This test is achieved by analyzing the eigenvalues of the second derivative or *Hessian matrix*:

$$\mathbf{H}_{ij} \equiv \frac{\partial^2 E}{\partial x_i \partial x_j} \tag{12}$$

For a true minimum, the eigenvalues of $\mathbf{H}$ must be all positive except for the six zeros corresponding to three translations and three rotations of the cluster (which don't change the energy in our case). This test is important because it can enable us to check postulated atomic structures to see if they are stable and also enables us to distinguish minima from saddle points and other more complex stationary points to be discussed in the next section.

The kind of accuracy possible in solving the Schrödinger equation today can be clearly seen by the impressive agreement of many ab initio results with experimental data. As an example, Figure 1 gives a number of results comparing ab initio interatomic bond lengths with experimental values. There is no question that with the present capabilities we can tap into the actual potential

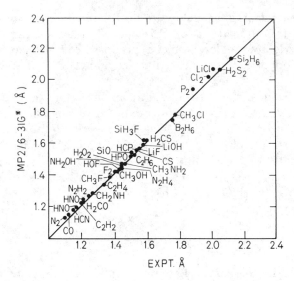

Figure 1. Plot of calculated bond lengths versus experimental bond lengths for a variety of molecules. (Hehre et al., 1986).

surfaces, which account for the chemical behavior of a wide range of systesms. Our next task is to extend this capability to understand the dynamics of geochemical reactions.

3.　TRANSITION-STATE THEORY AND MINERAL KINETICS

The ability to predict accurate potential surfaces means that we are in a position to investigate the nature of the kinetic barriers, including the activated complexes, of key surface processes. In fact, Born–Oppenheimer potential surfaces can be used not only with the transition-state approach to kinetics but also with the much more general and exact collision theory (e.g., scattering S-matrix theory). While methods based on collision and scattering theory have pointed out deficiencies in the traditional transition-state theory (TST), they have also served to uphold many of TST's simple claims. In turn, new generalized transition state theories have been born. For complex systems, the transition-state approach, while admittedly approximate, has been well established.

Numerous recent papers have invoked transition-state theory in developing the mechanisms of low-temperature water–rock reactions (e.g., Lasaga, 1981; Aagard and Helgeson, 1984). It is clear that one of the central themes in our understanding of heterogeneous kinetics is the elucidation of the important activated complexes and their modification by both surface properties and the composition of the solution near the surface as well as by temperature and pressure. One of the great advantages of the ab initio methods discussed in this

Chapter is the remarkable ability to finally predict quantatively reaction pathways and transition states for geochemically interesting reactions. At the same time, the rigorous molecular theory proposed here also paves the way to a solid foundation of the treatment of transition-state theory in water–rock kinetics.

Contrary to some usage in the literature, an important point to make about transition-state complexes is that they are not true minima and hence are not normal (in the thermodynamic sense) chemical species. For example, the attack by H^+ on the surface of feldspars will lead to a surface which has exchanged alkalies for SiOH or $SiOH_2^+$ groups. While these latter species are surface complexes, *none* of these chemical moieties is an activated complex. Further attack by H^+ or by H_2O will hydrolyze the Si–O–Al and Si–O–Si bridging bonds as shown in the reaction

$$H_2O + \,\equiv Si\text{–}O\text{–}Si \equiv \,\Rightarrow 2 \equiv Si\text{–}O\text{–}H \tag{13}$$

Again, neither side of reaction 13 represents the activated complex. Both sides represent "stable" surface species.

While the activated complex is not a true minimum, its structure is *uniquely and precisely* determined by the Born–Oppenheimer potential surface. To ascertain the structure of the activated complex, we must appraise carefully the topography of the potential surface describing the reaction sketched in Eq. 13. It is one of the aims of this Chapter to show what can be done to unravel the details of the activated complex in reactions such as reaction 13.

Rimstidt and Barnes (1980) use the shorthand notation $(SiO_2 \cdot 2H_2O)^{\ddagger}$ for the "activated complex" in the dissolution and precipitation of silica. While this approach is satisfactory for the purposes in their paper, it is important to remember that it is only a heuristic label; we expect the reaction itself to proceed with several elementary reactions, as illustrated in Figure 2. The last example raises another serious misuse of the theory [not to be attributed to Rimstidt and Barnes (1980)]. Activated complexes are specific to a particular *elementary reaction*. One cannot discuss the activated complex without specifying the nature of the elementary reaction taking place; nor can one speak of the activated complex of an overall reaction.

The definition and the structure of an activated complex is fairly straightforward, if one has access to the ab initio molecular results that are possible today. In the last section we introduced the Hessian matrix of a molecular cluster. We also pointed out that an important test for a stable structure is the requirement that the Hessian matrix at the "equilibrium geometry" have all positive eigenvalues (i.e., be concave upward in "all" directions). The same Hessian matrix can define the classical activated complex (see Fig. 3). An activated complex is a stationary point in the multibody potential surface of all the atoms of the cluster such that the Hessian matrix has all positive eigenvalues *except for one negative eigenvalue* (ignoring the six zero eigenvalues for translations and rotations). In other words, the shape of the potential surface in the neighborhood of the

I. Reaction in water

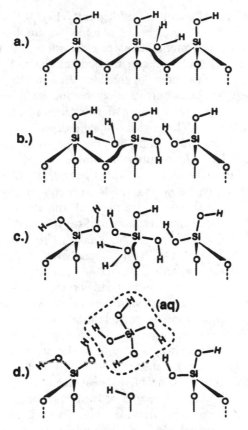

Figure 2. Reaction mechanism for the dissolution of quartz.

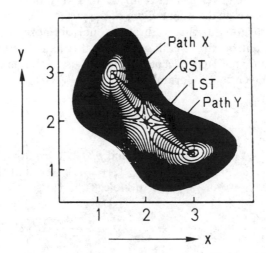

Figure 3. Hypothetical reaction surface illustrating two minima and one saddlepoint stationary point in between.

activated complex would look like a 'saddlepoint" (Fig. 3), if the potential surface were to depend on only two degrees of freedom. The saddlepoint arises because the activated complex is a minimum in one direction and a maximum in the perpendicular direction. Of course, for molecular clusters the potential surface typically contains tens of degrees of freedom. But in any of these cases, the requirement is that one and only one eigenvalue be negative. The positive eigenvalues would correspond to normal vibrations of the cluster around its minimum energy geometry. The single normal mode associated with the negative eigenvalue cannot be a vibration and becomes the motion associated with the *reaction coordinate*. The full reaction coordinate can, in fact, be obtained by beginning at the stationary point (activated complex) and initiating a motion using the negative eigenvalue normal mode. Subsequent moves are done using the method of steepest descent (i.e., we move all the atomic coordinates in the direction of $-\nabla V$). Note that with the theoretical tools available today, we can compute $E(\mathbf{R})$, $\partial E/\partial x_i$, and $\partial^2 E/\partial x_i \partial x_j$ completely within a near-exact solution of the Schrödinger equation (in the Born–Oppenheimer scheme). As a result, not only can the minima of the potential surface be obtained systematically but also any activated complexes can be obtained and verified from the ab initio Hessian. Furthermore, the same near-exact solutions allow a detailed analysis of the reaction coordinate both from the normal-mode analysis and from the subsequent steepest descent calculation.

Therefore, two rigorous concepts arise from the previous paragraph: (1) the activated complex is uniquely defined by the requirement of a stationary point with a Hessian that has a single negative eigenvalue and (2) the normal mode of the negative eigenvalue uniquely defines (at least near the activated complex) the reaction coordinate. There is no need for qualitative activated complexes, if the potential surfaces are available! Note also that for the complex multivariable potential surfaces that are of interest to us, it is very possible to find several "activated complexes" in different regions of the surface. The task, then, is to find the relevant elementary pathway through one (or more) of these activated complexes that controls the overall geochemical reaction being studied.

To illustrate the previous concepts, one could use a simple potential surface (Fig. 3) such as given by the formula

$$E = \cos(x)(e^y + e^{-y})$$

In this case the first derivatives, $\partial E/\partial x$ and $\partial E/\partial y$ are both set to zero at the stationary points: $(n\pi, 0)$, where $n = 0, \pm 1, \pm 2, \ldots$. To find the nature of the stationary points, however, we need to look at the Hessian:

$$\frac{\partial^2 E}{\partial x^2} = -\cos(x)(e^y + e^{-y}) \qquad \frac{\partial^2 E}{\partial y^2} = \cos(x)(e^y + e^{-y})$$

$$\frac{\partial^2 E}{\partial x \partial y} = -\sin(x)(e^y - e^{-y})$$

Evaluating the Hessian at the stationary point $(n\pi, 0)$ gives

$$\begin{bmatrix} (-1)^{n+1}2 & 0 \\ 0 & (-1)^n 2 \end{bmatrix}$$

Thus, *all* stationary points in this example are "activated complexes" with one positive and one negative eigenvalue. (The transition-state theory extension to this case would be interesting.)

Once the complex has been found, the familiar formulas of TST can be used to learn something about $\Delta S^{\ddagger}$, $\Delta H^{\ddagger}$ and the rate constant, k:

$$k = \frac{kT}{h} \frac{q^{\ddagger}/V}{q_{react}/V} \frac{\gamma^{react}}{\gamma^{\ddagger}} \exp\left(-\frac{\Delta E^{\ddagger}}{RT}\right) \tag{14}$$

where the partition functions can all be obtained from the geometry of the species and the vibrational frequencies (Lasaga, 1981). If we incorporate the zero-point energies of vibrations into the exponential, we have

$$\Delta H^{\ddagger} = \Delta E^{\ddagger} + \frac{1}{2}\sum_i h v_i^{\ddagger} - \frac{1}{2}\sum_i h v_i^{react} \tag{15}$$

We can rewrite Eq. 14 as

$$k = \frac{kT}{h} \frac{\gamma^{react}}{\gamma^{\ddagger}} e^{\Delta S^{\ddagger}/R} e^{-\Delta H^{\ddagger}/RT} \tag{16}$$

where

$$\Delta S^{\ddagger} = R \ln\left(\frac{q^{\ddagger}/V}{q^{react}/V}\right) \tag{17}$$

and the Arrhenius parameters can be obtained from

$$E_a = RT + \Delta H^{\ddagger} \tag{18}$$

and

$$A = e \frac{kT}{h} \frac{\gamma^{react}}{\gamma^{\ddagger}} e^{\Delta S^{\ddagger}/R} \tag{19}$$

4. MOLECULAR MECHANISMS OF WATER–ROCK KINETICS

With the vast increases in computational power, it is now feasible to calculate ab initio potential surfaces that incorporate bond-breaking and bond-forming processes. This means that we are now in a position to obtain the robust potential surfaces needed to map out the terrain wherein the dynamics of the geochemical reactions take place and begin to tackle the kinetic and structural questions that arise in the quantitative treatment of geochemical reactions. In particular, the reaction pathways and energetics of surface reactions involving oxides and silicates can be studied from the atomic point of view with these techniques.

One of the important conclusions from recent work (e.g., Gibbs, 1982; Lasaga and Gibbs, 1987, 1988; Dovesi et al., 1988) is that a major part of the structure and energetics of silicates and oxides can be accounted for by short-range forces, that is, what would be labeled "covalent" bonding in the traditional sense. As a result, large (but finite) atom clusters that mimic the local environment can provide very important insights into the atomic forces and bonding of minerals and glasses. As a corollary, we extend this local concept to the study of surface reactions.

One of the most important potential-energy surfaces to investigate from the point of view of geochemical kinetics, petrology, ceramics, and rock mechanics is the hydrolization of the Si–O–Si structural unit at the surface of silicates, as shown in the reaction

$$H_2O + \equiv Si-O-Si\equiv \ \Rightarrow 2\equiv Si-O-H \qquad (20)$$

This reaction is a key step in the dissolution processes occurring at the surfaces of silicates. Reaction 20 is also one of the essential aspects of the widely studied hydrolytic weakening in the area of rock mechanics. One of the current areas of active research in that field involves determining the molecular details of water or hydrogen defects in minerals. The incorporation of water by reactions such as Eq. 20 can affect the rates of diffusion or creep by many orders of magnitude. In addition, reaction 20 is the basic step by which water is initially incorporated into silicate melts, thereby significantly affecting both the phase diagrams and the transport properties such as viscosity and diffusion.

Figure 2 illustrates the elementary reactions that govern the dissolution of silica and similar silicates. The sequence depicts the sequential hydrolysis by water molecules of the three to four Si–O–Si bridges anchoring a surface silicon atom. At the end of the four hydrolysis steps, the silicon is surrounded by four OH groups and leaves the surface as orthosilicic acid. *None* of the steps illustrated in Figure 2 depict the activated complex. Figure 4 focuses on one of the hydrolysis steps. The attack of water molecules is decomposed into two molecular steps in Figure 4. The first step involves the adsorption of water near a Si-O-Si group. The second step involves the formation of a new Si–O bond by the oxygen of the adsorbed water and the cleavage of te Si–O–Si group. *It is the activated complex associated with this step that we believe accounts of the energetics of silica dissolution.*

4.1. Ab Initio Studies of Adsorption

Ab initio studies of adsorption have been done in a number of earlier papers (Sauer, 1987; 1989; Geerlings et al., 1984; Hobza et al., 1981; Mortier et al., 1984).

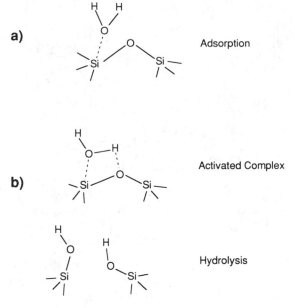

Figure 4. Elementary reactions involved in the hydrolysis reaction.

As discussed by Hobza et al. (1981), in studying the adsorption energetics of water on silica surfaces, the appropriate reaction to use is

$$H_2O \cdots H_2O + M \leftrightarrow M \cdots H_2O + H_2O \tag{21}$$

where M stands for the appropriate silica surface group. In other words, the energy gained in adsorbing the water must be weighed against the energy lost in breaking a hydrogen bond. In fact, a study of the hydrophilic or hydrophobic nature of a surface group can be done by comparing the ΔH_0 for adsorption to the ΔH_0 for the reaction

$$H_2O + H_2O \Rightarrow H_2O \cdots H_2O$$

in which the product is the water dimer (e.g., Hobza et al., 1981).

The first step in our ab initio study, therefore, is to ascertain the nature of water adsorption at silica surfaces. Note that the adsorption step in Figure 4 occurs close to the silicon atom. This type of adsorption, where a hydrogen bond is formed between the hydrogen of the incoming water and the oxygen of the silica surface, is termed "acceptor" adsorption. Normally the surface of silica is covered by numerous silanol (SiOH) groups reaching about 6–7 OH groups per 100 Å^2 (Iler, 1979). By far the most favorable mode of adsorption of water on silica is by "donor" adsorption onto a silanol group, that is, by the hydrogen bonding of the H in the silanol group to the oxygen of the incoming water molecule so that water "sits oxygen down on the SiOH groups" (Iler, 1979, p.

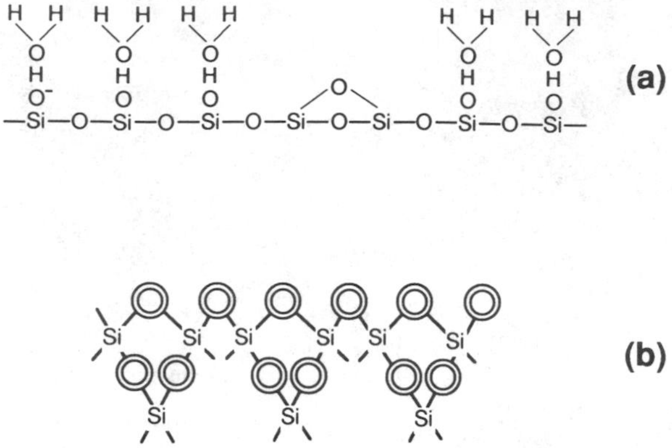

Figure 5. (*a*) Water adsorption onto a typical hydroxylated silica surface. (*b*) Surface of dehydroxylated silica.

627). This is shown in Figure 5. The major role of silanol groups in water adsorption is shown by the hydrophobic character of silica, once the silanol groups are eliminated during dehydroxylation (e.g., heating to several hundred degrees for several hours). The surface of the dehydroxylated silica is comprised of essentially Si–O–Si siloxane bonds (see Fig. 5*b*). We will discuss the adsorption onto silanol groups further below.

The key step in the kinetics of silica dissolution, however, requires a special type of water adsorption, specifically, the acceptor adsorption shown in Figure 4. This type of adsorption is required by the need to form a new SiO bond between the oxygen of the water molecule and the surface silicon atom in step 2 of Figure 4. Note that the kind of adsorption postulated in Figure 4 (and exhibited further below) leads to a surface silicon that is five-fold-coordinated. The existence of 5-fold coordinated silicon in glasses and melts has been postulated in several MD papers (see Kubicki and Lasaga, 1988). While there are no mineral structures that contain five-fold coordinated silicon, there is both theoretical and experimental evidence for its existence. In fact, five-fold coordinated silicon seems to play a key role in the high-pressure increase of diffusion in silica glass. Ab initio studies have shown that SiO_5 is indeed a *stable* structure (Damrauer et al., 1988). For example, Figure 6 shows the fully optimized (i.e., all 27 independent degrees of freedom were optimized) structure of $Si(OH)_5^-$ at the 3-21G* level. The geometry is that of a distorted trigonal bipyramid. As discussed earlier, the full Hessian is computed and all the 27 nonzero eigenvalues (frequencies of vibration) are indeed positive for the structure in Figure 6, verifying its stability. In fact, as the OH^- is brought in toward the four-fold-coordinated silicon, the energy monotonically decreases toward the minimum. The stability of five-fold Si in gas-phase complexes has also been recently verified in some time-of-flight experiments (Damrauer et al., 1988).

The situation with silicon is in interesting contrast to the behavior of its cousin, carbon. Numerous ab initio studies of the S_N2 nucleophilic reaction at

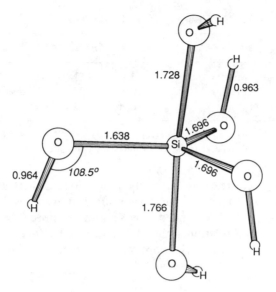

Figure 6. Fully optimized 3-21G* structure of Si(OH)$_5^-$

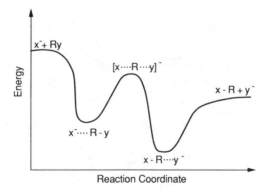

Figure 7. Schematic of the S$_N$2 nucleophilic reaction at a carbon center.

carbon centers, for instance, reactions such as

$$F^- + CH_3Cl \rightarrow CH_3F + Cl^- \tag{22}$$

have shown that the reaction pathway is schematically that shown in Figure 7 (Wolfe et al., 1981). Note that in this case, the five-fold-coordinated carbon moiety is *not* a stable species but rather an *activated complex*.

Table 2 gives our results for the adsorption of water on H$_3$SiOH. In each case, the structures and energies were obtained by optimizing all 21 degrees of freedom at the appropriate level of theory. This *full* minimization is important to verify

TABLE 2. 3-21G* Adsorption Energies (kcal mol^{-1})

	ΔE^a	ΔE^b	ΔH^a	ΔH^b
Donor adsorption	−14.46	−3.51	−11.94	−3.96
Acceptor adsorption	−10.91	+0.04	−7.76	+0.22
Water dimer	−10.96	—	−7.98	—

^{a} Energy change of reaction $A + H_2O \rightarrow A \cdot H_2O$.
^{b} Energy change relative to the water dimer (reaction 20).

whether the proposed structures are truly minima (see earlier), although the full minimization has not always been carried out in previous work. Calculations employed both the 3-21G* extended basis set and the more robust 6-31G* basis with added electron correlation at the MP2 level (see Hehre et al., 1986). Figures 8 and 9 give the H_3SiOH and H_2O optimized geometries for the two basis sets.

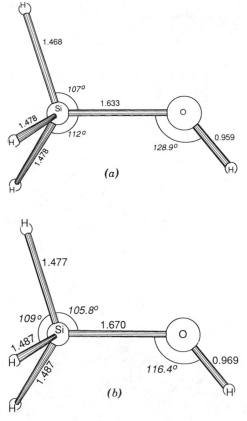

Figure 8. (a) Fully optimized 3-21G* geometry of H_3SiOH. (b) Fully optimized MP2/6-31G* geometry of H_3SiOH.

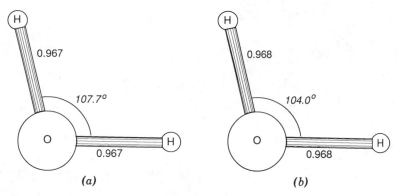

Figure 9. (a) Fully optimized 3-21G* geometry of H_2O. (b) Fully optimized MP2/6-31G* geometry of H_2O.

Note the generally very good agreement between the two calculations, showing the adequacy of the 3-21G* basis set. One important difference between the two sets is the narrowing of the SiOH angle in the higher level calculation. This narrowing gives a SiOH angle of 116°, in very good agreement with the experimentally derived angle in silanol, 113° (Hair, 1967). The SiO bond is also slightly longer in the higher level calculation. The water results at the MP2/6-31G* level are in excellent agreement with the experimental data on the water molecule ($r_{OH} = 0.958$ Å and HOH angle = 104.5°).

Figure 10 gives the results for the water dimer. The optimized water dimer structure also compared well with recent gas-phase experimental data on the gas water dimer by Dyke (1984); $r_{OO} = 2.98$ Å and tilt angle = 123°. The energy obtained in forming the hydrogen bond is given in Table 2.

Figure 11 gives the geometry of the adsorbed water molecule in the case of donor adsorption onto the silanol group. Note the strong hydrogen bond formed. The oxygen–oxygen distance in Figure 11 is 2.66 Å, which is actually shorter than the oxygen–oxygen distance (2.80 Å) calculated for the water dimer

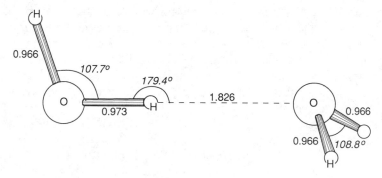

Figure 10. Fully optimized 3-21G* geometry of the water dimer.

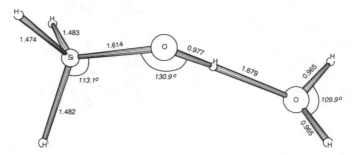

Figure 11. Fully optimized 3-21G* donor adsorption complex geometry.

(Fig. 10). Analysis of the Hessian matrix verifies that the adsorbed structure is a local minimum. In fact, Table 2 indeed confirms that water adsorption onto the hydrogens of silanol groups on silica surfaces is energetically favorable, that is, the energy change for the reaction

$$\equiv\text{SiOH} + \text{H}_2\text{O} \cdots \text{H}_2\text{O} \rightarrow \equiv\text{SiOH} \cdots \text{OH}_2 + \text{H}_2\text{O} \tag{23}$$

from Table 2 is $\Delta E = -3.5 \text{ kcal mol}^{-1}$.

Figure 12 gives the analogous adsorption for the more relevant (to us) acceptor adsorption. In agreement with experiment, this adsorption is not as stable as the one onto the silanol group. However, an analysis of the Hessian matrix indeed shows that the geometry in Figure 12 corresponds to a full minimum in the Born–Oppenheimer surface. Note that for the 3-21G* calculation the distance from the silicon to the oxygen of the water molecule is 2.79 Å, which corresponds roughly to the O–O distance in the water dimer. Furthermore, the distance between the hydrogen of the adsorbed water and the silanol oxygen is 1.91 Å, slightly larger than the OH distance (1.83 Å) in the water dimer. The MP2/6-31G* result for the acceptor adsorption (Fig. 12b) has the water molecule even further from the Si atom, although the hydrogen bonded OH distance is quite similar, 1.98 Å. In both cases, the OH bond length of the water molecule lengthens for the hydrogen atom involved in the H bond. Likewise, the original SiO bond length increases upon adsorption of the water molecule, a precursor state to the ultimate rupture of the bond in the subsequent dynamics.

The energetics of the reaction (reaction 21) for the case of the acceptor adsorption (see Table 2) are not favorable (i.e., $\Delta E > 0$). Because this adsorption is precisely the type of adsorption onto siloxane bonds (see Fig. 15, below), the unfavorable ΔE confirms the hydrophobic nature of the dehydroxylated disiloxane-rich surface of silica.

The energetics of the H_2O acceptor adsorption are of interest in monitoring the molecular dynamics of the adsorption process. We can obtain an interesting profile by fixing the bond distance between the silicon atom and the oxygen of the incoming water molecule and then optimizing all the remaining degrees of freedom. Figure 13 shows the results for several SiO bond distances. Note that

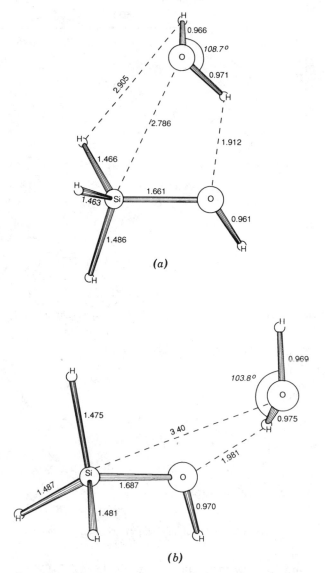

Figure 12. (a) Fully optimized 3-21G* acceptor adsorption complex geometry. (b) Fully optimized MP2/6-31G* acceptor adsorption complex geometry.

the energy rise to the right of the minimum is quite gradual so that the chemical intereaction between a water molecule and the silica surface would extend out to nearly 7 Å away from the surface. This result has important implications for the treatment of the adsorbed layer in silicates. In particular, it suggests that the so-called Stern layer may be several water molecules thick, even in "pure" water.

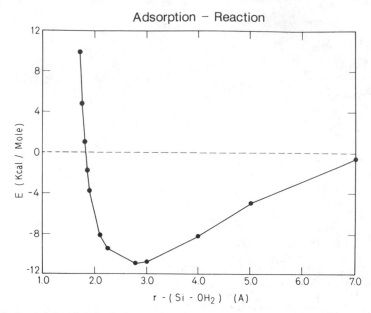

Figure 13. Potential surface for the interaction of an incoming water molecule with the silica surface (3-21G* results).

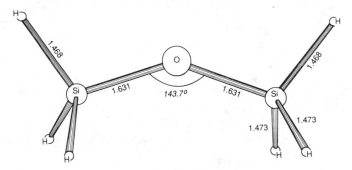

Figure 14. 6-31G* optimized geometry of disiloxane.

Figures 14 and 15 show the analogous results for a 6-31G* *full* optimization calculation on disiloxane and the acceptor adsorption of water onto the disiloxane bridge. In agreement with our localization hypothesis, the SiO and OH bond distances in Figure 15 agree closely with the corresponding distances in Figure 12b.

The energetics given in Table 2 must be corrected for zero-point energy of vibration to convert the energies to enthalpies:

$$\Delta H_{0\,\mathrm{K}} = \Delta E + \sum_{i} \frac{1}{2} h v_{i}^{\mathrm{P}} - \sum_{i} \frac{1}{2} h v_{i}^{\mathrm{R}} \tag{24}$$

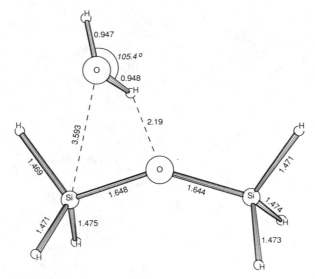

Figure 15. Fully optimized 6-31G* acceptor adsorption of water onto disiloxane.

where P and R stand for product and reactant, respectively, that is, the adsorbed complex and the infinitely separated water + surface in one case or the water dimer and the adsorbed water complex in the other. To calculate this correction, a full normal-mode analysis of each stable configuration is carried out. The frequencies are obtained in a fully ab initio manner by using the analytically derived Hessian matrix from the electronic wavefunctions. The corrected ΔH are given in Table 2. Note that part of the correction to the ΔE for the case with the infinitely separated reactants arises from the conversion of a translational degree of freedom into a vibrational degree of freedom. The more relevant corrections are the corrections to the ΔE for reaction 21 or 23 shown in Table 2 (which don't have this loss of degree of freedom). The donor adsorption is stabilized (lower ΔH, -3.96) by the zero-point correction while the acceptor adsorption is destabilized by the correction (ΔH increases to 0.22 kcal mol^{-1}).

4.2. Transition State of the Hydrolysis Reaction

Having obtained the adsorption data, we must now ascertain the kinetics of the hydrolysis reaction, which requires computation of the activated complex. Returning to Figure 2, it is clear that the hydrolysis reaction proceeds in such a manner that the original SiO bond is broken and new Si–OH and OH bonds are formed. Figure 16 illustrates this reaction path. In the parlance of reaction dynamics, we term the initial configuration the "entrance channel" and the final (product) configuration, the "exit channel." *Both* end members in Figure 16 are stable species (true minima); the transition-state complex must lie somewhere in between.

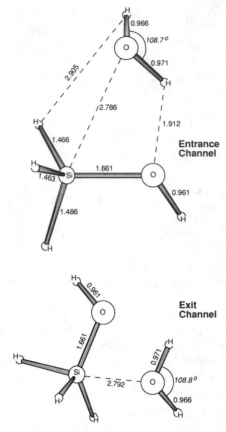

Figure 16. Reaction coordinate trajectory outlining the entrance and exit channels of the hydrolysis reaction.

To find the activated complex, a convenient method is the so-called linear synchronous transit (LST) method (Halgren and Lipscomb, 1977). In this method, the two stable end-member geometries are interpolated to produce a simple *reaction coordinate*. In the LST scheme the reaction coordinate, $f(0 \leqslant f \leqslant 1)$, is essentially an interpolation parameter between the interatomic distances, r_{ij}, of the two minima (i.e., the reactants and products). Thus for a given parameter, f, the corresponding interatomic distances between any pair of atoms i and j, is given by:

$$r_{ij} = (1-f)r_{ij}^{R} + f r_{ij}^{P} \qquad (25)$$

For a system with N atoms, there are $N(N-1)/2$ interatomic distances and only $3N - 6$ independent degrees of freedom (the x, y, z of atoms). Therefore, equation 25 is an overdetermined systems of equations. In the LST approach equation 25 is solved in the least-squares sense for each value of f (see Halgren and Lipscomb, 1977).

We can apply the LST scheme to the two minima shown in Figure 16. The results of applying the LST scheme to the hydrolysis reaction are shown in Figure 17. Note the very large energy barrier between the entrance and exit channels. If Figure 17 represented the true energetics of the hydrolysis reaction, then our mechanism would be quite at odds with experimental data on the activation energies of silicate–water reaction kinetics. However, the structure at the maximum in Figure 17 is not a stationary point, as required for a true activated complex. The next step, therefore, is to optimize further the LST structure. Because the complex involves the transfer of a hydrogen atom from the incoming water to the departing water (see Fig. 16), we expect (by symmetry) that the two OH bonds would be nearly equal in the activated complex. Similarly, the two SiO bonds (the one being formed and the one being broken) should be of similar length at the activated complex. Taking this into account, we optimized the $f = 0.5$ LST structure with respect to *all* degrees of freedom except with the constraint the two SiO bond lengths and the two particular OH bond lengths be equal. The resulting optimized geometry is shown in Figure 18. Note that the SiO bond lengths are both now lengthened to 1.84 Å (from the minimum geometry). The OH bond distances are 1.20 Å. The most interesting result is the drop in the activation barrier, ΔE, (which should be close to the activation energy barrier, $\Delta E^{\ddagger}$) to the value 21.87 kcal mol^{-1} (see Table 3). This value is remarkably close to the activation energy for silicate dissolution (Lasaga, 1984).

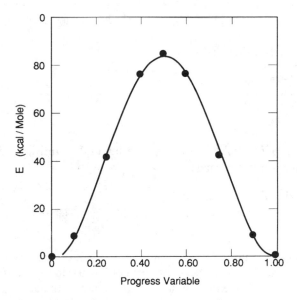

Figure 17. LST reaction coordinate potential energy (see text).

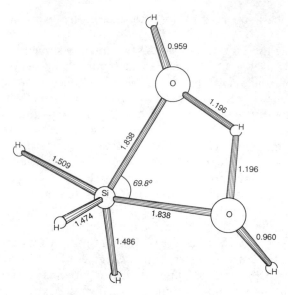

Figure 18. Approximate transition state at 3-21G*.

TABLE 3. Hydrolysis Transition State (kcal mol^{-1})

	ΔE^a	ΔE^b	ΔE^c	ΔE^d	ΔE^{expt}
E_a	21.87	22.74	16.70	~ 30.05	
Vibrational zero-point	-1.22	-1.30	-1.30	~ -1.25	
ΔH_0	20.65	21.44	15.40	~ 28.80	18.0^e

^{a} Approximate 3-21G* results.
^{b} Full 3-21G* transition state.
^{c} MP2/3-21G* results.
^{d} Approximate MP2/6-31G* results.
^{e} Quartz dissolution data.

Again, to compare the $\Delta E'$ values to experimental quantities, we must include the zero-point vibrational energies:

$$\Delta H^{\ddagger} = \Delta E^{\ddagger} + \sum_{i=1}^{3N-7} \frac{1}{2}h\nu_i^{\ddagger} - \sum_{i=1}^{3N-6} \frac{1}{2}h\nu_i^{ads} \tag{26}$$

This correction is quite nontrivial in this case and reduces the barrier by 1.22 kcal mol^{-1} (see Table 3). Note that part of the loss in the barrier is due to the fact that the activated complex is "loose" compared to the adsorption complex. Structures with many elongated bonds tend to have lower frequencies of vibration and hence smaller $\frac{1}{2}h\nu$ corrections. Therefore, the increase in zero-point energy of vibration is greater for the stable adsorption complex than for the activated complex. This correction has important consequences for the isotope effect in the dissolution and precipitation of quartz, which will be discussed in a future paper.

The approximate activated complex in Figure 18 was obtained by a reasonable constraint on the bond lengths. When a calculation of all the forces on the atoms is performed, one indeed obtains near zero values for all the forces (i.e. a near stationary point). Furthermore, the full frequency calculation on the structure in Figure 18 yields all positive eigenvalues (normal frequencies) except for *one* negative eigenvalue (imaginary frequency), as required for an activated complex. Therefore, this structure is very close to the true activated complex. In fact, the structure in Figure 18 can be input into a full search for a stationary point and the result, the true activated complex, is given in Figure 19. That this is a true activated complex follows from the exact satisfaction of the two key requirements (1) the structure in figure 19 is a stationary point for all degrees of freedom (Eq. 11) and (2) the eigenvalues of the second derivative matrix (Hessian) are all positive except for one negative and the usual six zeros. Interestingly enough, the symmetry of the two SiO bonds is closely maintained in the full transition state complex. However, the two OH distances (1.104 and 1.310 Å) are not symmetrical. This asymmetry arises from the symmetry breaking induced by the other groups in the activated complex and also manifests the malleable nature of the lengthened OH bonds.

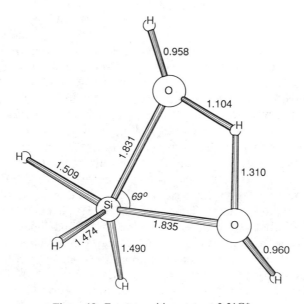

Figure 19. Exact transition state at 3-21G*.

The new values for $\Delta E^{\ddagger}$, $\Delta(\frac{1}{2}h\nu)$, and $\Delta H^{\ddagger}$ are given in Table 3. These values confirm the close approximation represented by the results from Figure 18. This close agreement is important for other calculations, because it is usually nontrivial to converge on the exact activated complex. The activation energy predicted from our exact activated complex is in excellent agreement with experimental values. One should add that the correct comparison, given the kinetic mechanism in Figures 2 and 4, should add the true enthalpy of adsorption for the water molecule (Lasaga, 1981, p. 37) and the RT correction in Equation 18:

$$E_a^{expt} = RT + \Delta H^{\ddagger} + \Delta H_{H_2O}^{ads} \tag{27}$$

However, we expect ΔH^{ads} to be either close to zero or a small negative number (e.g., several kilocalories per mole—see Table 2). The net effect of the corrections in Eq. 27 would most likely be to increase slightly the agreement.

It is important to analyze the frequency spectrum of the activated complex. In particular, we want to look at the normal mode with the negative eigenvalue (i.e., an imaginary frequency because $(2\pi\nu)^2 = \lambda$; for our case, $\nu^{\ddagger} = 884 \, i \, cm^{-1}$). As mentioned earlier, this normal mode is, in fact, the *reaction coordinate* for the hydrolysis reaction. If one analyzes the normal mode, the motion involves predominantly the transfer of the H atom, that is, the "vibration" has one OH distance increasing and the other OH distance decreasing. Interestingly, the two long SiO bonds (which ultimately must also become involved in the reaction) are *not* very active in the reaction coordinate normal mode. This result states that the H transfer dominates the hydrolysis reaction!

The dynamics of the activated complex can be nicely studied by an "ab initio movie," if the structure in Figure 19 is altered slightly along the negative eigenvalue normal mode (i.e., along the reaction coordinate) and then the energy allowed to drop along the steepest descent direction. This scheme will display, in movie fashion, the entire reaction coordinate (in an ab initio manner) as the activated complex slides right into one of the two minima in Figure 16 (depending on which way the complex was "tilted"). Indeed, if the OH bond in Figure 19 is slightly increased from 1.104 to 1.109 Å, the structure proceeds to move on to the exit channel depicted in Figure 16. On the other hand, if the process is started right at the structure in Figure 19, nothing will happen, which simply reiterates the stationary nature of the activated complex.

The energetics of the reaction can be improved by carrying out a calculation adding electron correlation. If we carry out calculations at the MP2 level (see Table 3) the activation energy changes to $\Delta E^{\ddagger} = 16.7 \, kcal \, mol^{-1}$. These results strongly indicate that we are indeed obtaining for the first-time a look at the key chemical moieties driving the dissolution of quartz and related minerals.

Even more extensive calculations were done optimizing both the adsorption and the activated complex structures at the MP2/6-31G* level. The results are given in Table 3. The approximate activated complex is given in Figure 20. Note that in this case a full optimization of the transitions state complex has not been

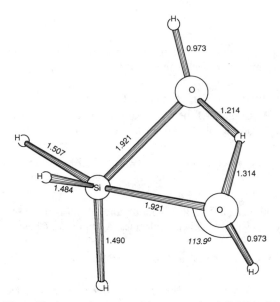

Figure 20. Approximate transition state using MP2/6-31G*.

done yet and hence the values are expected to be on the high side. Nonetheless, the value of $\Delta H^{\ddagger}$ is in the vicinity of the observed energetics of silicate reactions.

Future calculations will not only finish the optimization of the MP2/6-31G* transition state above but also carry out similar studies for the next higher cluster, disiloxane + water, namely, the hydrolysis reaction:

$$H_2O + H_3Si-O-SiH_3 \Rightarrow H_4SiO + H_4SiO \qquad (28)$$

One problem that will have to be addressed at some point is the transference of the potential surface data to complex situations such as water in contact with a silicate surface. While it is true that there are many water molecules present during the dissolution of quartz, it is clear that the close chemical interaction of one water molecule with a Si–O–Si bridge at the surface wil be dominated by the energetics that have been presented in this study. Therefore, while we cannot ignore the further complexities of liquid structure in the problem, the major features of the water reaction with Si–O bonds will emerge from the ab initio calculations that we are capable of carrying out at this stage. The close agreement of calculated and observed activation energies indicates that, indeed, these molecular moieties are the ones relevant to surface reactions. To be able to probe actual molecular mechanisms, the details of activated complexes and even obtain a first-principles "movie" of the adsorption and dissolution processes on surfaces is an exciting prospect. In turn, this result truly opens a new era for the study of surface reactions.

REFERENCES

Aagard, P., and H. C. Helgeson (1982), "Thermodynamic and Kinetic Constraints on Reaction Rates among Minerals and Aqueous Solutions, I. Theoretical Considerations," *Am. J. Sci.* **82**, 237–285.

Boys, S. F. (1950), *Proc. Roy. Soc.* (*London*) **A200**, 542.

Damrauer, R., L. W. Burggraf, L. P. Davis, and M. S. Gordon (1988), "Gas-Phase and Computational Studies of Pentacoordinate Silicon," *J. Am. Chem. Soc.* **110**, 6601–6606.

Dovesi, R., C. Pisani, C. Roetti, and B. Silvi (1988), "The Electronic Structure of α-Quartz: a Periodic Hartree–Fock Calculation," *Int. J. Quantum Chem.*

Dyke, Th. R., 1984, *Curr. Topics Chem.* **120**, 85.

Geerlings, P., N. Tariel, A. Botrel, R. Lissillour, and W. J. Mortier (1984), "Interaction of Surface Hydroxyls with Adsorbed Molecules. A Quantum Chemical Study," *J. Phys. Chem.* **88**, 5752–5759.

Gibbs, G. V. (1982), "Molecules as Models for Bonding in Silicates," *Am. Min.* **67**, 421–450.

Hair, M. L., (1967), *Infrared Spectroscopy in Surface Chemistry*, Dekker, New York, Interscience, London, p. 516.

Halgren, T. A., and W. N. Lipscomb (1977), "The Synchronous-Transit Method for Determining Reaction Pathways and Locating Molecular Transition States," *Chem. Phys. Lett.* **49**, 225–232.

Hehre, W. J., L. Radom, P. R. Schleyer, and J. A. Pople (1986), *AB INITIO Molecular Orbital Theory*, Wiley, New York.

Hobza, P., J. Saueer, C. Morgeneyer, J. Hurych, and R. Zahradnik (1981), "Bonding Ability of Surface Sites on Silica and Their Effect on Hydrogen Bonds. A Quantum-Chemical and Statistical Thermodynamic Treatment," *J. Phys. Chem.*, **85**, 4061–4067.

Iler, N. (1979), *The Chemistry of Silica*, Wiley, New York, 490 pp.

Kubicki, J. D., and A. C. Lasaga (1988), "Molecular Dynamics Simulations of SiO_2 Melt and Glass: Ionic and Covalent Models," *Am. Min.* **73**, 941–955.

Lasaga, A. C. (1981), "Transition State Theory," in A. C. Lasaga and R. J. Kirkpatrick, Eds., *Kinetics of Geochemical Processes, Rev. Mineral.*, Vol. 8, pp. 135–169.

Lasaga, A. C. (1984), "Chemical Kinetics of Water–Rock Interactions," *J. Geophys. Res.* **B6**, 4009–4025.

Lasaga, A. C., and G. V. Gibbs (1987), "Application of Quantum Mechanical Potential Surfaces to Mineral Physics Calculations," *Phys. Chem. Min.* **14**, 107–117.

Lasaga, A. C., and G. V. Gibbs (1988), "Quantum Mechanical Potential Surfaces and Calculations on Minerals and Molecular Clusters I. STO-3G and 6-31 G* Results," *Phys. Chem. Min.* **16**, 29–41.

Mortier, W. J., J. Sauer, J. A. Lercher, and H. Noller (1984), "Bridging and Terminal Hydroxyls. A Structural Chemical and Quantum Chemical Discussion," *J. Phys. Chem.* **88**, 905–912.

Rimstidt, J. D., and H. L. Barnes (1980), "The Kinetics of Silica–Water Reactions," *Geochim. Cosmochim. Acta* **44**, 1683–1699.

Sauer, J. (1987), "Molecular Structure of Orthosilicic Acid, Silanol, and $H_3SiO\overline{H}$ AlH_3 Complex: Models of Surface Hydroxyls in Silica and Zeolites," *J. Phys. Chem.* **91**, 2315–2319.

Sauer, J., 1989, Molecular Models in Ab Initio Studies of Solids and Surfaces: From Ionic Crystals and Semiconductors to Catalysts," *Chem. Rev.* **89**, 199–255.

Sauer, J., and R. Zahradnik (1984), "Quantum Chemical Studies of Zeolites and Silica," *Int. J. Quant. Chem.* **26**, 793–822.

Wolfe, S., J. Mitchell, and H. B. Schlegel (1981), "Theoretical Studies of S_N2 Transition States. 1. Geometries," *J. Am. Chem. Soc.* **103**, 7692–7694.

Zhidomirov, G. M., and V. B. Kazansky (1986), "Quantum-Chemical Cluster Models of Acid–Base Sites of Oxide Catalysts," *Adv. Catal.* **34**, 131–204.

10

ADSORPTION KINETICS OF THE COMPLEX MIXTURE OF ORGANIC SOLUTES AT MODEL AND NATURAL PHASE BOUNDARIES

Božena Ćosović

Rudjer Bošković Institute, Center for Marine Research, Zagreb, Croatia, Yugoslavia

1. INTRODUCTION

Adsorption processes at phase boundaries have an important role in biogeochemical cycles of elements in natural aquatic systems (Parks, 1975; Stumm and Morgan, 1981). The exchange of substances between the water and the atmosphere is influenced by mass-transport processes in the interfacial region at the air–water boundary. It is dependent on the chemical composition and physicochemical properties of the naturally occurring surface films, which consist of dissolved and particulate, surface-active organic molecules (Hunter and Liss, 1982). The elimination processes for most trace elements from the bulk water are regulated by adsorption on particles and scavenging mechanisms (Balistrieri et al., 1981; Sigg, 1987). The weathering of rocks and mineral dissolution processes are ruled by surface reactions and mediated by organic molecules that are present (Stumm and Furrer, 1987). Adsorption of organic substances at solid particles has a complex influence on the particle–particle interactions in establishing colloidal stability in aqueous systems. This matter is elaborated in more detail in Chapter 16 (O'Melia).

Surface film at the air–water interface and organic coatings at the mineral surface are formed by sorption of organic surface-active material from the aqueous phase. Natural aquatic systems contain a large number of organic substances with different functional groups and different hydrophobic properties. In adsorption at interfaces, depending on the adsorbate and the interface, there are hydrophilic, electrostatic, hydration, and hydrophobic interactions (Conway, 1976; Westall, 1987). Hydrophilic interactions depend on the presence of such functional groups as $-OH$, $-NH_2$, $>C=O$, $-COOH$, and $-NH-CH$ $=O$, as well as ionic groups that determine electrostatic hydration. Hydrophobic interactions arise from the reluctance of nonpolar molecules or molecular groups to be surrounded by polar H-bonded water due to the H-bond breaking that is induced by their presence. Hydrophobic and electrostatic (ionic) interactions are of great importance in determining the surface excess of solute species at the air–water, the water–mineral and the water–electrode interfaces.

The fact that organic matter in natural and polluted water contains a complex mixture of different naturally occurring substances and pollutants, most of them at very low concentrations, represents an important reason for the adsorption study of mixtures. These can be done at natural and model phase boundaries. Selected well-defined and easily controlled model interfaces have some advantages for the study of very complex systems such as adsorbable organic matter in natural waters.

1.1. Adsorption of Organic Substances at the Mercury Electrode–Solution Interface

Because of its nonpolar and hydrophobic character, the mercury–water may serve as a good model interface for the adsorption study and determination of the organic substances that are adsorbed primarily because of hydrophobic expulsion. There is generally a proportionality of adsorbability (free energy of adsorption) found at the mercury electrode to a number of $-CH_2$ groups in paraffinic hydrocarbon residues in nonpolar surfactants and a similar relation between the octanol water partition coefficient and chain length. This was recently also illustrated in the case of adsorption of aliphatic fatty acids (Ulrich et al., 1988).

The measured adsorption effect at the electrode is influenced by all dissolved and/or dispersed surface-active substances according to their concentration in the solution, adsorbability at the electrode, kinetics of adsorption, structure of the adsorbed layer, and some other factors. Adsorption of organic molecules on electrodes causes a change of the electrode double-layer capacitance. It is the result of an exchange between the counterions and water molecules from solution, followed by changes in the dielectric properties and the thickness of the double layer on the electrode surface, that is, parameters that determine the electrode capacitance (Bockris et al., 1963; Damaskin and Petrii, 1971).

By use of phase-sensitive alternating-current (AC) polarography one can measure selectively either the capacitive current (i.e., current needed to build up a certain potential on the electrode) or the Faradaic current of the oxidation or reduction processes that occur at the interface.

Capacitive current has also been successfully measured by the Kalousek commutator technique (Ćosović and Branica, 1973). This technique has some advantages for the study of adsorption of mixtures of adsorbable solutes, because it enables separation of adsorption effects of components as was shown for the mixture of nonionic and anionic detergents (Kozarac et al., 1976).

In the case of adsorption of more different adsorbable solutes, the capacitive behavior of the Hg–water interface can be described by the model of parallel capacitors (Jehring, 1974). For two substances, it follows

$$C = C_0(1 - \Theta_1' - \Theta_2') + C_1\Theta_1' + C_2\Theta_2' + (q_1 - q_0)\frac{d\Theta_1'}{dE} + (q_2 - q_0)\frac{d\Theta_2'}{dE} \quad (1)$$

where C represents the measured total differential capacitance (μF cm^{-2}), while C_0 is capacitance of the interface covered by water molecules, C_1 and C_2 that of the interface totally covered by organic molecules of individual components of the mixture; q_0, q_1, and q_2 are corresponding charge densities; and Θ_1' and Θ_2' are the values of surface coverage for the components in the mixture. Surface coverage is defined as $\Theta = \Gamma/\Gamma_{max} = \Gamma_{A max}$ is surface concentration at saturation level ($\Theta = 1$) and $A = 1/\Gamma_{max}$, surface area occupied by one adsorbed molecule. In the potential range of maximum adsorption, adsorption is potential independent, $do/dE = 0$ and the former equation is reduced to

$$C = C_0(1 - \Theta_1' - \Theta_2') + C_1\Theta_1' + C_2\Theta_2' \quad (2)$$

The total decrease of the double-layer capacitance caused by adsorption of two different organic molecules in the mixture is defined as

$$\Delta C = C - C_0 = \Theta_1'(C_1 - C_0) + \Theta_2'(C_2 - C_0) = \Theta_1'\Delta C_1 + \Theta_2'\Delta C_2 \quad (3)$$

This scheme can be used for three or more components of the mixture.

1.2. Adsorption Equilibrium

The relationship between the amount of substance on the electrode per unit area, $\Theta_1\Gamma_i$, the concentration of bulk solution, c_i, and the electrical state of the system, potential (E) or charge (q) at a given temperature, is given by the adsorption isotherm. This is obtained from the condition of equality of electrochemical potentials for bulk and adsorbed species i at equilibrium (Bard and Faulkner, 1981).

In ideal situations, if concentrations c_1 and c_2 as well as surface coverages Θ_1 and Θ_2 of the components of the mixture are very low, the competition and the interaction between organic molecules in the adsorbed layer can be neglected so that adsorption isotherms can be defined as

$$\Theta_1 = B_1 c_1, \quad \Theta_2 = B_2 c_2' c_2$$

Thus, adsorption of both types of organic molecules occurs independently. The measured decrease of the capacitance is additive in relation to values obtained for adsorption of individual components.

If two species are adsorbed competitively, the appropriate Langmuir isotherms are

$$\frac{\Theta'_1}{1-\Theta'_1-\Theta'_2}=B_1c_1 \tag{4}$$

$$\frac{\Theta'_2}{1-\Theta'_1-\Theta'_2}=B_2c_2 \tag{5}$$

These equations can be derived from a kinetic model assuming independent coverages Θ'_1 and Θ'_2, with the rate of adsorption of each species proportional to the free area $(1-\Theta'_1-\Theta'_2)$ and the solution concentrations c'_1, c_1 and c'_2, c_2 respectively. The rate of desorption is assumed to be proportional to Θ'_1 and Θ'_2. For $\Theta \ll 1$, the Langmuir isotherms are changed to the ideal case of independent adsorption.

Interactions between adsorbed species complicate the problem by making the energy of adsorption a function of surface coverage. Attractive forces in the adsorbed layer increase surface concentrations and cause formation of a more condensed layer, while repulsion results in decrease of surface concentration and a formation of a more loose adsorption layer. In the adsorption of mixture of two organic substances, one should take into consideration all possible interactions between the molecule of the same type and between different molecules.

The frequently used isotherm that includes the possibility of taking into account interactions between adsorbed species is the Frumkin isotherm. The adsorption of two components in the mixture is described by the equations

$$B_1c_1=\frac{\Theta'_1}{(1-\Theta'_1-\Theta'_2)}\exp(-2a_1\Theta'_1-2a_{12}\Theta'_2) \tag{6}$$

$$B_2c_2=\frac{\Theta'_2}{(1-\Theta'_1-\Theta'_2)}\exp(-2a_2\Theta'_2-2a_{12}\Theta'_1) \tag{7}$$

where a_1, a_2, and $a_{1,2}$ are interaction coefficients. For equilibrium adsorption of mixtures of adsorbable solutes, the model of parallel capacitors gives the relation $C=f(\Theta'_1\Theta'_2)$. At low surface coverages the measured capacitance decrease ΔC is additive with respect to individual components of the mixture. At higher surface concentrations the capacitance decrease is lower than the sum of individual effects because of competition and limited surface area. If the interactions between adsorbed molecules take place, and if $a_{12}>a_1$ or a_2, the capacitance decrease of the mixture could exceed the value that corresponds to the sum of individual components.

1.3. Adsorption Kinetics

The adsorption of a species from solution on the creation of fresh electrode surface proceeds in several steps. The inherent rate of adsorption, at least on mercury from aqueous solution, usually is rapid, so that the overall rate is frequently governed by mass transfer, diffusion, and convection (Ružić, 1987). The equilibration time can be very long. For example, equilibrium adsorption of humic substances in the concentration range below 1 mg dm^{-3} is established within 500–1000 s with convective mass transport (Raspor and Valenta, 1988). For strongly adsorbable surfactants, the adsorption equilibrium is completely on the side of adsorption. In very dilute solutions of surfactants, adsorption processes are controlled by transport of the surfactant from the bulk solution toward the electrode surface as a result of the concentration gradient formed in the diffusion layer, until surface saturation, $\Theta = 1$, is obtained. For non-equilibrium adsorption the apparent isotherms can be constructed for different time periods of adsorption that are shifted with respect to the true adsorption isotherm in the direction of higher concentrations. This is because the adsorption equilibrium can exist only at the surface of the electrode and not within the bulk of solution. The adsorption process is then controlled by kinetics of mass transport and measurable surface coverages can be obtained only at concentrations higher than those corresponding to the adsorption equilibrium. This is illustrated in Figure 1 for the diffusion controlled adsorption of Triton-X-100 at the Hg–water interface. Theoretical predictions of the adsorption process were obtained in this case by digital simulation, based on the finite difference method, assuming a Frumkin type of adsorption isotherm. Reasonably good agreement between experiment and theory was obtained for a lateral interaction factor

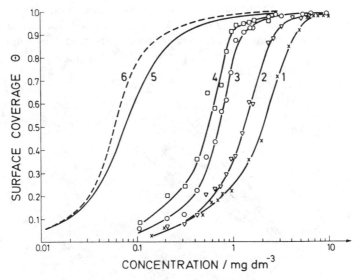

Figure 1. Apparent adsorption isotherms of Triton-X-100 in 0.55 mol dm^{-3} NaCl obtained with various accumulation times: (1) 30, (2) 60, (3) 180, (4) 300 s; and theoretical Frumkin adsorption isotherms of Triton-X-100 with interaction factors $a = 1.0$ (curve 5) and 1.25 (curve 6). Batina et al., 1985.)

between 1.0 and 1.25, the adsorption coefficient $B = 3 \times 10^6$ mol dm^{-3}, and the rate constant of adsorption $k_a = 10^4$ mol dm^{-3} s^{-1} (Batina et al., 1985).

During the process of adsorption of a mixture of adsorbable solutes the uncovered parts of the surface area are free for the attachment of other adsorbable molecules. If the diluted solution of strongly adsorbable surfactant also contains a higher concentration of a less adsorbable one, the latter will cover the electrode surface first and will then be replaced by the diffusion process of the stronger adsorbable substance. If the adsorption of both components is diffusion-controlled, then for $\Theta < 1$ the total capacitance decrease is in first approximation additive:

$$\Delta C = k \left[\Delta C_1 \frac{D_1^{1/2}}{\Gamma_{m1}} \cdot c_1 + \Delta C_2 \frac{D_2^{1/2}}{\Gamma_{m2}} \cdot c_2 \right] t^{1/2} \tag{8}$$

The total capacitance decrease is linear with half life $t^{1/2}$. Partial contributions to C of individual components depend on their capacitance decrease per mole (C/Γ_m) and concentration c. Linear dependence of the surface coverage, Θ, on the square root of time of adsorption of Triton-X-100 is presented in Figure 2, from which the value $\Gamma_m / D^{1/2} = 3.5 \times 10^{-8}$ mol s$^{-1/2}$ cm^{-3} was determined using the basic principle of Koryta's method (Heyrovsky and Kuta, 1966).

Nonequilibrium adsorption at the mercury has been studied in a number of mixtures of surface active substances of interest for natural aquatic systems.

Experiments were performed in two ways: (1) the ratios between the components of the mixture were maintained constant throughout the measurement and (2) one of the components had a constant concentration and for the other the concentration increased from zero to a maximum. Results obtained with mixtures prepared in different ways were in good agreement. The total adsorp-

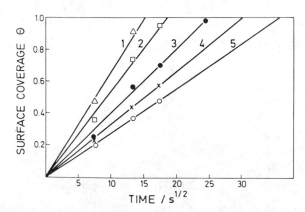

Figure 2. Surface coverage vs. square root of accumulation time for adsorption of Triton-X-100 at $E = -0.6$ V versus SCE. Concentration of Triton-X-100 in 0.55 mol dm^{-3} NaCl: (1) 1.25; (2) 0.94; (3) 0.73; (4) 0.63; (5) 0.52 mg dm^{-3}. (Batina et al., 1985.)

tion effect of different mixtures of selected surface active substances was compared with the values calculated from individual isotherms. Assuming that C_0, C_1, and C_2 do not change with the surface coverage and if components of the mixture do not influence one another in the process of adsorption, then Θ_1' and Θ_2' in equation 3 become Θ_1 and Θ_2; that is, the values determined by the adsorption isotherms for the substances investigated. The additivity of adsorption under nonequilibrium conditions is illustrated for the selected mixtures of surface active substances in Figure 3 (Ćosović et al., 1980). The total adsorption effect was found to correspond to the calculated one (dashed lines) for low values of surface coverage ($\Theta < 0.7$). In the region of high surface coverage, interactions at the electrode between similar molecules, as well as between different molecules, were more pronounced and usually caused a decrease in the adsorption effect of the mixture in comparison with the calculated value. Additional effects in the mixture could also result from the chemical interactions between molecules in the adsorption layer and in the solution, steric influences, reorientation of molecules at the surface, and so on. For example, high separation between the measured and the calculated isotherms was observed for all the values of Θ relating to the adsorption of the mixture containing sodium dodecyl sulfate and protein albumin, because of the binding ability of protein for anionic detergents (Ćosović et al., 1980). Knowledge of protein detergent interactions is also important in understanding the pollution effects of detergents in natural waters.

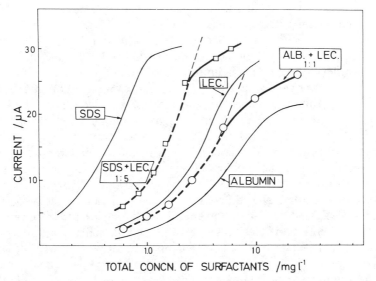

Figure 3. Adsorption isotherms of model surfactants (sodium dodecyl sulphate (SDS), egg albumin and lecithin) and their mixtures (SDS and leicthin, concentration ratio 1:5, and albumin and lecithin, concentration ratio 1:1) in seawater. Dashed lines correspond to calculated isotherms. (Ćosović et al., 1980.)

Detergents solubilize water insoluble substances. They merge into the inter-boundary layers of naturally occurring proteins at various phase boundaries, thus influencing mass-transport processes in natural systems. Hence the toxic effects of detergents to aquatic organisms are closely related to their interactions with lipids and proteins in biological membranes.

1.4. Effect of the Adsorbed Layer on the Mass- and Charge-Transfer Processes at the Electrode–Solution Interface

Very useful additional information on the adsorption behavior of the mixture and the structure of the adsorbed layer can be obtained by investigation of the influence of the adsorbed layer on the electrode processes of other ions and molecules, which are used as a probe. The choice of the probe depends, among other things, on the potential of the electrode reaction. The best results obtained if the potential of the electrode reaction is within the potential region of maximum adsorption. Polarographic oxidation–reduction waves of Cd(II) and nitrophenol are very suitable for such investigation. At pH values of natural seawater and freshwater samples (pH 7–8) cadmium bears a positive charge while nitrophenol is dissociated and present as negatively charged nitropheno-late. Since mass and charge-transfer processes at the covered electrode surface depend on the porosity of the adsorbed layer and possible interactions between electroactive species and adsorbable solutes, one can obtain valuable informa-tion about the structure and on the mechanisms of the exchange reactions at interfaces. Such investigations can be done at a surface completely covered by organic molecules ($\Theta > 1$) when additional information is not available from the measurement of capacitive effects (saturation level).

The adsorption behavior of different mixtures of albumin and lecithin at the mercury electrode and the influence of the adsorbed layer on the mass and charge transfer processes of cadmium were studied (Kozarac and Ćosović, 1984). The mixtures were composed of a constant concentration of one compo-nent, either lecithin or albumin, and an increasing concentration of the other component of the mixture. As shown in Figure 4, until a concentration ratio of approximately 1:1 is achieved, the inhibition effect is changed with every small addition of either albumin or lecithin, indicating that no preferential adsorption of any component of the mixture occurred. When the concentration of the increasing component in the mixture reaches the value of the constant compo-nent there are considerable differences in the behavior of the two types of mixtures investigated. The inhibition effect for cadmium caused by the mixture in which the concentration of lecithin is constant and the concentration of albumin increases remains more or less constant or changes very slowly with further addition of albumin. The inhibition effect of the mixture in which albumin is the constant component and lecithin the increasing one continuously changes with the addition of lecithin until the effect of the mixture coincides with the effect of lecithin at the given concentration. If lecithin is in excess, it displaces the albumin molecules from the electrode surface, while when the albumin is present in excess,

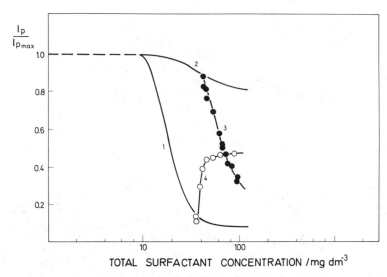

Figure 4. Dependence of the normalized reduction peak current of 10^{-4} mol dm^{-3} Cd(II) in seawater on the total concentration of surface-active substances. (1) Lecithin; (2) egg albumin; (3) mixture of constant concentration of egg albumin (42 mg dm^{-3}) + variable amounts of lecithin; (4) mixture of constant concentration of lecithin (35 mg dm^{-3}) + variable amounts of egg albumin. (Kozarac and Ćosović, 1984.)

it combines with the adsorption layer of lecithin. The permeability of the adsorbed layer depends on the ratio between the lipid and protein components in the mixture.

We have generally observed that adsorbed layers formed in neutral and alkaline solutions of natural biopolymers (proteins) and geopolymers (humic substances) are permeabile for Cd(II) ions (Kozarac et al., 1986). On acidification these layers show a strong inhibition effect on the oxidoreduction processes of Cd(II) ions. There are two possible explanations for this; either (1) repulsive forces of the negatively charged functional groups of the polymer cause a formation of a more loose or porous adsorption layer in alkaline solution and a more condensed one after protonation in the acidic solution, or (2) the charge-transfer process of cadmium can proceed via a bridging mechanism on negatively charged groups of the organic coating at the electrode surface.

When the applied probe was a negatively charged ion, as, for example, nitrophenolate ion, in accordance with our expectations, strong inhibition effects were observed for adsorbed layers of negatively charged proteins and humic substances. On the contrary, adsorbed layers that possess positively charged functional groups showed a positive effect on the charge-transfer processes of nitrophenolate ion. The effect of trimethyloctadecyl ammonium bromide (TOMA) on the cathodic wave of p-nitrophenol (PNP) is presented in Figure 5. Increased peak height of PNP above the value obtained in the absence of

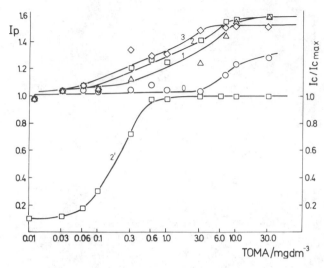

Figure 5. Effect of trimethyloctadecylammonium bromide (TOMA) on the cathodic current of 1 $\times 10^{-4}$ mol dm^{-3} p-nitrophenol (PNP) at the mercury electrode in 0.5 mol dm^{-3} NaCl. Accumulation time with stirring of solution: (0) 0, (1) 30, (2) 60, (3) 180 s. Curve (2') is apparent adsorption isotherm of TOMA for adsorption time $t_a = 60$ s with stirring of solution.

surfactant resulted due to accumulation of PNP in the adsorbed layer of TOMA at the electrode surface.

The orientation of the adsorbed organic molecules at the surface may also play a role in the accessibility of the functional groups for surface interactions with other ions and molecules. The surface charge of the mercury electrode can be adjusted as either positive or negative by the applied polarization potential. The mercury surface is changing from the uncharged to the negatively charged state in the potential region where reduction processes of cadmium and nitrophenol take place. However, in the adsorption of negatively charged humic substances and proteins at the negatively charged mercury electrode it is reasonable to expect that the functional groups are oriented toward solution. In the adsorption of positively charged surfactant such as TOMA the hydrophobic effect previals so that the hydrophilic functional groups, although of charge opposite to that of the mercury surface, are oriented toward solution and are accessible for interaction with negatively charged nitrophenolate.

Similar adsorption behavior was observed at the air–solution interface, as shown recently for interactions of PNP with lipids at the air–water interface, which were investigated by monolayer studies and light reflection spectroscopy (Kozarac et al., 1989, in press). The nitrophenolate ion does not accumulate at the air–water interface as well as in the neutral lipid monolayer. Its attachment to these hydrophobic interfaces is mediated by electrostatic interaction with positively charged functional groups of hydrophobic surfactants.

Adsorption of organic solutes at the surface of suspended particles, that is, the mineral–water interface, can be also characterized by specific coordinative interaction. Such coordinative interaction determines the free energy of adsorption and the orientation of organic molecules at the surface. These can be very much different from the adsorption of the same molecules at the hydrophobic interfaces (Ulrich et al., 1988).

1.5. Transformations of the Adsorbed Layer

Special examples of mixture adsorption are competitive adsorption of the different forms of the same substance, such as pH-dependent ionic and undissociated molecular forms, monomers, and associates of the same substance, as well as potential-dependent adsorption of the same compound in two different orientations in the adsorbed layer. Different orientations on the electrode surface—for example, flat and vertical—are characterized with different adsorption constants, lateral interactions, and surface concentrations at saturation. If there are strong attractive interactions between the adsorbed molecules, associates and micellar forms can be formed in the adsorbed layer even when bulk concentrations are below the critical micellar concentration (CMC). These phenomena were observed also at mineral oxide surfaces for isomerically pure anionic surfactants and their mixtures and for mixtures of nonionic and anionic surfactants (Scamehorn et al., 1982a–c).

Transformation of the adsorbed layer can be a slow process, as was demonstrated for the adsorption of valeric acid on the mercury electrode at pH 2 (Fig. 6) (Ružić et al., 1988). After the first step of adsorption, leading to a kind of semistable state, very slow and persistent changes in the adsorption layer occur. The overall adsorption equilibrium of the final state is somewhere between 10^3

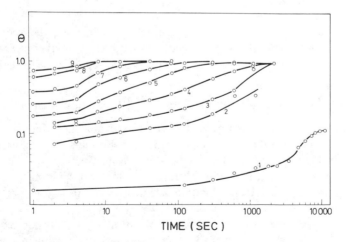

Figure 6. Progress of the adsorption process of valeric acid at pH 2, expressed in terms of the logarithmic increase of Θ versus the logarithm of the adsorption time. Concentrations of valeric acid: (1) 0.18, (2) 0.92, (3) 1.28, (4) 1.66, (5) 2.20, (6) 3.11, (7) 4.59, (8) 7.35, (9) 9.20 mmol dm^{-3} (Ružić et al., 1988.)

and 5×10^3 mol dm^{-3}, which is more than an order of magnitude higher than one could estimate from the experiments at the short adsorption time (~ 1 s). The change of state of adsorbed molecules was supported by experiments of inhibition of Cd(II) reduction by adsorbed molecules of valeric acid.

Transformation of the adsorbed layer can be catalyzed by the presence of a small amount of some other adsorbable compound. Effects of dodecyl alcohol (DOH) on the physicochemical properties of the adsorbed layer of sodium dodecyl sulfate (SDS) were studied in detail (Batina and Ćosović, 1987; Batina et al., 1988). Time dependence of the diffusion controlled adsorption of the mixture containing only 0.5% of DOH is presented in Figure 7. Up to a concentration of 2×10^{-4} mol dm^{-3} of SDS in the mixture, the capacity decreases with time and with increasing bulk concentration of surfactants as a result of the formation of the adsorption layer. In a higher concentration range, from 3×10^{-4} to 7×10^{-4} mol dm^{-3} of SDS in the mixture, after an initial decrease of the electrode double-layer capacity at the potential of maximum adsorption, a further slow increase is observed. At concentrations above 10^{-3} mol dm^{-3} of SDS in the mixture, the capacity value becomes constant again, which is the result of the virtually instantaneous and complete process of adsorbed layer formation. Resulting apparent adsorption isotherms for different mixtures and a selected time of adsorption (300 s) are given in Figure 8 together with the corresponding effects of the adsorbed layer on the reduction current of Cd(II). As observed, by increasing the SDS concentration in solution, up to about 2×10^{-4} mol dm^{-3}, the inhibition action of the adsorbed layer upon the reduction process of cadmium is demonstrated. A small increase in the reaction rate, which is more visible at lower frequences of the applied AC signal (Batina and Ćosović, 1987), is observed at higher concentrations. In mixtures of SDS and DOH, in the concentration range where the minimum of ΔC on the adsorption isotherms exists (Fig. 8a), a sudden increase of the peak height, that is, of the reaction rate due to the effect of the adsorbed layer, is observed. Dodecyl alcohol, which itself shows a strong inhibition (blocking) effect for the reduction of cadmium, when incorporated (solubilized) in the adsorption layer of SDS causes such changes of the adsorbed layer; that is, it becomes permeabile for Cd(II) ions.

1.6. Adsorbable Solutes of Natural Marine and Freshwater Samples

Natural organic matter in aquatic media are complex mixture of substances such as polysaccharides, proteins, peptides, lipids, and humic substances (Duursma and Dawson, 1981; Buffle, 1984). Humic material is itself a mixture of polymers of a wide range of molecular weights (Buffle, 1988). Besides naturally occurring substances, various artificial compounds are introduced in natural water as the result of human activities.

The adsorption behavior of organic molecules in natural samples at the mercury electrode can be described by competitive adsorption in which less-adsorbable materials are present at higher concentrations and strongly adsorbable substances are at lower concentrations. Since the adsorption process

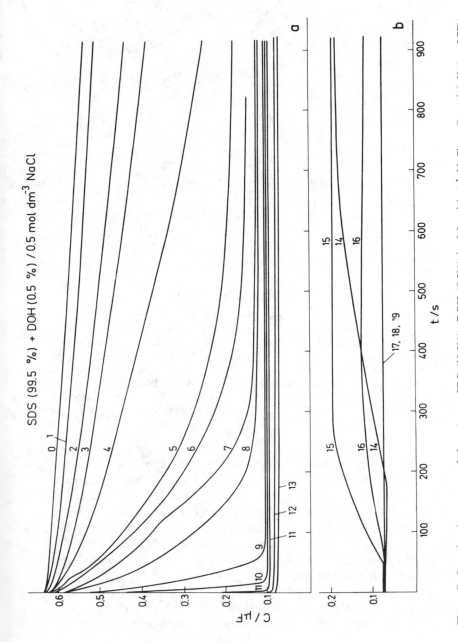

Figure 7. Capacity–time curves of the mixture SDS (99.5%) + DOH (0.5%) in $0.5\ mol\,dm^{-3}$ NaCl at $E = -0.6$ V (vs. SCE). Concentrations of SDS in the mixture: (a) (0) 0; (1) 2.08×10^{-7}, (2) 4.16×10^{-7}, (3) 6.24×10^{-7}, (4) 1.04×10^{-6}, (5) 1.46×10^{-6}, (6) 2.08×10^{-6}, (7) 3.12×10^{-6}, (8) 4.16×10^{-6}, (9) 1.04×10^{-5}, (10) 2.08×10^{-5}, (11) 4.16×10^{-5}, (12) 1.4×10^{-4}, (13) 2.08×10^{-4}, (b) (14) 3.12×10^{-3}, (15) 4.16×10^{-4}, (16) 7.29×10^{-4}, (17) 1.04×10^{-3}, (18) 2.08×10^{-3}, (19) $5.20 \times 10^{-3}\ mol\,dm^{-3}$. Frequency 230 Hz. (Batina and Ćosović, 1987.)

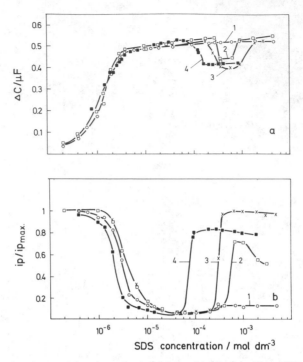

Figure 8. Comparison between the adsorption behavior and inhibition effects of SDS in the presence of DOH. (a) Adsorption isotherms of (1) recrislistallized SDS, (2) mixture of SDS (99.5%)+ DOH (0.5%), (3) mixture of SDS (98.%) + DOH (1.4%), and (4) SDS commercial chemical in 0.5 mol dm^{-3} NaCl. Accumulation time 300 s at $E = 0.6$ V vs. SCE. Frequency 230 Hz. (b) Dependence of the normalized reduction peak current of 10^{-4} mol dm^{-3} Cd^{2+} on the presence of surfactants in 0.5 mol dm^{-3} NaCl measured at frequency of 500 Hz: (1) recrystallized SDS, (2) mixture of SDS (99.5%) DOH (0.5%), (3) mixture of SDS (98.6%) + DOH (1.4%), and (4) SDS commercial. chemical. Accumulation time 300 s at $E = 0.4$ V vs. SCE; $i_{p,max}$ is the peak current of the AC voltammogram of Cd^{2+} in the absence of surfactant. (Batina et al., 1988.)

depends on both the adsorption constant and the adsorption kinetics, which include the mass transfer of surfactant molecules from the bulk solution toward the electrode surface, the measured electrochemical signal is influenced by the qualitative and quantitative composition of the complex mixture of adsorbable solutes in the sample. According to the standard procedures, for electrochemical measurement, adsorption effect is produced by all adsorbable organic molecules that reach the electrode surface during the accumulation period of up to 120 s with a continuous stirring of solution.

Humic substances were found to be predominant surface-active material in freshwater samples, while electrochemical studies have proved the role of lipid material in adsorption processes in the marine environment (Ćosović, 1985).

In comparison with the behavior of the total surfactants and the dissolved organic matter, the hydrophobic surfactants were found to be more enriched at the natural phase boundaries (Ćosović and Vojvodić, 1989). Thus, as a result of the adsorption processes in aquatic systems, the fractionation of organic matter occurs in the interfacial region, resulting in a different organic chemical composition in the interfacial layers in relation to the aqueous bulk phase. This has been demonstrated for the sea surface microlayer samples and for the organic layer formed at the halocline of the stratified estuary, as well as for the uptake of organic material by sorption on the mineral–water interface in the freshwater system. From the total amount of adsorbable solutes of natural-water samples, simple, low-molecular-weight compounds such as amino acids, short-chain mono- and dicarboxylic acids, and simple sugars, are not adsorbed at the mercury electrode, because of the low adsorption constant and high competition of other substances. However, because of specific interaction they can adsorb on the mineral surface of suspended solids. Their contribution is not visible from the electrochemical measurements.

Separation of natural organic matter into hydrophobic and hydrophilic fractions is possible by using the macroreticular resin XAD-8. We have used the modified procedure (Leenheer, 1981; Vojvodić and Ćosović, manuscript in preparation) for the investigation of selected natural aquatic samples. Surfactant activity was measured by the electrochemical method in influent and effluent samples for XAD-8 sorption at natural and acidic pH values. The hydrophobic (base + neutral) fraction is adsorbed on the resin at the natural pH of the sample, the hydrophobic (acid) fraction at pH 2, while the hydrophilic fraction is nonadsorbable on the resin and remains in the effluent. The results, the surfactant activity of the original sample and the percentage of surface active substances in different fractions, are presented in Table 1. Adsorption characteristics of separated fractions were studied by AC polarography. It showed that natural aquatic samples contain complex mixtures of organic compounds with different adsorption and dielectric properties, among which the hydrophobic fraction usually dominates in the adsorption of organic matter at model and natural phase boundaries.

Adsorption behavior of the natural sample was simulated by a model mixture containing the following components:

0.5 mg dm^{-3} Triton-X-100 (nonionic, strongly adsorbable surfactant, molecular weight 600, completely adsorbed at neutral pH on the XAD-8 resin)

5 mg dm^{-3} fulvic acid (predominantly adsorbed on XAD-8 in acidic solution)

5 mg dm^{-3} Dextran T-500 (polysaccharide, molecular weight 500.000, hydrophilic and nonadsorbable on XAD-8 resin)

Results of the fractionation of the mixture are presented in Table 2A. Rough characterization of the fractions was made by comparing their capacity current–potential curves with those of the model mixtures. The shape of the curve obtained in effluent samples after separation of the hydrophobic neutral component corresponds to that of the mixture of fulvic acid and Dextran. After removal of the hydrophobic acid fraction, the remaining fraction resembles the

TABLE 1. Fractionation of Organic Solutes in Natural Samples by Sorption on XAD-8 Resin

Sample	Date	SA (mg Triton-X-100 dm^{-3})	Fractions (%)		
			Hydrophobic Base + Neutral	Hydrophobic Acid	Hydrophilic
River water	June 1987	1.32	18.3	65.4	16.3
	September 1987	3.10	19.4	35.5	45.1
	March 1988	0.85	42.4	34.0	23.6
Estuarine water, S = 22‰, collected at the halocline	May 1987	0.44[a]	80.0		20.0
Dunaliella tertiolecta culture medium	March 1988	1.95	46.0	36.7	17.3

Ćosović and Vojvodić, 1989.

[a] Surfactant activity measured at pH 2.

TABLE 2. Fractionation of Organic Matter by Sorption on XAD-8 Resin[a]

A. Surfactant Activity (%) of Fractions		
Hydrophobic Base + Neutral (1)	Hydrophobic Acid (2)	Hydrophilic (3)
20.8	57.7	21.5

B. Effects of Different Fractions on Cathodic Current of Cd^{2+}

		Normalized Peak Current[b]		
Sample	pH	15 s	60 s	120 s
(1)+(2)+(3)	7	0.25	0.13	0.07
	2	0.23	0.08	0.06
(2)+(3)	7	0.21	1.03	1.06
	2	0.22	0.06	0.04
(3)	7	1.04	0.43	0.37
	2	0.93	0.40	0.38

[a] Composition of mixture: (1) 0.5 mg T-X-100 + (2) 5 mg FA + (3) 5 mg Dextran T-500.

[b] Normalized peak current = $\dfrac{i_p \text{ (in presence of surfactant)}}{i_p \text{ (in absence of surfactant)}}$.

adsorption behavior of Dextran at the mercury electrode. Further inspection was made using the Cd^{2+} reduction process as the probe for the characterization of the structure of the adsorbed layer. Results are presented in Table 2B. In the three component mixture the adsorption layer formed exhibits a strong inhibition effect for reduction of cadmium at neutral and acidic pH values, which corresponds mainly to the influence of Triton-X-100 in the adsorbed layer. In the mixture of fulvic acid and Dextran the organic coating formed is permeabile for Cd^{2+} ions at neutral pH; even a small accumulation of Cd^{2+} in the adsorbed layer is observed at short adsorption time (normalized peak current > 1). In acidic medium strong inhibition effect is observed as it is expected for the adsorption of fulvic acid. The hydrophilic fraction, that is, Dextran, itself shows an inhibition effect at both pH values, which increases with adsorption time. Conclusively, it can be summarized that the physicochemical properties of the adsorbed organic layer formed in the solution of the complex mixture are dynamically changed with time of adsorption. The contributions of the individual components depend more on their adsorption constants than on the weight percentage in the mixture. Small amounts of strongly adsorbable species, especially of low molecular weight, may have tremendous effects on the properties of the adsorbed layer. Triton-X-100 is not a naturally occurring surface-active substance. It was used in the model mixture as a representative for strongly adsorbable, low-molecular-weight compounds, such as lipids.

2. CONCLUSIONS

Adsorption processes at natural phase boundaries occur from the aqueous phase containing complex mixtures of different organic and inorganic compounds and they are not in equilibrium. Adsorption behavior of organic molecules can be described by competitive adsorption of the mixture, in which very often less adsorbable materials are present at higher concentrations and strongly adsorbable substances are present at lower concentrations. Since the adsorption process depends on both the adsorption constant and the adsorption kinetics, which include the mass transfer of adsorbable molecules from the bulk phase toward the surface as well as the intrinsic rate of attachment to the surface, the adsorption layer formed is influenced by qualitative and quantitative composition of the complex mixture of adsorbable solutes.

The focus of this chapter has been the nonequilibrium adsorption of a number of mixtures of surface-active substances of interest for natural aquatic systems at the model phase nonpolar hydrophobic mercury electrode. Mass transfer of adsorbable molecules was controlled by diffusion and by convective processes. At the concentration level of naturally occurring organic substances diffusion controlled adsorption takes place within 10–1000 min, while convective movements accelerate it up to 100–1000 s. Secondary transformations of the adsorbed layer can be very slow and sometimes take place within hours, as was shown in the case of valeric acid. Useful information on the adsorption behavior of the mixture and the structure of the adsorbed layer can be obtained by investigation of the influence of the adsorbed layer on the electrode processes of other ions and molecules, which are used as a probe. Since mass- and charge-transfer processes at the covered electrode surface depend on the porosity of the adsorbed layer and possible interaction between electroactive species and adsorbed molecules, the observed adsorption effect is indicative of the structure of the adsorbed layer as well as of the mechanisms of the exchange reactions at interfaces. Electrochemical processes of the positively charged cadmium(II) ion and negatively charged nitrophenolate ion were used as the probe of choice in this study.

Acknowledgment

The author wishes to acknowledge the financial support of the Authority for Scientific Research of the SR Croatia and the National Bureau of Standards, Washington, DC.

REFERENCES

Balistrieri, L., P. G. Brewer, and J. W. Murray (1981), "Scavening Residence Times of Trace Metals and Surface Chemistry of Sinking Particles in the Deep Ocean," *Deep Sea Res.* **28A**, 101–121.

Bard, A. J., and L. R. Faulkner (1981), *Electrochemical Methods*, Wiley, New York, pp. 488–553.

Batina, N., and B. Ćosović (1987), "Adsorption of a Mixture of Sodium Dodecyl Sulphate and Dodecyl Alcohol on a Mercury Electrode and Its Effect on the Electrochemical Processes of Cadmium(II) in Sodium Chloride Solution," *J. Electroanal. Chem.* **227**, 129–146.

Batina, N., I. Ružić, and B. Ćosović (1985), "An Electrochemical Study of Strongly Adsorbable Surface Active Substances. Determination of Adsorption Parameters for Triton-X-100 at the Mercury/Sodium Chloride Interface," *J. Electroanal. Chem.* **190**, 21–32.

Batina, N., B. Ćosović, and N. Filipović-Vinceković (1988), "The Effect of Trace Amounts of Dodecyl Alcohol on the Physicochemical Properties of Sodium Dodecyl Sulphate: Electrochemical Study at the Mercury/Sodium Chloride Interface," *J. Colloid. Interface Sci.* **125**, 69–79.

Bockris, J. O'. M., M. A. V. Devanathan, and K. Muller (1963), "The Structure of Charged Interfaces," *Proc. Phys. Soc. (London)* **A274**, 55–79.

Buffle, J. (1984), "Natural Organic Matter and Metal–Organic Interactions in Aquatic Systems," in H. Sigel, Ed., *Metal Ions in Biological Systems*, Vol. 18, *Circulation of Metals in the Environment*, Dekker, New York.

Buffle, J. (1988), *Complexation Reactions in Aquatic Systems*, Elis Horwood, Chichester.

Conway, B. E. (1976), "Hydrophobic and Electrostatic Interactions in Adsorption at Interfaces: Relation to the Nature of Liquid Surface," *Croat. Chem. Acta* **48**, 573–596.

Ćosović, B. (1985), "Aqueous Surface Chemistry· Assessment of Adsorption Characteristics of Organic Solutes by Electrochemical Methods," in W. Stumm, Ed., *Chemical Processes in Lakes*, Wiley, New York.

Ćosović, B., and M. Branica (1973), "Study of the Adsorption of Organic Substances at a Mercury Electrode by the Kalousek Technique," *J. Electroanal. Chem.* **46**, 63–69.

Ćosović, B., and V. Vojvodić (1989), "Adsorption Behaviour of the Hydrophobic Fraction of Organic Matter in Natural Waters," *Mar. Chem.* **28**, 183–198.

Ćosović, B., N. Batina, and Z. Kozarac (1980), "Adsorption of Some Mixtures of Surface Active Substances at the Mercury Electrode, Kalousek Commutator Measurements," *J. Electroanal. Chem.* **113**, 239–248.

Damaskin, B. B., O. A. Petrii, and V. V. Batrakov (1971), *Adsorption of Organic Compounds on Electrodes*, Plenum press, New York.

Duursma, E. K., and R. Dawson (1981), *Marine Organic Chemistry*, Elsevier, Amsterdam.

Heyrovsky, J., and J. Kuta (1966), *Principles of Polarography*, Academic Press, New York, pp. 287–335.

Hunter, K. A., and P. S. Liss (1982), "Organic Sea Surface Films," in E. K. Duursma and R. Dawson, Eds., *Marine Organic Chemistry*, Elsevier, Amsterdam, pp. 259–298.

Jehring, H. (1974), *Elektrosorptionanalyse mit der Wechsel Strompolarographie*, Akademie Verlag, Berlin.

Kozarac, Z., and B. Ćosović (1984), "Interaction of Cadmium with the Adsorbed Layer of Biogenic Surface Active Substances at the Mercury Electrode," *Bioelectrochem. Bioenerg* **12**, 353–363.

Kozarac, Z., V., Žutić, and B. Ćosović (1976), "Direct Determination of Nonionic and Anionic Detergents in Effluents," *Tenside* **13**, 260–265.

Kozarac, Z., B. Ćosović, and V. Vojvodić (1986), "Effects of Natural and Synthetic Surface Active Substances on the Electrochemical Reduction of Cadmium in Natural Waters," *Water Res.* **20**, 295–300.

Kozarac, Z., A. Dhathathreyan, and D. Mobius (1989), "Interaction of Nitrophenols with Lipids at the Air/Water Interface," *Colloid Polym. Sci.* **267**, 722–729.

Kozarac, Z., B. Ćosović, B. Šarić, A. Dhathathreyan, and D. Mobius (in press), "Interaction of Para-nitrophenol with Lipids at Hydrophobic Interfaces," submitted.

Leenheer, J. A. (1981), "Comprehensive Approach to Preparation, Isolation and Fractionation of Dissolved Organic Carbon from Natural Waters and Wastewaters," *Environ. Sci. Technol.* **15**, 578–587.

Parks, G. A. (1975), "Adsorption in the Marine Environment," in J. P. Riley and G. Skirrow, Eds., *Chemical Oceanography*, Vol. 2, Academic Press, New York.

Raspor, B., and P. Valenta, (1988), "Adsorption of Humic Substances Isolated from Marine and Estuarine Sediments," *Mar. Chem.* **25**, 211–226.

Ružić, I. (1987), "Time Dependence of Adsorption at Solid Liquid Interfaces," *Croat. Chem. Acta* **60**, 457–475.

Ružić, I., H. Ulrich, and B. Ćosović (1988), "Time Dependence of the Adsorption of Valeric Acid at the Mercury–Sodium Chloride Interface," *J. Colloid Interface Sci.* **126**, 525–536.

Scamehorn, J. F., R. S. Schechter, and W. H. Wade (1982a), "Adsorption of Surfactants on Mineral Oxide Surfaces from Aqueous Solutions. I. Isomerically Pure Anionic Surfactants," *J. Colloid Interface Sci.* **85**, 463–478.

Scamehorn, J. F., R. S. Schechter, and W. H. Wade (1982b), "Adsorption of Surfactants on Mineral Oxide Surfaces from Aqueous Solutions. II. Binary Mixtures of Anionic Surfactants," *J. Colloid Interface Sci.* **85**, 479–493.

Scamehorn, J. F., R. S. Schechter, and W. H. Wade (1982c), "Adsorption of Surfactants on Mineral Oxide Surfaces from Aqueous Solutions. III. Binary Mixtures of Anionic and Nonionic Surfactants," *J. Colloid Interface Sci.* **85**, 494–501.

Stumm, W., and G. Furrer (1987), "The Dissolution of Oxides and Aluminium Silicates; Examples of Surface-Coordination-Controlled Kinetics," in W. Stumm Ed., *Aquatic Surface Chemistry*, Wiley, New York.

Stumm, W., and J. J. Morgan (1981), *Aquatic Chemistry*, 2nd ed., Wiley, New York.

Ulrich, H. J., W. Stumm, and B. Ćosović (1988), "Adsorption of Aliphatic Fatty Acids on Aqueous Interfaces. Comparison between Two Model Surfaces: The Mercury Electrode and Al_2O_3 Colloids," *Environ. Sci. Technol.* **22**, 37–41.

Vojvodic, V., and B. Ćosović, "Isolation and Fractionation of Surface Active Substances from Natural Waters by Sorption on XAD-8 Resin," manuscript in preparation.

Westall, J. C. (1987), "Adsorption Mechanisms in Aquatic Surface Chemistry," in W. Stumm Ed., *Aquatic Surface Chemistry*, Wiley, New York, pp. 3–33.

11

REDOX REACTIONS OF METAL IONS AT MINERAL SURFACES

Bernhard Wehrli

Lake Research Laboratory, Institute for Water Resources and Water Pollution Control (EAWAG), Kastanienbaum, Switzerland; Swiss Federal Institute of Technology (ETH), Zürich, Switzerland

1. INTRODUCTION

1.1. Reactions at Geochemical Redox Boundaries

The presence of dissolved molecular oxygen in natural waters establishes a low level of chemically reactive electrons. As a consequence, dissolved and adsorbed metal ions are found in their higher oxidation states in oxic waters. The geochemical cycling of electrons on a global scale is dominated by photosynthesis and respiration with an electron flux in the order of 42 moles of electrons per square meter per year at the sea surface (Stumm, 1978). The reduced products of such photosynthetic activity accumulate at the bottom of rivers and lakes. Fast degradation of organic material and slow supply of dissolved oxygen produce steep redox gradients at the sediment–water interface. Similar oxic–anoxic transition zones are found at the boundaries of polluted groundwater plumes and at reduced mineral layers in soils.

Unknown redox kinetics often limit the predictive value of calculated redox equilibria in the aquatic environment. Thermodynamics indicate that manganese should be present in oxic waters as Mn(IV) oxide. However, in a long-term experiment Diem and Stumm (1984) have shown that Mn^{2+} is not oxidized by O_2 within several years (Fig. 1a). Microorganisms usually catalyze such slow redox reactions if the concentrations involved are high enough. Recently experimental evidence has been accumulated that aqueous mineral surfaces provide additional accelerated pathways for redox processes such as the oxygenation of metal ions (Davies and Morgan, 1989; Wehrli and Stumm, 1988) and the oxidation of organic pollutants (Stone, 1986). The manganese oxygenation is accelerated by iron oxide particles by a factor of more than 10^4

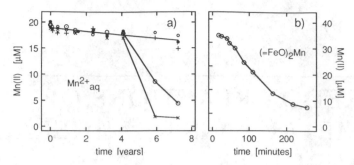

Figure 1. Oxygenation of Mn(II): (*a*) homogeneous solution at pH 8.4, data from Diem and Stumm (1984); (*b*) 10 m*M* goethite suspension at pH 8.5, data from Davies and Morgan (1989). The surface complex reacts within hours, whereas the homogeneous Mn(II) solutions are stable for years.

(Fig. 1*b*). Iron and manganese particles often accumulate at redox boundaries (Davison, 1985; De Vitre et al., 1988) and supply reactive mineral surfaces in zones with high chemical electron fluxes. Oxidation–reduction reactions of metal ions at mineral surfaces thus play a key role in the "geochemical cycling of electrons". The distribution and fate of many inorganic species is coupled with the reductive dissolution and the oxidative precipitation of manganese and iron at redox boundaries.

Applications of redox reactions on mineral surfaces in areas such as water treatment have been reviewed by Segal and Sellers (1984); Voudrias and Reinhard (1986) discussed the organic transformations at the solid–liquid interface. Here the discussion is confined to selected inorganic redox reactions. Table 1 lists some of the more important heterogeneous processes of inorganic

TABLE 1. Examples of Inorganic Redox Reactions on Mineral Surfaces[a]

Process	Oxidant	Reductant	Reference[b]
Heterogeneous equilibrium	FeOOH, MnOOH	Fe^{2+}, Mn^{2+}	
Oxidative adsorption	MnO_x	Fe(II), Co(II), Sn(II)	1
Reductive dissolution	MnO_x, FeOOH	U(IV), Cr(III), organics	2
Oxidation of pyrite	O_2, Fe(III)	FeS_2	3
Heterogeneous oxygenation	O_2	Adsorbed Fe(II), Mn(II), V(IV)	4

[a] Manganese and iron oxides act as electron acceptors; sulfides and reduced metal ions adsorbed to particles are heterogeneous reductants.
[b] References: (1) see Dillard and Schenk (1986) and references, cited therein. (2) Gordon and Taube (1962), Van der Weijden and Reith (1982), Stone (1986), Sulzberger (Chapter 14, this volume), (3) Lowson (1982), Luther (Chapter 6, this volume), (4) Tamura et al. (1976), Davies and Morgan (1989), Wehrli and Stumm (1988).

species in natural waters. Solid-dissolved equilibria such as

$$Fe^{III}OOH_{(s)} + 3H^+ + e^- \Leftrightarrow Fe^{2+} + 3H_2O \tag{1a}$$

are easily calculated from thermodynamic data. However, a mechanistic discussion of heterogeneous reaction rates should be based on the redox potential of elementary reaction steps such as the reduction of a surface Fe(III) center:

$$=Fe^{III}OH_2^+ + e^- \Leftrightarrow =Fe^{II}OH_2 \tag{1b}$$

In this chapter I will propose a kinetic estimate for the thermodynamics of reactions like Eq (1b). The solid phases listed in Table 1 may act as a reductant or an oxidant. One of the prominent geochemical electron donors is pyrite. From an estimate of global pyrite weathering of 36 Tg y^{-1} (Garrels et al., 1973) we may deduce an average electron flux on the land surface in the order of $0.02\ mol\ m^{-2}\ y^{-1}$. At redox boundaries in salt marshes and in lake sediments microbial sulfate reduction will intensify this "electron cycling." Luther (Chapter 6, this volume) discusses the details of sulfide redox mechanisms.

In contrast to pyrite, the hydroxides of iron and manganese act as electron acceptors. An adsorbed reducing agent may follow two different pathways after electron transfer to these mineral surfaces. Oxidation products with high particle affinity such as Fe(III) (Koch, 1957), Cr(III) (Zabin and Taube, 1964), Co(III) (Crowther et al., 1983), Sn(IV) (Rapsomanikis and Weber, 1985) will form strong surface complexes. Such an oxidative adsorption consumes the available mineral surface. If the oxidation products are anions such as Cr(VI) (Van der Weijden and Reith, 1982), As(V) (Oscarson et al., 1981) or organic compounds, which desorb more easily, the reactive interface is regenerated continuously and the reductive dissolution of the solid phase dominates the process. Stone (1986) studied the dissolution of manganese oxides in presence of reducing organic compounds. An account on the influence of light on this process is given by Sulzberger (Chapter 14, this volume). A last group of heterogeneous redox reactions involves species adsorbed to "innocent" surfaces such as Al_2O_3 or SiO_2 and silicates. Davies and Morgan (1989) observed acceleration of the Mn(II) oxygenation (Fig. 1b) also in the presence of silica and alumina, which have no accessible lower oxidation states. They attributed the accelerating kinetic effect to the change in coordination, when Mn^{2+} becomes adsorbed.

1.2. Two Paradigms

Mineral surfaces are rather complex reaction media. Two complementary paradigms have served so far as starting points for mechanistic discussions of reactions at the mineral–water interface: The concept of the solid–aqueous interface as an electrode and the picture of the mineral surface as a two-dimensional array of surface complexes. Both points of view are exploited in this chapter.

Marcus' theory (1965) has long been used to compare outer-sphere redox reactions in homogeneous solution with corresponding electrode kinetics. Recent theoretical developments by Astumian and Schelly (1984) allow the general comparison of homogeneous and heterogeneous rate constants. Such simple kinetic models account only for the change in geometry that accompanies the adsorption from solution to a two-dimensional surface and the electrostatic contribution from the diffuse double layer surrounding the mineral grains.

A discussion of specific catalytic effects of mineral surfaces must be based on thermodynamic and structural information of the reactive surface species. The first part includes a brief outline of the emerging picture of mineral surfaces as two-dimensional arrays of surface complexes. This part assembles also the kinetic tools that are useful for comparison of the reactivity of aqueous metal ions and their adsorbed surface complexes. The second part presents a reevaluation of the most extensively studied inorganic redox reactions in natural waters: the oxygenation of VO^{2+}, Mn^{2+}, Fe^{2+} and Cu^+ in homogeneous and heterogeneous systems.

2.   HETEROGENEOUS ELECTRON TRANSFER

A vast literature exists on the kinetics and mechanisms of electron-transfer reactions between dissolved metal ions. A recent review was written by Sutin (1986). Physical chemists, however, have dealt so far almost exclusively with reactions in aqueous solution of very low pH or high ligand concentrations. Such studies have shown that the *homogeneous* electron-transfer between couples such as Fe(III)/Fe(II) proceeds via three distinct steps. First the two reactants diffuse together and form a reactive intermediate called the *precursor complex*. The electron transfer occurs after an appropriate reorganization of the nuclear configuration. This yields a short-lived product called the *successor complex*. Finally the successor decomposes to the separated products of the redox reaction:

$$FeOH_{aq}^{2+} + Fe_{aq}^{2+} \xrightarrow{\text{precursor formation}} [Fe^{III}-OH-Fe^{II}]^{4+} \tag{2}$$

$$[Fe^{III}-OH-Fe^{II}]^{4+} \xrightarrow{\text{electron transfer}} [Fe^{II}-OH-Fe^{III}]^{4+} \tag{3}$$

$$[Fe^{II}-OH-Fe^{III}]^{4+} \xrightarrow{\text{dissociation of successor}} Fe_{aq}^{2+} + FeOH_{aq}^{2+} \tag{4}$$

The rate constant of the overall reaction at 25°C is $k = 3.1 \times 10^3 \, M^{-1} \text{s}^{-1}$ (Silverman and Dodson, 1952). In very acidic solution a pH-independent reaction with Fe^{3+} as a reactant is observed. This self-exchange between the aquo complexes proceeds much slower with a rate constant of $4 \, M^{-1} \text{s}^{-1}$. In this case the precursor complex consists probably of a simple encounter complex without a bridging ligand between Fe^{2+} and Fe^{3+}. The almost thousandfold acceleration of the electron transfer when $FeOH^{2+}$ replaces Fe^{3+} is typical for reactions with OH^- as an electron bridge in the precursor complex (Haim, 1983).

The ferric ion occurs only in trace concentrations in the neutral pH range of natural waters. Therefore, we should realize that the relevant Fe(III)/Fe(II) equilibrium in most aquatic environments involves *heterogeneous* electron transfer between dissolved Fe(II) and the surface centers of iron oxihydroxide particles. A similar three-step mechanism can be written for this process:

$$=Fe^{III}OH + Fe^{2+} \xrightarrow{\text{adsorption}} (=Fe^{III}-O-Fe^{II})^+ + H^+ \qquad (5)$$

$$(=Fe^{III}-O-Fe^{II})^+ \xrightarrow{\text{electron transfer}} (=Fe^{II}-O-Fe^{III})^+ \qquad (6)$$

$$(=Fe^{II}-O-Fe^{III})^+ + H_3O^+ \xrightarrow{\text{desorption}} =Fe^{II}OH_2 + Fe^{III}OH^{2+} \qquad (7)$$

An extensive numerical study on such heterogeneous three-step processes has been given by Stone and Morgan (1987). The adsorption–desorption kinetics of divalent metal ions is fast: Yasunaga and Ikeda (1986) report relaxation times in the order of milliseconds to seconds. The pH as a master variable governs the adsorption of Fe(II) in the preceding example. The elucidation of adsorption equilibria and the structure of precursor complexes such as $(=Fe^{III}-O-Fe^{II})^+$ at the mineral surface is therefore a prerequisite for the study of heterogeneous redox kinetics.

2.1. Adsorption

Alkali ions and inorganic anions such as NO_3^-, ClO_4^- adsorb electrostatically to mineral surfaces of opposite charge. Such surface species are analogs to outer-sphere complexes (ion pairs) in solution. Transition-metal ions, however, adsorb even against the electrostatic repulsion of a positively charged mineral surface. Figure 2 depicts the adsorption equilibria of vanadyl(IV) and ferrous iron to TiO_2. Both cations adsorb to the positively charged surface (the zero point of charge of anatase is $pH_{ZPC} \sim 6.4$). To explain such observations, Schindler and Stumm (1987) have developed a surface complexation model. The adsorption process is treated as a complex formation reaction, where surface $=M-OH-$ groups replace coordinated water molecules at the adsorbed metal center. The adsorption of a ferrous iron to an anatase surface may be given by equilibria such as

$$2 =Ti-OH + Fe^{2+} \overset{K_2}{\Leftrightarrow} (=Ti-O)_2Fe + 2H^+ \qquad (8)$$

From correlations between hydrolysis constant and surface complex formation

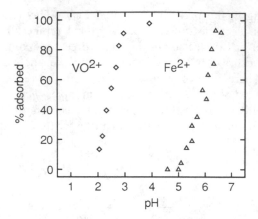

Figure 2. Adsorption of vanadyl and Fe(II) to TiO$_2$ (anatase). Both cations adsorb specifically to the positively charged surface. Conditions: 25°C, $I = 0.1$ (NaClO$_4$); [VO^{2+}] = 50 μM, 2 g L^{-1} TiO$_2$; [Fe^{2+}] = 100 μM, 10 g L^{-1} TiO$_2$.

constants (Schindler and Stumm, 1987) we expect that vanadyl (p$K_1^* \sim 6$) adsorbs stronger to an oxide surface than ferrous iron (p$K_1^* \sim 9.5$). Such equilibrium measurements quantify the concentration of reactive surface species. The structural interpretation, however, has to be based on spectroscopic evidence.

Recent ENDOR studies on the adsorption of Cu^{2+} (Rudin and Motschi, 1984) and VO^{2+} (Motschi and Rudin, 1984) have verified the conceptual model of inner-sphere coordination by surface ligands =M–OH. These authors found that the surface groups enter the coordination sphere of the adsorbed metal centers and replace one, two, or more water molecules. Molecular modeling techniques help visualize the possible structural arrangements on mineral surfaces. Figure 3 presents the two-dimensional array of surface ligands on the anatase surface. The available =TiOH groups are lined up in zigzag chains. The local geometry allows the formation of mono-, bi- and even tridentate surface complexes. The coordination of these surface ligands changes the reactivity of surface metal centers in different ways. These ligands act as σ donors and increase the electron density at the metal center, thus stabilizing higher oxidation states. This thermodynamic effect results in a lower redox potential. Such a change in $E°$ may directly affect electron-transfer rates (see below). The well-documented role of oxygen as an electron bridge suggests specific mechanistic effects: bridging surface =M–O groups may mediate electron transfer in a similar way as coordinated OH$^-$ in homogeneous solution (compare Eqs. 3 and 6).

2.2   Two Electron-Transfer Mechanisms

Electron transfer between metal ions may occur either as inner-sphere (is) or as outer-sphere (os) reaction. The first case involves a ligand exchange and the

Figure 3. The anatase surface shows zigzag chains of Ti surface octahedra. Each Ti site carries a =TiOH ligand that may bind adsorbed metal centers. The two surface groups to the right represent a bidentate (above) and a tridentate surface complex (below).

TABLE 2. Proposed Mechanisms of Reductive Dissolution

Solid	Reductant	Mechanism[a]	Evidence	Reference
α-Fe_2O_3	$V^{II}(pic)_3^-$	os	Diffusion-controlled	Segal and Sellers (1984)
MnO_2	Cr_{aq}^{II}	is	^{18}O tracer	Zabin and Taube (1964)
PbO_2	U_{aq}^{IV}	is	^{18}O tracer	Gordon and Taube (1962)
MnO_2	Fe_{aq}^{II}	is	Saturation effects	Koch (1957)

[a] os = outer-sphere; is = inner-sphere.

coordination of a bridging ligand such as in Eq. 2. An outer-sphere reaction occurs without ligand substitution. Reductive dissolution of oxides with different metal complexes as reductants illustrates the two mechanisms for heterogeneous systems (Table 2). The os mechanism has been assigned to many V^{2+} electron-transfer reactions that proceed faster than the ligand exchange at the V^{2+} center. The "bulky" negatively charged V(II) complex adsorbs electrostatically to the positive hematite surface. The outer-sphere oxidation of the V(II) center (Segal and Sellers, 1984) may be written in analogy to homogeneous systems as

$$=Fe^{III}OH_2^+ + (V^{II}L_3)^- \longrightarrow =Fe^{III}OH_2^+(V^{II}L_3)^-$$

adsorption: ion pair formation

$$=Fe^{III}OH_2^+(V^{II}L_3)^- \longrightarrow =FE^{II}OH_2(V^{III}L_3) \tag{9}$$

outer-sphere electron transfer

The ^{18}O-tracer studies of Gordon and Taube (1962) on the oxidation of U(IV) on PbO_2 have shown that both oxygen ions in the product UO_2^{2+} are derived from the oxide lattice. This result indicates an inner-sphere mechanism and is compatible with a binuclear U(IV) surface complex:

$$2=Pb^{IV}OH + U^{IV} \longrightarrow \begin{array}{c}=Pb^{IV}O\\ =Pb^{IV}O\end{array}\!\!>\!\!U^{IV} + 2H_3O^+$$

surface complexation

$$\begin{array}{c}=Pb^{IV}O\\ =Pb^{IV}O\end{array}\!\!>\!\!U^{IV} \longrightarrow \begin{array}{c}=Pb^{II}O\\ =Pb^{IV}O\end{array}\!\!>\!\!U^{VI} \tag{10}$$

inner-sphere electron transfer

In the last step the U(VI) desorbs from the surface as a uranyl ion, UO_2^{2+}. Two oxygen ions from the PbO_2 surface remain coordinated to the high valent uranyl. Recently Combes (1989) has shown by EXAFS (extended x-ray absorption fine-structure spectroscopy) that uranyl indeed forms bidentate surface complexes on goethite. The local coordination sites on α-FeOOH are structurally very similar as the PbO_2 sites with a rutile structure.

2.3. The Marcus Relations

Outer-sphere electron transfer is one of the simplest reaction types because no bonds are broken or formed. It is therefore not surprising that this class of reactions was the subject of early kinetic theories. More than 30 years ago Marcus (1965) derived a predictive theory for the rate constants of os redox reactions in homogeneous and heterogeneous systems. A didactic introduction was later given by the same author (Marcus, 1975), and Sutin (1986) reviewed modern refinements of the theory.

The second-order rate constant for electron transfer in solution can be given in terms of the Arrhenius equation

$$k = A \exp(-E_a/RT) \tag{11a}$$

where A stands for the preexponential factor ($M^{-1} s^{-1}$) and E_a refers to the activation energy (kJ mol^{-1}). Marcus replaced the factor A by the collision frequency $Z = \kappa k_B T/h$, where κ refers to the transmission coefficient, k_B is Boltzmann's constant, and h is Planck's constant. If every collision leads to a reaction ($\kappa = 1$) and $T = 298$ K, then the collision frequency in homogeneous solution is $Z = 10^{11}$ $M^{-1} s^{-1}$. In Marcus' theory the activation energy is split into two parts:

1. The two reactants must diffuse together. The work w (kJ mol^{-1}) required for this process is determined by the electrostatic forces between the two reactants. If one reactant (such as O_2) is uncharged the work term can be neglected.
2. Bond distances and bond angles between ions and solvent molecules change on electron transfer. The iron–oxygen bond distances for the aquo complex of Fe^{3+} are 0.13 Å shorter than in the case of Fe^{2+}. Prior to os electron transfer coordinated water molecules in such redox couples are rearranged at an intermediate position. The corresponding reorganization energy ΔG^* contributes as the second term to the activation energy:

$$k = Z \exp[-(w + \Delta G^*)/RT] \tag{11b}$$

Because Z can be calculated from collision theory and w is determined by simple electrostatics, Eq. 11b would allow a prediction of rate constants if a theoretical derivation can be found for the reorganization energy ΔG^*. Based on free-energy surfaces Marcus (1965) derived such an expression for ΔG^*. Figure 4a–c represent some schematic "cartoons" of one-dimensional energy surfaces. Many simplifications are made implicitly in these diagrams. First, a full quantum-mechanical treatment of reaction kinetics requires the calculation of the "Born–Oppenheimer" surface in $N - 1$-dimensional space, where N represents the number of all relevant coordinates, orientations, vibrational modes, and so on of the reactants and their surrounding solvent molecules. Here we consider only one general reaction coordinate, labeled as the "solvent reorganization." Second, we assume that the reaction is nonadiabatic, that is, that no mixing between the reactant states and the product states occurs. In this case the intersection of free-energy curves can be derived by straightforward geometry. In adiabatic reactions the free-energy profile between the reactant and product curves in Fig. 4a–c would be smoothed and the corresponding barriers lowered. Third, we approximate the free-energy curves by parabola of identical shape. This simplifies the mathematics. On the basis of these assumptions, Fig. 4a

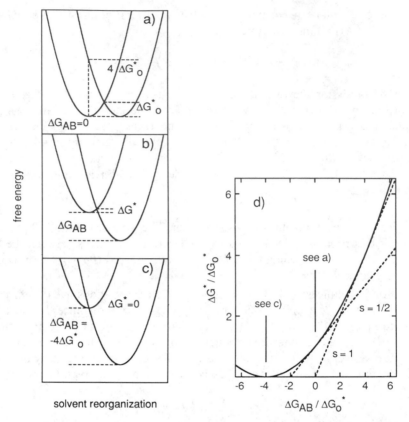

Figure 4 (*a*)–(*d*). Parabolic free-energy curves for outer-sphere electron-transfer reactions. The "general" reaction coordinate represents the reorganization of coordinated water molecules. (*a*) Electron transfer in a system without free-energy change such as reaction 12. In a photochemical reaction without solvent reorganization the energy barrier is $4\Delta G_0^*$. If thermal activation reorganizes the hydration shell, the energy barrier is four times smaller. (*b*) In exergonic reactions ($\Delta G_{AB} < 0$) the reorganization energy ΔG^* is smaller than the "intrinsic barrier" ΔG_0^*. (*c*) Limit of activationless transfer. (*d*) The parabolic Marcus relation (Eq. 13a) in normalized form. The relation describes the dependence of the energy barrier ΔG^* as a function of the free-energy change ΔG_{AB} and the intrinsic barrier ΔG_0^*. The slope is close to unity for very endergonic reactions, approaches 0.5 at $\Delta G_{AB} \sim \Delta G_0^*$, and is zero for activationless transfer (*c*).

represents the situation of a self-exchange reaction such as

$$Fe^{2+} + Fe^{3+} \xrightarrow{\;k_{11}\;} Fe^{3+} + Fe^{2+} \tag{12}$$

The free-energy change between the precursors and the successors of the electron transfer is zero in this case, $\Delta G_{AB} = 0$. If no thermal activation occurs (such as in photochemical reactions), an energy of $4\Delta G_0^*$ is required to bring the system from the left parabola to the product curve one the right-hand side of Fig. 4*a*. If thermal activation reorganizes the solvent shell, the energy barrier is only ΔG_0^*. Figure 4b shows the same parabolic curves for an exergonic reaction, $\Delta G_{AB} < 0$. The reorganization barrier is much smaller in this case. The special case of activationless electron transfer is shown in Fig. 4c. In the parabolic model the reorganization energy ΔG^* vanishes if the driving force of the reaction is $\Delta G_{AB} = -4 \Delta G_0^*$. Based on parabolic reaction diagrams Marcus (1965) derived a relation between the different free energy terms, which can be given in the following form:

$$\Delta G^* = \Delta G_0^* [1 + (\Delta G_{AB}/4 \Delta G_0^*)]^2 \tag{13a}$$

Figure 4d depicts a normalized version of this parabolic function. The calculation of the reorganization energy ΔG^* requires knowledge of the "intrinsic" barrier ΔG_0^*, which can be obtained from self-exchange experiments such as reaction 12. The free-energy change between precursor and successor ΔG_{AB} can often be replaced by the free-energy change of the overall redox step. Linear free-energy plots of $RT \ln k$ versus ΔG_{AB} for a series of reactions with similar 'intrinsic' energy barriers ΔG_0^* will show different slopes; the slope approaches unity for very exergonic reactions in the limit $\Delta G_{AB} = 4 \Delta G_0^*$. It decreases to 0.5 in the more usual range of near-equilibrium conditions $\Delta G_{AB} \sim \Delta G_0^*$ and is zero for activationless transfer at $\Delta G_{AB} = 4 \Delta G_0^*$ (Fig. 4c). In the "inverted region" beyond this point a decrease in the reaction rates with extremely exergonic potential is predicted. This feature of the Marcus relation as been debated for two decades. Only recently Closs and Miller (1988) presented experimental evidence for an "inverted region" in intramolecular redox kinetics of organic molecules. Transport processes will hide such effects in intermolecular reactions in natural waters. A more intuitive form of Eq. 13a is known as the *Marcus cross-relation*:

$$k_{12} \sim (k_{11} k_{22} K_{12} f_{12})^{1/2} \tag{13b}$$

with

$$\log f_{12} = \frac{(\log K_{12})^2}{\log (k_{11} k_{22}/Z^2)}$$

Here k_{11} and k_{22} are the rate constant for the self-exchange between the reduced and oxidized form of the two reactants and K_{12} refers to the equilibrium constant. This approximation is valid if the work term w in Eq. 11b cancels or can be neglected.

2.4. Comparison of Homogeneous and Heterogeneous Rate Constants

Parsons (1975) used the concept of particle surfaces as electrodes to compare homogeneous and heterogeneous reactions in the marine environment. Two factors are important in this comparison: (1) the change in geometry affects collision frequencies, steric interactions, and so on; and (2) adsorbed species may react with different activation energy. In gas-phase reactions the geometric effects tend to slow down the reaction rates. At 300 K the surface process occurs at competitive rates only if its activation energy is about 70 kJ mol^{-1} lower than in the homogeneous case (Laidler, 1987). Marcus (1965) compared the bimolecular electron transfer, k_{hom} ($M^{-1}\text{s}^{-1}$), with the first-order electrochemical reaction k_{het} (m s^{-1}) on an electrode of surface S:

$$A + B \xrightarrow{k_{\text{hom}}} C + D \tag{14}$$

$$A \xrightarrow{k_{\text{het}}} C \tag{15}$$

Simple collision theory (see Laidler, 1987) predicts the collision frequencies Z as preexponential factors in Eq. (11b):

$$Z_{\text{hom}} = N_A (8\pi kT/m^*)^{1/2} \sigma_{AB}^2 \qquad (\sim 10^{11}\ M^{-1}\text{s}^{-1}) \tag{16}$$
$$Z_{\text{het}} = (kT/2\pi m)^{1/2} \qquad (\sim 10^2\ \text{m s}^{-1}) \tag{17}$$

where m^* is the reduced mass $m_A m_B / m_A + m_B$. The term in brackets represents the relevant velocity of the species, and σ_{AB} stands for the effective collision cross section. If every collision leads to a reaction and the molecules follow hard-sphere dynamics, the effective cross section approaches $\sigma_{AB} = (r_A + r_B)^2$. Marcus (1965) normalized the rate constants from bimolecular exchange and electrode processes with the above Z values and derived the comparative relation

$$(k_{\text{hom}}/Z_{\text{hom}})^{1/2} \sim k_{\text{het}}/Z_{\text{het}} \tag{18}$$

Table 3 confronts measured electrokinetic rate constants k_{het} with values predicted from Eq. 18. The experimental rate constants $k_{\text{hom}} = k_{11}$ (see Eq. 12) from bimolecular self-exchange processes between species such as Fe^{3+} and Fe^{2+} were used. The agreement is usually better than a factor of 10 (except for the Co^{3+}/Co^{2+} with its low-spin–high-spin transition). This fair agreement confirms what we would expect: specifically, the activation energy of these os transfer reactions remains unchanged if a reactant is replaced by an electrode. Collision theory approximates the geometric changes correctly.

Not many redox reactions on mineral surfaces follow the two basic assumptions of Eq. 18: outer-sphere transfer and regeneration of the surface during the reaction. Fast reductive dissolutions with powerful reducing agents may potentially lead to an additional test of Eq. 18. In many cases the surface species are consumed during the reaction. Astumian and Schelly (1984) developed a theory to compare a second-order reaction of two reactants in solution (Eq. 14) with a heterogeneous reaction of a dissolved species A and an adsorbed species B. The two different reaction environments are outlined in Figure 5. Here we denote the surface complex as $=$MO–B:

$$A + \ =\text{MO–B} \xrightarrow{k_s} C + \ =\text{MO–D} \tag{19}$$

TABLE 3. Homogeneous Self-Exchange and Electrode Reactions

Redox Couple	$k_{hom}{}^a$ [$M^{-1}s^{-1}$]	k_{het}(meas.)^b [cm s^{-1}]	k_{het}(calc.)^c [cm s^{-1}]
V^{3+}/V^{2+}	1.0×10^{-2}	4.0×10^{-3}	3.2×10^{-3}
Mn^{3+}/Mn^{2+}	3.0×10^{-4}	1.0×10^{-5}	5.5×10^{-4}
Fe^{3+}/Fe^{2+}	4.2	5.0×10^{-3}	6.5×10^{-2}
Co^{3+}/Co^{2+}	3.3	2.0×10^{-7}	5.7×10^{-2}
Cu^{2+}/Cu^{+}	1.0×10^{-5}		1.0×10^{-4}
MnO_4^{-}/MnO_4^{2-}	7.1×10^{2}	$> 10^{-2}$	0.84
O_2/O_2^{-}	1.0×10^{3}		1.0

^{a} Experimental rate constants of homogeneous self-exchange (compare reaction 12). Data from Sutin (1986).
^{b} Electrokinetic constants of the reaction at electrode surface [from Marcus (1975); the Mn values are from Parsons (1975)].
^{c} Calculated constants for the heterogeneous process using Marcus' theory (Eq. 18). The agreement between theory and experiment is within an order of magnitude. (The deviation of the Co couple has been ascribed to its electronic structure.)

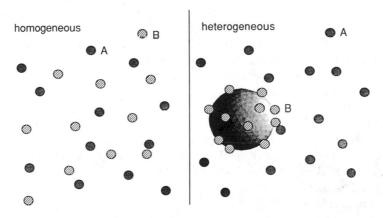

Figure 5. Geometric difference between homogeneous and heterogeneous reactions. The reduction in dimensionality that occurs when the reactants B are all adsorbed to the surface of a mineral particle slows down the reaction rate. This geometric effect is described by Eq. 22 using collision theory.

The authors derived the collision frequency for such a reaction as

$$Z_s = N_A (8kT/\pi m_A)^{1/2} r_P^2 \tag{20}$$

where the velocity is a function of the mass m_A of A, but the effective cross section is determined by the mean radius of the particles, r_P. For the case that the particles are completely covered with reactants B the relation

$$4\pi r_P^2 N_P = (2r_B)^2 N_B \tag{21}$$

holds, where N_P and N_B refer to the number of particles and species B. Combining the Eqs. 16, 20, and 21, Astumian and Schelly (1984) obtained the ratio

$$k_s/k_{hom} = \pi^{-1} [m_B/(m_A + m_B)]^{1/2} [r_B/(r_A + r_B)]^2 \exp(-\Delta E_a/RT) \tag{22}$$

The first three factors on the right-hand side of this relation are smaller than unity. The reduction in dimensionality that accompanies the transfer of reactant B from solution to an interface slows down the reaction rate. Using realistic values of the masses m_i and the radii r_i, one may predict a maximum estimate of a 50-fold reduction in the rate constant k_s. This geometric "disadvantage" is sometimes compensated by a lower activation energy E_a at the surface. Astumian and Schelly (1984) estimate that E_a of the heterogeneous reaction must be 2.5 to 7.5 kJ mol^{-1} lower to compensate for the geometric effects.

3.   APPLICATION: OXYGENATION KINETICS

The oxidation of metal ions by O_2 has been extensively studied because of the relevance of these reactions in geochemical cycles and in water-treatment technology. For different reasons the process lends itself to a conceptual discussion of heterogeneous redox reactions: (1) a large kinetic database of oxygenation kinetics in solution is available (Fallab, 1967; Davison and Seed, 1983; Millero et al., 1987), (2) heterogeneous oxygenation has recently been measured in spectroscopically well-characterized systems (Wehrli and Stumm, 1988), and (3) a report from Taube's group (Stanbury et al., 1980) shows that the oxygenation of Ru(II)–amine complexes follows an outer-sphere mechanism. This opens the perspective to apply Marcus theory to metal oxygenations in solution and at mineral surfaces.

3.1.   The Oxygenation of V(IV), Fe(II), Mn(II), and Cu(I)

The reduced species Fe^{2+} and Mn^{2+} have been detected electrochemically in anoxic waters (De Vitre et al., 1988). Vanadyl (VO^{2+}) is known to be incorporated in geoporphyrins in organic-rich sediments (Eckstrom et al., 1983). Moffett and Zika (1988) measured reduced Cu(I) photometrically in surface waters of the open ocean. The oxidation of these four metal species involves a simple one-electron transfer step. Haber and Weiss (1934) proposed a kinetic mechanism for the oxygenation of the ferrous ion, in which the first step in the four-electron reduction of the dioxygen molecule determines the rate. The redox potentials for the corresponding oxygen couples support this view: they are plotted in

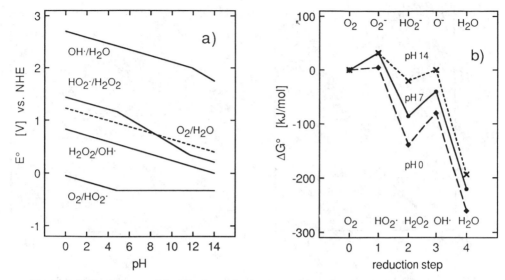

Figure 6. Reduction potentials of the four reduction steps of O_2. (*a*) pH dependence calculated form standard $E°$ data with $p_{O_2} = 1$ atm as reference state and pK values of the intermediates. References: Bielski et al. (1985), Sawyer and Valentine (1981), George (1965). Superoxide is a strong reducing agent. (*b*) Reaction path diagram for the reduction of O_2. The the formation of radicals in the first and third step is endergonic.

Figure 6*a* for the whole pH range. Figure 6*b* presents the four reduction steps in a reaction path diagram that was inspired by Schneider (1988). Both graphs show that the first reduction step form the dioxygen molecule to the superoxide radical is an "uphill" (endergonic) reaction. The superoxide anion (O_2^-) is a powerful reducing agent (Sawyer and Valentine, 1981), which is scavenged in natural systems by a variety of processes including the reduction of Cu(II) and even S_N2-type nucleophilic substitutions on organic compounds.

Several reports confirmed the finding of Stumm and Lee (1961) that the oxidation of Fe(II) by O_2 at neutral pH is accelerated hundredfold if the pH is raised by one unit (Davison and Seed, 1983, Millero et al., 1987). The empirical rate law of the ferrous ion oxygenation at neutral pH is

$$R = -d[Fe^{II}]/dt = k[Fe^{II}][O_2][H^+]^{-2} \qquad (23)$$

Experiments in the acidic pH range are more time-consuming, and kinetic data are quite rare. The work of Singer and Stumm (1970) is in agreement with earlier experiments by Holluta and Kölle (1964), which indicate a change in the pH dependence of the reaction from log $R \propto [H^+]^{-2}$ to $[H^+]^{-1}$ below pH = 5. The rates approach a pH-independent value below pH 3. Figure 7*a* summarizes the kinetic findings in terms of an observed first-order rate constant k at 25°C and $p_{O_2} = 1$ atm. This kinetic "titration curve" was plausibly interpreted by Millero

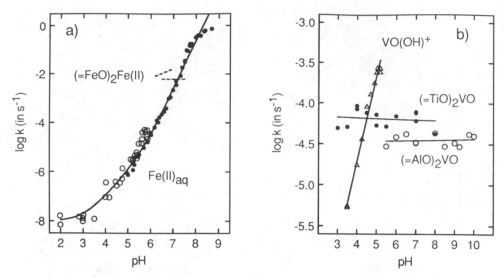

Figure 7. Oxygenation kinetics for 1 atm O_2. (a) Oxidation of Fe(II). Open circles represent data by Singer and Stumm (1970), dots are data from Millero et al. (1987). The solid line was calculated with Eq. 26. Small dotted lines represent heterogeneous rate constants for Fe(II) adsorbed to Fe(OH)$_3$ (upper line) and α-FeOOH (lower); data from Tamura et al. (1976). (b) Oxygenation of vanadyl in solution and as surface complexes; data from Wehrli and Stumm (1988).

(1985) as a parallel oxidation of the ferrous ion and its hydroxo complexes:

$$Fe^{2+} + O_2 \longrightarrow Fe^{3+} + O_2^-$$
$$Fe(OH)^+ + O_2 \longrightarrow Fe(OH)^{2+} + O_2^- \tag{24}$$
$$Fe(OH)_2 + O_2 \longrightarrow Fe(OH)_2^+ + O_2^-$$

Millero (1985) based his kinetic model on the assumption that the first step in the sequential reduction of O_2 to H_2O is rate-limiting. Since the hydrolysis constants of ferrous iron are $pK_1^* = 9.5$ and $p\beta_2^* \approx 20.6$, we may approximate the speciation of Fe(II) in these laboratory experiments the following way:

$$[Fe^{2+}] \approx [Fe^{II}]$$
$$[Fe(OH)^+] \approx K_1^*[Fe^{II}]/[H^+] \tag{25}$$
$$[Fe(OH)_2] \approx \beta_2^*[Fe^{II}]/[H^+]^2$$

The pseudo-first-order rate law for excess dissolved oxygen can then be given as the sum of the three parallel oxidation pathways:

$$\begin{aligned} R &= k_0[Fe^{2+}] + k_1[Fe(OH)^+] + k_2[Fe(OH)_2] \\ &= \{k_0 + k_1 K_1^*/[H^+] + k_2\beta_2^*/[H^+]^2\}[Fe^{II}] \\ &= k[Fe^{II}] \end{aligned} \tag{26}$$

The solid line in Fig. 7a was calculated according to Eq. 26 with the constants $k_0 = 1.0 \times 10^{-8}$, $k_1 = 3.2 \times 10^{-2}$ and $k_2 = 1.0 \times 10^4$ (s^{-1}) for 1 atm partial pressure of oxygen. The corresponding second-order rate constants $k_i' = k_i / K_H$ are given in Table 4. (The Henry's law constant for oxygen at 25°C is $K_H = 1.26 \times 10^{-3}$ M atm^{-1}.)

The empirical rate law in Eq. 23 holds only for the initial rates. Tamura et al. (1976) observed an autocatalytic effect of the ferric precipitates produced in the reaction. Sung and Morgan (1980) identified γ-FeOOH as the primary oxidation product at neutral pH and confirmed its autocatalytic effect. Adsorbed Fe(II) seems to compete in an additional parallel reaction with the dissolved ferrous species. Fast surface reaction rates resulted from a fit of the kinetic data. Examples of these constants are included in Fig. 7a for comparison. They represent only estimates of an order of magnitude because Tamura et al. (1976) did not determine the surface concentration of Fe(II). However, Figure 2 shows qualitatively that the ferrous ion is adsorbed specifically to mineral surfaces.

TABLE 4. Thermodynamics and Kinetics of Metal-Ion Oxygenation[a]

Redox Couple	$\log \beta_i$[b] (ox)	(red)	$E°$[c] (V)	$\log K$[d] First Step	$\log k'$ (M^{-1} s^{-1})	Reference[e]
O_2/O_2^-			-0.16			1
O_2/HO_2			0.12			1
Fe^{3+}/Fe^{2+}			0.771	-15.7	-5.1	2
$Fe(OH)^{2+}/Fe(OH)^+$	-2.19	-9.50	0.34	-8.45	1.4	2
$Fe(OH)_2^+/Fe(OH)_2$	-5.70	-20.6	-0.02	-3.04	6.9	3
$(=FeO)_2Fe^+/(=FeO)_2Fe$					0.7	4
Cu^{2+}/Cu^+			0.159	-5.41	4.3	5
$CuCl^+/CuCl$	0.4	3.1	0.32	-8.11	2.9	5
VO_2^+/VO^{2+}			1.000		<-5	6
$VO_2^+/VO(OH)^+$		-5.8	0.72	-10.1	0.02	6
$(=AlO)_2VO^+/(=AlO)_2VO$					-1.6	6
$(=TiO)_2VO^+/(=TiO)_2VO$					-1.3	6
$(=FeO)_2Mn^+/(=FeO)_2Mn$					-0.16	7
$(=AlO)_2Mn^+/(=AlO)_2Mn$					-1.55	7

[a] Figure 8 plots the linear free-energy relation between $\log K$ of the equilibrium $M_{red} + O_2 \Leftrightarrow M_{ox} + O_2^-$ and $\log k$ from experimental oxygenation rates.
[b] Stability constants of the oxidized (ox) and reduced (red) species from Smith and Martell (1979) except for CuCl (Ref. 5).
[c] The $E°$ values of the aquo complexes were taken from Bard et al. (1985). Reduction potentials of the hydroxo and chloro complexes were calculated from the listed stability constants.
[d] Equilibrium constants were calculated from the reduction potentials of the metal and oxygen couples using $\log K = \Delta E°/0.059$. The assumed product was O_2^- in the case of Fe and Cu and HO_2 in the case of vanadyl (see text).
[e] References: (1) Sawyer and Valentine (1981) for 1 M O_2 as standard state. (2) calculated from Singer and Stumm (1970); (3) Millero et al. (1987); (4) estimated value for goethite from Tamura et al. (1976); (5) Sharma and Millero (1988); (6) Wehrli and Stumm (1988), (7) Davies and Morgan (1989).

Vanadyl (VO^{2+}) is an ideal cation for the study of heterogeneous oxidations for several reasons: (1) the electron-transfer behavior of VO^{2+} is in many aspects similar to that of Fe^{2+} (Rosseinsky, 1972, Wehrli et al., in press), (2) the experimental conditions can be chosen so that vanadyl is completely adsorbed at pH >4 (Fig. 2), (3) the adsorbed V(IV) species have been characterized by ENDOR spectroscopy as inner-sphere surface complexes $(=MO)_xVO$ (Motschi and Rudin, 1984). Adsorption experiments are compatible with $x=2$. The oxygenation rates of VO^{2+} adsorbed to anatase and δ-Al_2O_3 follow the empirical rate law

$$-d\{(=MO)_2VO\}/dt = k'\{(=MO)_2VO\} \cdot [O_2] \qquad (27)$$

where $\{\ \}$ represent surface concentrations in moles per square meter (Wehrli and Stumm, 1989).

The rates are independent of pH, which suggests that the solution composition exerts only negligible effects on the speciation of adsorbed vanadyl. The oxygenation rate of dissolved V(IV), on the other hand, increases by an order of magnitude as the pH is increased by one unit. This observation indicates that the hydroxo complex $VO(OH)^+$ acts as the precursor of the oxidation step. Figure 7b compares the pseudo-first-order rate constants of the homogeneous and the heterogeneous reaction at 25°C and $p_{O_2}=1$ atm. Geometric effects of the reduction in dimensionality as summarized in Eq. 22 slow down the heterogeneous rate constant. These effects can be taken into account as follows. If we insert the relative molar mass of vanadyl and dioxygen and a radius $r_A=0.12$ mm for O_2 and $r_B=0.4$ mm for $VO(OH)^+_{aq}$ into Eq (22), we estimate a ratio $k_s/k_{hom}=0.17$ or a sixfold decrease in the rate of the surface reaction due the geometric effect. We may therefore compensate the geometric slowing down by calculating a corrected surface rate constant $k_s^c \sim 6k_s$. The resulting rate constants of dissolved and adsorbed vanadyl are remarkably similar:

$$VO(OH)^+ + O_2 \xrightarrow{\ k'\ } k'=1.07 \qquad (M^{-1}s^{-1})$$

$$(=TiO)_2VO + O_2 \xrightarrow{\ k'_s\ } k'_s=0.052; \quad k_s^c=0.31 \qquad (M^{-1}s^{-1}) \quad (28)$$

$$(=AlO)_2VO + O_2 \xrightarrow{\ k'_s\ } k'_s=0.028; \quad k_s^c=0.17 \qquad (M^{-1}s^{-1})$$

These corrected values correspond to an oxygenation of surface complexes extrapolated to a bimolecular reaction in solution. The close agreement in the oxygenation kinetics of dissolved $VO(OH)^+$ and with that of the surface complexes $(=MO)_2VO$ supports the evidence for inner-sphere surface coordination from spectroscopic and thermodynamic experiments.

The one-electron redox couples Mn^{2+}/Mn^{3+} and Cu^+/Cu^{2+} complete the emerging picture of metal oxygenations: Cu^+ shows low affinity for surfaces, and its oxidation occurs predominantly in solution. The oxidation of Mn(II), on the other hand, has so far been quantified only for surface complexes. The extremely slow reaction of dissolved Mn^{2+} (Diem and Stumm, 1984) (see Fig. 1a) may be a consequence of the prohibitive thermodynamics of the first electron-transfer step

$$Mn^{2+} + O_2 \Leftrightarrow Mn^{3+} + O_2^-, \qquad \Delta G^\circ = +177 \text{ kJ mol}^{-1} \qquad (29)$$

which results from the high reduction potential of $E^\circ = 1.5$ V for the Mn(III)/Mn(II) couple and the positive free energy of formation for the superoxide radical ($\Delta G^\circ \sim 31.9$ kJ mol^{-1}). Solid surfaces, however, seem to stabilize the products of reaction 29. Davies and Morgan (1989) found a kinetic behavior of adsorbed Mn(II) that is in line with the findings for vanadyl. The data are listed in Table 4 for comparison. The authors described the heterogeneous rate law as

$$-d[Mn^{II}]/dt = k_s' <(\equiv MO)_2 Mn > A[O_2] \qquad (30)$$

where $< >$ stands for the surface concentration in moles per gram of solid and A represents the solids concentration in grams per liter. Manganese was only partially adsorbed in these experiments. An increase in the solids concentration accelerates the reaction rate. The above rate law collapses to a form such as Eq. 27 in the limiting case of complete adsorption.

The oxygenation kinetics of Cu(I) in different electrolyte solutions has been measured by Sharma and Millero (1988). Contrary to Fe(II) and V(IV), the rates were found to be independent of $[H^+]$ in the range $5.3 < pH < 8.6$. The chloride ion, however, exerts a strong inhibitory effect. Millero (1985) interpreted the observations with a similar pseudo-first-order rate law as in the case of ferrous iron. In presence of excess oxygen the rate is given as

$$-d[Cu^I]/dt = k_0[Cu^+] + k_1[CuCl] + k_2[CuCl_2^-] + k_3[CuCl_3^{2-}] \qquad (31)$$

Sharma and Millero (1988) determined the corresponding second-order rate constants $k_0' = 2.1 \, 10^4$ and $K_1' = 8.7 \, 10^2 \, M^{-1} \text{s}^{-1}$ in sea water. The di and trichlorocomplexes were not sufficiently reactive to produce detectable rate constants. Thus the chloride ion, which stabilizes the soft reactant Cu(I) inhibits the oxygenation, whereas OH^-, which stabilizes the product Fe(III), accelerates the rate of Fe(II) oxidation. The reaction of Cu(I) with O_2 represents an interesting test case because the reverse reaction has been measured by pulse radiolysis. We may therefore apply the principle of microscopic reversibility to the electron-transfer step:

$$Cu^+ + O_2 \underset{k_-}{\overset{k_+}{\Leftrightarrow}} Cu^{2+} + O_2^- \qquad (32)$$

where k_+ was given above as k_0'. The review of Bielski et al. (1985) lists $k_- = 8 \times 10^9 \, M^{-1}\text{s}^{-1}$. From $K = k_+/k_-$ we obtain the kinetic estimate of

$\Delta G° = 31.9 \text{ kJ mol}^{-1}$ for reaction 32. This value should be compared to the thermodynamic redox potentials for the process with $1 M O_2$ as the standard state. The relevant redox potentials are 0.158 and -0.16 V for the reduction of Cu(II) and O_2 (Sawyer and Valentine, 1981), respectively. From these thermodynamic data we calculate $\Delta G° = -nFE° = 30.7 \text{ kJ mol}^{-1}$. The close agreement indicates that the redox kinetics of copper in natural waters is, indeed, governed by reaction 32 as the rate-limiting step.

3.2.   Outer-Sphere Reduction of Molecular Oxygen

The distinction between inner- and outer-sphere oxygenation evokes the question: "Does the dioxygen molecule enter the coordination sphere of the metal complex?" An inner-sphere reduction of O_2 would therefore involve a dioxygen complex as precursor and a coordinated superoxide radical in the successor complex. It should be possible to detect metal-bound superoxide by ESR spectroscopy. For an os reaction, on the other hand, only the encounter complex forms. The coordination shell of the metal center remains unchanged. Because of the transient nature of the os species, such mechanisms are difficult to prove directly. An indirect approach exploits the linear free-energy relation (LFER) based on Eq. 13. Because the rate-limiting step in the reduction of oxygen is endergonic (Fig. 7b) we expect a slope of a LFE plot $\log k$ versus $\log K$ of unity in the case that the "intrinsic" exchange barrier is smaller than the free energy of activation: $\Delta G_0^* < \Delta G^*$ (see Eq. 13a and Fig. 4d). The relevant redox potentials for the reactions discussed in the previous section were calculated from standard redox potentials (Bard et al., 1985) and the available complex formation constants (Smith and Martell, 1979). Based on the reversible equilibrium in reaction 32, the author assumes that the O_2/O_2^- couple determines the rate-limiting step. The only exception is vanadyl. The hypothetical product of reaction 33, $V^VO(OH)^{2+}$, would be very acidic. A protonation constant of VO_2^+ is not known. The relevant redox potential is therefore estimated from reaction 34:

$$V^{IV}O(OH)^+ + O_2 \longrightarrow V^VO(OH)^{2+} + O_2^- \tag{33}$$

$$VO(OH)^+ + O_2 \longrightarrow VO_2^+ + HO_2 \tag{34}$$

The thermodynamic and kinetic data are summarized in Table 4. The equilibrium constants as listed in Table 4 are calculated from $\log K = \Delta E°/0.059$ and $\Delta E° = -E°_{(M)} + E°_{(O_2)}$ and involve the standard reduction potentials of the metal couple and of oxygen at $[O_2] = 1 M$. This standard state is more suitable for kinetic calculations than the more widely used convention $p_{O_2} = 1$ atm. The calculation of *thermodynamic* equilibrium constants for the surface complexes is more difficult (Sposito, 1983) and requires further work. Figure 8 displays the resulting LFER. A theoretical line of slope one fits the data over a broad range of $13 \log k'$ units. This analysis supports an outer-sphere mechanism for the

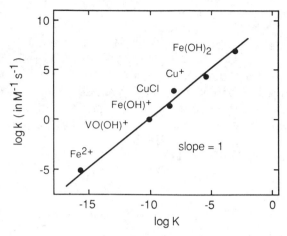

Figure 8. Linear free-energy relation for the oxygenation of metal ions. Data are listed in Table 4. The slope of unity is predicted by Marcus theory for endergonic outer-sphere electron-transfer steps.

oxygenation of V(IV), Fe(II), and Cu(I). Stanbury et al. (1980) derived a similar result for Ru(II)–amine complexes. They found a slope of $\frac{1}{2}$, which may indicate that the intrinsic barrier in Eq. 13a is approximately $\Delta G_0^* \approx \Delta G^*$ in this case. This group also published the first estimate of the self-exchange rate of the O_2/O_2^- couple, which is listed in Table 3. If the oxidation of Fe^{2+} occurs by an outer-sphere mechanism, then the kinetic observations of Singer and Stumm (1970) can be predicted from the self-exchange rates of the reactants. Application of the Marcus cross-relation (Eq. 13b) to the data of the self-exchange constants for iron and oxygen in Table 3 yields the prediction $k_0'(\text{calc.}) = 1.6 \times 10^{-5}$, which agrees well with the observed value $k_0' = 1.0 \times 10^{-5} \ M^{-1} s^{-1}$.

The three metal centers that follow the LFER for outer-sphere electron transfer in Figure 8 share a common aspect of their electronic structure. In the oxidation of d^1-V(IV), d^6-Fe(II), and d^{10}-Cu(I) an electron from a t_{2g} or a t_2 orbital is removed. Luther (Chapter 6, this volume) shows that these orbitals with π symmetry may overlap with the antibonding π^* orbital of O_2 in an encounter complex. In d^5-Mn(II), however, the leaving electron occupies an e_g orbital that points toward the coordinated ligands. The Mn(II) center is therefore a candidate for an inner-sphere electron transfer as proposed by Davies and Morgan 1989).

3.3. Kinetic Estimate for the Redox Potential of Adsorbed Metal Ions

Thermodynamic parameters such as the redox potential are difficult to measure for discrete surface species at the mineral–water interface. If the surface complexes are also oxidized by an os mechanism, we may use the LFER in Figure 8 to estimate E° from observed rate constants. The LFER plot translates to the numerical relation

$$\log k' = \log k'(0) + \Delta E^\circ / 0.059 \tag{35}$$

with the intercept at zero potential of $\log k'(0) = 10.2 \, (k/M^{-1} \, s^{-1})$, which is quite close to the theoretical limit of $\log k' = 11.0$ from Eq. 16. In order to compare the heterogeneous rate constants with those measured in solution, we correct for the reduction in dimensionality according to the approximation derived from Eq. 22 and use $k_s^c \sim 6 k_s'$. Applying the LFER in Eq. 35 to the oxygenation rates of vanadyl on anatase and of Fe(II) on goethite form Table 4, we calculate

$$(=FeO)_2 Fe^+ + e^- \Leftrightarrow (=FeO)_2 Fe^{II} \qquad (E^\circ = 0.36 \text{ V})$$

$$(=TiO)_2 VO^+ + e^- \Leftrightarrow (=TiO)_2 V^{IV}O \qquad (E^\circ = 0.73 \text{ V}) \tag{36}$$

These estimates are close to the reduction potentials for the monohydroxo complexes: $E^\circ = 0.34$ and 0.72 V for Fe(II) and V(IV), respectively. If we apply the procedure to the manganese data of Davies and Morgan (1989), we obtain an estimate of $E^\circ = 0.41$ V for the surface complex on goethite that contrasts with the value of $E^\circ = 0.9$ V for the couple $MnOH^{2+}/MnOH^+$. This large discrepancy in the case of Mn(II) is indirect evidence for an inner-sphere oxygenation of the Mn(II) surface complex. Thermodynamic and kinetic approaches are possible to test the extrapolation of redox potentials of surface species with the help of the LFER in Eq. 35: a thermodynamic calculation of E° may be based on the adsorption equilibria of the oxidized and reduced species. Kinetic measurements of other os redox processes may verify (or falsify) the estimates. The characterization of the reducing or oxidizing power of mineral surface species remains a challenge.

3.4. Half-Life Values in Natural Waters

A tentative answer to the question "What benefit may environmental science gain from such mechanistic arguments?" is given in Figure 9. The oxygenation kinetics of reduced metal centers in air-saturated waters vary over many orders of magnitude depending on (1) the redox potential of the aqueous metal couple and (2) the coordinated ligands. The large difference in the half-life of the aquo ions of Fe(II) and Cu(I) illustrates this point; the half-life of the weaker reducing agent Fe^{2+} is in the order of 10 years, while Cu^+ is oxidized within seconds. Coordinated oxygen donors drastically change the reduction potential. Two OH^- groups bound to Fe(II) shift its redox potential by ~ 0.77 V. As a consequence, the half-life of the Fe(II) species changes by 12 orders of magnitude or from 10 yr to less than 1 ms. Adsorption acts like hydrolysis: the two-dimensional array of oxygen donor ligands at mineral surfaces (Fig. 3) binds adsorbed metals in an inner-sphere coordination. Our kinetic analysis has shown that adsorption induces a change in redox potential that is similar to the effect of one coordinated OH^-. As a consequence surface complexes of Fe(II) and V(IV)

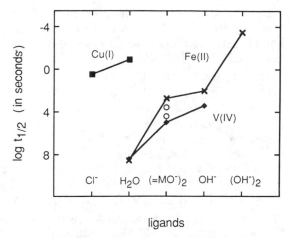

Figure 9. Half-life of metal complexes at $p_{O_2} = 0.21$. Fe(II) spans a range of 10 years to milliseconds. The surface complexes with the ligands $=MO^-$ show a reactivity similar to monohydroxo complexes in solution. Open circles represent Mn(II).

exhibit a half-life close to that of the monohydroxo complexes. If the change in redox potential on adsorption can be veryfied independently, the simple relation between $E°$ and log k in Figure 8 will allow us to treat surface oxygenation just like an electrode reaction.

It was the goal of this kinetic analysis to distinguish between the individual reactivities of different metal species. Only a mechanistic understanding at the molecular level allows us to find predictive correlations such as Eq. 35. In our heterogeneous environment, however, we face a massive parallelism in reaction rates. Within the kinetic framework of Hoigné's chapter (Chapter 2, this volume), the individual reactivities can be aggregated into "environmental factors," which allow the quantitative assessment of reaction rates measured in the field.

Acknowledgment

I thank Werner Stumm, Walter Schneider, and James J. Morgan. Their advice, support, and encouragement in front of different activation barriers made this journey into the world of rust and heterogeneity possible. A nice linear free-energy plot by René Schwarzenbach convinced me to calculate Figure 8.

REFERENCES

Astumian, R. D., and Z. A. Schelly (1984), "Geometric Effects of Reduction of Dimensionality in Interfacial Reactions," *J. Am. Chem. Soc.* **106**, 304–308.

Bard, A. J., R. Parsons, and J. Jordan (1985), *Standard Potentials in Aqueous Solutions*, IUPAC, Dekker, New York, p. 834.

Bielski, B. H. J., D. E. Cabelli, and R. L. Arudi (1985), "Reactivity of HO_2/O_2^- Radicals in Aqueous Solution," *J. Phys. Chem. Ref. Data.* **14**, 1041–1100.

Closs, G. L. and J. R. Miller (1988), "Intramolecular Long-Distance Electron Transfer in Organic Molecules," *Science* **240**, 440–447.

Combes, J. M. (1989), Ph.D. thesis, University of Paris, p. 7.

Crowther, D. L., J. G. Dillard, and J. W. Murray (1983), "The Mechanism of Co(II) Oxidation on Synthetic Birnessite," *Geochim. Cosmochim. Acta.* **47**, 1399–1403.

Davies, S. H. R., and J. J. Morgan (1989), "Manganese(II) Oxidation Kinetics on Oxide Surfaces," *J. Colloid. Interface Sci.* **129**, 63–77.

Davison, W. (1985), "Conceptual Models for Transport at a Redox Boundary," in W. Stumm, Ed., *Chemical Processes in Lakes*, Wiley-Interscience, New York, pp. 31–53.

Davison, W. and G. Seed (1983), "The Kinetics of the Oxidation of Ferrous Iron in Synthetic and Natural Waters," *Geochim. Cosmochim. Acta.* **47**, 67–79.

De Vitre, R. R., J. Buffle, D. Perret, and R. Baudat (1988), "A Study of Iron and Managanese Transformations at the $O_2/S(-II)$ Transition Layer in a Eutrophic Lake (Lake Bret, Switzerland): A Multimethod Approach," *Geochim. Cosmochim. Acta.* **52**, 1601–1613.

Diem, D. and W. Stumm (1984), "Is Dissolved Mn^{2+} Being Oxidized by O_2 in Absence of Mn-bacteria or Surface Catalysts?," *Geochim. Cosmochim. Acta.* **48**, 1571–1573.

Dillard, J. G., and C. V. Schenk (1986), "Interaction of Co(II) and Co(III) Complexes on Synthetic Birnessite: Surface Characterization," in J. A. Davis and K. F. Hayes, Eds., *Geochemical Processes at Mineral Surfaces*, American Chemical Society, Washington DC, pp. 503–522.

Eckstrom, A., C. J. R. Fooks, T. Hambley, H. J. Loeh, S. A. Miller, and J. C. Taylor (1983), "Determination of the Crystal Structure of a Porphyrine Isolated from Oil Shale," *Nature* **306**, 173–174.

Fallab, S. (1967), "Reactions with Molecular Oxygen," *Angew. Chem., Int. Ed.* **6**, 496–507.

Garrels, R. M., F. T. Mackenzie, and C. Hunt (1973), *Chemical Cycles and the Global Environment*, William Kaufmann, Los Altos, CA.

George, P. (1965), "The Fitness of Oxygen," in T. E. King, H. S. Mason, and M. Morrison, Eds., *Oxidases and Related Redox Systems*, Wiley, New York, pp. 3–36.

Gordon, G., and H. Taube (1962), "Oxygen Tracer Experiments on the Oxidation of Aquecus Uranium(IV) with Oxygen-Containing Oxidizing Agents," *Inorg. Chem.* **1**, 69–75.

Haber, F., and J. Weiss (1934), "The Catalytic Decomposition of Hydrogen Peroxide by Iron Salts," *Proc. Royal Soc. (London)* **A147**, 332–351.

Haim, A. (1983), "Mechanisms of Electron Transfer Reactions: The Bridged Activated Complex," *Progr. Inorg. Chem.* **30**, 273–357.

Holluta, J., and W. Kölle (1964), "Über die Oxydation von zweiwertigem Eisen durch Luftsauerstoff," *Das Gas- und Wasserfach.* **105**, 471–474.

Koch, D. F. A. (1957), "Kinetics of the Reaction between Manganese Dioxide and Ferrous Iron," *Aust. J. Chem.* **10**, 150–159.

Laidler, K. J. (1987), *Chemical Kinetics*, Harper, New York.

Lowson, R. T. (1982), "Aqueous Oxidation of Pyrite by Molecular Oxygen," *Chem. Rev.* **82**, 461–497.

Marcus, R. A. (1965), "On the Theory of Electron-Transfer Reactions. VI Unified Treatment for Homogeneous and Electrode Reactions," *J. Chem. Phys.* **43**, 679–701.

Marcus, R. A. (1975), "Electron Transfer in Homogeneous and Heterogeneous Systems," in E. D. Goldberg, Ed., *The Nature of Seawater*, Dahlem Konferenzen, Berlin, pp. 477–503.

Millero, F. (1985), "The Effect of Ionic Interactions on the Oxidation of Metals in Natural Waters," *Geochim. Cosmochim. Acta* **49**, 547–554.

Millero, F. J., S. Sotolongo, and M. Izaguirre (1987), "The Oxidation Kinetics of Fe(II) in Seawater," *Geochim. Cosmochim. Acta* **51**, 793–801.

Moffett, J. W., and R. G. Zika (1988), "Measurement of Copper(I) in Surface Waters of the Subtropical Atlantic and Gulf of Mexico," *Geochim. Cosmochim. Acta* **52**, 1849–1857.

Motschi, H., and M. Rudin (1984), "^{27}Al ENDOR Study of VO^{2+} Adsorbed on δ-Alumina," *Colloid Polym. Sci.* **262**, 579–583.

Oscarson, D. W., P. M. Huang, C. Defosse, and A. Herbillon (1981), "Oxidative Power of Mn(IV) and Fe(III) Oxides with Respect to As(III) in Terrestrial and Aquatic Environments," *Nature* **291**, 50–51.

Parsons, R. (1975), "The Role of Oxygen in Redox Processes in Aqueous Solution," in E. D. Goldberg, Ed., *The Nature of Sea Water*, Dahlem Konferenzen, Berlin, pp. 505–522.

Rapsomanikis, S., and J. H. Weber (1985), "Environmental Implications of Methylation of Tin(II) and Methyltin(IV) Ions in the Presence of Manganese Dioxide," *Environ. Sci. Technol.* **19**, 352–356.

Rosseinsky, D. R. (1972), "Aqueous Electron-Transfer Reactions. Vanadium(IV) as Reductant Compared with Iron(II)," *Chem. Rev.* **72**, 215–229.

Rudin, M., and H. Motschi (1984), "A Molecular Model for the Structure of Copper Complexes on Hydrous Oxide Surfaces: An ENDOR Study of Ternary Cu(II) Complexes on δ-Alumina," *J. Colloid Interface Sci.* **98**, 385–393.

Sawyer, D. T., and J. S. Valentine (1981), "How Super is Superoxide?" *Acc. Chem. Res.* **14**, 393–400.

Schindler, P. W., and W. Stumm (1987), "The Surface Chemistry of Oxides, Hydroxides and Oxide Minerals," in W. Stumm, Ed., *Aquatic Surface Chemistry*, Wiley-Interscience, New York, pp. 83–110.

Schneider, W. (1988), "Iron Hydrolysis and the Biochemistry of Iron—The Interplay of Hydroxide and Biogenic Ligands," *Chimia* **42**, 9–20.

Segal, M. G., and R. M. Sellers (1984), "Redox Reactions at Solid–Liquid Interfaces," *Adv. Inorg. Bioinorg. Mech.* **3**, 97–129.

Sharma, V. K., and F. J. Millero (1988), "Effect of Ionic Interactions on the Rates of Oxidation of Cu(I) with O$_2$ in Natural Waters," *Mar. Chem.* **25**, 141–161.

Silverman, J., and R. W. Dodson (1952), "The Exchange Reaction between the Two Oxidation States of Iron in Acid Solution," *J. Phys. Chem.* **56**, 846–852.

Singer, P. C., and W. Stumm (1970), "Acidic Mine Drainage: The Rate-Determining Step," *Science* **167**, 1121–1123.

Smith, R. M., and A. E. Martell (1979), *Critical Stability Constants*, Plenum Press, New York.

Sposito, G. (1983), "On the Surface Complexation Model of the Oxide–Aqueous Solution Interface," *J. Colloid Interface Sci.* **91**, 329–340.

Stanbury, D. M., O. Haas, and H. Taube (1980), "Reduction of Oxygen by Ruthenium(II) Ammines," *Inorg. Chem.* **19**, 518–524.

Stone, A. T. (1986), "Adsorption of Organic Reductants and Subsequent Electron Transfer on Metal Oxide Surfaces," in J. A. Davis and K. F. Davis and K. F. Hayes, Eds., *Geochemical Processes at Mineral Surfaces*, American Chemical Society, Washington DC, pp. 446–461.

Stone, A. T., and J. J. Morgan (1987), "Reductive Dissolution of Metal Oxides," in W. Stumm, Ed., *Aquatic Surface Chemistry*, Wiley-Interscience, New York, pp. 221–254.

Stumm, W. (1978), "What is the $p\varepsilon$ of the Sea?," *Thalassia Jugoslavica* **14**, 197–208.

Stumm, W., and G. F. Lee (1961), "Oxygenation of Ferrous Iron," *Indust. Eng. Chem.* **53**, 143–146.

Sung, W., and J. J. Morgan (1980), "Kinetics and Product of Ferrous Iron Oxygenation in Aqueous Systems," *Environ. Sci. Technol.* **14**, 561–567.

Sutin, N. (1986), "Theory of Electron Transfer," in J. J. Zuckerman, Ed., *Inorganic Reactions and Methods*, VCH, Weinheim, pp. 16–46.

Tamura, H., K. Goto, and M. Nagayama (1976), "The Effect of Ferric Hydroxide on the Oxygenation of Ferrous Ions in Neutral Solutions," *Corrosion Sci.* **16**, 197–207.

Van der Weijden, C. H., and M. Reith (1982), "Chromium(III)–Chromium(VI) Interconversions in Seawater," *Mar. Chem.* **11**, 565–572.

Voudrias, E. A., and M. Reinhard (1986), "Abiotic Organic Reactions at Mineral Surfaces," in J. A. Davis and K. F. Hayes, Eds., *Geochemical Processes at Mineral Surfaces*, American Chemical Society, Washington DC, pp. 462–486.

Wehrli, B., and W. Stumm (1988), "Oxygenation of Vanadyl(IV). Effect of Coordinated Surface Hydroxyl Groups and OH^-," *Langmuir* **4**, 753–758.

Wehrli, B., and W. Stumm (1989), "Vanadyl in Natural Waters: Adsorption and Hydrolysis Promote Oxygenation," *Geochim. Cosmochim. Acta* **53**, 69–77.

Wehrli, B., B. Sulzberger, and W. Stumm, "Redox Processes Catalyzed by Hydrous Oxide Surfaces," *Chem. Geol.* (in press)

Yasunaga, T. and T. Ikeda (1986), "Adsorption–Desorption Kinetics at the Metal-oxide–Solution Interface studied by Relaxation Methods," in J. A. Davies and K. F. Hayes, Eds., *Geochemical Processes at Mineral Surfaces*, American Chemical Society, Washington, DC, pp. 230–253.

Zabin, B. A., and H. Taube (1964), "The Reaction of Metal Oxides with Aquated Chromium(II) Ion," *Inorg. Chem.* **3**, 963–968.

12

MODELING OF THE DISSOLUTION OF STRAINED AND UNSTRAINED MULTIPLE OXIDES: THE SURFACE SPECIATION APPROACH

Jacques Schott

Laboratoire de Géochimie, Université Paul-Sabatier, Toulouse, France

1. INTRODUCTION

During the last decade there have been many major advances in the study of the kinetics of water–rock interactions. Indeed, one should remember that only 10 years ago most studies devoted to the interactions between aqueous solutions and minerals focused solely on the changes in the chemical composition of the solution. In the late seventies Berner's group first demonstrated that a detailed study of the surface chemistry of dissolving minerals was a prerequisite for understanding dissolution mechanisms (Petrovic et al., 1976; Holdren and Berner, 1979; Schott et al., 1981; Schott and Berner, 1983). From the examination of the surface chemistry of reacted feldspars, pyroxenes, and amphiboles with XPS, this group concluded that dissolution was controlled by chemical reactions at the mineral surface as opposed to the old idea of transport of solutes through a thick altered layer. Besides, as extensive SEM observations began to be carried out by the same group (Berner et al., 1980; Berner and Schott, 1982), the formation of crystallographically controlled etch pits initiated on dislocation outcrops was found to be a major and ubiquitous feature of the surface of weathered silicates. These findings not only supported the "surface control" hypothesis but also they led to the idea that the dissolution of silicate minerals was nonuniform and occurred preferentially at discrete sites on the surface such as dislocations, edges, and microfractures.

Concurrently, the chemical processes occurring at the solid–solution interface have been modeled by Lasaga (1981) and Aagaard and Helgeson (1982) using the transition-state theory [or activated-complex theory, (Eyring, 1935)]. Although many aspects of the formation and the stoichiometry of the activated complex remain uncertain, it is clear that a major step is the adsorption on the reactive sites of the important reactants, H_2O, H^+, OH^- and complex building ligands that weaken the critical metal–oxygen bonds. Among these species H^+ (or H_3O^+) play a central role and deserve special attention.

An important advance in understanding in detail the role of hydrogen during hydrolysis has been the use of ion-beam techniques, and particularly resonant nuclear reactions (RNR), which allow direct depth profiling of important elements such as H, Na, Al, and O. For instance, using RNR, Petit et al. (1987) and Schott and Petit (1987) have shown for the first time that pyroxene, olivine, and feldspar surfaces become protonated and/or hydrated to depths of several hundred angstroms during hydrolysis.

However, the major recent advance in the modeling of mineral dissolution has been the presentation by Stumm and coworkers (Furrer and Stumm, 1986; Zinder et al., 1986; Stumm and Furrer, 1987) of a surface coordination approach to explain the proton- and ligand-promoted dissolution of simple oxides (Al_2O_3, BeO, α-FeOOH). This approach, which relies on surface titrations and double layer concepts for characterizing the chemical speciation on mineral surfaces, offers several advantages:

Unlike XPS, and ion-beam techniques, it provides information about the interaction between the surface and solution.

It permits characterization of the precursor of the activated complex, perhaps making it possible for the first time to fill the gap between transition-state theory concept and experimental observation.

The treatment of surface titration data within the framework of transition-state theory (TST) thus appears to be one of the most promising tracks for elucidating and unifying the mechanisms of mineral dissolution. The goals of this chapter are to further explore this approach and to show that it can be applied to model the dissolution of complex oxides having several types of surface metal cation sites.

This chapter is organized under three topical areas. First, by analyzing the dissolution and speciation data of a basalt glass and various oxides we show that the surface characteristics and the dissolution behavior of complex oxides can be modeled from the properties of their constituent oxide components. Then we examine the main features of the steady-state dissolution of multiple oxides and their implications for the application of the surface coordination theory. Finally, we discuss the problem of the nature of the active dissolution sites by analyzing recent data on the dissolution of strained minerals.

2. SURFACE SPECIATION AND THE INITIAL DISSOLUTION OF MULTIPLE OXIDES

There is clear evidence that the dissolution of oxide minerals is promoted by the specific sorption of solutes at the mineral–solution interface. Moreover, it has been found that comparatively simple rate laws are obtained if the observed rates are plotted against the concentrations of adsorbed species and surface complexes (Pulfer et al., 1984; Furrer and Stumm, 1986). For example, in the presence of ligands (anions and weak acids) surface chelates are formed that are strong enough to weaken metal–oxygen bonds and thus to promote rates of dissolution proportional to their surface concentrations. Simple rate laws have been also observed with H^+—or OH^-—promoted dissolution of oxides in a manner that can be predicted from knowledge of the oxide composition and the surface concentrations of protons and hydroxyl radicals.

This surface coordination approach has been very successful in prediction of the rate of dissolution of simple oxides. However, in the case of crystalline mixed oxides like kaolinite (Webb and Walther, 1988), olivine, and feldspars (Blum and Lasaga, 1988; Amrhein and Suarez, 1988), it has not been possible to unify rate laws, and conflicting interpretations have been made from speciation data. This is likely to occur because of crystallographic constraints (i.e., different structures, lattice energies, silicate site occupancies), and the difficulty in characterizing the surface chemistry of crystalline mixed oxides. In order to determine whether dissolution mechanisms can be unified by surface chemistry concepts we have analyzed the dissolution of a tholeiite basalt glass ($Si_3AlFe_{0.5}Ca_{0.7}Mg_{0.77}Na_{0.33}K_{0.03}O_{10}$) in the hope of avoiding interference from structural ordering.

2.1. Dissolution and Speciation Data on $Si_3AlFe_{0.5}Ca_{0.7}Mg_{0.77}Na_{0.33}K_{0.03}O_{10}$

The main features of basalt glass dissolution are illustrated on Figure 1. For all elements one can observe after an ephemeral period of fast dissolution a linear steady-state dissolution period. Looking at the stoichiometry of cation release with respect to silica, we can recognize two different groups of cations:

The glass network formers (Al and Fe) exhibit a stoichiometric release over the entire pH range (this is reflected in Fig. 1b by a value of 1 for $R_X = [X/Si]_{aq}/(X/Si)_{solid}$, with $X = Al$ or Fe concentration).

The glass network modifiers (Na, Ca, Mg) show incongruent dissolution in acidic and neutral solutions ($R_X > 1$) and stoichiometric release in alkaline solutions ($R_X = 1$).

In agreement with hydrogen depth profiling and XPS analyses, these results show that the basalt glass surface is depleted in network modifying cations that

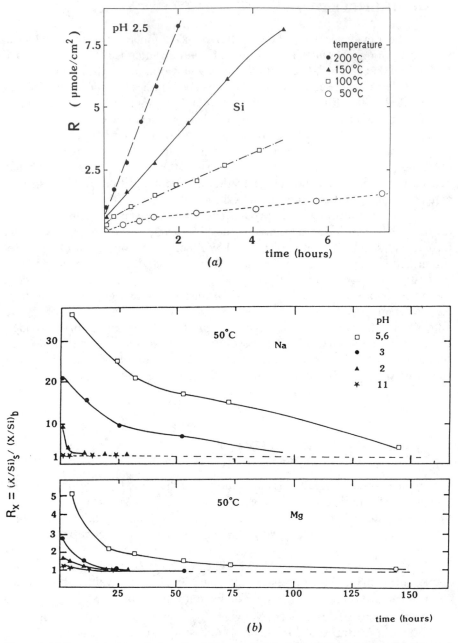

Figure 1. (*a*) Examples of basalt glass dissolution at pH 2.5 plotting silica release in solution as a function of time and temperature. (*b*) R_X of Na and Mg to Si at 50°C versus solution pH and time.

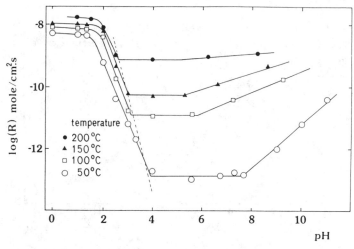

Figure 2. Plot of the dissolution rate of basalt glass versus solution pH at 50, 100, 150, and 200°C.

are instantaneously exchanged with protons (or H_3O^+) and are never re-adsorbed.

The steady-state basalt glass dissolution rates (moles of Si released per square centimeter per second) are shown on Figure 2 as a function of temperature and pH. The typical features of aluminosilicates can be recognized: dissolution decreases with increasing pH in the acid region; it is pH independent near the neutral region and increases with increasing pH at pH > 6–8, depending on temperature.

The chemical speciation of the basalt glass surface was deduced from surface titration experiments. Finely ground basalt powder was suspended in an aqueous solution of an inert electrolyte and titrated with acid and base. The total amount of H^+ or OH^- adsorbed on the surface was calculated by subtraction of a titration blank. Concentrations of dissolved species were monitored during the titrations (a charge balance equation should be written in order to compute the production or consumption of H^+ due to the release of charged species), and special care was taken to avoid contamination by atmospheric CO_2. Figure 3 shows the log of the surface concentration of H^+ and OH^- as a function of pH. Linear regressions are observed in the acid and alkaline regions, with slopes that are higher both in acid regions and at lower temperatures. The adsorption of reactants as a function of pH decreases with increasing temperature while the pH_{zpc} (zpc denotes the zero point of charge) of basalt glass shifts from about 6.8 at 25°C to 6.1 at 50°C. [This is in qualitative agreement with the results of Fokkink (1987) on hematite.]

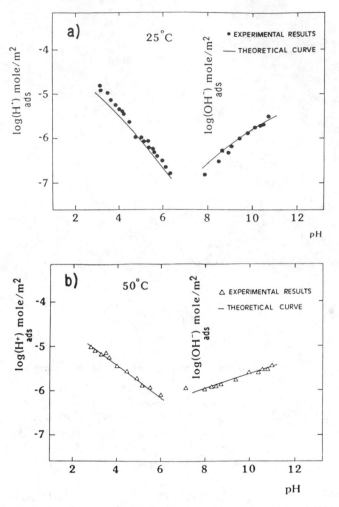

Figure 3. Surface titration of a basalt glass suspension at 25°C (*a*) and 50°C (*b*).

2.2. Modeling of the Surface Speciation of the Basalt Glass

For this purpose it is possible to extend to a multiple oxide the one-site model of Johnson (1984), which provides a thermodynamic description of the double layer surrounding simple hydrous oxides. Briefly, in this model the double layer charge is divided into the charge inside the slip plane, $\sigma[s]$, and that outside the slip plane $\sigma[d]$. While $\sigma[s]$ is determined from the occupied sites, $\sigma[d]$ is obtained from the Poisson–Boltzmann equation. Note that unlike the triple-layer model (Davis et al., 1978) which allows ions to form surface complexes at two different planes (0 or β) instead of at the slip plane only, this model does not distinguish between inner- and outer-sphere complexes. Expression of the electrical neutrality of the system solid solution ($\sigma[d] = \sigma[s]$) and in the bulk solution allow calculation of the concentration of the surface species solely from knowledge of the total number of surface sites and the intrinsic constants. Indeed, in this method the surface potential, which is considered as a dependent variable, is varied using an iterative method until $\sigma[d] + \sigma[s] = 0$.

To carry out surface speciation calculations for a multiple oxide, one must know the metal cations susceptible to adsorption. In the case of the basalt glass it is very tempting to assume that modifying cations (Na, Mg, Ca) are not susceptible to adsorption since the solid surface is depleted in them. (This is only true in acidic and neutral solutions, but in alkaline solutions modifier cations cannot provide adsorption sites for OH⁻ because their pH_{zpc} is too high; in other words, the concentration of surface species like $>CaO^-$ and $>MgO^-$ is insignificant even at high pH.) To verify this hypothesis we have performed experimental titrations with two mixtures of oxides having the same stoichiometry as the glass. The first mixture contained all the glass components, including network modifier cations, while the second one contained only network-forming cations. The results plotted in Figure 4 show that the titration curves of the basalt glass and of the mixture of oxides of former cations are the same, which confirms that network-modifying cations cannot provide adsorption sites. Thus, in our calculations, we have assumed that only Si, Al, and Fe sites were susceptible to adsorption by H^+, OH^-, and ligands, and that these sites acted independently of each other. The values of parameters used in the calculations are summarized in Table 1.

In Figure 3 the calculated surface charge of the basalt glass is compared with our experimental titrations. The superposition of the experimental and theoretical curves shows that the surface speciation of amorphous multiple oxides can be

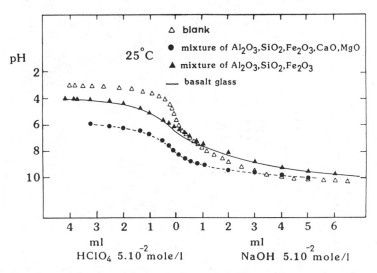

Figure 4. Experimental titration of the basalt glass and of different mixtures of constituent oxides.

TABLE 1. Intrinsic Acidity Constants and Number of Sites for Different Oxides

Oxides	Reference	$pK_{a_1 \text{ (intr)}}$	$pK_{a_2 \text{ (intr)}}$	N_s (sites/nm^2)
$Al_2O_3/NaCl$	Huang (1971)	5.7	11.5	8
$FeOOH/NaCl$	Yates (1975)	4.2	10.8	16.8
$Fe(OH)_3/NaNO_3$	Davis et al. (1978)	5.1	10.7	11
TiO_2/KNO_3	Yates (1975)	2.6	9.0	12
SiO_2/KCl	Abendroth (1970)	-2.0	7.2	5

easily deduced from the values of the equilibrium constants of the adsorption reactions at the different metal surface sites.

Using the same model we have calculated for the basalt glass the distribution of superficial sites as a function of solution pH. The results plotted in Figure 5 show that the dominant species are:

$>SiOH$, $>AlOH_2^+$, and $>FeOH_2^+$ in acid solutions.

The neutral species in solutions near neutral pH.

$>SiO^-$, $>AlOH$, and $>FeOH$ in alkaline solutions.

Let us hypothesize that the dissolution reaction is promoted by the adsorption of

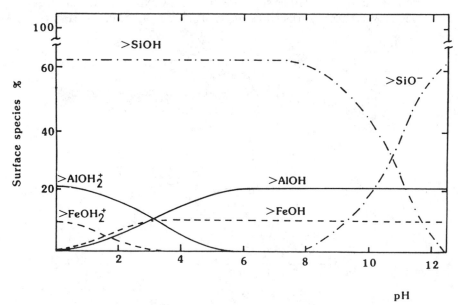

Figure 5. Predicted distribution of surface species on basalt glass as a function of solution pH at 25°C.

H^+ and OH^-, which weaken and break the metal cation framework bonds; then using the above calculations showing the distribution of surface species, basalt dissolution is promoted by the absorption of H^+ to Al and Fe sites (mainly Al) in acidic solutions, whereas it is controlled by the adsorption of OH^- at Si sites in alkaline solutions. The combination of the two distinct types of surface sites, Al and Fe on one hand, and Si on the other hand, results in a dissolution rate minimum at a pH value between the pH_{zpc} of the two groups of oxide components.

The prediction by our speciation model that in multiple oxides (as in quartz) Si sites are not susceptible to adsorption by H^+ is confirmed by the results of surface titrations of albite, labradorite, and anorthite (Fig. 6). One can see that the net adsorption of H^+ at the feldspar–solution interface increases markedly with the relative number of Al sites to Si sites, which is reflected by the stoichiometry of Al and Si in the different feldspars. These results help clarify the typical features of the dependence on pH of silicate dissolution (Table 2):

In alkaline solutions the same order of dependence on a_{OH^-} of dissolution rate of most silicates ($n = 0.3$) is consistent with OH^- adsorption occurring only at Si sites.

In acid solutions, by contrast, the variable order of dependence on a_{H^+} of dissolution rate of different silicates ($0.3 < n < 1$) is consistent with H^+ adsorption being metal-cation-specific (i.e., note the different reaction orders exhibited by aluminum and magnesium silicates).

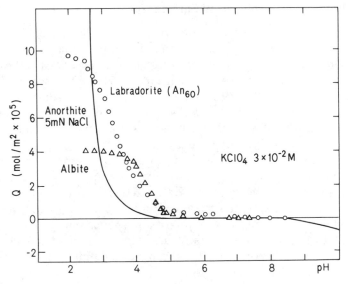

Figure 6. Experimental titrations at 25°C of ground powders of albite, labradorite (An_{60}), and anorthite [data for anorthite are from Amrhein and Suarez (1988)].

TABLE 2. Order of Dependence on H^+ Activity of the Rate of Dissolution of Various Oxides and Silicates ($T = 25°C$)

Mineral	Reaction Order		References
	Acid Solutions	Alkaline Solutions	
K-feldspar	0.33		Wollast (1967)
Albite	0.49	-0.30	Wollast and Chou (1985)
Anorthite	0.54		Fleer (1982)
Sr-feldspar	1.0		Fleer (1982)
Nepheline	1.0	-0.20	Lasaga (1984)
Augite	0.7		Schott and Berner (1985)
Enstatite	0.6		Schott et al. (1981)
Diopside	0.5		Schott et al. (1981)
Olivine	0.6	-0.28	Blum and Lasaga (1988)
	1.0		Grandstaff (1981)
Almandine	0.25	-0.20	Schott and Petit (1987)
Quartz	0.0	-0.30	Brady and Walther (in press)
Aluminum oxide	0.4	-1.0	Furrer and Stumm (1986) Webb and Walther (1988)
Iron oxide	0.48	—	Sigg and Stumm (1981)

2.3.   Dependence of Dissolution on Surface Charge

Because adsorption reactions are very fast compared to the metal detachment process, one expects the rate of dissolution to be proportional to the concentration of adsorbed H^+ or OH^- at the solid surface. This is verified in the case of a basalt glass. Linear regressions with a slope of 3.7–3.8 are observed in both acid and alkaline solutions in logarithmic plots of the rate of dissolution versus the surface charge (Fig. 7); this indicates that the order of the reaction is 3.8 with respect to C_H^s and C_{OH}^s. It is interesting to address the significance of the reaction order (n). Looking at the reaction order of proton-promoted dissolution in the case of δ-Al_2O_3, BeO, α-FeOOH, and SiO_2, Furrer and Stumm (1986), Zinder et al. (1986), and Guy and Schott (1989) have proposed that n represents the number of protonation steps required for metal detachment and is equal to the oxidation number of the central metal cation in the oxide. Indeed, they found $n = 2$ for BeO, $n = 3$ for Al_2O_3, and $n = 4$ for amorphous silica and quartz. Thus within this coordination approach, the order of the reaction observed in the case of a basalt glass ($n = 3.8$) seems to reflect the mean oxidation value of the network former metals of the glass

$$n = \Sigma \alpha_i Z_i \qquad (1)$$

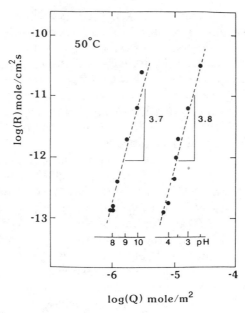

Figure 7. Logarithmic plot of basalt rate of dissolution versus surface charge.

where α_i and Z_i represent the mole fraction and the oxidation number of a network former metal i.

These findings allow us to describe the proton- and hydroxyl-promoted dissolution of an amorphous multiple oxide by the equation

$$R = k_H(C_H^s)^n + k_{OH}(C_{OH}^s)^n + k_{H_2O} \tag{2}$$

where C_H^s and C_{OH}^s can be either determined experimentally or predicted from the knowledge of the acidity constants of the surface hydroxyl groups of the constituent network forming oxides. k_H, k_{OH} and k_{H_2O} are the reaction constants for H^+, OH^-, and water-promoted dissolution, respectively.

The same order of reaction measured in acid and alkaline solutions, together with the observation at all pH values of stoichiometric dissolution of network-forming metal cations, may appear inconsistent with the initiation of dissolution at two distinct types of surface sites in acid and alkaline solutions. In fact, the stoichiometry of dissolution is more strongly related to the bonding characteristics of solids than to their surface speciation. One can understand that dissolution is more congruent in glasses, where network-forming cations are randomly distributed and where the energy of the metal–oxygen bonds is the same for Al, Fe, and Si (Zarzycki, 1982), than in crystalline silicates, where bonding characteristics and the surface chemistry of the different faces can vary greatly. For instance, this is the case of kaolinite, where the functional groups that control the initial reaction steps are $>$AlOH on the planar gibbsite layer and the edge faces and $>$SiOH at the planar siloxane layer (Wieland and Stumm, 1989).

Thus, one understands that the global speciation approach used above is inappropriate to kaolinite and that dissolution processes at edge faces, gibbsite layers, and siloxane layer should be considered separately.

By contrast, in the case of feldspars such as albite, a separate treatment of the different crystal faces is not required; a close agreement between experimental and theoretical titration curves is observed when using our speciation model and assuming that only Al and Si sites are susceptible to charging by H^+ and OH^- (Fig. 8). If, using the experimental data of Chou and Wollast (1984), one plots the log of the rate of dissolution of albite versus $\log C_H^s$, a linear regression is observed with a slope of about 3, which is consistent with protonation occurring solely at surface aluminum sites (Fig. 9). A similar reaction order has been measured by Amrhein and Suarez (1988) for the dissolution of anorthite. However, it should be noted that Blum and Lasaga (1988) have reported a first-order dependence of the dissolution rate of albite on concentrations of specific surface species. A rigorous characterization of the reaction orders for silicate dissolutions should probably wait for more experimental data and particularly very precise determinations of the surface speciation of feldspars as a function of dissolution progress. Indeed, because aluminum is preferentially leached during initial dissolution of feldspar in acid solutions, the surface charge decreases markedly as a function of time. This is illustrated in Figure 10, which

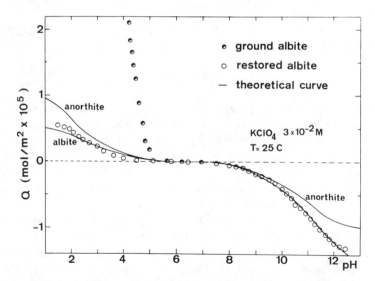

Figure 8. Comparison of experimental surface titrations of ground and restored albite (ground albite annealed for 80 h at 1000°C) with theoretical titrations of albite and anorthite ($N_s = 12$ adsorption sites nm^{-2}). Note the good superposition of theoretical and experimental curves for "restored" albite. Possible explanations for the important surface charge exhibited by ground albite are that (1) surface defects produced by prolonged grinding can introduce extra adsorption sites and (2) weakening and breaking of surface Si–O bonds due to grinding can increase $pK_{Si, 1 \text{ (intr.)}}$ and thus allows proton adsorption on Si sites even in midly acid solutions.

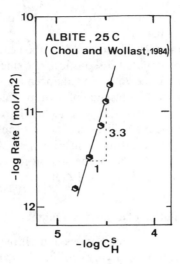

Figure 9. Logarithmic plot of albite dissolution rate versus its surface charge.

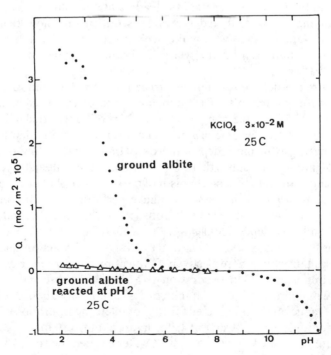

Figure 10. Comparison of the surface titration curves of freshly ground albite and of albite that has been reacted at pH $2(10^{-2} M \ HClO_4)$ during 10 days.

compares the surface charge of fresh albite and that of albite reacted during 10 days at pH 2 ($\sim 10^{-2} M$ $HClO_4$).

Surface complexation models like those described can also be used to predict the ligand- and cation-protonated dissolution of oxides. As an illustration, we have analyzed within this framework the enhancement of the rate of dissolution of amorphous silica due to the presence of sodium chloride in solution (Wirth and Gieskes, 1979). The adsorption of Na on silanol groups has been calculated from the value of pK for the Na^+–silica gel reaction, and has been reported as a function of pH and NaCl concentration in Figure 11. Again, linear regressions were observed when plotting the log of the dissolution rate against the log of Na surface concentration (Guy and Schott, 1989).

2.4. Steady-State Dissolution of Complex Oxides

We have shown above that dissolution rates of multiple oxides can be related to the abundance and speciation of hydrogen and hydroxyl radicals at different metal centers at the surface. Since dissolution of most complex oxides is nonstoichiometric, the identity of these centers varies as a function of time and experimental conditions. The selective removal of some cations from the solid surface creates a reacted layer that is depleted in those elements that dissolve rapidly (i.e., modifying cations during basalt dissolution or sodium, calcium, and aluminum in the case of feldspars). As steady-state dissolution is controlled by the dismantling of these altered layers, it is critical to know their chemical characteristics and to identify the main mechanisms that control their formation. Two important findings obtained via microbeam techniques will be presented here.

First, hydrogen-depth profiling performed on reacted aluminosilicates shows that the preferential removal of some metals proceeds via an exchange reaction with H^+ or H_3O^+/H_2O (i.e. compare Na and H profiles reported in Fig. 12). As a result, silicate surfaces become protonated and/or hydrated to depths of several hundred angstroms or more depending on pH and temperature (Fig. 12).

A second important result is that the formation of these altered layers does not involve a simple protonation mechanism of the surface. Following the method of Pederson et al. (1986) with glasses, we have performed dissolution runs with albite and olivine in ^{18}O-labeled solutions (50% ^{18}O) at pH 2 and $T = 200°C$. Hydrogen and ^{18}O depth profiles were measured on a 7-MeV Van de Graaff accelerator by means of the resonant ^1H (^{15}N, $\alpha\gamma$)^{12}C reaction and ^{18}O (p, α) ^{15}N nonresonant reaction, respectively. Typical results for albite are shown on Figure 13. Important ^{18}O incorporation (~ 5 atom %) is observed to depths that can reach 2 or 4 μm. The $^{18}O:H$ uptake ratios (Fig. 14) are about 3, which is well outside the range of values expected for the complete incorporation of any simple molecular or ionic form of water (0.33 for hydromium, 0.50 for molecular water, 0.00 for bare protons). Whatever forms of water entered the solid, they have not been immobilized and conserved upon reaction. Oxygen-18 uptakes in excess of H are consistent with multiple silanation–condensation reactions only within the

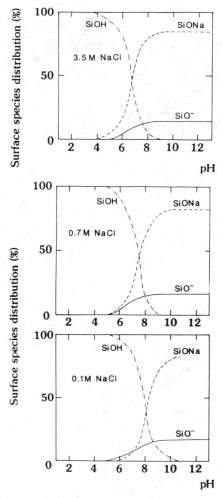

Figure 11. Calculated surface speciation of amorphous silica in NaCl solutions at 25°C (pK for Na–silica gel reaction = 7.0).

surface layer:

$$\text{Si}-^{16}\text{O}-\text{Si} + \text{H}_2{}^{18}\text{O} \longrightarrow \text{Si}-^{16}\text{O}-\text{H} + \text{Si}-^{18}\text{O}-\text{H}$$

$$\text{Si}-^{16}\text{O}-\text{H} + \text{Si}-^{18}\text{O}-\text{H} \begin{array}{c} \xrightarrow{\ 50\%\ } \text{Si}-^{18}\text{O}-\text{Si} + \text{H}_2{}^{16}\text{O} \\ \xrightarrow{\ 50\%\ } \text{Si}-^{16}\text{O}-\text{Si} + \text{H}_2{}^{18}\text{O} \end{array}$$

These results show that multiple reactions of water with the albite matrix, rupturing and reforming many network bonds (Si–Al–O), occur simultaneously with Na and Al leaching. Casey et al. (1988b) recently reached similar conclusions in the study of the reaction of labradorite feldspar in acid solutions at 25°C

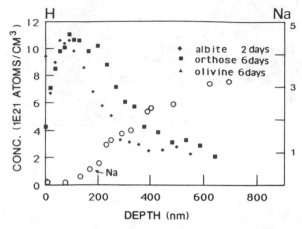

Figure 12. Hydrogen and sodium (relative to albite) depth profile of several silicates leached at 200°C in deionized water. [After Petit et al., (1989).]

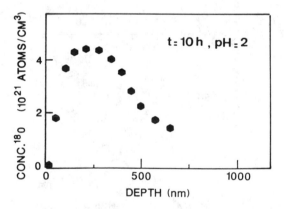

Figure 13. ^{18}O depth profile of albite reacted 10 h at 200°C (pH 2).

in which they showed that the amount of hydrogen added to the feldspar was much less than one would expect from ion exchange. Obviously, caution must be exercised when proposing reaction mechanisms solely on the basis of a comparison of hydrogen uptake and cation extraction.

As shown above, surface titration curves of reacted albite exhibit notable differences, particularly in the acid pH range, relative to those obtained with fresh albite (but they are very close to those obtained with silica gels). Thus, one understands that surface titrations performed on fresh silicates can only help elucidate initial dissolution mechanisms.

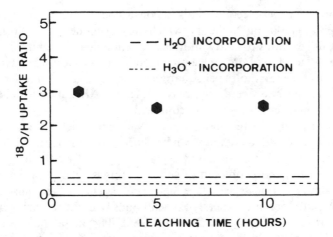

Figure 14. Ratios of ^{18}O and H uptake in albite versus leaching time. Note that the data do not correspond to any simple molecular or ionic form of water.

2.5.   Surface Speciation Approach and Transition-State Theory (TST)

Transition-state theory may be useful in testing the dissolution mechanisms presented above. According to TST, for any elementary chemical reaction the reactants should pass through a free-energy maximum, labeled the "activated complex", before they are converted to products. It is assumed that the reaction rate-determining step is related to the decomposition of this activated complex:

$$r = k_r C^* \tag{3}$$

where C^* is the concentration of the activated complex, k_r is the rate constant, which can be expressed (Lasaga, 1981; Aagaard and Helgeson, 1982) as

$$k_r = \frac{kTK^*}{h\sigma\gamma_j^*} \tag{4}$$

k and h represent the Boltzmann and Planck constants, respectively; σ refers to the ratio of the rate of decomposition of the activated complex to that of the overall reaction; γ_j^* is the activity coefficient of the jth activated complex; and K^* represents the equilibrium constant for the jth reaction forming the activated complex. The constant K^* can be written in the usual thermodynamic way:

$$K^* = e^{-(\Delta G^*/RT)} = e^{(\Delta S^*/R)}e^{-(\Delta H^*/RT)} \tag{5}$$

where ΔG^*, ΔH^*, and ΔS^* are the standards Gibbs free energy, enthalpy, and entropy change from reactants to activated complex. It should be noted that k_r has units of second^{-1}, whereas the experimental data are reported as moles per square centimeter per second. This inconsistency of units demonstrates a potential pitfall in using a theory which has been developed primarily for homogeneous reactions to study interfacial phenomena. To maintain consistent units in the calculation of ΔS^*, measured reaction rates should be divided by the number of surface sites (moles per square centimeter) to determine a k_r with units of second^{-1} (Dove and Crerar, in press). Note that ΔH^* does not change when the rate is normalized to give units consistent with TST.

As emphasized by Furrer and Stumm (1986) and Wieland et al. (1988), the precursor of the activated complex can be characterized from the determination of the surface speciation of the solid.

In the case of ligand-promoted dissolution, this is easily characterized since the ligands directly attack the surface metal center. The surface chelate formed is strong enough to loosen the metal–oxygen bonds without requiring any further proton adsorption (Furrer and Stumm, 1986). The metal ligand is thus the precursor of the activated complex, and the rate of dissolution should be proportional to its surface concentration if the detachment step is rate-determining:

$$r = k_L \{ML\} = k_L C_L^* \tag{6}$$

The H^+- and OH^--promoted dissolution of oxides involves a different mechanism. Protons and OH^- are bound to the surface hydroxyl groups or to oxide ions closest to the solid surface. In this configuration they polarize the bonds between oxygen and metal ions. In order to obtain a sufficient weakening of critical metal-oxygen bonds and thus detachment of a surface central metal ion, a sufficient number (n) of oxide or hydroxide species neighboring the surface metal have to be protonated or deprotonated. Like the adsorption of ligands, that of H^+ or OH^- at the surface is very fast compared to metal detachment. Therefore, the dissolution rate should be proportional to the surface concentration of the metal ions that present the coordination arrangement corresponding to the last protonation (or deprotonation) step before metal detachment. These surface species can be regarded as the precursors of the activated complex. Thus

$$r = k_H C_H^* = k_H \{>MOH_n\} \tag{7}$$

where $>MOH_n$ represents the surface concentration of metal centers surrounded with n-protonated oxide or hydroxide radicals.

As only C_H^s or C_{OH}^s are experimentally accessible, one has to derive a relationship between these quantities and C^* in order to apply Eq. 3. Furrer and Stumm (1986) and Wieland et al. (1988) have shown that if the fraction of surface sites occupied by adsorbed species is $\ll 1$, the probability of finding a surface site in the coordination arrangement of the precursor of the activated complex is proportional to the nth power of C_H^s and C_{OH}^s. For example, the proton-promoted dissolution can be expressed as

$$r = k_H C_H^* = k_H (C_H^s)^n \tag{8}$$

If n is an integer for a simple oxide, noninteger values may be assigned to n for the case of complex oxides since several types of metal centers are involved in the protonation or hydroxylation processes.

In the case of the basalt glass studied above, $n = 3.7$ (mean valence of the network-forming metals). The stoichiometry of the activated complex formed in acid solution is likely to be the same as that of the network-forming metals since modifying cations are instantaneously exchanged with H^+:

$$(Si_{1.1}Al_{0.37}Fe_{0.18}Ca_{0.26}Mg_{0.28}Na_{0.12}O_{3.7}) + 1.2H^+$$
$$\longrightarrow (Si_{1.1}Al_{0.37}Fe_{0.18}O_{3.7}H_{1.2})$$
$$(Si_{1.1}Al_{0.37}Fe_{0.18}O_{3.7}H_{1.2}) + 3.7H^+$$
$$\longrightarrow (Si_{1.1}Al_{0.37}Fe_{0.18}(OH)_{3.7}H_{1.2})^{*3.7+}$$

In alkaline solutions the stoichiometry of the activated complex is the same as that of the glass.

Similarly, in the case of albite, the formation of the activated complex under acidic conditions can be written in a general form:

$$(Na_{0.5}Al_{0.5}Si_{1.5}O_4) + 0.5H^+ \longrightarrow (H_{0.5}Al_{0.5}Si_{1.5}O_4)$$
$$(H_{0.5}Al_{0.5}Si_{1.5}O_4) + xH^+ \longrightarrow (H_{0.5+x}Al_{0.5-y}Si_{1.5}O_4)^{*(x-3y)+}$$

The activation enthalpy ΔH^* and entropy ΔS^* for the formation of the activated complex can be derived from Eqs. 4 and 5 if one assumes that the overall dissolution of the basalt glass is controlled by the decomposition of the activated complex. Such a calculation is very rough since it requires several simplifications ($\Delta C_p^* = 0$, $\gamma_j^* = 1$) and assumes that the overall dissolution reaction involves only one elementary reaction. Values of ΔH^* and ΔS^* for the dissolution of the basalt glass and other silicates are reported in Table 3. Examination of the ΔS^* values is interesting since the entropy of activation is widely seen as an indicator of the configuration of the activated complex (e.g., see Benson, 1976). Values of ΔS^* derived from dissolution runs are negative, which reflects the restricted motion imposed by the binding of the reactant species (loss of translational and rotational freedom). As a result, the decrease in entropy will be larger if the activated complex is tightly bounded than if it is loosely bonded. We found that ΔS^* decreases with increasing pH from the acid pH to the near neutral pH region. This signifies that the activated complex formed during the H_2O-promoted dissolution of the basalt glass is more tightly bound than those formed during the H^+- or OH^--promoted dissolution, which polarize and weaken the

TABLE 3. Apparent Activation Entropies and Enthalpies of Overall Dissolution for Various Silicates

Mineral	pH	$-\Delta S^*$ $(J\,K^{-1}\,mol^{-1})$	ΔH^* $(kJ\,mol^{-1})$	Reference
Enstatite	2–6	130	47.5	Schott et al. (1981)
Diopside	4–6	170	47.5	Schott et al. (1981)
	1–2	60	78.2	
Wollastonite	4	60	72	Murphy (1985)
Olivine	3–5	140	35	Grandstaff (1981)
Quartz	4–6	160	75	Murphy (1985)
	9–10	130	105	Knauss and Wolery (1988)
Albite	1–3	30	86	Aagaard and Helgeson (1982)
	3–8	250	36	Aagaard and Helgeson (1982)
K-feldspar	1–3	50	79.5	Aagaard and Helgeson (1982)
Kaolinite	2	120	51	Webb and Walther (in press)
	5	230	21.2, 61(1)	(1) Wieland et al. (1988)
	11	120	50.7	
Basalt glass	3	150	–	
	5	212	69	This study
	7	200	75	
	8	170	75	

metal oxygen bonds exposed at the solid surface. Similar trends and orders of magnitude of ΔS^* are observed with other silicates.

3. DISLOCATIONS AND THE NATURE OF "ACTIVE SITES" FOR DISSOLUTION

We have seen above that the kinetics of mineral dissolution is well explained by transition-state theory. The framework of this theory and kinetic data for minerals have shown that dissolution is initiated by the adsorption of reactants at "active sites." Until now these active sites have been poorly characterized; nevertheless, there is a general consensus that the most active sites consist of dislocations, edges, point defects, kinks, twin boundaries, and all positions characterized by an excess surface energy. Also these concepts have been strongly supported by the results of many SEM observations which have shown that the formation of crystallographically controlled etch pits is a ubiquitous feature of weathered silicates.

Concurrently, after reviews of the theories of etch-pit formation, Lasaga (1983) and Helgeson et al. (1984) concluded that the rate of mineral dissolution can be greatly enhanced by the presence of dislocations intersecting the solid surface. This can have enormous implications because dislocation densities can vary by many orders of magnitude between crystals and within different parts of single crystals.

Thus, experimental studies relating dissolution rates to defect concentrations are needed. This has been done recently in two different types of experiments where the importance of line defects has been measured. Casey et al. (1988) and Holdren et al. (1988) have measured the rate of dissolution of rutile and labradorite powders that were shocked with an explosive charge to induce a high density of dislocations ($\geq 10^{11}$ cm^{-2}). Concurrently, in order to avoid artifacts due to fine particles, Schott et al. (1989) have performed with the rotating disc apparatus dissolution runs of single crystals of calcite in which dislocations were induced in a constant-strain apparatus. Examples of the results are shown in Fig. 15. Surprisingly, it can be seen that dislocations have only a small effect on the dissolution rates: the dissolution rates of samples with over 6 order of magnitude in dislocation density differ only by a factor less than 3.

So, why are rates of dissolution and dislocation densities so weakly related? To address this paradox, energetic and kinetic considerations are necessary.

3.1. Lattice Defects and Reaction Rates

Lattice defects may influence the rates of mineral dissolution in two ways: (1) by changing the bulk thermodynamic properties and (2) by creating sites of accelerated dissolution on the solid surface. Strain and core energies associated with dislocations contribute insignificantly to the total energy of minerals. For instance, even with the extremely high dislocations density of 10^{11} cm^{-2}, the free energy of calcite is increased by only 80 J mol^{-1}, which corresponds to a 25°C activity of 1.04. The same features are observed for quartz and silicate minerals.

In contrast, the second effect could play an important role because the core energy is highly localized. For example, if we could calculate a molar free energy for a core by assuming that this core energy disrupts one calcite formula unit of atoms, it follows that the core energy for a mole of such formula units would be ~ 40 kJ, which is comparable to the activation energy for dissolution.

Following this track, several approaches can be used to model the kinetics of dissolution of strained solids. Blum and Lasaga (1987) have performed Monte Carlo simulations where the dissolution rate constant at any surface site is a function of the enthalpy of dissolution of the crystal. Schott et al. (1989) have followed a similar approach that allows a much simpler analysis of experimental data.

If in the basic equation of TST, which describes the rate of dissolution of minerals under highly undersaturated conditions

$$r = k_+ S_e \Pi_i a_i^{n_i} \tag{9}$$

$k_+ S_e$ (k_+ dissolution rate constant and S_e = effective surface area) is replaced with

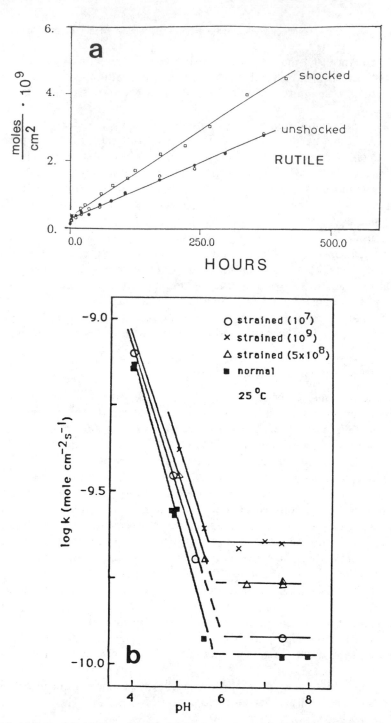

Figure 15. (*a*) Rate of dissolution of shocked (10^{11} dislocations/cm^2) and unshocked rutile powders in 1.06 N hydrofluoric acid solutions at 25°C [after Casey et al. (1988a)]. (*b*) Rate of dissolution of single crystals of calcite as a function of dislocation density and pH at 25°C [after Schott et al. (1989)].

a summation over surface sites with different activation energies toward dissolution, the following expression is obtained:

$$r = \Pi_i a_i \left[v \left\{ \underbrace{\Sigma S_i \; \exp\left(-\frac{\Delta G_p^+}{kT}\right)}_{\substack{\text{perfect} \\ \text{surface sites}}} + \underbrace{\Sigma S_i \; \exp\left(-\frac{\Delta G_d^+}{kT}\right)}_{\substack{\text{dislocated} \\ \text{surface sites}}} \right\} \right] \quad (10)$$

In this equation v is the frequency factor, k is Boltzmann's constant, ΔG_p^+ is the activation energy for dissolution of perfect crystal surface, and ΔG_d^+ is the activation energy for dissolution of dislocated crystal surface. In the case of a surface intersected only by dislocations, ΔG_p^+ differs from ΔG_d^+ by the magnitude of the strain energy U_{st} associated with the dislocation ($U_{st} = U_{line} \rho V$ with $U_{line} = (\tau b^2/4\pi)\ln(r_s/r_0) + \alpha \tau b^2$, where τ is the shear modulus, b is the Burgers vector, r_s is the radius of the strain field, r_0 is the radius of the core, and α is a factor describing the contribution of the core energy). If it is assumed that the activated complex is the same whether it occurs at a dislocation or a perfect surface site, then Eq. 10 can be written

$$r = k_e S + 2\pi \rho k_e S \int_{r_0}^{r} \exp\left(\frac{U_s}{RT}\right) r \, dr + 2\pi r^2 H \rho S \, k_e \exp\left(\frac{U}{RT}\right) \quad (11)$$

where k_e is the dissolution constant of unstrained material, S is the surface area, r and Hr = radius and depth of a hollow core centered about a dislocation.

Using Eq. 11, we can compare the effect of dissolution at dislocations to the overall dissolution rate. For instance, one predicts that the critical dislocation density ρ_c at which the dissolution rate due to dislocations and to the dislocation-free surface become equal is $\rho_c \simeq 2.10^9$, 3.10^9, and 10^{10} cm^{-2} for quartz, calcite, and rutile, respectively. These values compare well with experimental data and calculations of Blum and Lasage (1987) for quartz. Obviously, the effect of dislocation on the overall rate of dissolution become significant only for very high dislocation densities.

Moreover, numerical solutions of Eq. 11 for calcite, quartz, and feldspars show that after a short transient period (a few minutes for calcite), the rate of dissolution remains roughly constant. Our calculations (see Table 4) indicate that during this steady-state period the overall dissolution rate should increase by a factor of only 3 or 4 at maximum due to two competing effects: decreasing surface strain energy and increasing surface area, as dislocation cores dissolve and widen.

3.2. Dislocations and Active Sites

Possible explanations for the small effect of dislocations on dissolution rates, even at very high dislocation densities, are that (1) in contrast to intuitive

TABLE 4. Calculated Rate of Calcite Dissolution at $25°C^a$

r (Å)	Activity at Surface of Hallow Core	S_d/S_T	Rate of Dissolution
	$H = 50$		
15	9	0.07	2.3
20	4.6	0.13	2.3
40	1.55	0.5	2.5
60	1.1	1.1	2.9
	$H = 200$		
15	9	0.28	3.5
20	4.6	0.5	3.3
40	1.55	2.0	4.1
60	1.1	4	5

[a] Normalized to unstrained calcite. Dissolution rate calculation as a function of radius r of the hollow core, based on Eq. 11, for $\rho = 10^{10}$ cm^{-2} (S_T = total perfect surface area; S_d = total surface area of hollow cores).

expectations, the number of active sites per surface area could only be weakly related to dislocation density; and (2) active sites provided by dislocations may not be regenerated during the course of dissolution.

The first explanation can be easily understood if one notices that the dislocation densities, which can rarely exceed 10^{11} cm^{-2}, are negligible compared to the total number of surface sites in solids ($1 - 10 \times 10^{14}$ cm^{-2}). Note also that at very high defect densities, increasing overlap of dislocation strain fields will occur. However, our surface titrations in acid solutions of shocked (10^{11} dislocations cm^{-2}) and undeformed labradorite powders (An$_{60}$) (Fig. 16) show a notable increase of the surface concentration of protons for shocked labradorite. (This is accompanied by a similar increase of Ca release in solution: five times in normalized conditions.) The interpretation of these surface titrations is not straightforward since shock deformations, besides dislocations, induce also all kind of defects (point defects, microcracks, surface defects, etc.). However, calculation of labradorite surface charge using our speciation model shows that the observed charge on shocked labradorite cannot be solely explained by an increase in the total number of adsorption sites. In fact, it is likely that the weakening and breaking of surface Si–O bonds due to shock deformations results in an increase of p$k_{Si, 1 \text{ (intr.)}}$ and allows proton adsorption on Si sites.

On the other hand, dissolution under steady-state conditions requires, as explained by Furrer and Stumm (1986), that after each metal detachment step the active site will be regenerated (i.e., the fraction of active sites to total surface sites

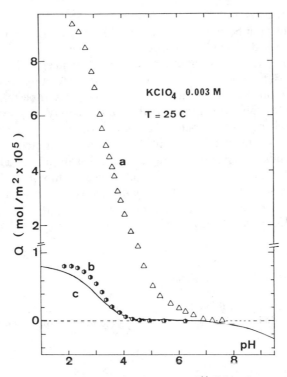

Figure 16. Surface titration of shocked labradorite (An_{60}, 10^{11} dislocations cm^{-2}): (a) unannealed sample; (b) sample annealed for 80 h at 1000°C; (c) theoretical titration curve for labradorite ($N_s = 12$ adsorption sites nm^{-2}).

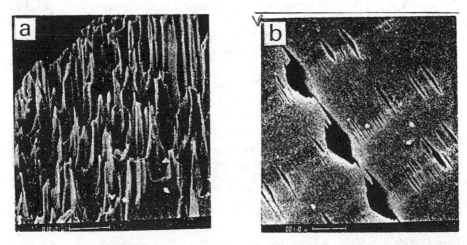

Figure 17. Examples of etching units found in silicates and resulting from (a) the specific adsorption of fluoride (teeth produced in the laboratory by treatment of hornblende with HF + HCl solutions) and (b) localized chemical impurities; lens-shaped etch pits in augite along amphibole exsolution lamella boundaries (likely to be produced by specific adsorption of reactants on Al sites).

remains roughly constant). This is not likely to occur in the case of dislocation-promoted dissolution because, as shown above, the strain energy is quickly and irreversibly released as hollow cores begin to open up, and also because transport control will tend to limit dissolution when steep etch pits are formed.

These findings, together with the observation that etch pits are developed in a similar manner on both deformed and undeformed samples of feldspar and calcite (e.g., see Murphy, 1989), indicate that etch pits may only be weakly related to dislocations. Probably, the dense etch pitting observed in natural samples of quartz and silicates must reflect their aqueous chemical environment (i.e., the presence of ligands, which considerably enhance dissolution) and the presence in these solids of localized chemical impurities such as aluminium, which favor the specific adsorption of F^- and organic ligands as oxalate, silicilate, and similar. This specific adsorption on chemical impurities may result in localized enhancements of dissolution as illustrated by Figure 17.

Acknowledgments

This is CNRS/INSU/DBT contribution number 103. Financial support was provided by the Institut National des Sciences de l'Univers (Thème "Fluides et Cinétique" of the Programme "Dynamique et Bilan de la Terre"). I am very grateful to Roland Wollast and David Crerar for their helpful comments and corrections on the manuscript.

REFERENCES

Aagard, P. and H. C. Helgeson (1982), "Thermodynamic and Kinetic Constraints on reaction among Minerals and Aqueous Solutions. I. Theoretical Considerations," *Am. J. Sci.* **282**, 237–285.

Abendroth, R. P. (1970), "Behavior of Pytogenic Silica in Single Electrolytes", *J. Colloid Interface Sci.* **34**, 591.

Amrhein, C. and D. L. Suarez (1988), "The Use of a Surface Complexation Model to Describe the Kinetics of Ligand-Promoted Dissolution of Anorthite," *Geochim. Cosmochim. Acta* **52**, 2795–2807.

Benson, S. W. (1976), *Thermochemical Kinetics*, 2nd ed., Wiley, New York, 320 pp.

Berner, R. A., and J. Schott (1982), "Mechanism of Pyroxene and Amphibole Weathering II. Observations of Soil Grains," *Am. J. Sci.* **282**, 1214–1231.

Berner, R. A., E. L. Sjöberg, M. A. Velbel and M. D. Krom (1980), "Dissolution of Pyroxenes and Amphiboles during Weathering," *Science* **207**, 1205–1206.

Blum, A. and A. C. Lasaga (1987), "Monte Carlo Simulations of Surface Reaction Rate Laws," in W. Stumm, Ed., *Aquatic Surface Chemistry*, Wiley-Interscience, New York, pp. 255–292.

Blum, A. and A. C. Lasaga (1988), "Role of Surface Speciation in the Low Temperature Dissolution of Minerals," *Nature* **331**, 431–433.

Brady, P. V. and J. V. Walther (in press), "Kinetics of Quartz Dissolution at Low Temperatures", *Chem. Geol.*

Casey, W. H., M. J. Carr and R. A. Graham (1988a), "Crystal Defects and the Dissolution Kinetics of Rutile," *Geochim. Cosmochim. Acta* **52**, 1545–1556.

Casey, W. H., H. R. Westrich, and G. W. Arnold (1988b), "Surface Chemistry of Labradorite Feldspar Reacted with Aqueous Solutions at pH = 2,3 and 12," *Geochim. Cosmochim. Acta* **52**, 2795–2807.

Chou L., and R. Wollast (1984), "Study of the Weathering of Albite at Room Temperature and Pressure with a Fluidized Bed Reactor," *Geochim. Cosmochim. Acta* **48**, 2205–2217.

Davis, J. A., R. O. James and J. O. Lechie (1978), "Surface Ionization and Complexation at the Oxide/Water Interface. I. Computation of Electrical Double Layer Properties in Simple Electrolytes," *J. Colloid Interface Sci.* **63**, 480–499.

Dove P., and D. A. Crerar (in press), "Kinetics of Quartz Dissolution in Electrolyte Solutions Using a Hydrothermal Mixed Flow Reactor," *Geochim. Cosmochim. Acta.*

Dugger, D. L., J. H. Stanton, B. N. Irby, B. L. McConnel, W. W. Cummings, and R. W. Maatman (1964), "The Exchange of Twenty Metal Ions with the Weakly Acid Silanol Group of Silica Gel," *J. Phys. Chem.* **68**, 757–760.

Eyring, H. (1935), "The Activated Complex in Chemical Reactions," *J. Chem. Phys.* **3**, 107–115.

Fleer, V. N. (1982), "The Dissolution Kinetics of Anorthite ($CaAl_2Si_2O_8$) and Synthetic Strontium Feldspar ($SrAl_2Si_2O_8$) in Aqueous Solutions at Temperatures below 100°C: With Applications to the Geological Disposal of Radioactive Nuclear Wastes," Ph.D. thesis, Pennsylvania State University, University Park, PA.

Fokkink L. G. L. (1987), "Ion Adsorption on Oxides," Ph.D. thesis, University of Wageningen, Netherlands.

Furrer, G., and W. Stumm (1986), "The Coordination chemistry of weathering· I Dissolution kinetics of δ-Al_2O_3 and BeO," *Geochim. Cosmochim. Acta* **48**, 2405—2432.

Grandstaff, D. E. (1981), "The Dissolution Rate of Forsteritic Olivine from Hawaiian Beach Sand," Third International Symposium on Water–Rock Interaction, *Proceedings*, Alberta Research Council, Edmonton, pp. 72–74.

Guy, C., and J. Schott (1989), "Multisite Surface Reaction versus Transport Control during the Hydrolisis of a Complex Oxide," *Chem. Geol.* **78**, 181–204.

Helgeson, H. C., W. M. Murphy and P. Aagaard (1984), "Thermodynamic and Kinetic Constraints on Reaction Rates among Minerals and Aqueous Solutions. II. Rate Constants, Effective Surface Area and the Hydrolysis of Feldspar, "*Geochim. Cosmochim. Acta* **48**, 2405–2432.

Holdren, G. R. Jr., and R. A. Berner (1979), "Mechanism of Feldspar Weathering. I. Experimental Studies," *Geochim. Cosmochim. Acta* **43**, 1161–1171.

Holdren, G. R., Jr., W. H. Casey, H. R. Westrich, M. Carr, and M. Boslough (1988), "Bulk Dislocation Densities and Dissolution Rates in a Calcic Plagioclase," *Chem. Geol.* **70**, 72 (abstr.).

Huang, C. P. (1971) "The Chemistry of the Aluminum Oxide–Electrolyte Interface," Ph.D. thesis, Harvard University, Cambridge, MA.

Johnson, R. E., Jr. (1984) "Thermodynamic Description of the Double Layer Surrounding Hydrous Oxides," *J. Colloid Interface Sci.* **100**, 540–554.

Knauss, K. G., and T. J. Wolery (1988), "The Dissolution Kinetics of Quartz as a Function of pH and Time at 70°C," *Geochim. Cosmochim. Acta* **52**, 43–54.

Lasaga, A. C. (1981), "Transition State Theory," A. C. Lagasa and R. J. Kirkpatrick, Eds.,

Kinetics of Geochemical Processes, Reviews in Mineralogy, Mineralogical Society of America, Vol. 8, 135–169.

Lasaga, A. C. (1983), "Kinetics of Silicate Dissolution," in Fourth International Symposium on Water–Rock Interaction, *Proceedings,* Misasa, Japan, pp. 269–274.

Lasaga, A. C. (1984), "Chemical Kinetics of Water–Rock Interactions," *J. Geophys. Res.* **89,** (B6), 4009–4025.

Murphy, W. M. (1985), "Thermodynamic and Kinetic Constraints on Reaction Rates among Minerals and Aqueous Solution," Ph.D. thesis, University of California, Berkeley.

Murphy, W. M. (1989), "Dislocations and Feldspar Dissolution," *Eur. J. Mineral.* **1,** 315–326.

Pederson, L. R., D. R. Baer, G. L. McVay, and M. H. Engelhard (1986), "Reaction of Soda Lime Silicate Glass in Isotopically Labelled Water," *J. Non-Cryst. Solids* **86,** 369–380.

Petit, J. C., G. Della Mea, J. C. Dran, J. Schott, and R. A. Berner, (1987)," Diopside Dissolution: New Evidence from H-Depth Profiling with a Resonant Nuclear Reaction," *Nature* **325,** 705–706.

Petit, J. C., J. C. Dran, A. Paccagnella, and G. Della Mea (1989), "Structural Dependence of Crystalline Silicate Hydration during Aqueous Dissolution," *Earth and Planet. Sci. Lett.* **93,** 292–298.

Petrovic, R., R. A. Berner, and M. B. Goldhaber (1976), "Rate Control in Dissolution of Alkali Feldspar. I. Study of Residual Grains by X-Ray Photoelectron Spectroscopy", *Geochim. Cosmochim. Acta* **40,** 537–548.

Pulfer, K., P. W. Schindler, J. C. Westall, and R. Grauer (1984) "Kinetics and Mechanism of Dissolution of Bayerite (γ-Al(OH)$_3$) in HNO$_3$–HF solutions at 298.2 K," *J. Colloid Interface Sci.* **101,** 554–564.

Schott, J. and R. A. Berner (1983), "X-Ray Photoelectron Studies of the Mechanism of Iron Silicate Dissolution during Weathering," *Geochim. Cosmochim. Acta* **47,** 2333–2340.

Schott, J. and R. A. Berner (1985), "Dissolution Mechanisms of Pyroxenes and Olivines during Weathering", in J. I. Drever, Ed., *The Chemistry of Weathering,* Reidel, Dordrecht, pp. 35–53.

Schott, J., and J. C. Petit (1987), "New Evidence for the Mechanisms of Dissolution of Silicate Minerals," in W. Stumm, Ed., *Aquatic Surface Chemistry,* Wiley-Interscience, New York, pp. 293–312.

Schott J., R. A. Berner, and E. L. Sjöberg (1981), "Mechanism of Pyroxene and Amphibole Weathering: I. Experimental Studies of Iron-Free Minerals," *Geochim. Cosmochim. Acta* **45,** 2123–2135.

Schott, J., S. Brantley, D. Crerar, C. Guy, M. Borcsik, and C. Willaime (1989), "Dissolution Kinetics of Strained Calcite," *Geochim. Cosmochim. Acta* **53,** 373–382.

Sigg, L. and W. Stumm (1981), "The Role of Surface Weak Acids with the Hydrous Geothite Surface," *Colloids Surf.* **2,** 101–117.

Stumm, W. and G. Furrer (1987), "The Dissolution of Oxides and Aluminum Silicates; Examples of Surface-Coordination-Controlled Kinetics," in W. Stumm, Ed., *Aquatic Surface Chemistry,* Wiley-Interscience, New York, pp. 197–219.

Webb, S. C., and J. V. Walther (1988), "A Surface Complex Reaction Model for the pH-Dependence of Corundum and Kaolinite Dissolution Rates," *Geochim. Cosmochim. Acta* **50,** 1861–1869.

Webb, S. C., and J. V. Walther (in press) "Temperature Dependence of Kaolinite Dissolution," *Am. J. Sci.*

Wieland, E. and W. Stumm (1989), "The Dissolution Kinetics of Kaolinite" (Abstr.). Workshop on Aquatic Chemical Kinetics, Kartause Ittingen (Switzerland), March 19–23, 1989.

Wieland, E., B. Wehrli, and W. Stumm (1988), "The Coordination Chemistry of Weathering: III. A Generalization on the Dissolution Rates of Minerals," *Geochim. Cosmochim. Acta* **52**, 1969–1981.

Wirth, G. S., and J. M. Gieskes (1979), "The Initial Kinetics of the Dissolution of Vitreous Silica in Aqueous Media," *J. Colloid Interface. Sci.* **68**, 492–500.

Wollast, R. (1967), "Kinetics of the Alteration of K-Feldspar in Buffered Solutions at Low Temperature," *Geochim. Cosmochim. Acta* **31**, 635–648.

Wollast, R, and L. Chou (1985), "Kinetic Study of the Dissolution of Albite with a Continuous Flow-through Fluidized Bed Reactor," in J. I. Drever, Ed., *The Chemistry of Weathering*, Reidel, Dordrecht, pp. 75–96.

Yates, D. E. (1975), "The Surface of the Oxide/Aqueous Electrolyte interface," Ph.D. thesis, University of Melbourne, Melbourne, Australia.

Zarzycki, J. (1982), *Les verres et l'Etat vitreux*, Masson, Paris, 391 pp.

Zinder B., G. Furrer and W. Stumm (1986), "The Coordination Chemistry of Weathering. II. Dissolution of Fe(III) Oxides," *Geochim. Cosmochim. Acta* **50**, 1861–1869.

13

DISSOLUTION OF OXIDE AND SILICATE MINERALS: RATES DEPEND ON SURFACE SPECIATION

Werner Stumm and Erich Wieland

Institute for Water Resources and Water Pollution Control (EAWAG), Dübendorf, Switzerland; Swiss Fedral Institute of Technology (ETH), Zürich, Switzerland

1. INTRODUCTION

Most chemical reactions that occur in natural waters occur at solid–solution interfaces. Chemical weathering processes, essentially caused by the interaction of water and atmosphere with the earth's crust, transform primary minerals into solutes and soils, and eventually into sedimentary rocks. These processes participate in controlling the global hydrogeochemical cycles of many elements. Biota exerts its influence on the weathering reaction by providing, at local sites, protons and CO_2 (as a consequence of respiration) and surface complex-forming ligands and reductants. Such metabolic products are highly effective in promoting reductive and nonreductive dissolution of oxide minerals.

The objectives of this chapter are (1) to illustrate that the surface structure is important in characterizing surface reactivity and that kinetic mechanisms depend on the coordinative environment of the surface groups, (2) to derive a general rate law for the surface-controlled dissolution of oxide and silicate minerals and illustrate that such rate laws are conveniently written in terms of surface species, and (3) to illustrate a few geochemical implications of the kinetics of oxide dissolution.

1.1. Coordinative Properties of Surface

Figure 1 illustrates two pertinent models of the distribution of species at the mineral–water interface. In Figure 1*a* the surface is treated as if it were an

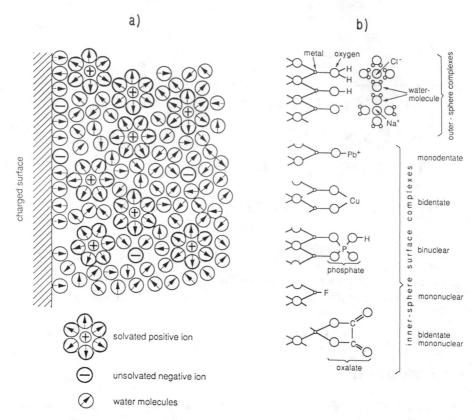

a)

b)

solvated positive ion

unsolvated negative ion

water molecules

Figure 1. Schematic representations of two models on the structure and ionic composition of the mineral–water interface. (*a*) The solid surface is treated like an electrode. The solvated ions present in a diffuse ion swarm neutralize the surface charge in a diffuse double layer. (*b*) The mineral surface, represented here by an oxide surface, is covered in the presence of water by amphoteric surface hydroxyl groups >M—OH that can interact with protons and metal ions. The underlying central ion in the surface layer of the oxide acting as a Lewis acid can exchange its structural OH ions against other ligands (anions or weak acids). The extent of surface coordination and its pH dependence can be quantified by mass-action equations and can be explained by considering the affinity of the surface sites for metal ions or ligands and the pH dependence of the activity of surface sites and ligands. The reactivity depends on the type of surface species present. The diffuse ion swarm and the outer-sphere surface complexation mechanisms of adsorption involve almost exclusively electrostatic interactions, whereas inner-sphere complex mechanisms involve ionic as well as covalent bonding.

electrode. The distribution of ions is influenced by electrostatic interaction; some ions may become specifically[†] adsorbed. This *electric double-layer* model (Stern–Gouy–Chapman) has been very useful in describing the distribution of charges on the solution side and in predicting many phenomena of colloid stability. However, the selectivity of interactions of hydrous oxide surfaces with

---

† The term "specific adsorption" is used whenever adsorption occurs by interactions that are not due solely to electrostatic interactions.

many solute species (H^+, OH^-, metal ions, and ligands) can be accounted for only by considering specific chemical interactions at the solid surface. In Fig. 1*b* the *surface complex-formation model* is represented schematically. Hydrous solid surfaces contain functional groups (e.g. hydroxo groups) Me—OH, at the surfaces of oxides of Si, Al, and Fe. These surface functional groups represent enormous facilities for the specific adsorption of cations and anions.

The pH-dependent charge of a hydrous oxide surface results from proton transfers at the surface. The surface OH groups represent σ-donor groups and are, like their counterparts in solution, able to form complexes with metal ions (Stumm et al., 1970, 1976; Schindler et al., 1976, Hohl and Stumm, 1976). The central ion acting as a Lewis acid can exchange its structural OH^- ion against other ligands (anions on weak acids, ligand exchange) (Stumm et al., 1980; Sigg and Stumm, 1981). The concept of surface complex formation has been extensively documented in recent reviews (Sposito, 1983; Schindler and Stumm, 1987). The extent of surface coordination and its pH dependence can be quantified by *mass-action equations*. The equilibrium constants are conditional stability constants—at constant temperature, pressure, and ionic strength–the values of which can be corrected for electrostatic interaction (Schindler and Stumm, 1987). The nature and bonding patterns of surface complexes have far-reaching implications for the mechanism of interfacial processes and their kinetics (Wehrli, Chapter 11, this volume; Stumm et. al., in press).

Figure 2 gives a scheme illustrating some applications in geochemistry and technology where surface reactivity (kinetics of dissolution, catalytic activity, photochemical activity) depends on surface structure, expecially on surface coordination. It has been shown by various spectroscopic techniques [electron-spin resonance (ESR), electron double-resonance spectroscopy (ENDOR), electron-spin echo modulation (e.g., see Motschi, 1987), Fourier transform infrared spectroscopy (Zeltner et al., 1986), and *in situ* X-ray absorption studies of surface complexes (EXAFS) (Hayes et al., 1987; Brown, 1989)] that inner-sphere

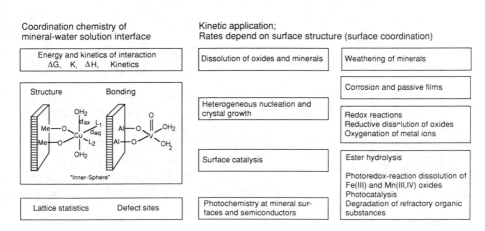

Figure 2. Kinetic applications of the coordination chemistry at the mineral–water interface.

complexes with metal ions and ligands are formed under many different conditions. Thus, a solute can become adsorbed as an inner-sphere complex, as an outer-sphere complex or within the ion swarm of the diffuse double layer and, as has been shown by Wehrli et al. (1989), and by Wehrli in this Chapter 11 of volume, the reactive properties of adsorbed species are critically dependent on the type of interaction.

2. THE RATE-DETERMINING STEP IN DISSOLUTION KINETICS; THE *SURFACE SPECIES* AS THE PRECURSOR OF THE ACTIVATED COMPLEX

We would like to provide the reader first with a *qualitative* understanding of the subject of dissolution kinetics.

The dissolution of a mineral is a sum of chemical and physical reaction steps. If the chemical reactions at the surface are slow in comparison with the transport (diffusion) processes, the dissolution kinetics is controlled by one step in the chemical surface processes; thus, rates of transport of the reactants from the bulk solution to the surface and of products from the surface into the solution can be neglected in the overall rate. It has been shown by Petrovic et al. (1976) and Berner and Holdren (1979) that the dissolution of many minerals, especially under conditions encountered in nature, are surface-controlled.

In the dissolution reaction of a metal oxide, the coordinative environment of the metal changes; for example, in dissolving an aluminum oxide, the Al^{3+} in the crystalline lattice exchanges its O^{2-} ligand for H_2O or another ligand L. The most important reactants participating in the dissolution of a solid mineral are H_2O, H^+, OH^-, ligands (surface complex building), reductants, and oxidants (in case of reducible or oxidizable minerals).

Thus the reaction occurs schematically in two sequences:

$$\text{Surface sites} + \text{reactants } (H^+, OH^- \text{ or ligands}) \xrightarrow{\text{fast}} \text{surface species} \quad (1)$$

$$\text{Surface species} \xrightarrow[\text{detachment of Me}]{\text{slow}} Me(aq) \quad (2)$$

Although each sequence may consist of a series of smaller reaction steps, the rate law of surface-controlled dissolution is based on the assumptions that (1) the attachment of reactants to the surface sites is fast and (2) the subsequent detachment of the metal species from the surface of the crystalline lattice into the solution is slow and thus rate-limiting. In the first sequence, the dissolution reaction is initiated by the surface coordination with H^+, OH^-, and ligands that polarize, weaken, and break the metal–oxygen bonds in the lattice of the surface. Figure 3 gives a few examples on the surface configurations that enhance or inhibit dissolution. Since reaction 2 is rate-limiting, the rate law on the dissolution reaction will show a dependence on the concentration (activity) of the

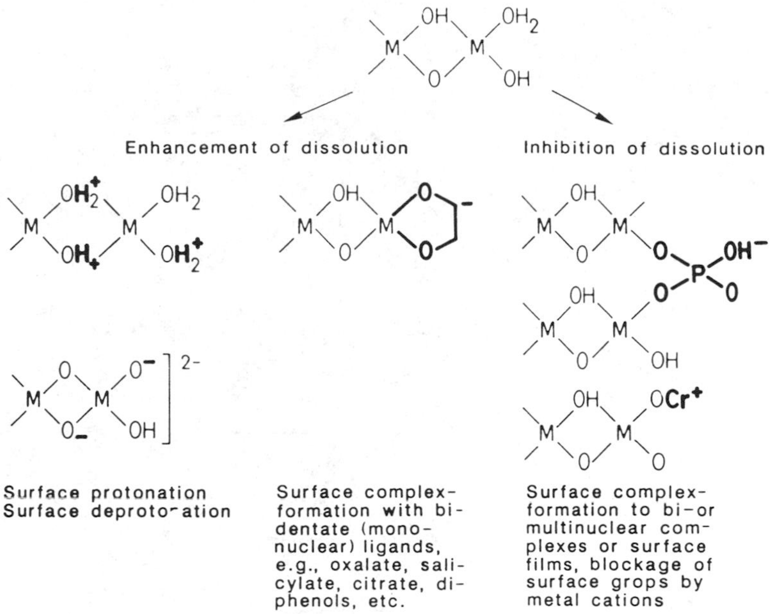

Figure 3. Possibilities for the dependence of surface reactivity and of kinetic mechanisms on the coordinative environment of the surface groups.

particular surface species.

$$\text{Dissolution rate} \propto \{\text{surface species}\} \tag{3a}$$

We reach the same conclusion (Eq. 3a) if we treat the reaction sequence according to the transition-state theory (see Stone and Morgan, Chapter 1 in this book). The particular surface species that has formed from the interaction of H^+, OH^-, or ligands with surface sites is the precursor of the activated complex (Fig. 4):

$$\text{Dissolution rate} \propto \{\text{precursor of the activated complex}\} \tag{3b}$$

The surface concentration of the particular surface species (which is equal to the concentration of the precursor of the activated complex) (Eq. 3) can usually be determined from the knowledge of the number of surface sites and the extent of surface protonation or surface deprotonation or the surface concentration of ligands. Surface protonation or deprotonation can be measured from alkalimetric or acidimetric surface titrations, and ligands bound to the surface sites can be determined analytically, from the change in the concentration of ligands in solution.

There is much experimental evidence (e.g., Hachiya et al., 1984, Hayes and Leckie, 1986) that surface complex formation reactions are usually fast reactions;

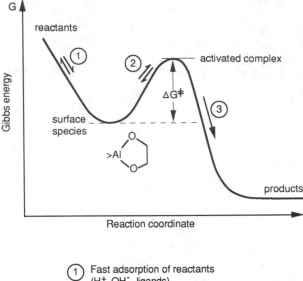

① Fast adsorption of reactants
   (H⁺, OH⁻, ligands)

② Fast activation of the surface species

③ Slow detachment of the activated
   surface complex

Figure 4. Schematic representation of the activated-state theory. The precursor to the activated state in dissolution reactions is a surface species. In the dissolution of Al_2O_3 (cf. Fig. 5) an oxalato surface complex is the precursor of the activated state.

the order of their reaction rates corresponds to that of the rate constants for the release of a water molecule from the hydrated metal ions in homogeneous metal complex systems.

A simple example.   The ligand-promoted dissolution of a metal oxide such as Al_2O_3 is given in Figure 5. The shorthand representation of a surface site as given in this figure is a simplification that takes into account either the detailed structural aspects of the oxide surface nor the oxidation state of the metal ion and its coordination number. The scheme in Figure 5 indicates that the ligand—for example, oxalate—is bound very rapidly, in comparison to the dissolution reaction:

$$\text{Me} \begin{array}{c} OH_2 \\ \\ OH \end{array} + C_2O_4^{2-} + H^+ \rightleftharpoons \text{Me} \begin{array}{c} O-C\diagup^O \\ \bigg| \\ O-C\diagdown_O \end{array} + 2H_2O \tag{4}$$

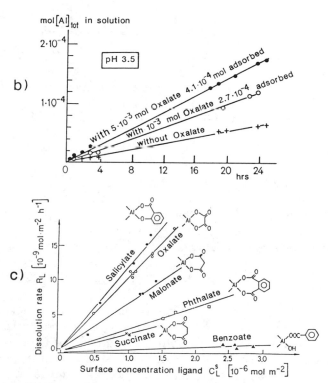

Figure 5. (*a*) The ligand-catalyzed dissolution reaction of a M_2O_3 can be described by three elementary steps: a fast ligand adsorption step (equilibrium), a slow detachment process, and fast protonation subsequent to detachment restoring the incipient surface configuration. (*b*) In accordance with the reaction scheme of (*a*) the rate of ligand-catalyzed dissolution of $\delta\text{-Al}_2O_3$ by the aliphatic ligands oxalate, malonate, citrate, and succinate, R_L (nmol $m^{-2} h^{-1}$), can be interpreted as a linear dependence on the surface concentrations of chelate complexes, C_L^s.

The scheme in Figure 5 corresponds to a steady-state condition; this also implies that the active sites are continuously regenerated after the metal detachment.

Different ring structures are differently effective in promoting the dissolution of Al_2O_3 (Furrer and Stumm, 1986):

$$k_{oxalate} > k_{malonate} > k_{succinate} \tag{5}$$

$$k_{salicylate} > k_{phthalate} > k_{benzoate} \tag{6}$$

The rate constants in Eqs. 5 and 6 have the same sequence as the corresponding stability contants with Al^{3+} in solution. It has been shown before (Sigg and Stumm, 1981) that the stability of surface complexes correlate with the stability of corresponding solute complexes; for instance, the equilibrium constants of the reactions

$$>Al-OH + HA \rightleftharpoons >AlA + H_2O; \quad *K_1^s \tag{7a}$$

$$AlOH^{2+} + HA \rightleftharpoons AlA^{2+} + H_2O; \quad *K_1 \tag{7b}$$

can be related in the sense of a linear free-energy relationship; a plot of $\log *K_1^s$ versus $\log *K_1$ for various complex formers, HA, gives a linear regression with a slope of 1 (Sigg and Stumm, 1981).

3. A GENERAL RATE LAW FOR DISSOLUTION

A general rate law for the dissolution of minerals is derived by considering, in addition to the surface coordination chemistry, established models of lattice statistics and activated-complex theory:

$$R = k \chi_a P_j S \tag{8}$$

where R is the proton- or ligand-promoted dissolution rate (mol $m^{-2} s^{-1}$); χ_a denotes the mole fraction of dissolution active sites, P_j represents the probability of finding a specific site in the coordinative arrangement of the precursor complex, and S is the surface concentration of sites (mol m^{-2}). Surface complexes (surface chelates and proton complexes with the central cation) are precursors in the rate-limiting detachment of a central metal ion from the surface into the solution. In Eq. 8 k represents the appropriate rate constant (second^{-1}) and is related to the activation energy of conversion of a suitable surface complex (precursor) to an activated surface complex and, in turn, to the crystal bond energy that has to be broken.

The precursor concentration C_p^s corresponds to

$$C_p^s = \chi_a P_j S \tag{9}$$

In ligand-promoted dissolution of oxides (Furrer and Stumm, 1986) the probability, P_j, of finding a surface site in the form of a surface–ligand complex, ML, is given by

$$P_j \propto \{>\text{ML}\} = C_L^s \tag{10}$$

and the ligand-promoted dissolution rate becomes

$$R_L = k'\{>\text{ML}\} = k'C_L^s \tag{11}$$

where $\{>\text{ML}\}$ is the concentration of surface-bound (adsorbed) ligands.

In acid (proton)-promoted dissolution of oxides the probability of finding a surface site in the form of a precursor configuration is proportional to the surface concentration of protonated sites to the power j

$$P_j \propto \{>\text{MOH}_2^+\}^j = (C_H^s)^j \tag{12}$$

and the rate of dissolution is

$$R_H = k'_H\{>\text{MOH}_2^+\}^j = k'_H(C_H^s)^j \tag{13}$$

where $\{>\text{MOH}_2^+\}$ is the concentration of protonated surface hydroxo group $[=C_H^s(\text{mol m}^{-2})$ in excess relative to zero proton charge], and j is an exponent that arbitrarily corresponds to the oxidation state of the central metal ion in the crystalline lattice, for example, $j=3$ for Al, Fe(III); $j=2$ for Be(II), Mg(II); $j=4$ for Si (Fig. 7, below).

Figure 6 schematically depicts the proton-promoted dissolution of hydrous oxides (e.g., Al_2O_3). In fast initial steps, the protons become bound to the surface hydroxyl groups or to the oxide ions closest to the metal center at the surface of the lattice. Subsequent to surface protonation, the detachment of the metal ion from the surface is the slowest of the consecutive steps. Therefore, the rate of the proton-promoted dissolution, R_H, is proportional to the concentration (activity) of the surface species D(Fig. 6).

Equation 8 needs some explanation. The particular surface species in the rate-determining step (or the precursor) (Eqs. 2, 3) of the proton-promoted dissolution depends on the geometric coordinative arrangement of the bound surface protons. The protons may be assumed to be relatively mobile in the surface layer; that is, they can be shifted from surface OH groups to neighboring oxygen bridges (as in tautomerism). A random distribution of surplus protons in the surface layer may be assumed. We need a quantitative relationship between P_j and χ_H, where χ_H is the mole fraction of surface sites that carry an additional bound proton and P_j denotes the probability of finding a reactive precursor configuration. The question is: How does the probability P_j of a metal center to be surrounded by $j(0,1,2,3,4)$ protonated functional groups depend on the concentration of bound protons or surface protonation? *Lattice statistics*

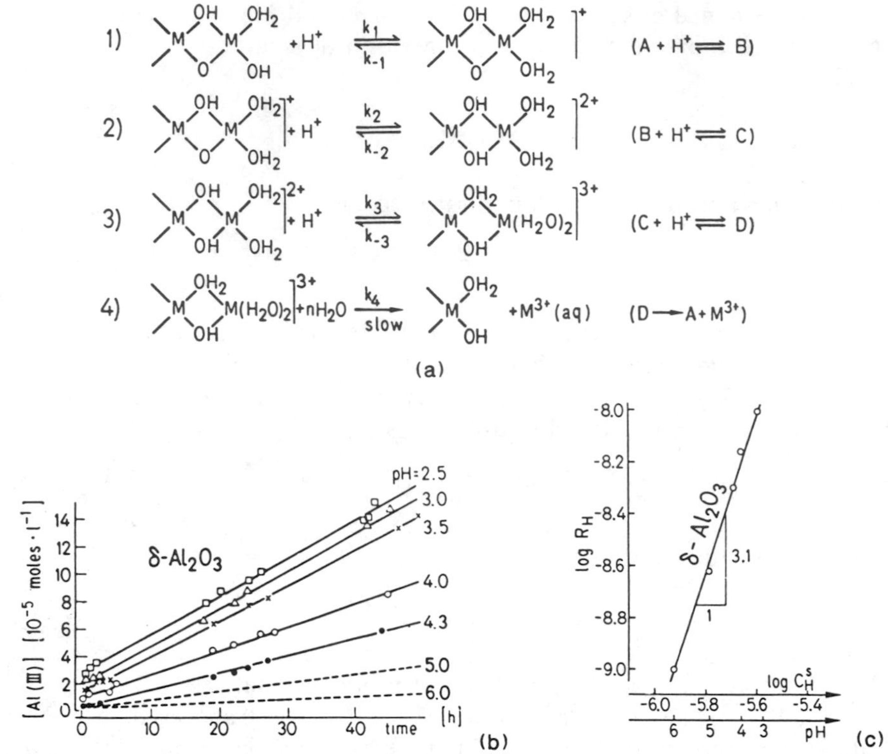

Figure 6. (a) Schematic representation of the proton-promoted dissolution process at a M_2O_3 surface site. Three preceding fast protonation steps are followed by a slow detachment of the metal from the lattice surface. (b) The reaction rate derived from individual experiments is proportional to the surface protonation to the third power.

provide a simple framework to estimate probabilities (mole fractions) of such geometric arrangements. The probability P_j is given by the Bernoulli scheme (De Finetti, 1974; Wieland et al., 1988):

$$P_j = \frac{4!}{j!(4-j)!} \chi_H^j (1 - \chi_H)^{4-j} \tag{14}$$

In the case of Al_2O_3 and BeO, where $\chi_H \ll 1$, Eq. 14 simplifies to

$$P_j \cong \chi_H^j; \quad P_j \propto (C_H^s)^j \tag{15}$$

and the general rate law expressed in Eq. 8 simplifies to the rate law derived by Furrer and Stumm (1986) (Eq. 13).

The dissolution rate of most oxides increases both with increasing surface protonation and with *decreasing deprotonation* (or hydroxylation). In this case, the probability of finding a precursor configuration is

$$P_j \propto \{>MO^-\}^i = (C_{OH}^s)^i \tag{16}$$

where $\{>MO^-\}$ is the concentration of deprotonated surface groups $(=C_{OH}^s \ (mol \ m^{-2})$ in excess relative to zero proton charge) and i is an exponent. The rate law of the hydroxide-promoted dissolution becomes:

$$R_{O-H} = k_{O-H} \{>MO^-\}^i = k_{OH}(C_{OH}^s)^i \tag{17}$$

As this equation illustrates, the rate is related to the negative charge imparted to the surface either by binding of OH^- (as a ligand) or by deprotonation. For possible mechanisms, see Casey et al. (1989).

The overall rate of dissolution is given by

$$R = k_H'(C_H^s)^j + k'(C_{O-H}^s)^i + k_L'(C_L^s) + k_{H_2O}' \tag{18}$$

which is the sum of the individual reaction rates, assuming that the dissolution

---

TABLE 1. Model Assumptions

---

1. Dissolution of slightly soluble hydrous oxides:
    Surface process is rate-controlling
    Back reactions can be neglected if far away from equilibrium
2. The hydrous oxide surface, as a first approximation, is treated like a cross-linked polyhydroxo- oxo acid:
    All functional groups are identical
3. Steady state of surface phase:
    Constancy of surface area
    Regeneration of active surface sites
4. Surface defects, such as steps, kinks, and pits, establish surface sites of different activation energy, with different rates of reaction:

    Active sites $\xrightarrow{\text{faster}}$ Me(aq)   (a)

    Less active sites $\xrightarrow{\text{slower}}$ Me(aq)   (b)

    Overall rate is given by (a):
    Steady-state condition can be maintained if a constant mole fraction, x_a, of active sites to total (active and less active) sites is maintained, i.e. if active sites are continuously regenerated
5. Precursor of activated complex:
    Metal centers bound to surface chelate, or surrounded by n protonated functional groups
    $(C_H^s/S) \ll 1$

---

occurs parallel at different metal centers (Furrer and Stumm, 1987). Table 1 gives the assumptions that are implied in the rate laws given.

4.   SURFACE MORPHOLOGY AND DISSOLUTION KINETICS

An assessment of linear free-energy relations (Wieland et al., 1988) has shown that the Madelung energy is the most promising parameter for estimating activation energies of the dissolution process. The Madelung energy is defined as the energy that would be required to separate a particular ion (point charge) from its equilibrium position in a crystalline structure to an infinite distance. Concerning the effects of various reactive surface sites (kinks, steps, edges) on the dissolution, Wehrli (1989) investigated the interdependence on surface morphology and dissolution kinetics with Monte Carlo simulations. He found that a simple partition function describes the distribution of different types of surface sites as a function of $\Delta H^{\#}$, the activation enthalpy of dissolution. The same parameter determines the surface roughness. These results indicate a zero-order dissolution rate and that the steady-state morphology of dissolving mineral particles may be determined by the kinetic process.

5.   CASE STUDIES

The ligand-promoted dissolution is illustrated by the effect of bidendate chelate formers seen in Figure 5. An example of the acid-promoted dissolution of Al_2O_3 and the dependence of some types of oxides on surface protonation is given in Figures 6 and 7.

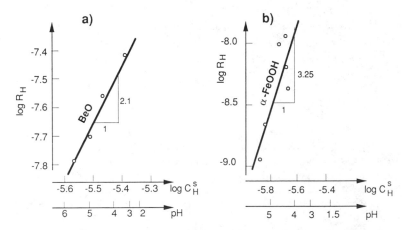

Figure 7. Rate of dissolution depends on surface protonation or surface deprotonation (Eqs. 13, 18). For Al_2O_3, see Figure 6*b*.

5.1.   Silica, Quartz, Basalt Glass, and UO$_2$

Most oxides show, in accordance with the general rate equation (Eq. 18) the same trend with regard to the rate dependence on pH: a decrease in pH in the acid range and an increase in pH in the alkaline range. Although the dissolution of SiO$_2$(s)

$$SiO_2 + 2H_2O = H_4SiO_4(aq) \tag{19}$$

does not involve protons or OH$^-$ ions, the rate of solution is dependent on pH, as shown in Figure 8. The pH of ZPC (zero point charge due to H$^+$ or OH$^-$) is around pH = 3. As Figure 8 illustrates, both the positive surface charge, due to bound proton, and the negative surface charge, due to deprotonation (equivalent to bound OH$^-$), enhance the dissolution rate. Guy and Schott (in press) plotted careful titration curves on amorphous silica and determined the sites that are protonated or deprotonated at each pH level and showed that four successive protonation steps in the acid range and four deprotonation steps in the alkaline range are necessary to detach Si from the silica structure. This interpretation reinforces the idea that the reaction order of proton-promoted dissolution of oxides reflects the oxidation number of the central ion in the oxide.

Schott (Chapter 12, this volume) applies the speciation approach to the dissolution of *basalt glass*. As Schott points out, such glasses have little or no structural ordering and thus interpretation may be rendered more simple than in crystalline mixed oxides (e.g., olivine and feldspars). The dissolution behavior of

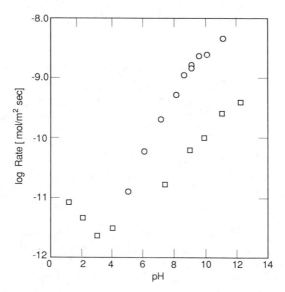

Figure 8. Rate of dissolution of silica as a function of pH: ○, vitreous silica in 0.7 M NaCl solution (Wirth and Gieskes. 1979); □, Quartz in 0.2 M NaCl solution (Wollast and Chou, 1985).

complex oxides can be modeled from the properties of their constituent oxide components. The chemical speciation of the basalt glass surface was deduced from surface titration data. As Schott's Figure 7 (in Chapter 12) illustrates, linear regressions with a slope of 3.7–3.8 are observed in both acid and alkaline solutions in logarithmic plots versus the surface charge given by C_H^s in the acid or by C_{OH}^s in the alkaline range. The numerical values of these slopes are representative for a charge of 4 for Si.

Bruno (personal communication, 1989) points out that the dissolution of $UO_2(c)$ under reducing conditions $(H_2(g)/Pd)$ at $25°C$ showed a rate-order dependence on $[H^+]^{-0.3}$, which suggests a hydroxo-promoted dissolution reaction via the formation of a surface complex with OH^- at the UO_2 surface (or a corresponding deprotonation of surface OH groups).

5.2. HCO_3^- Enhances the Dissolution Rate of Hematite

Hydrogen carbonate and carbonate from complexes with Fe(III) in solution— $(Fe^{III}(CO_3)_2^-$ —predominates in seawater and at the surface of iron(III) (hydr)oxides (Bruno, personal communication, 1989). Figure 9 shows the dependence of the dissolution rate as a function of the bicarbonato surface complex:

$$Rate \propto \{ >Fe_2O_3 - HCO_3^- \} \tag{20}$$

5.3.   Silicates: Olivine and Albite

The weathering of silicates has been investigated extensively in recent decades. Interpretation of the dissolution reaction is rendered more involved in the case of

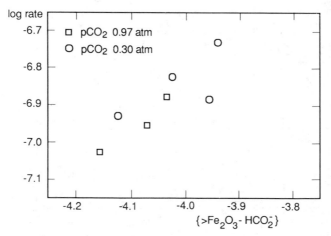

Figure 9. Dependence of the dissolution rate of hematite, α-Fe_2O_3, on the surface concentration of the HCO_3^- complex.

crystalline mixed oxides, such as olivine and feldspars because, understandably, it is more difficult to characterize the surface chemistry of crystalline mixed oxides. Furthermore, in many instances the dissolution of a silicate mineral is incipiently incongruent. As Schnoor explains in this volume, (Chapter 17) this initial incongruent dissolution step is followed by a congruent dissolution controlled surface reaction. Wollast and Chou (1985) and Blum and Lasaga (1988) determined the influence of pH on the dissolution rate of albite and olivine, respectively. This rate dependence is typical and illustrates the enhancement of dissolution rate by surface protonation and surface deprotonation. A zero-order dependence on $[H^+]$ has often been reported near the pH_{ZPC}; this is generally interpreted in terms of a hydration reaction of the surface (see Eq. 18).

5.4. Silicates: Kaolinite and Muscovite

All studies concerning the weathering of phyllosilicates provide clear evidence that the dissolution of clay minerals and micas is very slow under natural conditions (Correns, 1963; Polzer and Hem, 1965; Calvera and Talibudeen, 1978; t'Serstevens et al., 1978; Lim and Clemency, 1981; Carroll-Webb and Walther, 1988). Hence they are characteristic secondary weathering products occurring in soils and sediments.

The main feature of clay minerals and micas is the layered crystallographic structure. Muscovite is a 2:1[tetrahedral–octahedral–tetrahedral(T–O–T)] phyllosilicate. In an ideal structure, aluminum exists in the octahedral sheet ($\equiv O$) between two tetrahedral sheets ($\equiv T$), whose cations are composed of 25% Al and 75% Si. Interlayer K^+ cations balance the resulting negative charge (see schematic representation in Fig. 12, below).

Kaolinite is a 1:1 (T–O) phyllosilicate. The fundamental unit of its structure is an extended sheet of two constituents: a silica-type layer of composition $(Si_4O_{10})^{4-}$ and a gibbsite-type layer of composition $(OH)_6Al_4(OH)_2O_4$ (see schematic representation in Fig. 10). Ideally, kaolinite crystals are not permanently charged. However, due to isomorphic substitution of Si by Al at the siloxane surface, kaolinite platelets carry a small, permanently negative charge (Van Olphen, 1977). Lim et al. (1980) and Talibudeen (1984) postulate that the permanent charge of kaolins is caused by contamination with small amounts of 2:1 phyllosilicates rather than a consequence of isomorphic substitution.

5.4.1. The Ligand-Promoted Dissolution

Only a few studies are published concerning the effect of complex-forming ligands, such as low-molecular-weight organic acids and fulvic acids, on the dissolution of phyllosilicates. Schnitzer and Kodama (1976) showed that the decomposition of mica is not significantly enhanced in the presence of fulvic acid. Complex-forming ligands (e.g., aspartic, citric, salicylic, tartaric and tannic acids) promote the dissolution of clay minerals and change the stoichiometry of Al and Si release (Huang and Keller, 1971).

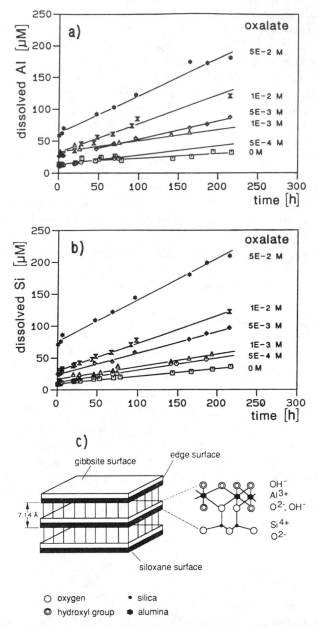

Figure 10. The oxalate-promoted dissolution of kaolinite. Both Al and Si detachment is promoted in the presence of oxalate. The linear increase of Al and Si concentrations represents the zero-order kinetics indicating the homogenity of surface sites. Figure 10c displays a schematic representation of the kaolinite structure. It reveals the 1:1 structure due to the alternation of silica-type (black) and gibbsite-type layers (white).

Carroll–Webb and Walther (1988) studied the dissolution of kaolinite in the presence of ligands such as phthalate, succinate, malonate, and tris. As the scatter in the data was of the same order as the effect of the increased dissolution, the authors concluded that in their experiments the long-term dissolution of kaolinite was not affected by organic ligands. However, as was shown in Figure 5, organic ligands may form surface chelate complexes that enhance the dissolution rate of δ-Al_2O_3. Thereby oxalate, which forms five-membered rings, and salicylate, which forms six-membered rings with surface Al centers, are the most effective ligands in the homologous series of dicarboxylic acids and aromatic acids, respectively, and have been selected for our studies. The dissolution of kaolinite at pH 4 and in the presence of various concentrations of oxalate is shown in Figure 10. The rates of kaolinite dissolution R_{Al} and R_{Si} were determined from the increase of Al and Si concentrations in solution, respectively. The concentrations increase linear with time and, hence, follow a zero-order kinetics. Increasing concentrations of oxalate enhance the detachment of both Al and Si. The question is whether the proton- and the ligand-promoted dissolution reactions occur simultaneously on different surface sites. However, a better correlation between the ligand-promoted dissolution rate, R_{Ox}, and the surface concentrations of oxalate, C_{Ox}^s, is achieved if we assume that the proton- and ligand-promoted dissolution are competitive reactions occurring at the same surface sites:

$$R_{Ox} = R_T - k_n(C_j^s - C_{Ox}^{'s}) \tag{21}$$

where R_T and R_{Ox} are the overall and the ligand-promoted dissolution rate, respectively; k_n denotes the rate constant of the proton-promoted dissolution reactions: and C_j^s and C_{Ox}^s represent the surface concentrations of precursor (protonated) configurations and of oxalate, respectively. In Eq. 9 C_j^s is given by $C_j^s = SP_j$.

Figure 11 gives the correlations between the rate of oxalate promoted dissolution and the surface oxalate concentrations. It indicates that there is a linear correlation between the detachment of Si from the kaolinite surface and the oxalate concentration; hence, a pH-independent rate constant $k_{Ox,Si}'$ can be evaluated (Table 2). Although the rates of the oxalate-promoted detachment of Al, $R_{Ox,Al}$, are linearly proportional to increasing surface oxalate concentrations, the evaluated rate constants $k_{Ox,Al}$ significantly depend on pH (Fig. 11). The ability of oxalate to remove Al from the kaolinite surface decreases with decreasing H^+ activity in solution. It is interesting to notice that the apparent rate constant k_{Ox} of oxalate-promoted dissolution of δ-Al_2O_3 takes a pH-independent value of 10.8×10^{-3} h^{-1} (Table 2) (Furrer and Stumm, 1986).

The effect of salicylate was explored in a few experiments at pH 2, 4, and 6. Salicylate promotes the dissolution of kaolinite at pH > 4. At pH 2 it reveals no catalytic effect due to protonation of the surface-bound chelates (Furrer and Stumm, 1986). The rate constant of salicylate-promoted dissolution was determined from Si release for pH > 4 (Table 2).

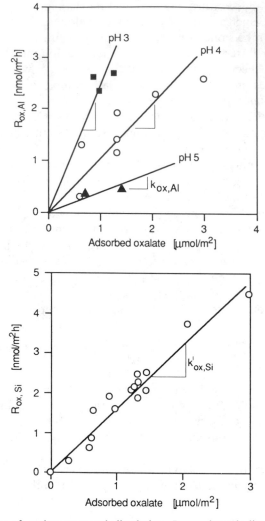

Figure 11. The rate of oxalate-promoted dissolution, $R_{Ox,Al}$, depends linearly on the surface concentration of oxalate. The pH-independent rate constant, $k_{Ox,Si}$, determined from the oxalate-promoted detachment of Si, is $1.71 \times 10^{-3} \pm 0.22 \times 10^{-3} \, h^{-1}$. The rate constant, $k_{Ox,Al}$, increases with increasing H^+ activity: $0.34 \times 10^{-3} \, h^{-1}$ (pH 5), $0.88 \times 10^{-3} \, h^{-1}$ (pH 4), $2.31 \times 10^{-3} \, h^{-1}$ (pH 3).

The results of preliminary dissolution experiments with muscovite are displayed in Figure 12. They illustrate that oxalate increases the dissolution rate of muscovite at pH 3. However, the catalytic effect of oxalate is hardly measurable and, hence, an evaluation of data leading to a more mechanistic understanding of muscovite dissolution appeared to be difficult.

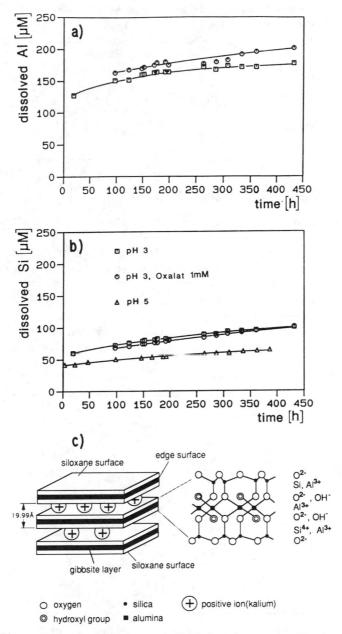

Figure 12. The proton- and oxalate-promoted dissolution of muscovite. The slow weathering kinetics is a characteristic of micas. Oxalate affects the stoichiometry of Al and Si release, but has not a significant catalytic effect. Figure 12c displays a schematic representation of the muscovite structure. It reveals the 2:1 structure. For example, an Al layers (black) exists in an octahedral sheet between two tetrahedral sheets (white) whose cations are composed of 25% Al and 75% Si. Siloxane and edge surfaces are exposed to solution.

TABLE 2. Rate Constants of Oxalate- and Salicylate-Promoted Dissolution

Ligand	$\delta\text{-Al}_2\text{O}_3$ $k_L\,(\text{h}^{-1})$	Kaolinite $k_L\,(\text{h}^{-1})$
Oxalate	10.8×10^{-3}	1.71×10^{-3} $\pm 0.22 \times 10^{-3}$
Salicylate (pH > 4)	12.5×10^{-3}	2.37×10^{-3} $\pm 0.56 \times 10^{-3}$

5.4.2. Stoichiometry of Dissolution Processes

The coexistence of various structure-forming metal centers is a characteristic of silicates distinguishing them from simple oxides such as aluminum oxide, quartz, and iron oxide. In weathering processes, the structure-forming cations may be detached from the mineral surface at the same rate (stoichiometric process) or at different rates (nonstoichiometric process).

The pH-dependent rates of kaolinite dissolution are presented in Figure 13. Within the experimental time range ($t = 10$–15 days), the proton-promoted dissolution of kaolinite occurs nonstoichiometrically, that is, the detachment of Si is faster than the release of Al from the kaolinite surface ($R_{H,Si} > R_{H,Al}$).

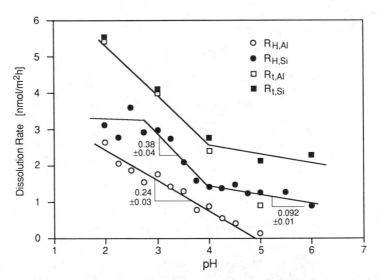

Figure 13. The dependence of the rates of proton- and oxalate-promoted dissolution on solution pH. The rate of the proton-promoted Al release linearly decreases to higher pH. Three sequences of pH dependence may be identified for the proton-promoted detachment of Si, $R_{H,Si}$. Proton- and oxalate-promoted dissolution are superimposed and the total dissolution rate $R_{T,Al}$ and $R_{T,Si}$ are composed of two additional rates: $R_T = R_H + R_{Ox}$.

However, the presence of oxalate significantly changes the dissolution characteristic: at pH < 3, the overall dissolution rates $R_{T,Si}$ and $R_{T,Al}$ are equal indicating the stoichiometric detachment of Al and Si. At pH = 4 and 1 mM oxalate, the dissolution occurs almost stoichiometrically but even at high oxalate concentrations, $R_{T,Al}$ is always smaller than $R_{T,Si}$. At pH > 5, the dissolution process is nonstoichiometric. It is worth mentioning that the rate of Al detachment never exceeds the rate of Si detachment in the pH range 2–6 and various oxalate concentrations. This indicates that Al and Si detachments are not independent surface reactions. These findings are also confirmed by experiments with salicylate.

5.4.3. *Proton-Promoted Dissolution*

The pH dependence of kaolinite dissolution can be discussed by expressing proton-promoted rates R_H in terms of pH:

$$\log R_H = \log k_H - s \cdot \text{pH} \tag{22}$$

where k_H denotes the apparent rate constant. Figure 13 reveals that the rate of proton-promoted Al detachment, $R_{H,Al}$, linearly depend on pH in the pH range 2–5 ($s = 0.24 \pm 0.03$). Three sequences may be clearly identified for $R_{H,Si}$: (1) at pH < 3, $R_{H,Si}$ is constant within the experimental errors; (2) between pH 3 and 4, $R_{H,Si}$ strongly depends on pH ($s = 0.38 \pm 0.04$); and (3) at pH > 4, $R_{H,Si}$ is only weakly affected by pH ($s = 0.092 \pm 0.01$). An explanation of the pH dependence of dissolution rates is obtained by considering the effect of surface protonation. Figure 14 illustrates the pH-dependent protonation of the distinct surface sites of a kaolinite platelet. The functional groups acting as coordinative counterparts in protonation and deprotonation equilibria are: Al–OH–Al groups at the gibbsite surface (=SOH), AlOH– and/or SiOH– groups at the edge surface, (=MOH), and (Al–O–Si)$^-$ groups at the siloxane layer due to isomorphic substitution (=XO) (Fig. 10)

The permanent negatively charged surface sites at the siloxane layer (XO) are accessible to ion-exchange reactions with cations, such as Na$^+$, K$^+$, Ca^{2+}, and Al^{3+} (inset in Fig. 14) (Schindler et al., 1987).

The protonation of surface hydroxyl groups at the gibbsite and edge surfaces is displayed in Figure 14. The surface proton concentration $\Gamma_{H,V}$ denotes the excess proton density with respect to pH$_{ZPC}$ = 7.5 of the hydroxyl groups at the edge face. The total excess proton density (solid line) may be assigned to two successive protonation equilibria at the kaolinite surface (broken lines).

Gibbsite Surface (25°C)

$$\text{SOH}_2^+ \rightleftharpoons \text{SOH} + \text{H}^+, \qquad pK_{a1}^s \text{(intr.)} = 4.04$$

Edge Surface

$$\text{MOH}_2^+ \rightleftharpoons \text{MOH} + \text{H}^+, \qquad pK_{a1}^s \text{(intr.)} = 6.31$$

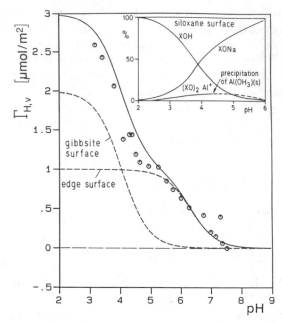

Figure 14. Surface protonation and ion exchange equilibria at the kaolinite surfaces. The inset represents the protonation and ion-exchange reactions at the permanent negatively charged surface sites of the siloxane layer (0.1 M NaNO$_3$, [Al] = 1.6 × 10^{-4} M, [XO]$_t$ = 1.46 × 10^{-3} M). The excess proton density, $\Gamma_{H,V}$, at the surface hydroxyl group is displayed as a function of pH. Surface protonation is interpreted as a successive protonation of two distinct types of OH groups localized at the gibbsite and edge surfaces. The pH$_{ZPC}$ of the edge surface is about 7.5.

The surface protonation equilibria are interpreted according to the constant-capacitance model of Schindler and Stumm (1987). The two model parameters, the "intrinsic" constant pK^s_{a1}(intr.) at zero surface charge and the integral capacitance of the flat electric double layer $C(=2\ \mathrm{F\,m^{-2}})$ are determined from titration curves. The methods of the acidimetric titration of kaolinite suspensions are discribed by Wieland (1988).

The intrinsic acidity constants of MOH groups at the edge surface are comparable to the values evaluated for δ-Al$_2$O$_3$ (Kummert and Stumm, 1980). The SOH groups localized at the gibbsite surface are more acid than AlOH groups of aluminum oxides. We postulate that Al–OH–Al groups at the gibbsite surface and AlOH groups at the edge surface are protonated. The higher acidity of Al–OH$_2^+$–Al groups may be explained according to the different acidity constants of dissolved species Al$_2$(OH)$_2^{4+}$ (pK = 7.7) and Al(OH)$_2^+$ (pK = 9.3).

5.4.4. A Mechanistic Interpretation of the Dissolution Process

In agreement with the results presented, a mechanistic interpretation of kaolinite and muscovite dissolution is suggested:

1. *The dissolution is controlled by the detachment of Al.* Since the dissolution of silica is not promoted in presence of oxalate and salicylate (Bennett et al., 1988; Wieland, 1988), we may conclude that Si centers do not form stable surface complexes with these ligands. Hence, the siloxane layer of kaolinite and muscovite is not reactive with respect to dissolution reactions. Therefore, the detachment of both Al and Si is a consequence of the formation of surface complexes with Al sites.

2. *Surface protonation of the edge and of the gibbsite surface promote the dissolution process.* The pH dependence of Si detachment (Fig. 13) reflects the protonation of the kaolinite surface (Fig. 14). At pH < 7.5, AlOH groups at the edge surface are protonated and dominate the overall weathering process. At pH < 5, both the proton density and the dissolution rate of the edge surface remain constant. In the pH range 3–5.5, Al–OH–Al groups at the gibbsite surface are protonated and, consequently, the dissolution process is dominated by the detachment of Al and Si centers from the gibbsite surface.

3. *Reconstitution of a secondary Al phase.* The pH dependence of the Al detachment (Fig. 13) may not be explained simply by surface protonation reactions. Aluminum(III) may adsorb on the siloxane layer (inset in Fig. 14) and reconstitute a secondary precipitate. Hence, the pH dependence of Al detachment reflects the release of Al from this Al-rich precipitate rather than the dissolution process at the kaolinite surface. In the presence of oxalate, the Al phase is dissolved and the dissolution process occurs stoichiometrically at low pH. The accumulation of Al on mica surfaces has already been postulated (t'Serstevens et al., 1978). Figure 15 reveals that the Al center occurring at the surface mainly affects the dissolution characteristic of kaolinite and muscovite.

Generally, the protonation of Al sites promotes the dissolution process with increasing H^+ activity in acid solution (Al_2O_3, kaolinite, muscovite), whereas the rate of silica dissolution even decreases or remains constant (pH $\leq$ 3). Obviously the more Al centers are exposed per unit surface area, the higher the proton-promoted dissolution rate and the more effective are surface chelates in catalyzing the weathering process.

5.5. Reductive and Catalytic Dissolution of Fe(III) (Hydr)oxides

Changes in oxidation state affect the solubility of metal (hydr)oxides. The oxides of transition elements become more soluble on reduction, whereas other oxides such as Cr_2O_3 or V_2O_3 become more soluble on oxidation. The reduction of surface metal centers in reducible metal oxides typically leads to easier detachment of the reduced metal ions from the lattice surface. This is readily accounted for by the larger lability of the reduced metal–oxygen bond in comparison to the nonreduced metal–oxygen bond; for example, the Madelung energy of the

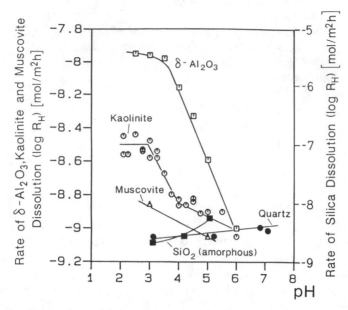

Figure 15. The pH dependence of the proton-promoted dissolution rates of kaolinite, muscovite, and their constitutent oxides of Al_2O_3 and amorphous SiO_2 or quartz, respectively. With increasing H^+ activity, the rate of Al detachment is promoted whereas the rate of Si detachment is slowed down.

Fe^{II}–O bond in a crystalline lattice is much smaller than that of the Fe^{III}–O bond.

As Stone and Morgan (1987) comprehensively surveyed the reductive dissolution of metal oxides, we need only some examples of recent work in our laboratory on the reductive dissolution of Fe(III) (hydr)oxides (Banwart et al., in press; Sulzberger, 1988; Suter et al., 1988; Zinder et al., 1986). In each case investigated, the dissolution rate was controlled by surface chemical reactions. The rate law can be accounted for by the following reaction sequence: (1) surface complex formation with reductant, (2) electron transfer leading to some Fe(II) in the lattice surface, and (3) detachment of Fe(II) from the mineral surface into the solution. The detachment (step 3) is rate-determining; that is, the precursor of the activated complex is a surface >Fe(II) species (with surrounding bonds that are partially protonated), whose concentration is proportional to that of the preprecursor, the surface complex. Correspondingly, the rate of dissolution was found to be proportional to the concentration of the surface bound reductant:

$$R_{reductive} \propto \{ >Fe-O-reductant \} \qquad (23)$$

Figure 16 illustrates the reaction scheme that accounts for the reductive dissolution of Fe(III) (hydr)oxides by ascorbate. Figure 17 gives experimental results illustrating the zero-order dissolution rate with varying ascorbate

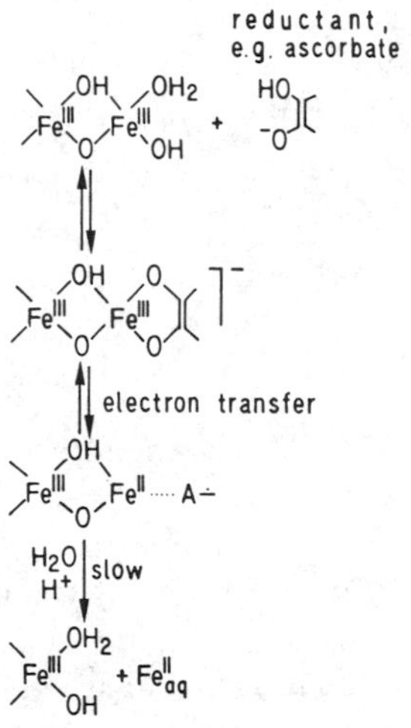

reductant,
e.g. ascorbate

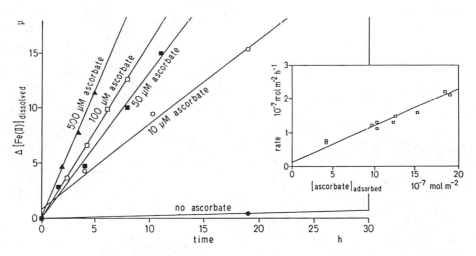

Figure 16. Reaction sequence for the dissolution of Fe(III) (hydr)oxides by a reductant such as ascorbate. The fast adsorption of the reductant is followed by steps that involve the electron transfer; $>$Fe(III) is reduced to $>$Fe(II) and ascorbate oxidized to a radical that is desorbed form the surface. The resulting Fe(II)–O bond at the surface is more labile than the Fe(III)–O bond. Then Fe(II) becomes detached from the surface and the original surface structure is reconstituted.

Figure 17. Representative results on the dissolution of hematite by ascorbate at pH 3. As shown by the inset, the rate of dissolution is proportional to the surface concentration of ascorbate (from Banwart et al., in press).

concentrations and its dependence on the concentration of the ascorbate surface complex.

Iron(III) (hydr)oxides can be dissolved [as Fe(III)] most effectively by a catalytic mechanism that involves a bridging ligand and Fe(II) (for simplicity, changes in charge are omitted):

$$>Fe^{III}OH + HL \rightleftharpoons >Fe^{III}LH + OH^-$$

$$>Fe^{III}LH + Fe^{2+} \rightleftharpoons >Fe^{III}L\,Fe^{II} + H^+$$

$$>Fe^{III}L\,Fe^{II} \overset{ET}{\rightleftharpoons} >Fe^{II}L\,Fe^{III} \qquad (24)$$

$$>Fe^{II}L\,Fe^{III} + H_2O \rightleftharpoons >Fe^{II}OH_2 + Fe^{III}L$$

$$>Fe^{II}OH_2 \rightleftharpoons > + Fe^{II}(aq)$$

Thus, as exemplified in Figure 18, ligands such as oxalate bind via ligand exchange at the surface $>Fe^{III}OH$. Subsequently, a ternary complex $>Fe^{III}LFe^{II}$ is formed. The efficiency of a bridging ligand depends on its surface complex formation properties, the stability of the ternary complex as a function of pH, the electronic properties of the bridging ligand (a potentially conjugatable electronic arrangement facilitates the electron transfer), and its ability to transport the successor complex away from the surface before the reverse process occurs (Suter et al., 1988; Wehrli et al., in press). It is interesting to note that Fe(II) binds to hematite and goethite at pH 3–4 only if a ligand such as oxalate is present. The effect of the ligand is caused by an increase in concentration of adsorbed Fe(II); furthermore, oxalato–Fe(II) is, thermodynamically speaking, a much stronger

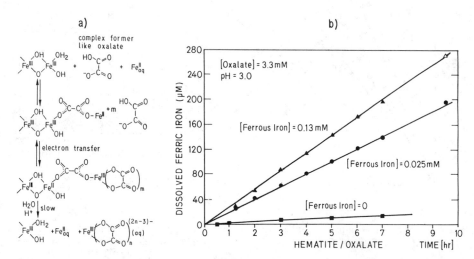

Figure 18. (*a*) Catalytic dissolution of Fe(III) (hydr)oxides. In the presence of a suitable bridge-building ligand, Fe(II) catalyzes the dissolution of Fe(III). Ligands such as oxalate adsorb via ligand exchange at the surface. Subsequently a ternary complex with Fe(II) is formed. The electron transfer from the Fe(II) on the solution side to the $>Fe(III)$ at the lattice surface reduces the latter to $>Fe(II)$, which then becomes detached, replacing the Fe(II) in solution. The Fe(III), now bound to the successor complex, goes into solution. (*b*) Representative experimental results (from Suter et al., 1989).

reductant than Fe(II). In addition, oxalate binds the reductant Fe(II) at a greater distance from the surface than OH^-; this facilitates the detachment of the oxidized surface complex. A detailed discussion of similar inner-sphere electron transfer reactions at the pyrite–water interface is given by Luther (Chapter 6 in this volume).

Figure 18 gives representative experimental results (Suter et al., 1988) for the dissolution of hematite in the presence of oxalate and Fe(II). The total amount of Fe(III) dissolved during an experiment is much higher than the amount of Fe(II) present; the concentration of Fe(II) remains constant. There is no reduction, although the reduction of the surface Fe(II) has a strong effect on the kinetics of the reaction. As shown by Sulzberger (Chapter 14, this volume) the effect described can become important by establishing an autocatalytic mechanism in the light-induced dissolution of Fe(III) oxides in the presence of surface ligands such as oxalate; the Fe(II) produced photochemically acts as a catalyst for the further dissolution.

6. A FEW GEOCHEMICAL AND TECHNOLOGICAL IMPLICATIONS

6.1. Weathering

In weathering, primary minerals become transformed into solutes and soils and eventually into sedimentary rocks (see Schnoor, Chapter 17, this volume). Weathering is an important feature of the hydrogeochemical cycle of elements. Partners in weathering processes are (1) the *lithosphere* (primary minerals, aluminum silicates, oxides, carbonates, clays, humus); (2) the *atmosphere* (CO_2 as a source of carbonic acids, oxidants, organic acids, oxides of S and N that convert in atmospheric depositions into acid rain); (3) the *biota*, which produces ligands and reductants (biogenic acids such as oxalic, succinic, tartaric, ketogluconic, *p*-hydroxybenzoic acid, in addition to fulvic and humic acids, are typically found in top soils and in surface waters; and (4) *water*, which operates both as a reagent and a transporting agent of dissolved and particulate components from land to sea (Stumm and Wollast, 1990).

Calcareous minerals and evaporate minerals (halides and sulfates) contribute approximately 38 and 17%, respectively, of the total solutes in world's rivers. About 45% are due to the weathering of silicates. The global weathering rate is influenced by temperature and the extent of the land covered by vegetation; the latter dependence results from the CO_2 production in soils, which is a consequence of plant respiration and the decay of organic matter as well as the release of complex-forming ligands. The silicate weathering is more important than the carbonate weathering in the long-term control of atmospheric CO_2. The HCO_3^- and Ca^{2+} ions produced by the weathering of $CaCO_3$ (see Wollast, Chapter 15, this volume),

$$CaCO_3 + CO_2 + H_2O \rightleftharpoons Ca^{2+} + 2HCO_3^- \qquad (25)$$

are precipitated in the ocean (through incorporation into marine organisms) as $CaCO_3$. The CO_2 consumed in the carbonate weathering is released again on formation of $CaCO_3$ in the ocean (reversal of the reaction given above); on the other hand, the weathering of calcium silicates, represented in a simplified way by

$$CaSiO_3 + 2CO_2 + 3H_2O = Ca^{2+} + 2HCO_3^- + H_4SiO_4 \qquad (26)$$

produces also Ca^{2+} and HCO_3^-, which form $CaCO_3$ in the sea, but only half of

TABLE 3. Rate of Dissolution of Silicates at pH 5 (in mol SiO_2 $m^{-2}s^{-1}$)

	O:Si	Log Rate	Reference[a]	Comments
Framework Silicates				
Quartz	2	−13.39	1	In pure water
Quartz	2	−11.85	2	In 0.2 M NaCl
Opal	2	−12.14	1	In pure water
Opal	2	−10.92	3	In 0.1 M NaCl
Albite	2.67	−11.32	4	
Albite	2.67	−11.37	5	$P_{CO_2}=1$ atm
Orthoclase	2.67	−11.77	5	$P_{CO_2}=1$ atm
Oligoclase	2.67	−11.59	5	$P_{CO_2}=1$ atm
Anorthite	4	−11.49	6	
Nepheline	4	−8.55	7	
Sheet Silicates				
Kaolinite	2.5	−12.55	8	
Muscovite	3.33	−12.70	8	
Chain Silicates				
Enstatite	3	−10.00	9	
Diopside	3	−9.85	9	
Tremolite	2.88	−10.47	9	
Augite	3	−10.60	10	
Wollastonite	3	−7.70	11	
Ring and Orthosilicates				
Olivine	4	−9.49	12	

From Wollast and Stumm (in preparation)

[a] References: (1) Rimstidt and Barnes (1986), (2) Wollast and Chou (1987), (3) Wirth and Gieskes (1979), (4) Wollast and Chou (1985), (5) Busenberg and Clemency (1976), (6) Amrhein and Suarez (1988), (7) Tole et al. (1986), (8) Wieland and Stumm (in preparation), (9) Schott et al. (1981), (10) Schott and Berner (1985), (11) Rimstidt and Dove (1988), (12) Blum and Lasaga (1988).

the CO_2 consumed in the weathering is released and returned to the atmosphere on $CaCO_3$ formation. Of course, the net loss of CO_2 may be made up ultimately by metamorphic and magnetic breakdown of $CaCO_3$ deep in the earth with the help of SiO_2, basically by a reaction such as $CaCO_3 + SiO_2 = CaSiO_3 + CO_2$.

Table 3 presents a survey of rates of dissolution of various silicates at pH = 5. This table is taken from a recent compilation by Wollast and Stumm (in preparation). It is remarkable that the experimentally determining sequence of increasing rate of silica release is in the same order as the chemical weathering sequence proposed by Goldich (1938) on the basis of field observations.

6.2. The Role of Ligands in the Natural Cycling of Iron in Natural Environments

The cycling of iron in natural environments is of great importance to the geochemical cycling of other reactive elements. The oxidation of Fe(II) to Fe(III) (hydr)oxides is accompanied by the binding of reactive compounds [heavy metals, silicates, phosphates, and other oxyanions of metalloids such as As(III,V) and Se(III,V)] to the hydrous Fe(III) oxide surface, and the reduction of the hydrous Fe(III) oxides to dissolved Fe(II) is accompanied by the release of these substances.

The light-induced processes on the dissolution of hydrous Fe(III) oxides are discussed in this volume by Sulzberger (Chapter 14). Here, we present our views on the reactions of reductive dissolution of Fe(III) (hydr)oxides that occur at the oxic–anoxic boundary in oceans and lakes. This is schematically presented in Figure 19. As shown by Sulzberger et al. (1989) the important processes are:

1. The principal reductant is the biodegradable biogenic material that settles in the deeper portions of the water column.

2. Electron transfer becomes more readily feasible if, as a consequence of fermentation processes—typically occurring around a redox potential of 0–200 V—molecules with reactive functional groups such as hydroxy and carboxyl groups are formed.

3. Within the depth-dependent redox gradient, concentration peaks of solid Fe(III) and of dissolved Fe(II) develop (Davidson, 1985), the peak of Fe(III) overlying the peak of Fe(II).

4. As illustrated in Figure 18, the Fe(II), forming complexes with these hydroxy and carboxy ligands, encounter in their upward diffusion the settling Fe(III) (hydr)oxides and interact with these according to catalytic mechanisms (Eq. 26), thereby dissolving rapidly the Fe(III) hydr(oxides).

The sequence of diffusional transport of Fe(II), oxidation to insoluble $Fe(OH)_3$ and subsequent settling and reduction to dissolved Fe(II) typically occurs within a relatively narrow redox-cline. The continuous cycle of oxidation–precipitation, binding of reactive compounds and subsequent reduction–dissolution and release of reactive elements, is of great importance for many marine metabolic

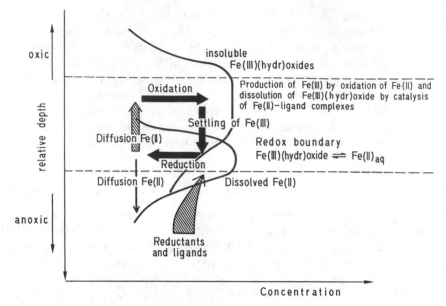

Figure 19. Transformations of Fe(II, III) at an oxic–anoxic boundary in the water or sediment column (modified from Davidson, 1985). Peaks in the concentration of solid Fe(III) (hydr)oxides and of dissolved Fe(II) are observed at locations of maximum Fe(III) and Fe(II) production, respectively. The combination of ligands and Fe(II) produced in underlying anoxic regions are most efficient in dissolving Fe(III) (hydr)oxides. Redox reactions of iron—oxidation accompanied by precipitation, reduction accompanied by dissolution—constitute an important cycle at the oxic-anoxic boundary which is often coupled with transformations (adsorption and desorption) or reactive elements such as heavy metals, metalloids, and phosphates.

transformations and for the coupling of the iron cycle with that of other elements.

6.3.   Soils

Biogenic ligands (dicarboxylic and hydroxy carboxylic acids and phenols) have a pronounced effect in enhancing the dissolution rate of minerals. The complexation in solution tends to raise the concentration of Al and Fe in solution and to extend the domain of congruent dissolution; higher concentrations of soluble Al(III) and Fe(III) can be built up before a new phase is formed. The biogenic solutes aid in the formation of very small particles and in stabilizing negatively charged colloids. These organic solutes assist the downward mobilization of Al and Fe observed in the podsolidation of soils, by increasing the dissolution of the Al(III) minerals and of hydrous Fe(III) oxides. In the latter case the enhancement of dissolution is also caused by reductive mechanisms.

Many geologists and soil scientists have repeatedly suggested the important influences of biological processes on the weathering reactions. The synthesis of terrestrial biomass is accompanied by acidification of the surrounding soil. This acidification results from the fact that plants take up more nutrient cations than anions; thus H^+ ions are released via roots. The ashes of plants and wood are alkaline ("potash"). Every temporal and spatial imbalance between production and mineralization of biomass leads to a modification of the H^+ ion balance of the environment. Concerning the effects of microorganisms, it is often difficult to distinguish between direct effects (enzymatic processes) and indirect effects (nonenzymatic processes). The latter effects are due to the release of metabolic intermediate or end products (H^+, CO_2, organic acids and other ligands, reductants, and ammonia). In the case of oxides of Mn(III,IV) and of Fe(III), direct mediating effects of autotrophic microorganisms have been documented (Arnold et al., 1986; Nealson, 1983; Ghiorse, 1986).

6.4. Corrosion and Passive Oxides

Corrosion rates of many metals are often determined or influenced by the dissolution kinetics of the film of corrosion products, often hydrous oxides, that form on top of the corroding surface (Grauer and Stumm, 1982). For example, it is well known that the corrosion rate of passive Fe in HNO_3 can be accelerated markedly by the addition of phosphates. Most likely H_3PO_4 on $H_2PO_4^-$ forms bidentate ligands with the functional groups of the oxide surface that increase the dissolution rate in a similar way as oxalic acid does (Zinder et al., 1986). On the other hand, phosphate at higher pH values may act as an inhibitor, because under these conditions it may form binuclear phosphate complexes (much more free energy is needed to detach simultaneously two metal centers), or a film with an iron(III) phosphate plane may be formed. Zutić and Stumm (1984) studied the dissolution of aluminum oxide with an oxide film covered-rotating disk aluminum electrode.

Acknowledgment

Our research on the surface chemistry of minerals has been supported by the Swiss National Foundation; it profited from data and ideas contributed by Steven Banwart, Gerhard Furrer, Jerald Schnoor, Christophe Siffert, Laura Sigg, Barbara Sulzberger, Daniel Suter, Bernhard Wehrli, and Bettina Zinder.

REFERENCES

Arnold, R. G., T. J. Di Christina, and M. R. Hoffmann (1986), "Dissimilative Fe(III) Reduction by Pseudomonas sp 200—Inhibitor Studies," *Appl. Environ. Microbiol.* **52**, 281–294.

Banwart, S., S. Davies, and W. Stumm (in press), "The role of Oxalate in Accelerating the Reductive Dissolution of Hematite (α-Fe$_2$O$_3$) by Ascorbate," *Colloids Surf.*

Bennett, P. C., M. E. Melcer, D. I. Siegel and J. P. Hassett (1988), "The Dissolution of Quartz in Dilute Aqueous Solutions of Organic Acids at 25°C," *Geochim. Cosmochim. Acta* **52**, 1521–1530.

Berner, R. A., and R. R. Holdren, Jr. (1979), "Mechanism of Feldspar Weathering. II. Observations of Feldspars from Soils," *Geochim. Cosmochim. Acta* **43**, 1173–1186.

Blum, A. E., and A. C. Lasaga (1988), "Role of Surface Speciation in the Low-Temperature Dissolution of Minerals," *Nature* **331**, 431–433.

Brown, G. E. (in press), "In-situ X-Ray Absorption Spectroscopic Studies of Ions at Oxide Water Interfaces," *Chimia*.

Calvera, F. and O. Talibudeen (1978), "The Release of Aluminium from Aluminosilicate Minerals. I. Kinetics," *Clays Clay Min.* **26**, 434–440.

Carroll-Webb, S. A., and J. V. Walther (1988), "A Surface Complex Reaction Model for the pH-Dependence of Corundum and Kaolinite Dissolution Rates," *Geochim. Cosmochim. Acta* **52**, 2609–2623.

Casey, W. H., H. R. Westrich, G. W. Arnold, and J. F. Banfield (1989), "The Surface Chemistry of Dissolving Labradorite Feldspar," *Geochim. Cosmochim. Acta* **53**, 821–832.

Correns, C. W. (1963), "Experiments on the Decompositio of Silicates and Discussion of Chemical Weathering," *Clays Clay Min.* **10**, 443–459.

Davidson, W. (1985), "Conceptual Models for Transport at a Redox Boundary," in W. Stumm, Ed., *Chemical Processes in Lakes*, Wiley-Interscience, New York.

De Finetti, B. (1974), *Theory of Probability*, Vol. 1, Wiley-Interscience, New York, 148 pp.

Furrer, G., and W. Stumm (1986), "The Coordination Chemistry of Weathering I: Dissolution Kinetics of δ-Al_2O_3 and BeO," *Geochim. Cosmochim. Acta* **50**, 1847–1860.

Ghiorse, W. C. (1986), "Microbial Reduction of Manganese and Iron," in A. J. B. Zehnder, Ed., *Environmental Microbiology of Anaerobes*, Wiley-Interscience, New York.

Goldich, S. S. (1938), "A Study in Rock Weathering," *J. Geol.* **46**, 17–58.

Grauer, W., and W. Stumm (1982), "Die Koordinationschemie oxidischer Grenzflächen und ihre Auswirkung auf die Auflösungskinetik oxidischer Festphasen in wässrigen Lösungen," *Colloid Polym. Sci.* **260**, 959–970.

Guy, Ch. and Schott, J. (in press), "Multisite Surface Reaction versus Transport Control during the Hydrolysis of a Complex Oxide," *Chem. Geol. Kinet.* (spec. issue).

Hachiya, K., M. Sasaki, Y. Saruta, N. Mikami, and T. Yasunaga (1984), "Static and Kinetic Studies of Adsorption–Desorption of Metal Ions on a γ-Al_2O_3 Surface, I. Static Study of Adsorption–Desorption II. Kinetic Study by Means of Pressure Jump Technique," *J Phys. Chem.* **88**, 23–31.

Hayes, K. F., and J. O. Leckie (1986), "Mechanism of Lead Ion Adsorption at the Goethite–Water Interface," in J. A. Davis and K. F. Hayes, Eds., *Geochemical Processes at Mineral Surfaces*, American Chemical Society, Washington, pp. 114–141.

Hayes, K. F., A. L. Roe, G. E. Brown, Jr., K. O. Hodgson, J. O. Leckie, and G. A. Parks (1987), "In situ X-Ray Absorption Study of Surface Complexes: Selenium Oxyanions on α-FeOOH," *Science* **238**, 783–786.

Hohl, H. and W. Stumm (1976), "Interaction of Pb^{2+} with Hydrous γ-Al_2O_3," *J. Colloid Interface Sci.* **55**, 281–288.

Huang, W. H., and W. D. Keller (1971), "Dissolution of Clay Minerals in Dilute Organic Acids at Room Temperature," *Am Min.* **56**, 1082–1095.

Lim, C. H., M. L. Jackson, R. D. Koons, and P. A. Helmke (1980), "Kaolins: Sources of

Differences in Cation-Exchange Capacities and Cesium Retention," *Clays Clay Min.* **28**, 223–229.

Lin, F. Ch., and Ch. V. Clemency (1981), "The Kinetics of Dissolution of Muscovites at 25°C and 1 atm CO_2 Partial Pressure," *Geochim. Cosmochim. Acta* **45**, 571–576.

Motschi, H. (1987), "Aspects of the Molecular Structure in Surface Complexes; Spectroscopic Investigations," in W. Stumm, Ed., *Aquatic Surface Chemistry*, Wiley-Interscience, New York, pp. 111–125.

Nealson, K. H. (1983), "The Microbial Iron Cycle," and "The Microbial Manganese Cycle," in W. E. Krumbein, Ed., *Microbial Geochemistry*, Blackwell, Oxford, pp. 159–221.

Petrovic, R., R. A. Berner, and M. B. Goldhaber (1976), "Rate Control in Dissolution of Alkali Feldspars, Studies of Residual Feldspar Grains by X-Ray Photo-electron Spectroscopy," *Geochim. Cosmochim. Acta* **40**, 537–577.

Polzer, W. L., and J. D. Hem (1965), "The Dissolution of Kaolinite," *J. Geophys. Res.* **70**, 6233–6240.

Schindler, P. W., and W. Stumm (1987), "The Surface Chemistry of Oxides, Hydroxides and Oxide Minerals," in W. Stumm, Ed., *Aquatic Surface Chemistry*, Wiley-Interscience, New York, pp.83–110.

Schindler, P. W., P. Liechti, and J. C. Westall (1987), "Adsorption of Copper, Cadmium and Lead from Aqueous Solution to the Kaolinite/Water Interface," *Neth. J. Agric. Sci.* **35**, 219–230.

Schnitzer, M. and H. Kodama (1976), "The Dissolution of Mica by Fulvic Acid," *Geoderma* **15**, 381–391.

Sigg, L., and W. Stumm (1981), "The Interaction of Anions and Weak Acids with the Hydrous Goethite (α-FeOOH) Surface," *Colloids Surf.* **2**, 101–117.

Sposito, G. (1983), "On the Surface Complexation Model of the Oxide–Aqueous Solution Interface," *J. Colloid Interface Sci.* **91**, 329–340.

Stone, A. T., and J. J. Morgan (1987), "Reductive Dissolution of Metal Oxides," in W. Stumm, Ed., *Aquatic Surface Chemistry*, Wiley-Interscience, New York, pp. 222–254.

Stumm, W., C. P. Huang, and S. R. Jenkins (1970), "Specific Chemical Interaction Affecting the Stability of Dispersed Systems," *Croat. Chem. Acta* **42**, 223–245.

Stumm, W., H. Hohl, and F. Dalang (1976), "Interaction of Metal Ions with Hydrous Oxide Surfaces," *Croat. Chim. Acta* **48**, 491–504.

Stumm, W., R. Kummert, and L. Sigg (1980), "A Ligand Exchange Model for the Adsorption of Inorganic and Organic Ligands at Hydrous Oxide Interfaces," *Croat. Chem. Acta* **53**, 291–312.

Stumm, W., J. Sinniger, and B. Sulzberger (in press), *Croat. Chim. Acta*.

Stumm, W., and R. Wollast (1990). "Coordination Chemistry of Weathering I: Kinetics of the Surface-Controlled Dissolution of Oxide Minerals," *Reviews of Geophysics* **28**.

Sulzberger, B. (1988), "Oberflächen-Koordinationschemie und Redox-Prozesse: Zur Aulfösung von Eisen(III)-oxiden unter Lichteinfluss," *Chimia* **42**, 257–261.

Sulzberger, B., D. Suter, Ch. Siffert, S. Banwart, and W. Stumm (1989), "Dissolution of Fe(III) (hydr)oxides in Natural Waters; Laboratory Assessment on the Kinetics Controlled by Surface Coordination," *Mar. Chem.* **28**, 127–144

Suter, D., C. Siffert, B. Sulzberger, and W. Stumm (1988), "Catalytic Dissolution of Iron(III) (hydr)oxides by Oxalic Acid in the Presence of Fe(II)," *Naturwissenschaften* **75**, 571–573.

Talibudeen, O. (1984), "Change Heterogenity and the Calorimentry of K–Ca Exchange–Adsorption in Clays and Soils," *Adsorpt. Sci. Tehcnol.* **1**, 235–246.

t'Serstevens, A., P. G. Rouxhet, and A. J. Herbillon (1978), "Alteration of Mica Surfaces by Water and Solutions," *Clays Clay Min.* **13**, 401–410.

van Olphen, H. (1977), *An Introduction to Clay Colloid Chemistry*, Wiley-Interscience, New York.

Wehrli, B. (1989), "Monte Carlo Simulations of Surface Morphologies during Mineral Dissolution," *J. Colloid. Interface Sci.* **132**(1), 230–242.

Wehrli, B., B. Sulzberger, and W. Stumm (1989), "Redox Processes Catalyzed by Hydrous Oxide Surfaces," *Chem. Geol.* **78**, 167–179.

Wieland, E. (1988), "Die Verwitterung schwerlöslicher Mineralien—ein koordination-schemischer Ansatz zur Auflösungskinetik," Ph.D. thesis No. 8532, ETH, Zürich, Switzerland.

Wieland, E., B. Wehrli, and W. Stumm (1988), "The Coordination Chemistry of Weathering: III: A Generalization on the Dissolution Rates of Minerals," *Geochim. Cosmochim. Acta* **52**, 1969–1981.

Wollast, R., and L. Chou. (1985), "Kinetic Study of the Dissolution of Albite with a Continuous Flow-through Fluidized Bed Reactor," in J. I. Drever, Ed., *The Chemistry of Weathering*, NATO ASI SERIES C **149**, pp. 75–96.

Wollast, R., and W. Stumm (in preparation), "Coordination Chemistry of Weathering: II. Experimental Approaches and Case Studies (in preparation).

Zeltner, W. A., E. C. Yost, M. L. Machesky, M. I. Tejedor-Tejedor, and M. A. Anderson (1986), "Characterization of Anion Binding on Goethite Using Titration Calorimetry and Cylindrical Internal Reflection-Fourier Transform Infrared Sepectroscopy," in J. A. Davis and K. F. Hayes, Eds., *Geochemical Processes at Mineral Surfaces*, American Chemical Society, Washington, PP. 142–161.

Zinder, B., G. Furrer, and W. Stumm (1986), "The Coordination Chemistry of Weathering. II. Dissolution of Fe(III) Oxides," *Geochim. Cosmochim. Acta* **50**, 1861–1870.

Zutic, V. and W. Stumm (1984), "Effect of Organic Acids and Fluoride on the Dissolution Kinetics of Hydrous Alumina. A Model Study Using the Rotation Disc Electrode," *Geochim. Cosmochim. Acta* **48**, 1493–1503.

14

PHOTOREDOX REACTIONS AT HYDROUS METAL OXIDE SURFACES: A SURFACE COORDINATION CHEMISTRY APPROACH

Barbara Sulzberger

Institute for Water Resources and Water Pollution Control (EAWAG), Dübendorf, Switzerland; Swiss Federal Institute of Technology (ETH) Zürich, Switzerland

1. INTRODUCTION

The motivation for studying photoredox reactions occurring at metal hydroxide surfaces originates from different research and application fields: examples are photochemical conversion and storage of solar energy with artificial systems, photochemical wastewater treatment, and, more recently, research related to aquatic geochemistry and atmospheric chemistry. In the field of photochemical solar energy conversion, the photochemical processes of main interest are those that lead to the storage of light energy in the form of chemical potential. In nature, the most prominent such process is photosynthesis. Hence, there have been many attempts to mimic some features of photosynthesis for artificial solar energy conversion. In heterogeneous photoredox systems, aimed at the conversion of solar energy or at the photocatalytic degradation of organic pollutants, metal hydroxides (either in colloidal form or as electrodes) are often investigated as photocatalysts. (The term "metal hydroxides" or "hydrous metal oxides" used in this chapter includes oxides, hydroxides, and oxide hydroxides.) One requirement for a metal hydroxide to be used as a photocatalyst is its stability toward dissolution. In nature, on the other hand, the light-induced dissolution of minerals such as iron(III) and manganese (III, IV) hydroxides plays an important role in the cycling of iron and manganese, which is coupled to the biogeochemical cycling of many other chemical compounds such as heavy metals and nutrients, including phosphate (Zinder and Stumm, 1985). The light-induced dissolution of these minerals is coupled to the photochemical oxidation of an electron donor, which may be a biogenic or a xenobiotic substance, and thus is also of importance in the abiotic degradation of pollutants in natural waters. The photocatalytic oxidation of compounds, where metal oxides act as photocatalyts, may be accompanied by formation of hydrogen peroxide (Frank and Bard, 1977; Kormann et al., 1988), an important redox component in natural waters.

The electronic structure of hydrous metal oxides with semiconductor properties, such as the crystalline phases of hydrous iron(III) oxides, are often described with the band energy model. The band energy model, however, is just one limiting model. As Goodenough (1971) has stated, the electronic structure of the outer electrons of such crystalline solids can alternatively be described by the crystal- or ligand-field theory. Crystal-field theory assumes that the interactions between neighboring atoms are so weak that each electron remains localized at a discrete atomic position; band theory assumes that the interactions are so large that each electron is shared between the nuclei and thus corresponds to a delocalized model. Molecular-orbital theories are capable of integrating both limiting models and are more appropriate for the description of the chemical bond (effective charges, covalent vs. ionic bonds, participation of different atomic orbitals in the bond, etc.) than is a nearly free electron approximation of the energy band theory (Marfunin, 1979). Within the band energy model, semiconductors are characterized by energetically nonoverlapping bands. If a semiconductor is exposed to a redox couple in solution, the bulk Fermi level moves to its equilibrium position while the surface band edge position remains fixed. As a consequence band bending occurs. In an n-type semiconductor the bands are bent upward whereas in a p-type semiconductor they are bent downwards. (Fig. 1). On absorption of light with energy equal or higher than the band-gap energy of the semiconductor, a band-to-band transition occurs, moving an electron from the filled valence band to the vacant conduction band. In the electrical field caused by the band bending, photoelectrons move in opposite direction from holes. The charge carriers that reach the surface can undergo redox reactions with adsorbed species at the solid/liquid interface. The thermodynamic requirement for such a redox reaction to occur is that the energy of the highest occupied molecular orbital of the electron donor is less negative than the energy of the valence band edge and that the energy of the lowest unoccupied moleculer orbital of the electron acceptor is more negative than the energy of the conduction band edge (Fig. 1). The efficiency of the redox reaction at the solid/liquid interface—which is in competition with the recombination of the photogenerated electron–hole pair—depends on how rapidly the minority carriers reach the surface of the solid and how rapidly they are captured through interfacial electron transfer by a thermodynamically appropriate electron donor or acceptor. [For a recent review of photocatalysis on semiconductors, see

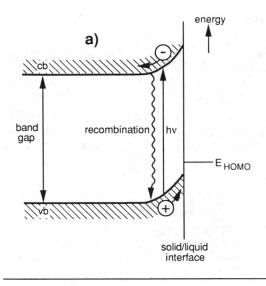

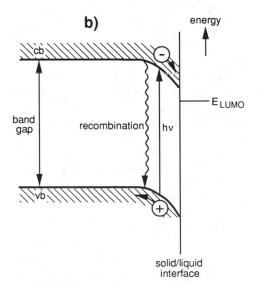

Figure 1. Light absorption of an *n*-type (*a*) and a *p*-type (*b*) semiconductor leading to a photoelectron–photohole pair. The photoholes in an *n*-type semiconductor and the photoelectrons in a *p*-type semiconductor are the minority carriers. (Key: vb = valence band; cb = conduction band; E_{HOMO} = energy of the highest occupied molecular orbital of an electron donor; E_{LUMO} = energy of the lowest unoccupied molecular orbital of an electron acceptor.)

Pichat and Fox (1988)]. As will be discussed later in this chapter, in many cases the surface structure, especially the coordinative interactions taking place at the surface, determines the efficiency of a photoredox reaction at the solid–liquid interface of a hydrous metal oxide.

The aqueous surfaces of metal oxides present functional hydroxyl groups (>MOH). Interactions of these groups with dissolved protons, metal ions, and organic and inorganic anions occur by inner-sphere or outer-sphere coordination or association with the diffuse double layer. The inner-sphere surface coordination has been described by surface ligand exchange reactions and with the constant-capacitance model (Dzombak and Morel, 1990; Schindler and Stumm, 1987, Stumm et al., 1980), which is a special case of "mean field theory" (Sposito, 1984). Spectroscopic investigations with magnetic resonance methods (EPR, ENDOR, ESEEM) (Motschi, 1987), with EXAFS (Hayes et al., 1987) and with FTIR (Zeltner et al., 1986) have confirmed the inner-sphere structure of different surface complexes. Possible structures of an inner-sphere surface complex are shown schematically in Figure 2 for the case of the specific adsorption of oxalate on an iron(III) hydroxide surface. Mononuclear, bidentate surface complexes are preferably formed at edges and corners of a hydrous oxide surface, whereas binuclear, bidentate surface complexes can be formed at an ideal surface of a metal hydroxide, if the interatomic distance between neighboring metal centers is of the order of a few angstroms. Under certain circumstances, mononuclear surface complexes may be formed (Blesa et al., 1987).

1.1.   Objectives

The author will first review some general concepts of photoredox reactions related to homogeneous and to heterogeneous photoredox systems and will then discuss the possible roles of surface-bound ligands and of surface compounds

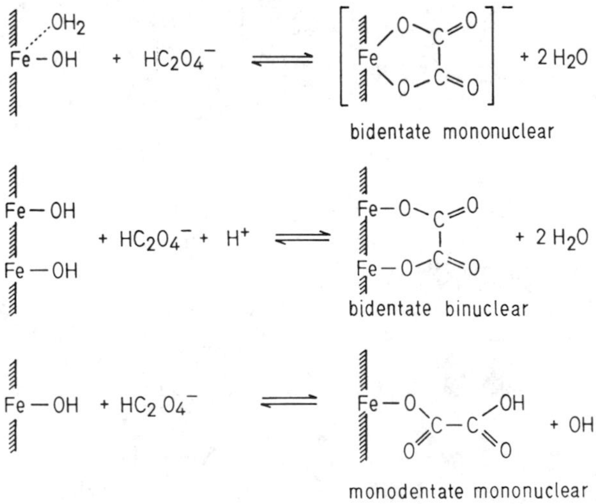

Figure 2. Schematic representation of oxalate adsorption equilibria on an iron(III) hydroxide surface.

acting as electron donors and in some cases as chromophores in the photocatalytic degradation of organic pollutants, in the spectral sensitization of metal oxides, and in the photochemical reductive dissolution of iron(III) hydroxides. A case study on the light-induced dissolution of iron(III) hydroxides with model compounds will be presented and the different possible reaction pathways explored. Finally, the application of these studies to natural water systems will be discussed.

2. PHOTOREDOX REACTIONS: SOME GENERAL CONCEPTS

A photoredox reaction is a redox reaction that occurs after electronic excitation of one or several reaction partners. By electronic excitation of a molecule or a solid such as a semiconductor a strongly reducing photoelectron and a strongly oxidizing photohole may be formed (Fig. 1). The ionization potential of an electronically excited species decreases, and the electron affinity increases approximately by the excitation energy.

Photoredox reactions occur either via *inter*molecular or via *intra*molecular processes. The most general scheme of an intermolecular photoredox reaction (Rehm and Weller, 1970) is shown in reaction 1: A compound C undergoes a transition from the ground state to an electronically excited state by absorption of light, $C \rightarrow C^*$. If such an excited compound encounters a reaction partner Q (either an electron donor or an electron acceptor), electron transfer may take place, leading to $C^{\mp}$ and $Q^{\pm}$:

$$C^* + Q \underset{k_{-1}}{\overset{k_1}{\rightleftharpoons}} C^*..Q \overset{k_2}{\rightarrow} C^{\mp}..Q^{\pm} \underset{k_{-3}}{\overset{k_3}{\rightleftharpoons}} C^{\mp} + Q^{\pm} \tag{1}$$

$$hv \downarrow\uparrow 1/\tau_0 \qquad\qquad\qquad k_b \downarrow$$

$$C \qquad\qquad\qquad\qquad C + Q$$

where τ_0 is the mean lifetime of the excited state, $C^*..Q$ stands for the encounter complex, which may also be an excited complex, before the electron-transfer step and $C^{\mp}..Q^{\pm}$ for the ion pair after electron transfer. Deactivation of the excited state that does not involve the reaction partner Q can occur via radiative or via radiationless transition. The radiationless, thermal deactivation of an excited state, leading to the ground state, $C^* \xrightarrow{\text{thermal}} C$, is called *internal conversion*. Both, the radiative deactivation and the internal conversion of C^* are in competition with chemical quenching by electron transfer leading to the ion pair, $C^{\mp}..Q^{\pm}$. It is, however, often the degree of ionic separation from the ion-pair $C^{\mp}..Q^{\pm}$, which is in competition with reverse electron transfer within the ion pair to form the ground-state reactants $C + Q$, that determines the overall efficiency of light-induced charge separation (Harriman et al., 1983). Reaction scheme 1 can also be applied to heterogeneous photoredox systems. In this case

C* represents an electronically excited solid species, for example, a semi-conductor, and C*..Q, a solute Q that is adsorbed at the surface of C*. If $C^{\mp}$ is reconverted to its original oxidation state, either by oxidation or by reduction, C can be considered as a photocatalyst according to the following definition: A photocatalyst, PC, is a light-absorbing species that enables the redox reaction but that remains unchanged after the overall process:

$$PC \xrightarrow{hv} PC*$$

$$PC* + A \longrightarrow PC + B \qquad (2)$$

The term "photocatalyst" is used independent of the sign of the free energy of the overall process. For heterogeneous systems it is convenient to distinguish between "heterogeneous photocatalysis," where the rate of a thermodynamically favorable reaction ($\Delta G < 0$) is increased by the presence of the illuminated solid, and "heterogeneous photosynthesis," where a thermodynamically unfavorable reaction ($\Delta G > 0$) is caused to occur by the presence of an illuminated solid, leading to conversion of radiant to chemical energy (Bard, 1979).

A homogeneous, bimolecular photoredox system, which has been extensively investigated, involves ruthenium(II) trisbipyridyl [$Ru(bpy)_3^{2+}$] as photocatalyst and methylviologen (MV^{2+}) as electron acceptor, where triethanolamine (TEOA) or ethylenediamintetraacetate (EDTA) are used as sacrificial electron donors for the reduction of $Ru(bpy)_3^{3+}$ (Tazuke et al., 1987):

$$\{Ru(bpy)_3^{2+}\}* + MV^{2+} \underset{k_{-1}}{\overset{k_1}{\rightleftharpoons}} \{Ru(bpy)_3^{2+}\}* MV^{2+} \xrightarrow{k_2} Ru(bpy)_3^{3+} \cdot\cdot MV^{+}$$

$$\underset{k_{-3}}{\overset{k_3}{\rightleftharpoons}} Ru(bpy)_3^{3+} + MV^{+}$$

$$+$$

$$TEOA \qquad (3)$$

$$Ru(bpy)_3^{2+} + TEOA_{ox}$$

$$hv \qquad 1/\tau_0$$

$$k_b$$

$$Ru(bpy)_3^{2+} \qquad Ru(bpy)_3^{2+} + MV^{2+}$$

$$MV^{2+} = H_3C-\overset{+}{N}\!\!\diagcirc\!\!-\!\!\diagcirc\!\!\overset{+}{N}-CH_3$$

Increasingly systems are investigated in which the photocatalyst, the electron acceptor, and/or the electron donor are linked through molecular bridges. In case of molecular systems, such "an appropriate assembly of suitable molecular components capable of performing light-induced functions can be called a photochemical molecular device" (Balzani et al., 1987). Such photochemical molecular devices allow efficient electron transfer through bridging ligands and the local separation of the primary photochemical products, or "a photoinduced charge separated state of sufficient lifetime and chemical stability for driving repetitively a catalytic reaction" (Lehn, 1987). In these systems the photochemistry of coordination compounds plays an important role.

Many phytoredox reactions occur as *intramolecular* processes via charge-transfer transitions. This is especially true for the photochemistry of coordination compounds. A charge-transfer transition leads to a major reorganization of the electron density distribution between the metal center and the ligand(s). For the following discussion ligand-to-metal, metal-to-ligand, and metal-to-metal charge-transfer transitions are of particular interest.

A metal-to-ligand charge-transfer (MLCT) transition causes an increase in the electron density on the ligand(s) and a decrease on the metal center. As an example, Fe(II) pentacyanide complexes with an additional aromatic ligand exhibit absorption bands in the visible that are due to a MLCT transition and that can be interpreted in terms of a quantitative transfer of an electron, localized

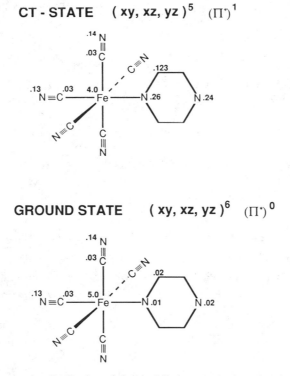

Figure 3. Electron density distribution of the six "d"-electrons in the ground state and the first excited state, which is a charge-transfer state, of $[Fe^{II}(CN)_5 \text{ pyrazine}]^{3-}$, as calculated by an extended Hückel calculation (Calzaferri and Grüniger, 1979; Calzaferri, 1986).

on the iron center, to the aromatic ligand (Calzaferri and Grüniger, 1979):

$$(L)_n M–AR \underset{}{\overset{h\nu}{\rightleftharpoons}} (L)_n M^+ – AR^- \tag{4}$$

Such an example is shown in Figure 3, where the aromatic ligand is pyrazine.

A ligand-to-metal charge transfer (LMCT) transition, on the other hand, causes an increase in the electron density on the metal center and a decrease on the ligand(s). The intense color of the permanganate ion, MnO_4^-, for example, is due to a charge-transfer transition from the oxygen ligand to the metal center (Ballhausen and Gray, 1965). Another example of a LMCT transition is the absorption band in the near UV of iron(III) oxalato complexes. Irradiation of $Fe^{III}(C_2O_4)_n^{(2n-3)-}$ within this absorption band leads to reduction of the iron center and to oxidation of the ligand(s) with high quantum yields (Vincze and Papp, 1987).

Photoredox reactions induced by charge-transfer transitions are not restricted to the homogeneous phase but can also occur at the surface of metal hydroxides. As will be discussed in the next section, a ligand-to-metal charge-transfer transition is involved in the light-induced dissolution of iron(III) hydroxides, and a metal-to-metal charge-transfer transition is proposed to be responsible for the sensitization of TiO_2 by specifically adsorbed $Fe^{II}(CN)_6^{4-}$ (Vrachnou et al., 1987).

3.   ROLE OF SURFACE COMPOUNDS IN HETEROGENEOUS PHOTOREDOX REACTIONS

In photoredox reactions occurring at the surface of hydrous metal oxides, a surface compound can have different functions; it can act merely as electron donor or acceptor without being involved as the chromophore, or it can act as electron donor or acceptor *and* as the light absorbing species, relevant for the heterogeneous photoredox reaction.

3.1.   Adsorbed Ligand Acting Merely as Electron Donor

The activity of an adsorbed electron donor for the interfacial electron transfer depends on both its ability to be oxidized by the photogenerated hole (Fig. 1a) and its adsorption properties. The first is a thermodynamic, the second a kinetic requirement. Kinetic experiments performed at various concentrations of the electron donor can give information on its adsorption properties. Different authors have reported, that the rate of the photocatalytic oxidation of an electron donor at the TiO_2 surface varied as a function of the dissolved concentration of the electron donor according to a Langmuir-type isotherm (Fig. 4) (Herrmann et al., 1983; Al-Ekabi et al., 1989). Also the rate of the photochemical reduction of a dissolved electron acceptor has been reported to be a

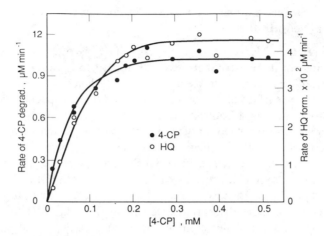

Figure 4. Effect of 4-chlorophenol (4-CP) concentration on the rate of degradation of 4-CP ($\bullet$) and the rate of formation of hydroquinone ($\circ$), which is a major intermediate of the 4-CP degradation. Conditions: initial pH 5.8; flow rate, 250 mL min^{-1}; temperature, 30 $\pm$ 2°C (Al-Ekabi et al., 1989).

Langmuir-type function of the concentration of the *electron donor* in presence of TiO$_2$, which is an *n*-type semiconductor (Henglein, 1984). These findings are in agreement with the assumption of an inner-spherically bound electron donor, since a Langmuir-type adsorption is consistent with the formation of a surface complex by surface ligand exchange (Stumm and Furrer, 1987):

$$>MOH + HL^- \rightleftharpoons >ML^- + H_2O \qquad K^s = \{>ML^-\}/(\{>MOH\}[HL^-])$$

(5)

$$\{>ML^-\} = \frac{K^s S[HL^-]}{1 + K^s[HL^-]} \qquad (\text{in mol m}^{-2})$$

(6)

where K^s is the conditional microscopic equilibrium constant of the adsorption equilibria and S is the concentration of the total surface sites that are available for the adsorption of the anion HL^-, $S = \{>MOH\} + \{>ML^-\}$. Equation 6 thus follows directly from Eq. 5.

The efficiency of interfacial electron transfer depends also on the type of surface complex formed between a specifically adsorbed electron donor and a surface metal center. Using methylviologen (MV^{2+}) as electron acceptor, Darwent and Lepre (1986) have reported that the yield of photochemical MV$^+$ formation was much higher in the presence of an electron donor where two functional groups are likely to be bound to the TiO$_2$ surface, compared to an electron donor that can only form a monodentate surface complex.

3.2. Ternary Surface Complex Acting as Chromophore in the Spectral Sensitization of TiO_2

Despite the relatively large band-gap energy, TiO_2 is often used as a photocatalyst because of its relative inertness with regard to dissolution. In order to render such materials more suitable as photocatalysts for the visible part of the solar spectrum, the surface of the solid phase is modified by specific adsorption of a chromophore such as a coordination compound (Calzaferri, 1981). Vrachnou et al. (1987) have reported on the efficient visible light sensitization of TiO_2 by surface complexation with $Fe(CN)_6^{4-}$. Evidence of formation of an inner-sphere ternary surface complex at the surface of TiO_2, CN^- acting as a bridging ligand, is furnished by FTIR spectroscopy (Desilvestro et al., 1988). The surface modified TiO_2 exhibits an absorption spectrum with a maximum at 430 nm, which is different from the absorption spectrum of $K_4Fe(CN)_6$ in solution and is attributed to a charge-transfer complex between $Fe(CN)_6^{4-}$ and surface Ti^{4+} ions (Vrachnou et al., 1987). From these data I deduce the following scheme:

$$Ti^{IV}-CN-Fe^{II}(CN)_n \xrightarrow{\ hv\ } Ti^{III}-CN-Fe^{III}(CN)_n \qquad (n=1-5) \qquad (7)$$

The authors attribute the bleaching of this charge-transfer transition within the 20-ns duration of a laser pulse to a very rapid photoinduced electron injection in the conduction band of TiO_2. Another example of an electron transfer through a bridging ligand of a ternary surface complex will be given in Section 4.

3.3. Surface Complex Acting as Chromophore in the Photochemical Dissolution of Iron(III) Hydroxides

In the example given above a metal-to-metal charge-transfer transition through a bridging ligand of a ternary surface complex leads to the sensitization of a metal oxide by electron transfer. A ligand-to-metal charge-transfer transition involving the specifically adsorbed ligand and a surface metal ion may also induce redox reactions. This mechanism is discussed for photoredox reactions occurring at iron(III) hydroxide surfaces (Waite and Morel, 1984a), where surface complexes may undergo similar photoredox reactions as the corresponding iron(III) coordination compounds in solution. Faust and Hoffmann (1986), Litter and Blesa (1988), and Siffert (1989) have investigated the wavelength dependence of the relative rate of dissolved iron(II) formation in the photochemical dissolution of iron(III) hydroxides using hematite–bisulfite, maghemite–EDTA, and hematite–oxalate as model systems, respectively. Figure 5 shows that only light in the near UV ($\lambda < 400$ nm) leads to an enhancement of the dissolution of hematite in the presence of oxalate that is due to the redox process at the surface (Siffert, 1989). By comparing this "action spectrum" with the absorption spectrum of hematite (Marusak et al., 1980) and of dissolved $Fe^{III}(C_2O_4)_n^{(2n-3)-}$. complexes, Siffert (1989) concludes that either a ligand-to-

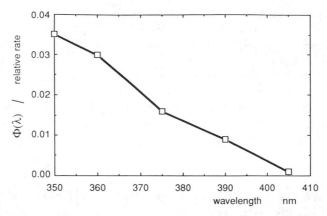

Figure 5. Relative rate of the light-induced reductive dissolution of hematite in the presence of oxalate as a function of the wavelength. Experimental conditions: 0.5 g L^{-1} hematite; initial oxalate concentration: 3.3 m mol L^{-1}; pH = 3.0; nitrogen atmosphere. The relative rate is the rate of hematite dissolution at constant incident light intensity. Under the assumption that the light intensity, absorbed by the oscillator that enables the photoredox reaction, corresponds to the incident light intensity, $I_{A\lambda} = I_{0\lambda}$, the relative rate equals the quantum yield, Φ_λ, of dissolved iron(II) formation. As is discussed in the text that follows, the photochemically formed dissolved iron(II) catalyzes the thermal dissolution of hematite. In order to keep the rate of this thermal dissolution constant, a sufficiently concentration of iron(II), $[Fe^{2+}] = 0.15$ m mol L^{-1}, was added to the suspension from the beginning. Thus, the relative rate corresponds to the dissolution rate due to the surface photoredox process (Siffert, 1989).

metal charge-transfer transition of the surface complex and/or a $Fe^{III} \leftarrow O^{-II}$ charge-transfer transition of the hematite surface lattice are the oscillators that drive the redox reaction leading to reductive dissolution of the solid phase. It is, to the author's knowledge, not yet experimentally proved which of these two mechanisms is mainly operative in the light-induced dissolution of crystalline iron(III) hydroxides. In the case of hematite, there is experimental evidence, however, that band-gap excitation of the 2.2-eV band gap, which corresponds to a d–d transition, does not lead to an enhancement of the dissolution of hematite in the presence of oxalate (Siffert, 1989). The $Fe^{III} \leftarrow O^{-II}$ charge-transfer transition of hematite corresponds to an electronic transition between the filled $2p$ oxygen band and an empty iron band in terms of semiconductor language and has been assigned the excitation of the "effective band gap" (3.3 eV) of hematite (Faust and Hoffmann, 1986).

Waite and Morel (1984b) have investigated the photodissolution of amorphous iron(III) hydroxides in organic-free medium. The authors suggest that in an organic free system a hydroxylated ferric surface species is the primary chromophore and that near-UV irradiation results in the photolysis of the $>Fe^{III}OH$-surface complex in analogy to photolysis of dissolved $FeOH^{2+}$ complexes (Faust and Hoigné, in press). Thus, in the light-induced dissolution of iron(III) hydroxides several charge-transfer transitions may play a role; a ligand-to-metal charge-transfer transition of a surface complex, and/or a lattice–oxygen to lattice–iron(III) charge-transfer transition of the solid phase. Irrespective of which of these two mechanisms is primarily operative, it is crucial that the electron donor form an inner-sphere surface complex at the surface of the iron(III) hydroxide for an efficient electron transfer to occur.

The examples given above illustrate that the coordinative interactions taking place at the surface of a hydrous metal oxide play an important role in these heterogeneous photoredox reactions. This is especially true for the light-induced dissolution of iron(III) hydroxides, as will be discussed in the next section.

4. CASE STUDY: LIGHT-INDUCED DISSOLUTION OF HEMATITE IN THE PRESENCE OF OXALATE

4.1. Motivation

Our motivation for studying the light-induced dissolution of hydrous iron(III) oxides in laboratory experiments is to elucidate the mechanisms of these processes and to establish general rate laws that allow prediction of the factors that are important in controlling the steady-state concentration of dissolved iron(II) in the photic zone of a natural water.

Figure 6 shows schematically the aquatic redox cycle of iron. Under the conditions usually encountered in natural aquatic systems, the reduction of iron(III) is accompanied by dissolution and the oxidation of iron(II) by precipitation. Reductive dissolution of iron(III) hydroxides occurs primarily at the sediment–water interface under anoxic conditions in the presence of reductants, such as products of the decomposition of biological material or exudates of organisms. Reductive dissolution of iron(III) hydroxides, however, can also occur in the photic zone in the presence of compounds that are metastable with respect to iron(III), that is, compounds that do not undergo redox reactions with iron(III) unless catalyzed by light. The direct biological mediation of redox processes may also influence the redox cycles of iron (Arnold et al., 1986; Price and Morel, Chapter 8, this volume). Dissolved oxygen is usually the oxidant of

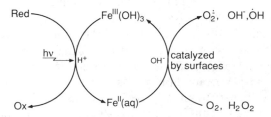

Figure 6. Aquatic redox cycle of iron.

iron(II). If the light-induced dissolution of iron(III) hydroxides occurs with a sufficiently high efficiency, a steady-state concentration of dissolved iron(II) may be maintained. Diurnal variation of the iron(II) concentration is observed in acidic surface waters (Collienne, 1983; McKnight et al., 1988; Sulzberger *et al.*, in press). The presence of iron in dissolved form is often a prerequisite for the uptake of this element by the photoplankton (Anderson and Morel, 1980; Finden et al., 1984; Rich and Morel, in press). It has been suggested that iron may be a limiting nutrient in marine systems (Martin and Fitzwater, 1988).

We have chosen hematite–oxalate as a model system, since the photochemical properties of colloidal hematite (Stramel and Thomas, 1986) and the photochemistry of iron(III) oxalato complexes in solution (Parker and Hatchard, 1959) have been studied extensively. The experiments presented in this section were carried out as batch experiments with monodispersed suspensions of hematite (diameter of the particles 50 and 100 nm), synthesized according to Penners and Koopal (1986) and checked by electron microscopy and X-ray diffraction. An experimental technique developed for the study of photoredox reactions with colloidal systems (Sulzberger, 1983) has been used. A pH of 3 was chosen to maximize the adsorption of oxalate at the hematite surface. This case study is described in detail by Siffert (1989) and Siffert et al. (manuscript in preparation).

4.2. Reaction Pathways Involved in the Light-Induced Dissolution of Hematite in the Presence of Oxalate

Apart from the photoredox reaction occurring at the surface of hematite and leading to dissolved iron(II), Fe^{II} is also produced through photolysis of dissolved iron(III) trioxalato complexes. Dissolved iron(III) is formed via thermal pathways, where Fe^{II} acts as a catalyst.

The various elementary steps involved in the surface photoredox reaction, leading to dissolution of hematite are outlined in Figure 7. The two-dimensional structure of the surface of an iron(III) hydroxide given in this figure is highly schematic. The charges indicated correspond to relative charges. An important step is the formation of a hypothetical bidentate, mononuclear surface complex. With pressure jump relaxation technique it has been shown that the adsorption equilibria at the mineral–water interface are generally established fast (Hayes and Leckie, 1986). Election transfer occurs via an electronically excited state (indicated by an asterisk) which is either a ligand-to-metal charge-transfer transition of the surface complex and/or a $Fe^{III} \leftarrow O^{-II}$ charge-transfer transition of hematite. The oxalate radical undergoes a fast decarboxylation reaction yielding CO_2 and the $CO_2^{\cdot-}$ radical, which is a strong reductant that can reduce a second surface iron(III) in a thermal reaction. Thus, two surface iron(II) and two CO_2 may theoretically be formed per absorbed photon. However, the quantum yield of this surface redox reaction is less than two (Siffert, 1989) because of loss reactions such as thermal deactivation from the excited state. For the sake of simplicity this thermal reaction of the $CO_2^{\cdot-}$ radical is omitted in Figure 7. We assume that detachment of Fe^{II} from the crystal lattice is the rate-limiting step of the overall reaction. After detachment of the surface group, the surface of hematite is reconverted into its original configuration; hence, the surface concentration of active sites and of adsorbed oxalate does not change throughout the experiment.

Part of the photochemically formed iron(II) is readsorbed on the hematite surface, with oxalate acting as a bridging ligand. Two possible structures of such a ternary surface complex are shown in reactions 8 and 9. The adsorbed Fe^{II} may

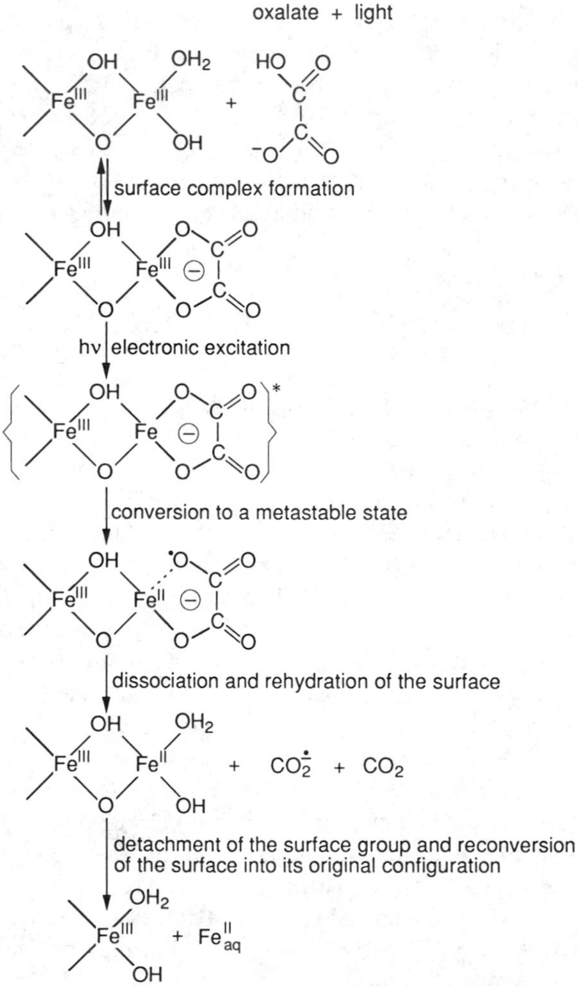

Figure 7. Schematic representation of the various steps involved in the light-induced reductive dissolution of hematite in the presence of oxalate.

be coordinated with one or two additional ligands, and is thermodynamically a suitable reductant of surface iron(III). We assume that electron transfer from the adsorbed iron(II) to the surface iron(III) occurs through the bridging ligand:

$$\qquad\qquad\qquad\qquad\qquad\qquad\qquad\qquad\qquad\qquad\qquad (8)$$

$$\qquad\qquad\qquad\qquad\qquad\qquad\qquad\qquad\qquad\qquad\qquad (9)$$

The surface iron(II) thus formed is detached from the surface of hematite as the rate-determining step (Suter et al., 1988). This pathway is a thermal reductive dissolution of hematite, which leads, however, to an increase of the concentration of dissolved iron(III) while the concentration of dissolved iron(II) remains constant; Fe^{II} acts as a catalyst for the dissolution of iron(III) hydroxides.

Dissolved iron(III) formed via this Fe^{II}-catalyzed thermal dissolution process is photochemically reduced through photolysis of iron(III) trioxalato complexes with high quantum yields (Vincze and Papp, 1987) according to the following stoichiometry:

$$2[Fe^{III}(C_2O_4)_3]^{3-} \xrightarrow[N_2]{h\nu} 2Fe^{2+} + 2CO_2 + 5C_2O_4^{2-} \qquad (10)$$

Because of the high quantum yields of this reaction in a broad spectral range, ferric oxalate is used as an actinometer for the determination of the absolute incident light intensity in photochemical experiments (Parker, 1968). The extent to which this homogeneous photoredox reaction occurs in a heterogeneous system depends on the incident light intensity and the attenuation of the light by the colloidal particles. In dissolution experiments, carried out at a relatively high light intensity ($I_0 \cong 4\,kW\,m^{-2}$) all dissolved iron was present as iron(II), whereas at a relatively low light intensity ($I_0 = 4W\,m^{-2}$) dissolved Fe^{III} could be detected in addition to Fe^{II} (as seen in Figs. 8 and 9).

4.3.   Rate Expression

As for all photochemical reactions, the rate, r_λ, of the preceding homogeneous and heterogeneous photoredox reactions depends, at a given wavelength, on the amount of light absorbed by the chromophore that is involved in the photoredox reaction, $I_{A\lambda}$, and on the quantum yield, Φ_λ, of the photoredox reaction:

$$r_\lambda = I_{A\lambda}\, \Phi_\lambda \tag{11}$$

According to the Beer–Lambert law, $I_{A\lambda}$ is a function of the concentration of the chromophore:

$$I_{A\lambda} = I'_{0\lambda}[1 - \exp(-\varepsilon_\lambda c L 2.3)] \tag{12}$$

$$I_{A\lambda} \cong I'_{0\lambda}(\varepsilon_\lambda c L 2.3) \qquad \text{for } \varepsilon_\lambda c L 2.3 \ll 1 \tag{13}$$

where $I'_{0\lambda}$ in moles of photons per liter per hour is the volume-averaged light intensity that is available to the primary chromophore, ε_λ, the molar extinction coefficient of the chromophore at a given wavelength in liters per mole per centimeter, L is the light path length in centimeters and c is the concentration of the chromophore in moles per liter. For a sufficiently small concentration of the chromophore and/or of the extinction coefficient, the rate of the photochemical reaction, r_λ (in moles per liter per hour), is then directly proportional to the concentration of the light-absorbing species involved in the photoredox reaction:

$$r_\lambda = kc \tag{14}$$

where the experimental rate constant, k is

$$k = I'_{0\lambda}\varepsilon_\lambda L 2.3 \Phi_\lambda \tag{15}$$

4.3.1. Rate Expression of the Surface Photoredox Reaction.

As discussed in the previous section, a ligand-to-metal charge-transfer transition of the surface complex (mechanism 1) and/or a $Fe^{III} \leftarrow O^{-II}$ charge-transfer of hematite (mechanism 2) are the oscillators involved in the surface photoredox reaction, leading to reductive dissolution of hematite in the presence of oxalate. The elementary steps and the derivations of the rate expressions of photochemical surface iron(II) formation of mechanism 1 and 2 are outlined in reactions 16–19, Eqs. 20–26, reactions 27–31, and Eqs. 32–37, respectively.

Mechanism 1

$$>Fe^{III}OH + HC_2O_4^- \rightleftharpoons\ >Fe^{III}C_2O_4^- + H_2O \text{ (adsorption equilibrium)} \tag{16}$$

$$>Fe^{III}C_2O_4^- \underset{k_{-1}}{\overset{h\nu_{CT}}{\rightleftharpoons}} >Fe^{II}C_2O_4^{\cdot -}$$
$$\text{(ligand-to-metal charge-transfer transition)} \tag{17}$$

$$>Fe^{II}C_2O_4^{\cdot -} + H_2O \xrightarrow{k_2} >Fe^{II}OH_2 + CO_2^{\cdot -} + CO_2$$
$$\text{(decarboxylation and rehydration of the reduced surface site)} \tag{18}$$

$$>Fe^{II}OH_2 \xrightarrow{k_3} Fe^{II}(aq) + > + H_2O$$

(detachment of the surface group)   (19)

where $>$ denotes the lattice surface of hematite.

Under continuous irradiation, the concentration of the charge-transfer state reaches rapidly a stationary value:

$$\frac{d\{>Fe^{II}C_2O_4^{\dot{-}}\}}{dt} = I_A - k_{-1}\{>Fe^{II}C_2O_4^{\dot{-}}\} - k_2\{>Fe^{II}C_2O_4^{\dot{-}}\} = 0 \tag{20}$$

$$I_A = \{>Fe^{II}C_2O_4^{\dot{-}}\}(k_{-1} + k_2) \tag{21}$$

The quantum yield is defined as follows (Balzani and Carassiti, 1970):

$$\Phi_\lambda = \frac{\text{number of molecules undergoing a photochemical primary process}}{\text{number of photons absorbed by the reactant}}$$

(22)

If we consider $>Fe^{II}OH_2$ to be a primary photoproduct, then the quantum yield is, in terms of rates:

$$\Phi_\lambda = \frac{k_2\{>Fe^{II}C_2O_4^{\dot{-}}\}}{I_A} .2 \tag{23}$$

and for the steady state of $\{>Fe^{II}C_2O_4^{\dot{-}}\}$:

$$\Phi_\lambda = \frac{k_2}{k_{-1} + k_2} .2 \tag{24}$$

The factor 2 accounts for the reduction of a second surface iron(III) by $CO_2^{\dot{-}}$ in a thermal reaction.

From Eqs. 14, 15, and 24 follows

$$r_\lambda = k\{>Fe^{III}C_2O_4^-\} \qquad \text{(in mol m}^{-2}\text{ h}^{-1}\text{)} \tag{25}$$

where

$$k = I'_{0\lambda}\varepsilon_\lambda L 2.3 \frac{k_2}{k_{-1} + k_2} .2 \qquad \text{(in h}^{-1}\text{)} \tag{26}$$

Mechanism 2

$$\alpha\text{-}Fe_2O_3 \underset{k_{IC}}{\overset{h\nu}{\rightleftharpoons}} \alpha\text{-}Fe_2O_3^*(e_{ss}^-, h_{ss}^+) \qquad \text{(electronic excitation)} \tag{27}$$

$$\alpha\text{-}Fe_2O_3^* (e_{ss}^-, h_{ss}^+) + C_2O_4^{2-}(\text{adsorbed}) \xrightarrow{k_1} >Fe^{II}C_2O_4^{\dot{-}}$$

(electron transfer)   (28)

$$>Fe^{II}C_2O_4^{\dot{-}} \xrightarrow{k_{-1}} >Fe^{III}C_2O_4^{-} \qquad \text{(back electron transfer)} \quad (29)$$

$$>Fe^{II}C_2O_4^{\dot{-}} + H_2O \xrightarrow{k_2} >Fe^{II}OH_2 + CO_2^{\dot{-}} + CO_2$$

$$\text{(decarboxylation and rehydration of the reduced surface site)} \quad (30)$$

$$>Fe^{II}OH_2 \xrightarrow{k_3} Fe^{II}(aq) + > + H_2O \qquad \text{(detachment of the surface}$$

$$\text{group)} \quad (31)$$

where (e_{ss}^{-}, h_{ss}^{+}) stands for the photoelectron localized on a surface iron center and the photohole localized on an oxygen ion of the surface lattice and k_{IC} is the rate constant of the internal conversion.

For a stationary value of the excited state follows

$$\frac{d\{\alpha - Fe_2O_3^*\}}{dt} = I_A - k_{IC}\{\alpha\text{-}Fe_2O_3^*\} - k_1\{\alpha\text{-}Fe_2O_3^*\}\{C_2O_4^{2-}\}_{ads} = 0 \quad (32)$$

$$I_A = \{\alpha - Fe_2O_3^*\}(k_{IC} + k_1\{C_2O_4^{2-}\}_{ads}) \quad (33)$$

$$\Phi_\lambda = \frac{k_2\{>Fe^{II}C_2O_4^{\dot{-}}\}}{I_A}.2 \quad (34)$$

For the steady state of $\{>Fe^{II}C_2O_4^{\dot{-}}\}$, Eq. 34 is transformed to

$$\Phi_\lambda = \frac{k_2 k_1\{C_2O_4^{2-}\}_{ads}}{(k_{-1} + k_2)(k_{IC} + k_1\{C_2O_4^{2-}\}_{ads})}.2 \quad (35)$$

From Eqs. 14, 15, and 35 follows

$$r_\lambda = k'\{\alpha - Fe_2O_3\}\{C_2O_4^{2-}\}_{ads} \qquad \text{(in mol m}^{-2}\text{h}^{-1}) \quad (36)$$

where

$$k' = I'_{0\lambda}\varepsilon_\lambda L2.3 \frac{k_2 k_1}{(k_{-1} + k_2)(k_{IC} + k_1\{C_2O_4^{2-}\}_{ads})}.2 \qquad \text{(in m}^2\text{ mol}^{-1}\text{h}^{-1}) \quad (37)$$

The rate expressions is Eqs. 25, 26 and 36, 37 show that (1) for both mechanisms the rate of photochemical surface Fe^{II} formation, $d\{Fe^{II}\}_{surf}/dt$, depends on the concentration of *adsorbed oxalate* and (2) for mechanism 2, the rate also depends on the solid concentration. Furthermore, the quantum yield of surface iron(II) formation depends on whether mechanism one or two is operative.

The rate of formation of dissolved iron(II), $d[Fe^{2+}]_{diss}/dt$, depends, in addition, on the efficiency of detachment of reduced surface iron ions from the crystal lattice. The detachment step is a key step in the overall dissolution kinetics of slightly soluble minerals, since it is assumed to be the rate-determining step (Stumm and Furrer, 1987). The efficiency of detachment depends primarily on the crystallinity, and thus the stability of the iron(III) hydroxide phase, and also on the coordinative surrounding of the reduced surface metal centers. It has been shown that a combination of a reductant and a ligand that forms stable surface complexes in the dark is especially efficient for the thermal reductive dissolution of hydrous iron(III) oxides (Banwart, et al. 1989). The role of a ligand as an electron donor *and* as a "detacher" in the photochemical dissolution of hydrous iron(III) oxides remains to be elucidated.

4.3.2. Rate Expression, Taking into Account the Other Reaction Pathways.

Since the rate of the photochemical dissolution via surface photoredox reaction depends on the surface concentration of the adsorbed oxalate and not the solution concentration, it is constant under the experimental conditions of this case study, whereas the rate of the homogeneous photoreduction of iron(III) depends on the concentration of dissolved iron(III) trioxalato complexes. The rate of the photochemical iron(II) production is the sum of the rates of the heterogeneous and the homogeneous photoredox reaction. Since, under the

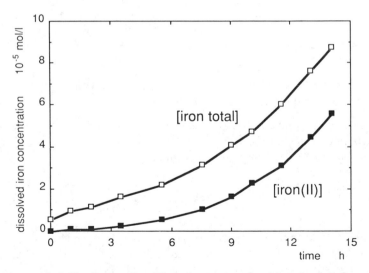

Figure 8. Autocatalytic dissolution of hematite in the presence of oxalate under the influence of light. Experiemntal conditions: 0.5 g L^{-1}, hematite; initial oxalate concentration 3.3 m mol L^{-1}; pH = 3.0; nitrogen atmosphere; incident light intensity $I_0 = 4$ W m^{-2}; $\lambda = 375$ nm; reaction volume 250 mL; irradiated surface 50 cm^2 (Siffert, 1989).

experimental conditions of the experiment shown in Figure 8 there is no sink for Fe^{II}, the rate of dissolved Fe^{II} formation is

$$r_{Fe^{2+}} = \frac{d[Fe^{2+}]}{dt} = k_1\{>Fe^{III}C_2O_4^-\} + k_2[Fe^{III}(C_2O_4)_3^{3-}]_{diss} \qquad (38)$$

Since the first term is constant:

$$r_{Fe^{2+}} = const. + k_2[Fe^{III}(C_2O_4)_3^{3-}]_{diss} \qquad (39)$$

where

$$k_2 = I'_{0\lambda}\varepsilon_\lambda L 2.3\Phi_\lambda \qquad (compare\ Eq.\ 15) \qquad (40)$$

The photochemically formed iron(II) catalyzes the thermal dissolution of hematite, which results in a gradual increase in the concentration of dissolved iron(III). The rate of this reaction pathway depends on the surface concentration of adsorbed iron(II) at a given oxalate concentration (Suter et al., 1988). The sink of dissolve iron(III) is the homogeneous photoredox reaction. The rate of Fe^{3+} formation is thus

$$r_{Fe^{3+}} = \frac{d[Fe^{3+}]}{dt} = k_3\{Fe^{II}\}_{ads} = k_3\frac{K^sS[Fe^{II}]_{diss}}{1+K^s[Fe^{II}]_{diss}} - k_2[Fe^{III}(C_2O_4)_3^{3-}]_{diss} \qquad (41)$$

where K^s is the equilibrium constant for the adsorption of Fe^{II} and S is the maximum capacity of the hematite surface for the adsorption of Fe^{II} through a bridging ligand. These coupled reaction pathways lead to an autocatalytic dissolution mechanism as seen in Figure 8.

4.4. Influence of Oxygen on the Light-Induced Dissolution of Hematite in the Presence of Oxalate

Figure 9 shows the appearance of dissolved iron(II) and the disappearance of oxalate in an experiment carried out under nitrogen atmosphere. The reaction occurs according to the following stoichiometry:

$$\alpha\text{-}Fe_2O_3 + C_2O_4^{2-}\ (ads.) + 6H^+ \xrightarrow[N_2]{h\nu} 2Fe^{2+} + 2CO_2 + 3H_2O \qquad (42)$$

This is no longer true in presence of oxygen, as shown in Figure 10. In this case, oxalate is oxidized much faster than under nitrogen and the dissolution of hematite is inhibited. [Since this experiment was carried out at low pH, the difference in dissolved iron under nitrogen and oxygen cannot be explained in terms of reprecipitation of iron(III) in an oxygen environment.] Obviously a reaction takes place that competes with the reduction and subsequent detachment of iron(II) groups from the surface. Plausibly, adsorbed molecular oxygen

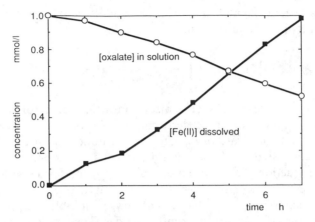

Figure 9. Light-induced dissolution of hematite in the presence of oxalate under *nitrogen atmosphere*. Experimental conditions: 0.5 g L^{-1} hematite; initial oxalate concentration 1 m mol L^{-1}; pH $= 3$; $I_0 \cong 4 \text{ kW m}^{-2}$; white light from a high-pressure xenon lamp after passing a Pyrex glass filter; irradiated surface $\cong 50 \text{ cm}^2$; reaction volume 250 mL (Siffert, 1989).

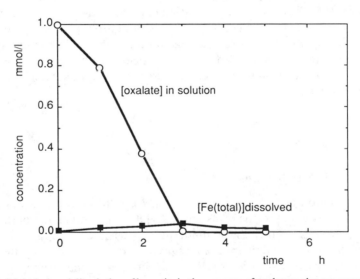

Figure 10. Light-induced dissolution of hematite in the presence of oxalate under *oxygen atmosphere*. Experimental conditions: 0.5 g L^{-1} hematite; initial oxalate concentration: 1 m mol L^{-1}; pH $= 3$; $I_0 \cong 4 \text{ kW m}^{-2}$; white light from a high-pressure xenon lamp after passing a Pyrex glass filter; irradiated surface $\cong 50 \text{ cm}^2$; reaction volume: 250 ml (Siffert, 1989).

reacts with the photoelectrons, e_{ss}^-, trapped on the iron(III) sites at the hematite surface:

$$O_2(\text{ads.}) + e_{ss}^- \xrightarrow[\text{hematite}]{h\nu} O_2^{\cdot -} \tag{43}$$

Thus, to a certain extent, hematite acts as a photocatalyst for the oxidation of oxalate in the presence of oxygen:

$$C_2O_4^{2-} (\text{ads.}) + 2H^+ + O_2 \xrightarrow[\text{hematite}]{hv} H_2O_2 + 2CO_2 \tag{44}$$

Oxidation of oxalate with OH radicals (formed, e.g. through the Fenton reaction) in addition to the heterogeneous photochemical oxidation may account for the higher rate of oxalate disappearance in an oxygen-saturated suspension, compared to the rate in a deaerated suspension.

Since the reduction of adsorbed molecular oxygen competes with detachment of the reduced surface iron from the crystal lattice, it is the efficiency of detachment that decides to what extent oxygen inhibits the photochemical reductive dissolution of hydrous iron(III) oxides. The efficiency of detachment depends above all on the crystallinity of the iron(III) hydroxide phase and is expected to be much higher with iron(III) hydroxide phases less crystalline and thus less stable than hematite. Not only does the efficiency of the light-induced dissolution of iron(III) hydroxides depend on their crystal and surface structure, but so does the efficiency of photoxidation of electron donors. Leland and Bard (1987) have reported that the rate constants of photooxidation of oxalate and sulfite varies by about two orders of magnitude with different iron(III) oxides. From their data they concluded that "this appears to be due to differences in crystal and surface structure rather than to difference in surface area, hydrodynamic diameter, or band gap."

5.   APPLICATION TO NATURAL SYSTEMS

5.1.   A Simple Steady-State Model

The accumulation of moles Fe^{II}, $V(d[Fe^{2+}]/dt)$, in the photic zone of a natural-water system, is equal to the mass inflow, $Q[Fe^{2+}]_{in}$, minus the outflow mass discharge, plus or minus the amount due to chemical (photochemical and thermal) reactions. In a first approximation the reduction of particulate iron(III) and the oxidation of dissolved iron(II) in the photic, oxygenated zone of a surface water can be expressed as

$$Fe(OH)_3(s) + \text{electron donor (ads.)} \xrightarrow{hv} Fe^{2+}(aq) + \text{oxidized electron donor} \tag{45}$$

$$Fe^{2+}(aq) + \tfrac{1}{4}O_2 + 2OH^- + \tfrac{1}{2}H_2O \xrightarrow{k_{ox}} Fe(OH)_3(s) \tag{46}$$

From the previous discussion and from experimental observations (Waite, 1986) it follows that the rate of the photochemical dissolution of hydrous iron(III) oxides is expected to depend on the concentration of the *specifically adsorbed electron donor*:

$$r = k_{red}\{>Fe^{III}L\} \tag{47}$$

where k_{red} is the empirically determined rate constant. The mass balance for dissolved iron(II) is then given by the following expression, applicable to a shallow lake, where sunlight penetrates to the bottom (Schnoor et al., 1989):

$$V\frac{d[Fe^{2+}]}{dt} = Q[Fe^{2+}]_{in} - Q[Fe^{2+}] - Vk_{ox}[Fe^{2+}][O_2(aq)][OH^-]^2$$

$$\text{accumulation} \qquad \text{inflow} \qquad \text{outflow} \qquad \text{oxidation/precipitation}$$

$$+ Vk_{red}\{>FeL\} \tag{48}$$

$$\text{reduction/dissolution}$$

where V = water volume
Q = volumetric flow rate
$[Fe^{2+}]_{in}$ = inflow concentration from seepage and/or streams
k_{ox} = oxidation rate constant
k_{red} = rate constant for light-induced reductive dissolution
$\{>Fe^{III}L\}$ = surface concentration of the specifically adsorbed electron donor

Assuming a steady state for Fe^{2+}, Eq. 48 can be transformed as follows:

$$[Fe^{2+}]_{ss} = \frac{[Fe^{2+}]_{in} + t_0 k_{red}\{>Fe^{III}L\}}{1 + t_0 k_{ox}[O_2(aq)][OH^-]^2} \tag{49}$$

where t_0 is the hydraulic retention time, $t_0 = (V/Q)$. This simple model assumes that the reduction of particulate iron(III) in the photic zone of a natural-water system occurs primarily via a photoredox reaction at the surface of iron(III) hydroxides and that the main oxidation pathway of iron(II) is the homogeneous oxidation by molecular oxygen. It neglects (1) the oxidation of Fe^{II} with oxidants other than O_2 (e.g., H_2O_2) and the surface catalyzed oxidation of iron(II) (Wehril, Chapter 11, this volume); (2) the reduction of iron(III) via homogenous photoredox reactions, which, as discussed, may be an important pathway of photochemical iron(II) formation in aquatic systems (Waite and Morel, 1984a, Miles and Brezonik, 1981); and (3) the biologically mediated reduction of iron(III) (Price and Morel, Chapter 8, this volume), a reaction that is also driven indirectly by light for it is coupled to the photochemical oxidation of water to $\dot{O}_2$ in the photosynthetic system of the phytoplankton cells. If important, these other reactions can be included in Eq. 48.

The kinetics of the homogeneous oxidation reaction of iron(II) under conditions typical of natural waters is well established (Stumm and Lee, 1961; Singer and Stumm, 1970). The rate constant, k_{red}, of the photochemical reductive dissolution of iron(III) hydroxides accounting for mechanism 1 is a function of wavelength-dependent and wavelength-independent factors:

$$k_{red} = \eta L 2.3 \int_{\lambda} I'_0(\lambda) \varepsilon(\lambda) \Phi(\lambda) \, d\lambda \qquad (50)$$

where η is the efficiency of detachment of reduced iron ions from the surface. As has been discussed, η is a key parameter in controlling the overall kinetics of the photochemical reductive dissolution of iron(III) hydroxides.

From this model one expects a light-intensity-dependent steady state concentration of dissolved iron(II) in the photic zone of a natural-water system. What steady-state concentration can be reached depends on the pH, the content and the type of dissolved organic matter, and above all the crystallinity of the iron(III) hydroxide phase.

5.2.   Natural System Study

In the photic zone of natural waters, the reactive authigenically formed iron(III) hydroxides are likely to be ferrihydrite. Such amorphous iron(III) phases may originate from mononuclear $Fe^{III}(OH)_i^{(3-i)}$, formed by oxygenation of Fe^{II} and subsequent hydrolysis. These amorphous iron(III) hydroxides are thermodynamically less stable than hematite and are expected to be more readily dissolved

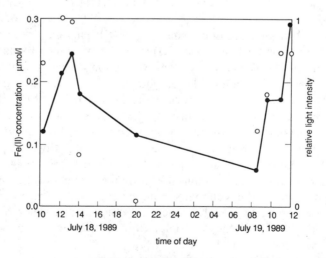

Figure 11. Diurnal variation of the concentration of dissolved Fe(II)● and of the incident light intensity ○ in Lake Cristallina. [The maximal measured light intensity is arbitrary set to one (Sulzberger et al., in press)].

than hematite in an oxic environment under the influence of light. An example is the X-ray amorphous iron(III) hydroxide phase in Lake Cristallina (in the southern Alps of Switzerland) that is formed by oxygenation and subsequent precipitation of ferrous iron. Lake Cristallina, which is an acidic lake at an elevation of 2400 m, has an average pH of 5.2. The primary source of iron(II) is biotite weathering (Giovanoli et al., 1988). Eventually iron(II) is oxidized and precipitated. Feldspar from the Cristallina catchment area is covered by a film of orange appearance and, at the bottom of the lake, areas of iron-rich sediment can be identified by the naked eye. Since the content of organic carbon is very low in Lake Cristallina (in the order of 0.5 mg L^{-1}), the light penetrates to the bottom. Figure 11 shows the diurnal variation of the concentration of dissolved iron(II) in Lake Cristallina. These results are interpreted in terms of photochemical reductive dissolution of the iron(III) hydroxide phase and oxidation and reprecipitation of iron(II). Addition of oxalate to a small enclosure within the lake had a tremendous effect on the photochemical iron(II) formation (Sulzberger et al., in press).

6. CONCLUSIONS

In photoredox reactions occurring at metal hydroxide surfaces the efficiency of the interfacial electron transfer depends on the type of surface coordination compound formed between a thermodynamically suitable ligand acting as electron donor or acceptor and a surface metal center. In the photochemical reductive dissolution of hydrous iron(III) oxide, it is, in addition, the efficiency of detachment of reduced iron ions from the surface that controls the overall kinetics of this process. The efficiency of detachment depends on the coordinative environment of reduced surface iron centers and on the crystallinity of the solid phase. Detachment of Fe^{II} from the surface is in competition with reduction of adsorbed molecular oxygen. With iron(III) hydroxide phases less crystalline than hematite, the inhibition of the photochemical reductive dissolution by oxygen is less pronounced. Since the rate of the photochemical reductive dissolution of iron(III) hydroxides depends on the concentration and the extinction coefficient of the chromophore involved in the surface photoredox reaction, more conclusive information is needed on the role of the surface complex as a chromophore. For such studies, a suitable ligand would be one that forms an inner-sphere surface complex with surface iron(III), exhibiting a ligand-to-metal charge-transfer band in a spectral window of the solid phase and being readily photolyzed. In the photic zone of a natural-water system dissolved iron(II) can be formed via different pathways: via heterogeneous and via homogeneous photoredox reactions. Also direct biological mediation of redox processes may lead to the formation of Fe^{II}. Furthermore, iron(II) acts as a catalyst in the thermal dissolution of iron(III) hydroxides. This Fe^{II}-catalyzed dissolution, of importance at the oxic-anoxic boundaries (Sulzberger et al., 1989), may also be operative in the photic zone of a natural-water system and lead to autocatalytic dissolution of iron(III) hydroxides. The diurnal cycle of the iron(II) concentration found in acidic waters is interpreted in terms of the photochemical reductive dissolution of particulate iron and oxidation and precipitation of iron(II). It is expected that also at a higher pH, usually encountered in surface waters, a light-intensity-dependent steady-state concentration of iron(II) is maintained by photochemical processes.

Acknowledgments

I thank Gion Calzaferri, University of Bern, and François M. M. Morel, Massachusetts Institute of Technology, for reviewing this chapter. Useful comments from Jürg Hoigné, Janet G. Hering, EAWAG, and Michael M. Hoffmann, California Institute of Technology, are appreciated. I am grateful to Jerald L. Schnoor, The University of Iowa, for the fruitful collaboration in the field experiment. Above all, I thank Christophe Siffert for the collaboration and for allowing some of the results of his Ph.D. thesis to be published here. Last but not least I thank Werner Stumm for his support and his advice.

REFERENCES

Al-Ekabi, H., N. Serpone, E. Pelizzetti, C. Minero, M. A. Fox, and R. B. Draper (1989), "Kinetic Studies in Heterogeneous Photocatalysis. 2. TiO$_2$-Mediated Degradation of 4-Chlorophenol Alone and in a Three Component Mixture of 4-Chlopenol, 2,4-Dichlorophenol, and 2,4,5-Trichlorophenol in Air-Equilibrated Aqueous Media," *Langmuir* **51**, 250–255.

Anderson, M. A., and F. M. M. Morel (1980), "Uptake of Fe(II) by a Diatom in Oxic Culture Medium", *Marine Biology Letters* **1**, 263–268.

Arnold, R. G., T. J. DiChristina, and M. R. Hoffmann (1986), "Dissimilative Fe(III) Reduction by Pseudomonas sp. 200-Inhibitor Studies," *Appl. Environ. Microbiol.* **52**, 281–294.

Ballhausen, C. J., and H. B. Gray (1965), *Molecular Orbital Theory*, Benjamin, New York.

Balzani, V., and V. Carassiti (1970), *Photochemistry of Coordination Compounds*, Academic Press, London.

Balzani, V., L. Moggi, and F. Scandola (1987), "Towards a Supramolecular Photochemistry: Assembly of Molecular Components to Obtain Photochemical Molecular Devices", in V. Balzani, Ed., *Supramolecular Photochemistry*, NATO ASI Series, Ser. C, Vol. 214, Reidel, Dordrecht, Netherlands.

Banwart, S., S. Davies, and W. Stumm (1989), "The Role of Oxalate in Accelerating the Reductive Dissolution of Hematite (α-Fe$_2$O$_3$) by Ascorbate," *Colloids Surf.* **39**, 303–309.

Bard, A. J. (1979), "Photoelectrochemistry and Heterogeneous Photocatalysis at Semiconductors," *J. Photochem.* **10**, 59–75.

Blesa, M. A., H. A. Marinovich, E. C. Baumgartner, and A. J. G. Maroto (1987), "Mechanism of Dissolution of Magnetite by Oxalic Acid–Ferrous Ion Solutions", *Inorg. Chem.* **26**, 3713–3717.

Calzaferri, G. (1981), "Umwandlung von Lichtenergie in chemische Energie II. Experimente zu diesem Thema," *Chimia* **35**, 209–220.

Calzaferri, G. (1986), "Photoredox–Reaktionen", *Chimia* **40**, 74–93.

Calzaferri, G., and H. R. Grüniger (1979), "Charge-Transfer-Uebergänge im Pentacyano-benzonitril-ferrat (II)," *Helv. Chimica Acta* **62**, 1112–1120.

Collienne, R. H. (1983), "Photoreduction of Iron in the Epilimnion of Acidic Lakes", *Limnol. Oceanogr.* **28**, 83–100.

Cornell, R. M., and P. W. Schindler (1987), "Photochemical Dissolution of Goethite in Acid/Oxalate Solution," *Clays Clay Min.* **35**, 347–352.

Darwent, J. R., and A. Lepre (1986). "Interfacial Electron Transfer in Colloidal TiO$_2$ Accelerated by Surface Adsorption and the Electric Double Layer," *J. Chem. Soc., Faraday Trans. 2* **82**, 2323–2335.

Desilvestro, J., S. Pons, E. Vrachnou, M. Grätzel (1988), "Electrochemical and FTIR Spectroscopic Characterization of Ferrocyanide-Modified TiO$_2$ Electrodes Designed for Efficient Photosensitization," *J. Electroanal. Chem.* **246**, 411–422.

Dzombak, D., and F. M. M. Morel (1990), *Aquatic Sorption: Stability Constants for Hydrous Ferric Oxide*, Wiley-Interscience, New York.

Faust, B. C., and M. R. Hoffmann (1986), "Photoinduced Reductive Dissolution of α-Fe$_2$O$_3$ by Bisulfite," *Environ. Sci. Technol.* **20**, 943–948.

Faust, B. C., and J. Hoigné (in press), "Photolysis of Fe(III)–Hydroxy Complexes as Sources of OH Radicals in Clouds, Fog and Rain," *Atmosph. Environ.*

Finden, D. A. S., E. Tipping, G. H. M. Jaworski, and C. S. Reynolds (1984), "Light-Induced Reduction of Natural Iron(III) Oxide and Its Relevance to Phytoplankton," *Nature* **309**, 783–784.

Frank, S. N., and A. J. Bard (1977), "Heterogeneous Photocatalytic Oxidation of Cyanide and Sulfite in Aqueous Solutions at Semiconductor Powders," *J. Phys. Chem.* **81**, 1484–1488.

Giovanoli, R., J. L. Schnoor, L. Sigg, W. Stumm, and J. Zobrist (1988), "Chemical Weathering of Crystalline Rocks in the Catchment Area of Acidic Ticino Lakes, Switzerland," *Clays Clay Min.* **36**, 521–529.

Goodenough, J. B. (1971), "Metallic Oxides", in H. Reiss, Ed., *Progress in Solid-State Chemistry*, Vol. 5, Pergamon Press, Oxford.

Harriman, A., G. Porter, and A. Wilowska (1983), "Photoreduction of Benzo-1,4-quinone Sensitised by Metalloporphyrins," *J. Chem. Soc., Faraday Trans. 2* **79**, 807–816.

Henglein, A. (1984), "Catalysis of Photochemical Reactions by Colloidal Semiconductors," *Pure Appl. Chem.* **56**, 1215–1224.

Hayes, K. F., and J. O. Leckie (1986), "Mechanism of Lead Ion Adsorption at the Goethite/Water Interface," in *Geochemical Processes at Mineral Surfaces*, J. A. Davis and K. F. Hayes, Eds. (ACS Symposium Series No. 323), American Chemical Socieity, Washington, DC.

Hayes, K. F., A. L. Roe, G. E. Brown, Jr., K. O. Hodgson, J. O. Leckie, and G. A. Parks (1987), "In situ X-Ray Absorption Study of Surface Complexes: Selenium Oxyanions on α-FeOOH," *Science* **238**, 783–786.

Herrmann, J. M., M. N. Mozzanega, and P. Pichat (1983), "Oxidation of Oxalic Acid in Aqueous Suspensions of Semiconductors Illuminated with UV or Visible Light," *J. Photochem.* **22**, 333–343.

Kormann, C., D. W. Bahnemann, and M. R. Hoffmann (1988), "Photocatalytic Production of H$_2$O$_2$ and Organic Peroxides in Aqueous Suspensions of TiO$_2$, ZnO, and Desert Sand," *Environ. Sci. Technol.* **22**, 798–806.

Lehn, J. M. (1987), "Photophysical and Photochemical Aspects of Supramolecular Chemistry," in V. Balzani, Ed., *Supramolecular Photochemistry*, NATO ASI Series, Ser. C, Vol. 214, Reidel, Dordrecht, Netherlands.

Leland, K. L., and A. J. Bard (1987), "Photochemistry of Colloidal Semiconducting Iron Oxide Polymorphs," *J. Phys. Chem.* **91**, 5076–5083.

Litter, M. I., and M. A. Blesa (1988), "Photodissolution of Iron Oxides. I. Maghemite in EDTA Solutions," *J. Colloid Interface Sci.* **125**, 679–687.

Marfunin, A. S. (1979), *Physics of Minerals and Inorganic Materials*, Springer-Verlag, Berlin, Heidelberg, New York.

Martin, J. H., S. E. Fitzwater (1988), "Iron Deficiency Limits Phytoplankton Growth in the North-East Pacific Subarctic," *Nature* **331**, 341–343.

Marusak, L. A., R. Messier, and W. B. White (1980), "Optical Absorption Spectrum of Hematite, α-Fe_2O_3 Near IR to UV," *J. Phys. Chem. Solids* **41**, 981–984.

McKnight, D. M., B. A. Kimball, and K. E. Bencala (1988), "Iron Photoreduction and Oxidation in an Acidic Mountain Stream," *Science* **240**, 637–640.

Miles, C. J., and P. L. Brezonik (1981), "Oxygen Consumption in Humic-Colored Waters by a Photochemical Ferrous-Ferric Catalytic Cycle," *Environ. Sci. Technol* **15**, 1089–1095.

Motschi, H. (1987), "Aspects of the Molecular Structure in Surface Complexes; Spectroscopic Investigations", in W. Stumm, Ed., *Aquatic Surface Chemistry*, Wiley-Interscience, New York.

Parker, C. A. (1968), *Photoluminescence of Solutions*, Elsevier, Amsterdam, pp. 208–214.

Parker, C. A., and C. G. Hatchard (1959), "Photodecomposition of Complex Oxalates. Some Preliminary Experiments by Flash Photolysis," *J. Phys. Chem.* **63**, 22–26.

Penners, N. H. G., and L. K. Koopal (1986), "Preparation and Optical Properties of Homodisperse Hematite Hydrosols," *Colloids Surf.* **19**, 337–349.

Pichat, P., and M. A. Fox (1988), "Photocatalysis on Semiconductors," in M. A. Fox and M. Chanon, Eds., *Photoindenced Electron Transfer*, Part D, Elsevier, Amsterdam.

Rehm, D., and A. Weller (1970), "Kinetics of Fluorescence Quenching by Electron and H-Atom Transfer," *Israel J. Chem.* **8**, 259–271.

Rich, H. W., and F. M. M. Morel (in press), "Availability of Well-Defined Iron Colloids to the Marine Diatom Thalassiosira Weissflogii," *Limnol. Oceanogr.*

Schindler, P. W., and W. Stumm (1987), "The Surface Chemistry of Oxides, Hydroxides and Oxide Minerals," in W. Stumm, Ed., *Aquatic Surface Chemistry*, Wiley-Interscience, New York.

Schnoor, J. L., R. Giovanoli, L. Sigg, W. Stumm, B. Sulzberger, and J. Zobrist (1989), "Fate of Iron and Aluminum in Lake Cristallina (Switzerland)," *EAWAG News*, No. 26/27.

Siffert, C. (1989), "L'effet de la Lumière sur la Dissolution des Oxydes de Fer(III) dans les Milieux Aqueux," Ph.D. thesis, ETH, Zürich, No. 8852

Siffert, C., B. Sulzberger, and W. Stumm, manuscript in preparation.

Singer, P. C., and W. Stumm (1970), "Acidic Mine Drainage: The Rate-Determining Step," *Science* **167**, 1121–1123.

Sposito, G. (1984), *The Surface Chemistry of Soils*, Oxford University Press, New York.

Stramel, R. D., and J. K. Thomas, (1986), "Photochemistry of Iron Oxide Colloids," *J. Colloid Interface Sci.* **110**, 121–129.

Stumm, W., and G. Furrer (1987), "The Dissolution of Oxides and Aluminium Silicates; Examples of Surface-Coordination-Controlled Kinetics," in W. Stumm, Ed., *Aquatic Surface Chemistry*, Wiley-Interscience, New York.

Stumm, W., and G. F. Lee, (1961) "Oxygenation of Ferrous Iron," *Indust. Eng. Chem.* **53**, 143–146.

Stumm, W., R. Kummert, and L. Sigg (1980), "A Ligand Exchange for the Adsorption of Inorganic and Organic Ligands at Hydrous Oxide Interfaces," *Croat. Chem. Acta* **53**, 291–312.

Sulzberger, B. (1983), "Experimente zur photochemischen Energiespeicherung mit metall-beladenen Zeolithen," Ph.D. thesis, University of Bern, Bern, Switzerland.

Sulzberger, B., D. Suter, C. Siffert, S. Banwart, and W. Stumm (1989), "Dissolution of Fe(III)(hydr)oxides in Natural Waters; Laboratory Assessment on the Kinetics Controlled by Surface Chemistry," *Mar. Chem.* **28**, 127–144.

Sulzberger B., J. L. Schnoor, R. Giovanoli, J. G. Hering, and J. Zobrist (in press), "Biogeochemistry of Iron in an Acidic Lake," *Aquatic Sciences*.

Suter, D., C. Siffert, B. Sulzberger, and W. Stumm (1988), "Catalytic Dissolution of Iron(III) (hydr)oxides by Oxalic acid in the Presence of Fe(II)," *Naturwissenschaften* **75**, 571–573.

Tazuke, S., N. Kitamura, and H. B. Kim (1987), "Photoinduced Looping Electron Transfer. What Occurs Between Electron Transfer and Charge Separation?," in V. Balzani, Ed., *Supramolecular Photochemistry*, NATO ASI Series, Ser. C, Vol. 214, Reidel, Dordrecht, Netherlands

Vincze, L., and S. Papp (1987), "Individual Quantum Yields of $Fe^{3+} OX_n^{2-} H_m^+$ Complexes in Aqueous Acidic Solutions ($OX^{2-} = C_2O_4^{2-}$, $n = 1-3$, $m = 0, 1$)," *J. Photochem.* **36**, 289–296.

Vrachnou, E., N. Vlachopoulos, and M. Grätzel (1987), "Efficient Visible Light Sensitization of TiO_2 by Surface Complexation with $Fe(CN)_6^{4-}$," *J. Chem. Soc., Chem. Commun.* 868–870.

Waite, T. D. (1986), "Photoredox Chemistry of Colloidal Metal Oxides," in *Geochemical Processes at Mineral Surfaces*, J. A. Davis and K. F. Hayes, Eds. (ACS Symposium Series No. 323), American Chemical Society, Washington, DC.

Waite, T. D., and F. M. M. Morel (1984a), "Photoreductive Dissolution of Colloidal Iron Oxide: Effect of Citrate," *J. Colloid Interface Sci.* **102**, 121–137.

Waite, T. D., and F. M. M. Morel (1984b), "Photoreductive Dissolution of Colloidal Iron Oxides in Natural Waters," *Environ. Sci. Technol.* **18**, 860–868.

Zeltner, W. A., E. C. Yost, M. L. Machesky, M. I. Tejedor–Tejedor, and M. A. Anderson (1986), "Characterization of Anion Binding on Goethite Using Titration Calorimetry and Cylindrical Internal Reflection–Fourier Transform Infrared Spectroscopy," *Geochemical Processes at Mineral Surfaces*, J. A. Davis and K. F. Hayes, Eds. (ACS Symposium Series No. 323), American Chemical Society, Washington, DC.

Zinder, B., and W. Stumm (1985), "Die Auflösung von Eisen(III)-oxid; ihre Bedeutung im See und im Boden," *Chimia* **39**, 280–288.

15

RATE AND MECHANISM OF DISSOLUTION OF CARBONATES IN THE SYSTEM CaCO$_3$–MgCO$_3$

Roland Wollast

Laboratoire d'Océanographie, Université Libre de Bruxelles, Brussels, Belgium

1. INTRODUCTION

Carbonate minerals are among the most reactive minerals found at the earth's surface. The dissolution of calcite and dolomite, the two dominant carbonates exposed to weathering, represents roughly 50% of the chemical denudation of the continents (Wollast and Mackenzie, 1983; Meybeck, 1987). Calcium and magnesium bicarbonates are therefore the most abundant ions present in freshwaters. In the oceanic environment, dissolution of calcareous skeletons of organisms, in the water column or in the sediments after burial, drastically affects the accumulation of the carbonates in marine deposits. Only 20% of the CaCO$_3$ produced by marine organisms is preserved in the sediments in areas where seawater and porewater remain oversaturated with respect to the carbonate phase (Wollast, 1981). It is therefore not surprising that numerous kinetic studies have been devoted to the elucidation of the mechanism of dissolution of carbonates. Although a large variety of carbonate minerals has already been studied, we will discuss here mainly the case of the mineral system CaCO$_3$–MgCO$_3$ because of its practical importance.

It is a well-known fact that the rate of dissolution of carbonate minerals is considerably enhanced under strong acid conditions. In fact, the rate of the chemical reaction becomes so fast at low pH that the rate of dissolution is limited by the transport of the reacting species between the bulk of the solution and the surface of the mineral (Berner and Morse, 1974). The rate can then be described in terms of diffusion (molecular or turbulent) of the reactants and products through a stagnant boundary layer. The thickness of this layer is strongly dependent on the local turbulence generated near the solid–liquid interface and is related, for instance, to the stirring rate of a suspension of carbonate grains in the aqueous phase. However, the conditions required for observation of a diffusion-controlled mechanism seldom are found in nature and are of limited interest from a chemical point of view. We will thus focus on the kinetics of dissolution of carbonate minerals under relatively mild conditions, which is more characteristic of natural environments. Experimental results on the diffusion controlled mechanism and its theoretical interpretation can be found in Berner and Morse (1974), Plummer and Wigley (1976), Plummer et al., (1978, 1979), Rickard and Sjöberg (1983), and Sjöberg and Rickard (1984).

The earlier investigations of the dissolution kinetics of carbonates were mainly limited to the influence, on the rate of reaction, of pH or, more precisely, of the degree of undersaturation of aqueous solution with respect to the mineral. Also, the experiments were generally conducted in batch reactors where several variables change simultaneously with the progress of the reaction. Often only one chemical variable, either pH or p_{CO_2}, is more or less well controlled during the course of the dissolution. These unfavorable experimental conditions led to empirical expressions of the rate law, where the rate of dissolution was generally expressed as a function of the degree of undersaturation raised to an nth power [for a general review, see Morse, (1983)].

Recently, more systematic and rigorous approaches have been used to study the influence of various individual parameters on the rate of reaction. They allow one to formulate rate equations in terms of the effect of the concentrations of individual components. In turn, the rate dependence with respect to the concentration of reactants and products participating in the dissolution and reprecipitation reactions permits the identification of the mechanism and the elementary steps responsible for the rate control of the various kinetic processes involved.

2. SURFACE REACTION-CONTROLLED RATE OF DISSOLUTION FAR FROM EQUILIBRIUM

Under most of the natural conditions, the rate of dissolution of carbonate minerals is far less than that expected for rate control by diffusion. The chemical reaction at the water–mineral interface is then assumed to be the rate-determining step. This reaction consists in the attachment or interactions of reactants with specific surface sites where the critical crystal bonds are weakened, which, in turn, allows the detachment of anions and cations of the surface into the solution.

In terms of transition-state theory, the rate of the surface reaction can be related to that of decomposition of an activated surface complex. The rate of production of this activated complex is rapid enough to assume that it is in equilibrium with the reactants responsible for its formation. In homogeneous systems it is usually easy to identify the nature of the elementary steps leading to the formation of the activated complex by investigating the effect of the concentration of reactants and products on the reaction rate. For heterogeneous systems involving a solid phase, the problem is much more complicated since the formation of the activated complex depends to a large extent on the properties of the mineral–water interface, which are presently poorly understood.

It seems, however, that single carbonate minerals exhibit a rather simple behavior that allows a fruitful application of some elementary concepts of the kinetic theory. In the $CaCO_3$–H_2O system, the dissolution of calcium carbonate may occur according to the reactions

$$CaCO_3 \longrightarrow Ca^{2+} + CO_3^{2-} \tag{1}$$

$$CaCO_3 + H^+ \longrightarrow Ca^{2+} + HCO_3^- \tag{2}$$

$$CaCO_3 + 2H^+ \longrightarrow Ca^{2+} + H_2CO_3 \tag{3}$$

$$CaCO_3 + H_2CO_3 \longrightarrow Ca^{2+} + 2HCO_3^- \tag{4}$$

where Ca^{2+}, CO_3^{2-}, and HCO_3^- represent total dissolved species (free ions plus ion pairs). One can reasonably expect that one or more of these reactions constitute the elementary steps for the formation of the surface activated complex. The reaction rate depends then on the concentration of the reacting species appearing in the reactions involved in the formation of the surface complexes. This can be investigated experimentally by measuring the individual effect of the concentration of each reactant on the rate of dissolution. It is also preferable in a first approach to use conditions far from the dissolution equilibrium in order to avoid the influence of the backward precipitation reactions.

The first comprehensive study of the influence of the concentration of the reactants on the rate of dissolution of carbonates has been performed fairly recently by Plummer et al. (1978) and Busenberg and Plummer (1986) in the case of calcite and aragonite and by Busenberg and Plummer (1982) in the case of dolomite.

Figure 1 shows the influence of pH on the rate of dissolution of calcite single crystals in near absence of CO_2 and far from equilibrium. Below pH 5, the rate of dissolution is strictly proportional to the activity of H^+ and becomes independent of pH above this value. In other words, reaction 2 prevails under acidic conditions and reaction 1, under alkaline conditions. These authors have also shown (Fig. 2) that carbonic acid contributes markedly to the dissolution process under high p_{CO_2} conditions. The rate dependence on the concentration of H_2CO_3 is again simply first-order and corresponds to the stoichiometry of reaction 4. The overall rate of dissolution can thus be described by

$$r_d = k_1 a_{H^+} + k_2 a_{H_2CO_3} + k_3 \tag{5}$$

In terms of surface complexes, the first term in this equation corresponds to the protonation of the surface and the second one, to its carbonatation. The third

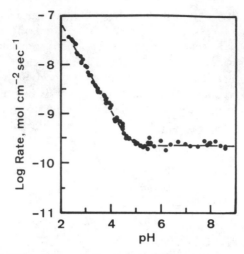

Figure 1. Influence of pH on the rate of dissolution of calcite at low p_{CO_2} at pH below 9. (After Busenberg and Plummer, 1986).

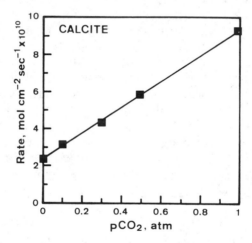

Figure 2. Influence of p_{CO_2} on the rate of dissolution of calcite at constant pH (5.6–5.7). Adapted from the experimental results, Table 6, of Busenberg and Plummer (1986).

one can be considered as the hydration of surface sites of the calcite minerals to form an activated complex that then decomposes rapidly. In Eq. 5, the first term prevails under acidic conditions and the second, under high p_{CO_2}. In most natural environments the third term, corresponding to a simple hydration of the surface of calcite and the detachment of Ca^{2+} and CO_3^{2-} ions, must be the dominant process. The rate dependence on the reactants for the dissolution of aragonite is similar to that of calcite (Busenberg and Plummer, 1986). On the other hand, Busenberg and Plummer (1982) found that the rate dependence of the dissolution of dolomite with respect to protons and to p_{CO_2} was only 0.5.

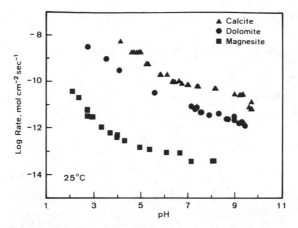

Figure 3. Influence of pH on the rate of dissolution of carbonates in the system $CaCO_3$–$MgCO_3$. (After Chou et al., in press.)

These results were confirmed to a large extent during a comparative study of the rate of dissolution of various carbonates performed with continuous-flow reactor (Chou et al., in press). The results of this study for calcite, dolomite, and magnesite are shown in Figure 3. Also, other single carbonates investigated during this study (aragonite and witherite) exhibit a kinetic behavior during dissolution similar to that of calcite. In the case of dolomite, the formation of the activated complex on the surface of the mineral is more complicated. In the study by Chou et al. (in press), the reaction order was also found to be fractional, but equal to 0.75 instead of 0.5 as found by Busenberg and Plummer (1982).

These authors have suggested that the fractional order reflects the existence of a two-step process. They found experimental evidences suggesting that there is a first rapid step corresponding to the reaction

$$CaMg(CO_3)_2 + H \overset{K_1}{\rightleftharpoons} MgCO_3 + Ca^{2+} + HCO_3^- \qquad (6)$$

The rate-limiting step is the protonation of the $MgCO_3$ site followed by the detachment of the corresponding ions:

$$H^+ + MgCO_3 \overset{k_2}{\longrightarrow} Mg^{2+} + HCO_3^- \qquad (7)$$

The rate of this reaction is given by

$$r = k_2 a_{H^+} X_{MgCO_3} \qquad (8)$$

where X_{MgCO_3} represents the mole fraction of the surface of dolomite covered by $MgCO_3$ sites. Since the first step of the dissolution of dolomite is very rapid, it may be assumed that reaction 6 is almost at equilibrium, and thus we have

$$K_1 = \frac{a_{Ca^{2+}} \, a_{HCO_3^-} X_{MgCO_3} \, \gamma_{MgCO_3}}{a_{H^+}} \tag{9}$$

where γ_{MgCO_3} is the activity coefficient for the $MgCO_3$ species at the surface of dolomite. Substituting X_{MgCO_3} in the rate equation gives

$$r = \frac{k_2 K_1 a_{H^+}^2}{a_{Ca^{2+}} \, a_{HCO_3^-} \gamma_{MgCO_3}} \tag{10}$$

The classical kinetic approach used for homogeneous reactions indicates that the reaction order should be 2 rather than 0.5 if such a mechanism is considered. However, there are drastic simplifications in this approach. If the surface sites do not behave ideally, γ_{MgCO_3} may, in turn, be a complicated function of the reactants and, for example, of a_{H^+}. This may reduce the overall reaction order with respect to a_{H^+}. Existence of electrical surface charges and the necessity to protonate successively a given metal center have been evoked in order to explain the fractional order usually observed in the case of oxides and silicates (see Schott, Chapter 12, and Stumm and Wieland, Chapter 13, both in this volume). A better understanding of the origin of the proton rate dependence thus requires a better knowledge of surface properties of the mineral surface.

Finally, it is interesting to compare the values of the rate constants obtained for the various carbonates (Table 1). If we first compare the various values of the

TABLE 1. Comparison of Rate Constants of Dissolution and Precipitation of Simple Carbonates Determined in Various Experimental Studies[a]

	$\log k_1$	$\log k_2$	$\log k_3$	$\log k_{-3}$	Reference
Calcite					
Powders	−4.29	−7.47	−9.93		Plummer et al. (1978)
Single crystal	−5.06	−7.32	−9.93		Busenberg and Plummer (1986)
Powders	−4.05		−10.19	−1.73	Chou et al. (1989)
Powders				−2.39	Nancollas and Reddy (1971)
Powders				−1.82 to −1.93	Inskeep and Bloom (1985)
Aragonite					
Single crystal	−5.06	−7.32	−9.46	−2.38[b]	Busenberg and Plummer (1986)
Powders	−3.92		−10.00	−1.78	Chou et al. (in press)
Magnesite					
Powders	−8.60		−13.35		Chou et al. (in press)

[a] Symbols according to Eq. (18) and rates expresses as moles per square centimeter per second.
[b] Value: calculated here from Figure 6.

forward rate constants obtained for the same mineral, for example, calcite, the agreement is usually very satisfactory except for k_1. At low pH the reaction rate is very rapid and the transport may become rate-limiting, especially in the case of single crystal or with the rotating-disk technique. The systematically low values of k_1 obtained under these conditions may be attributed to the mixed kinetic control, that is, diffusion and surface control, of the dissolution process (Rickard and Sjöberg, 1983). If we now consider the two polymorphs of $CaCO_3$, the rate of dissolution of aragonite is slightly higher than that of calcite, which reflects the small difference in bond energy between the two crystal structures. Witherite exhibits rates of dissolution similar to those of $CaCO_3$. However, the rate of dissolution of magnesite is several orders of magnitude lower, although the solubility of the $MgCO_3$ is of the same order of magnitude as those for the other single carbonates. This lower rate is probably due to the fact that it is very difficult to hydrate the compact $MgCO_3$ structure.

3. THE BACKWARD RATE OF DISSOLUTION NEAR EQUILIBRIUM

Many natural aquatic systems have a chemical composition close to saturation with respect to calcite or even dolomite. This is the case, for instance, for seawater, which is usually slightly oversaturated in the upper part of the water column and slightly undersaturated at greater depths. Under these conditions, the rates of both precipitation and dissolution contribute significantly to the overall rate of reaction. Even though the reaction paths may be very complex, there is a very direct and important link between the kinetic rate constants, according to which the rates of forward and reverse microscopic processes are equal for every elementary reaction. The fundamental aspect of this principle forms the essential aspect of the theory of irreversible thermodynamics (Prigogine, 1967).

For equilibrium to be reached, all elementary processes must have equal forward and backward rates. This differs from steady-state conditions, where only certain reactions and processes have balanced rates. Thus, at equilibrium, dissolution and precipitation rates of minerals should become necessarily equal. Note also that under equilibrium conditions the net flux of dissolved components at the water–mineral interface is equal to zero and the eventual limitation of the rate by the transport of reactants and products disappear. Close to equilibrium, the reaction kinetics always become entirely controlled by the surface reactions.

To illustrate this, let us consider the case of an aqueous system close to saturation with respect to calcite at rather high pH and low p_{CO_2}, corresponding to a classical natural situation. According to Eq. 5, the rate of dissolution is then controlled mainly by the hydration of the surface of calcite and the forward rate is a constant equal to k_3. The backward precipitation reaction corresponding to the stoichiometry of reaction 1 is given by

$$r_p = k_{-3} a_{Ca^{2+}} \, a_{CO_3^{2-}} \qquad (11)$$

and the overall dissolution rate is given by the difference between the forward and backward reactions:

$$r = r_d - r_p = k_3 - k_{-3} a_{Ca^{2+}} a_{CO_3^{2-}} \tag{12}$$

The equilibrium condition for reaction 1 specifies that

$$K_{eq} = a_{Ca^{2+}} a_{CO_3^{2-}} \tag{13}$$

corresponding to the condition that the net rate is equal to zero. According to Eq. 12, one thus obtains

$$a_{Ca^{2+}} a_{CO_3^{2-}} = \frac{k_3}{k_{-3}} = K_{eq} \tag{14}$$

It is interesting to point out that Eq. 12 can be rewritten as

$$r = k_3 \left(1 - \frac{k_{-3}}{k_{+3}} a_{Ca^{2+}} a_{CO_3^{2-}} \right) \tag{15}$$

or

$$r = k_3 \left(1 - \frac{a_{Ca^{2+}} a_{CO_3^{2-}}}{K_{eq}} \right) = k_3 (1 - \Omega) \tag{16}$$

where Ω represents conventionally the degree of saturation of the solution with respect to the mineral considered. This equation is similar to the empirical rate equation used by many authors:

$$r = k(1 - \Omega)^n \tag{17}$$

where n is often found to be a fractional number different from one.

If several parallel paths are leading to the dissolution of the carbonate minerals, the equilibrium condition requires that the net rates must cancel out. In the case of calcite, where we have three parallel reactions, we must thus have

$$r_{net} = k_1 a_{H^+} + k_2 a_{H_2CO_3} + k_3 - k_{-1} a_{HCO_3^-} a_{Ca^{2+}}$$
$$- k_{-2} a_{HCO_3^-}^2 a_{Ca^{2+}} - k_{-3} a_{Ca^{2+}} a_{CO_3^{2-}} = 0 \tag{18}$$

However, this condition is not sufficient to maintain the composition of the aqueous-phase constant, and each of the forward dissolution rates must also be compensated by the corresponding backward precipitation reaction (Lasaga, 1981). We must thus have simultaneously

$$k_1 a_{H^+} = k_{-1} a_{HCO_3^-} a_{Ca^{2+}} \tag{19}$$

$$k_2 a_{H_2CO_3} = k_{-2} a_{HCO_3^-}^2 a_{Ca^{2+}} \tag{20}$$

$$k_3 = k_{-3} a_{Ca^{2+}} \, a_{CO_3^{2-}} \tag{21}$$

These conditions must necessarily be fulfilled for any system reaching equilibrium regardless of the complexity of the reaction mechanism. However, as shown later, these constraints must be applied with great care, especially in the case of heterogeneous reactions. One must be sure that the system is really approaching equilibrium and that the solid phase precipitated has exactly the same composition and the same properties as the initial mineral.

Experimentally, the backward precipitation reaction has been investigated mainly at high pH and low p_{CO_2}. The decrease in the rate of dissolution of $CaCO_3$ observed at pH > 8 (Fig. 3) clearly shows that the rate of the backward reaction becomes important. This backward rate may be estimated by subtracting the net rate observed from the forward rate calculated according to Eq. 5. As shown here above, at high pH and low p_{CO_2} the backward rate should be proportional to the activity product $IAP_{CaCO_3} = a_{Ca^{2+}} \, a_{CO_3^{2-}}$ (Eq. 11). The results of these calculations applied to the experiments realized with the fluidized-bed reactor are shown in Figure 4. Despite the uncertainties involved in the experimental and calculation procedures, a good correlation between the calculated precipitation rate and the IAP_{CaCO_3} is observed. Furthermore, the slope of the regression line, corresponding to the value of k_{-3}, is equal to 1.9×10^{-2} mol cm^{-2} s^{-1}. Since the forward rate constant k_3 was found to be equal to 6.5×10^{-11} mol cm^{-2} s^{-1} in the same experiments, we have, according to Eq. 14

$$K_{calcite} = \frac{k_3}{k_{-3}} = 3.5 \times 10^{-9}$$

which is in very close agreement with the thermodynamic value equal to 3.3 $\times 10^{-9}$ (Plummer and Busenberg, 1982).

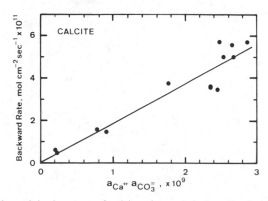

Figure 4. Backward precipitation rate of calcite observed during dissolution experiments in a fluidized-bed reactor. (After Chou et al., in press.)

Theoretically, one could consider that a more powerful approach for estimating the backward reaction rate and understanding its mechanism is to study directly the precipitation reaction from oversaturated solutions. However, the precipitation reactions are even more complicated than the dissolution reactions. They are very sensitive to the surface properties, the presence of impurities adsorbed on the interface, and the inclusions of foreign ions and molecules in the precipitated phase. As a consequence, the experimental results concerning the precipitation reactions are often inconsistent. In the field of calcite precipitation at low p_{CO_2} and high pH, a recent careful study of Inskeep and Bloom (1985) conducted using a pH-stat method at fixed p_{CO_2} indicates that the rate of crystal growth of this mineral is well described by the rate model and reaction mechanism proposed by Reddy and Nancollas (1971) and Nancollas and Reddy (1971). This mechanistic model assumes that the elementary reaction describing calcite precipitation under these conditions can be visualized as Ca^{2+} and CO_3^{2-} ions dehydrating on the surface of the mineral to form an activated complex. Adapting Eq. 12, the rate of precipitation may thus be written as

$$r = k_{-3} \left(a_{Ca^{2+}} a_{CO_3^{2-}} - \frac{k_3}{k_{-3}} \right)$$

$$= k_{-3} (a_{Ca^{2+}} a_{CO_3^{2-}} - K_{eq}) \tag{22}$$

Figure 5 shows the linear relationship obtained between the rate of growth of calcite and the oversaturation as expressed in Eq. 22. The value of the rate

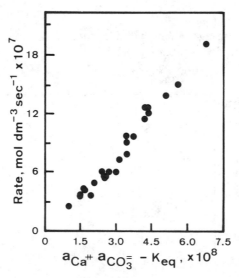

Figure 5. Rate of growth of calcite at high pH and low p_{CO_2} as a function of the departure from equilibrium. (After Inskeep and Bloom, 1985.)

constant k_{-3} determined in the study of Inskeep and Bloom (1985) was found to be between 1.5×10^{-2} and 1.9×10^{-2} mol cm^{-2} s^{-1}, which is in very good agreement with the value obtained from the dissolution experiments close to equilibrium (Chou et al., in press). This value is also coherent with the forward rate constant and the equilibrium constant, according to the relation defined in Eq. 14.

Busenberg and Plummer (1986) performed, on the other hand, experiments on the growth rate of calcite and aragonite at higher p_{CO_2} values using single crystals suspended in calcium bicarbonate solutions at very high degrees of over-saturation ($\geqslant 100$ times). These results differ from those described above, particularly in the case of aragonite, which we will discuss here in more detail. The backward reaction observed by these authors is much too slow in comparison with the theoretical rates calculated on the basis of the thermo-dynamic constraints (Eqs. 19–21).

Busenberg and Plummer (1986) usually express the rate of the backward reaction by the equation

$$r = k_4 a_{Ca^{2+}} a_{HCO_3^-} \tag{23}$$

with

$$k_4 = k_4' + k_4''(a_{HCO_3^-})_s + k_4'''(a_{OH^-})_s \tag{24}$$

This equation is similar to the backward reaction terms of Eq. 11 if it is remembered that the product $a_{OH} - a_{HCO_3^-}$ is related to $a_{CO_3^{2-}}$ by an equilibrium condition. There is, however, a fundamental difference in that the activities of HCO_3^- and OH^- in Eq. 24 are those at the surface of the mineral and not in the bulk of the solution. The relations between the activities of these ions at the surface of the solid and in the aqueous solution are given by Langmuir or Freundlich isotherms.

Let us apply to their experimental results the approach presented here. As shown previously, at low p_{CO_2} and high pH, the only important forward reaction rate is simply given by k_3 and the corresponding backward reaction rate, by $k_{-3} a_{Ca^{2+}} a_{CO_3^{2-}}$. Although the experiments were carried out with aragonite samples of various origins, there is a satisfactory relationship between the observed backward reaction rate and the IAP_{CaCO_3} (Fig. 6). However, the slope of the regression line, which corresponds to the rate constant k_{-3}, is equal only to 4.2×10^{-3}. If this value is compared to the constant of the corresponding forward reaction given by Busenberg and Plummer for aragonite ($k_3 = 3.5 \times 10^{-10}$), one obtains an IAP_{CaCO_3} constant for the dissolution of aragonite equal to 8.2×10^{-8} instead of 4.6×10^{-9} for the equilibrium constant. A similar conclusion is obtained when the results at high p_{CO_2} are considered. At p_{CO_2} equal to 1 atm, the reaction step involving H_2CO_3 is five times faster than the constant rate k_3. The corresponding backward reaction is then proportional to $a_{HCO_3^-}^2 a_{Ca^{2+}}$ (Eq. 20). The relation between the backward rate observed by Busenberg and Plummer (1986) at $p_{CO_2} = 1$ atm and this activity product gives a value for

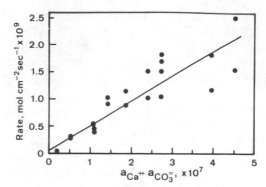

Figure 6. Rate of growth of aragonite single crystals as a function of the IAP_{CaCO_3} at low p_{CO_2} (2.85 $\times 10^{-3}$ atm). The solubility of aragonite corresponds to an IAP_{CaCO_3} equal to 4.6×10^{-9}. [Adapted from experimental results of Busenberg and Plummer (1986), Table 9.]

k_{-2} equal to 1.3×10^{-4}. Here again, the value of the backward reaction constant k_{-2} gives a solubility product for aragonite equal to 3.7×10^{-4} instead of 4.4×10^{-5} as predicted from the equilibrium constant. Similar observations could be made with calcite, although the discrepancies are less pronounced. Busenberg and Plummer (1986) attribute these observations to the fact that the concentration of $(H^+)_s$ at the surface of the mineral, which, in fact, controls the concentration of $(OH^-)_s$ and $(HCO_3^-)_s$ appearing in the expression of the backward reaction rate (Eq. 24), is different from the concentration of H^+ in the bulk solution. In fact, they use the thermodynamic constraints on the backward reaction rate constant to evaluate the concentration of protons at the surface of the mineral. As already pointed out by Inskeep and Bloom (1985), this term representing a pH difference between the surface and the bulk solution was not necessary in order to describe the precipitation rate observed in their experiments. Furthermore, close to equilibrium, the thermodynamic constraint applies necessarily to the bulk of the solution.

A likely alternative explanation is that the solid precipitated in the experiments of Busenberg and Plummer (1986) is rather disorded or has not been completely dehydrated. It is important to note that their experiments were conducted with single crystals while Inskeep and Bloom (1985) used powders. Also, the experiments of Busenberg and Plummer (1986) were conducted at much higher degrees of oversaturation. These conditions (low surface area and high rate of precipitation) favor a rapid growth of the precipitated layer, which does not have the chance to age during a sufficient time in contact with the aqueous solution and thus to recrystallize to the stable aragonite or calcite phase.

The influence of the crystal growth rate on the stability of the precipitated calcium carbonates has already been described by several authors. In his saturometry measurements, Weyl (1967) showed that the stability of the precipitating phase obtained by pumping oversaturated seawater through a column packed with carbonate seeds was a function of the flow rate and thus of the rate of precipitation. These results were later confirmed by Wollast et al. (1980) and Schoonmaker (1981), who conducted batch experiments with over-saturated seawater to which reagent-grade calcite powders were added. They observed an increase of the solubility of the precipitated phase with the increase of the thickness of the coating of the seeds. Schoonmaker (1981) further showed that aging of this carbonate coatings decreased their apparent solubility. These authors attributed these effects to the amount of $MgCO_3$ incorporated in the calcite crystallographic structure. The most recent study of this problem (Busenberg and Plummer, 1989) demonstrates that the increase of solubility with increasing rate of growth is also proportional to the concentration of Na_2SO_4 in the aqueous phase. Their conclusion is that the differences in solubilities must be attributed rather to the existence of point defects due to the presence of Na^+ and SO_4^{2-} in the crystal lattice.

It is also interesting to note that a similar phenomenon is observed with the growth of quartz in aqueous solution of silicic acid oversaturated with respect to this mineral (Stöber, 1967; Wollast, 1974). The solid deposited has a tendency to attain a more increasingly disordered structure as the precipitated layer increas-es. There is a further slow reorganization of the precipitated layer, leading finally to the formation of relatively well-crystallized quartz (Mackenzie, 1971). An evaluation of the solubility of the disordered phase initially deposited on quartz based on the backward reaction rate indicates that this rather amorphous silica has a solubility 10 times higher than quartz.

4. CONCLUSIONS

Dissolution kinetics of single carbonates such as calcite, aragonite, and magnes-ite exhibit simple dependence with respect to a limited number of reactants, specifically, H^+, H_2CO_3, and H_2O. It is thus easy to identify the elementary steps leading to the formation of the surface activated complex and the nature of the products of the detachment process following the decomposition of the activated complex.

On the other hand, the order of rate dependence of the dissolution of dolomite with respect to H^+ is fractional, as for many oxides and silicates, indicating more complex reactions at the mineral surface. A further elucidation of the origin of this fractional order requires a better knowledge of the surface properties of the mineral and of the protonation reaction of the various surface sites. Classic acid–base titration of the surface of carbonates is difficult to realize because of the high reactivity of these minerals and the complexity of the dissolved carbonate system.

Dissolution kinetics of a simple component close to saturation and the mechanism of the backward precipitation reaction are still subject to contro-versy. Minerals such as calcite and aragonite are known to reach rapidly a dissolution equilibrium when placed in closed aqueous systems. According to simple and classical thermodynamical concepts, this requires that each forward reaction be exactly balanced by a corresponding backward reaction. This condition introduces considerable constraints on the nature of the backward reactions and on the value of the corresponding kinetic constant. As shown here, this is not always observed experimentally because it is very likely that a rather disordered phase is first precipitated and that the system does not, in fact, approach a true equilibrium with the most stable solid phase on the time scale of the experiments.

On the other hand, various hypotheses concerning the properties of the solid–liquid interface and the composition of the double layer have been proposed in order to explain the observed kinetics. These hypothese are, however, often purely speculative. Here again, direct evaluations of properties of the mineral surface and of the interface are barely existing and severely lacking.

REFERENCES

Berner, R. A., and J. W. Morse (1974), "Dissolution Kinetics of Calcium Carbonate in Sea Water. IV. Theory of Calcite Dissolution," *Am. J. Sci.* **274**, 108–134.

Busenberg, E., and L. N. Plummer (1982), "The Kinetics of Dissolution of Dolomite in CO_2–H_2O Systems at 1.5 to 65°C and 0 to 1 atm P_{CO_2}," *Am. J. Sci.* **282**, 45–78.

Busenberg, E., and L. N. Plummer (1986), "A Comparative Study of the Dissolution and Crystal Growth Kinetics of Calcite and Aragonite," in F. A. Mumpton, Ed., *Studies in Diagenesis*, USGS Bulletin **1578**, U.S. Government Printing Office, Washington, D.C. pp. 139–168.

Busenberg, E. and L. N. Plummer (1989), "Thermodynamics of Magnesian Calcite Solid—Solutions at 25°C and 1 atm Total Pressure," *Geochim. Cosmochim. Acta* **53**, 1189–1208.

Chou, L., R. M. Garrels, and R. Wollast (in press), "Comparative Study of the Kinetics and Mechanisms of Dissolution of Carbonate Minerals," *Chem. Geol.*

Inskeep, W. P., and P. R. Bloom (1985), "An Evaluation of Rate Equation for Calcite Precipitation Kinetics at p_{CO_2} Less than 0.01 atm and pH Greater than 8," *Geochim. Cosmochim. Acta* **49**, 2165–2180.

Lasaga, A. C. (1981), "Rate Laws of Chemical Reaction," in A. C. Lasaga and R. J. Kirkpatrick, Eds., *Kinetics of Geochemical Processes* (*Rev. Min.* **8**, 1–68). Mineralogical Society of America, Washington, D.C.

Mackenzie, F. T. (1971), "Synthesis of Quartz at Earth-Surface Conditions," *Science* **173**, 533–535.

Meybeck, M. (1987), "Global Chemical Weathering of Surficial Rocks Estimated from River Dissolved Loads," *Am. J. Sci.* **287**, 401–428.

Morse, J. W. (1983), "The Kinetics of Calcium Carbonate Dissolution and Precipitation," in P. H. Ribbe, Ed., *Carbonates: Mineralogy and Chemistry.* (*Rev. Min.* **11**, 227–264). Mineralogical Society of America, Washington, D.C.

Nancollas, G. H., and M. M. Reddy (1971), "The Crystallization of Calcium Carbonate, II. Calcite Growth Mechanism," *J. Colloid Interface Sci.* **37**, 824–829.

Plummer, L. N., and E. Busenberg (1982), "The Solubilities of Calcite, Aragonite and Vaterite in CO_2–H_2O Solutions between 0 and 90°C, and an Evaluation of the Aqueous Model for the System $CaCO_3$–CO_2–H_2O," *Geochim. Cosmochim. Acta* **46**, 1011–1040.

Plummer, L. N., and T. M. L. Wigley (1976), "The Dissolution of Calcite in CO_2-Saturated Solutions at 25°C and 1 atmosphere Total Pressure," *Geochim. Cosmochim. Acta* **40**, 191–201.

Plummer, L. N., T. M. L. Wigley, and D. L. Parkhurst (1978), "The Kinetics of Calcite Dissolution in CO_2-Water System at 5° to 60°C and 0.0 to 1.0 atm CO_2," *Am. J. Sci.* **278**, 179–216.

Plummer, L. N., T. M. L. Wigley, and D. L. Parkhurst (1979), "Critical Review of the Kinetics of Calcite Dissolution and Precipitation," in E. A. Jenne, Ed., *Chemical Modeling in Aqueous Systems—Speciation. Sorption. Solubility, and Kinetics,* (ACS Symposium Series No. 93), American Chemical Society, Washington, DC, pp. 537–573.

Prigogine, I. (1967), *Introduction to Thermodynamics of Irreversible Processes,* Wiley, New York, 147 pp.

Reddy, M. M., and G. H. Nancollas (1971), "The Crystallization of Calcium Carbonate. I. Isotopic Exchange and Kinetics," *J. Colloid Interface Sci.* **36**, 166–172.

Rickard, D., and E. L. Sjöberg (1983), "Mixed Kinetic Control of Calcite Dissolution Rates," *Am. J. Sci.* **283**, 815–830.

Schoonmaker, J. E. (1981), "Magnesian Calcite–Seawater Reactions: Solubility and Recrystallization Behavior," Ph.D. Dissertation, Northwestern University, Evanston, IL.

Sjöberg, E. L., and D. Rickard (1984), "Temperature Dependence of Calcite Dissolution Kinetics between 1 and 62°C at pH 2.7 to 8.4 in Aqueous Solutions," *Geochim. Cosmochim. Acta* **48**, 485–493.

Stöber, W. (1967), "Formation of Silicic Acid in Aqueous Suspensions of Different Silica Modifications," in *Equilibrium Concepts in Natural Water Systems.* (W. Stumm, Ed.) (ACS Symposium Series No. 67), American Chemical Society, Washington DC, Chapter 7, pp. 161–182.

Weyl, P. K. (1967), "The Solution Behavior of Carbonate Minerals in Sea Water," *Proceedings of the International Conference on Tropical Oceanography,* Miami, FL, pp. 178–228.

Wollast, R. (1974), "The Silica Problem," in E. G. Goldberg, Ed., *The Sea,* Wiley, New-York Vol. 5, pp. 359–392.

Wollast, R. (1981), "Interactions in Estuaries and Coastal Waters," in B. Bolin and R. B. Cook, Eds., *The Major Biogeochemical Cycles and Their Interactions,* Wiley, New York, pp. 385–407.

Wollast, R., and F. T. Mackenzie (1983), "The Global Cycle of Silica," in S. E. Aston, Ed., *Silicon Geochemistry and Biogeochemistry,* Academic Press, London, pp. 39–76.

Wollast, R., R. M. Garrels, and F. T. Mackenzie (1980), "Calcite–Seawater Reactions in Ocean Surface Water," *Am. J. Sci.* **280**, 831–848.

16

KINETICS OF COLLOID CHEMICAL PROCESSES IN AQUATIC SYSTEMS

Charles R. O'Melia

*Department of Geography and Environmental Engineering,
The Johns Hopkins University, Baltimore, Maryland*

1. INTRODUCTION

Solid–water interfaces provide sites for important abiotic and biotic reactions that involve the uptake and transformation of pollutants. The availability of solid surfaces and the physical and chemical characteristics of solid–water interfaces in aquatic environments can be determined by particle aggregation and deposition reactions. The kinetics of these colloid chemical reactions play an important role in the transport, reactivity, fate, and impact of pollutants and other particle-reactive substances in natural waters.

This chapter is written with two objectives: (1) to discuss field, laboratory, and modeling results in which the kinetics of particle aggregation and deposition in natural aquatic systems are developed and tested and (2) to indicate approaches for studying the kinetics of such reactions in these and other aquatic systems in practice and in research. The chapter begins with a consideration of the kinetics of particle deposition in groundwater aquifers. This is followed by an assessment of particle aggregation and sedimentation in lakes. A modeling approach for the kinetics of particle–particle interactions in aquatic systems is then presented. Finally, experimental and theoretical directions for studying the kinetics of these reactions in natural systems are suggested.

2. PARTICLE PASSAGE AND RETENTION IN AQUIFERS

Extensive field studies of the transport of latex microspheres and bacteria in a sandy freshwater aquifer on Cape Cod, Massachusetts, were conducted by

Harvey et al. (1989). In one experiment, fluorescent latex microspheres and a chloride tracer were introduced as a pulse input into the aquifer through an injection well located in a plume of contaminated groundwater at a distance of 500 m from a treated wastewater infiltration bed. A mixture containing six types of latex particles differing in size and surface characteristics was used. Their passage in the aquifer was monitored with samples collected from multilevel sampling wells situated downgradient at a distance of 6.9 m from the injection well. The travel time from the injection well to the sampling wells was estimated to be 21 days with the chloride tracer; the corresponding average flow velocity is 0.33 m day^{-1}. The aquifer material consists largely of quartz and feldspar; mean grain size and aquifer porosity were reported to be ~ 0.05 cm and 0.38, respectively.

Results of the passage of three sizes of the microspheres (0.23, 0.91, and 1.35 μm in diameter) are presented in Figure 1. Actual particle number concentrations observed at a sampling well are presented as a function of time after injection in the main part of the figure. The inset contains these data as dimensionless concentrations normalized to the latex concentration in the input. Results for the chloride tracer (not shown) indicate substantial passage after 18 days and a peak concentration at 21 days. These data indicate that removal decreased with increasing particle size in the order 0.23 > 0.91 > 1.35 μm. The

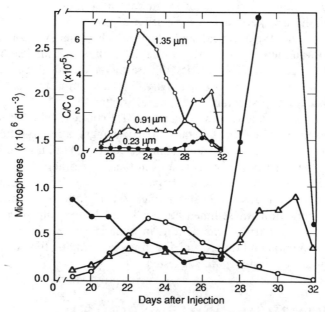

Figure 1. Transport and deposition of latex microspheres through a natural sandy aquifer, from Harvey et al. (1989). Particle concentration (number dm^{-3}) and dimensionless concentration (insert) as a function of time after injection for three sizes of carboxyl latex. Flow velocity ~ 0.33 m day^{-1} ($U \sim 0.125$ m day^{-1}), $L = 6.9$ m, $a_c = 0.025$ cm, $\varepsilon = 0.38$, $T = 11°$C, travel time ~ 21 days.

data also indicate that the 1.35-μm particles followed the chloride tracer fairly well, and they suggest an apparent delay or retardation in the passage of the 0.23- and 0.91-μm particles.

Discussion of these results is directed at the following question. How do these field observations compare with present theories and experiments for the kinetics of particle deposition in porous media? In answering this question, laboratory experiments and modeling results will be considered.

Some results of laboratory studies using a continuous input of non-Brownian latex particles (4 μm diameter) are presented in Figure 2, taken from the work of Tobiason and O'Melia (1988). Effluent concentration (dimensionless) is plotted as a function of time after the latex suspension was introduced to the system (also dimensionless). In these experiments the size of the particles in the influent suspension was constant while the chemistry of the aqueous phase was varied by changing the type and concentration of inorganic salts. Passage through the beds varied from ~ 98 to $\sim 20\%$, with removal increasing with increasing ionic strength and Ca^{2+} in the solution. Physical parameters such as media size (0.04 cm), bed depth (25 cm), bed porosity (0.40), suspended particle size (4 μm), flow rate (120 m day^{-1}), and temperature (25°C) were similar in all of these runs so that mass transport of the particles within the porous media was constant among the runs. The chemical compositions of the solid phases were also constant; the porous media were glass beads and the suspended particles were surfactant-free latex. The pH of the system was maintained at ~ 7. Differences in the kinetics of deposition among the curves shown in Figure 2 must arise from differences in chemical interactions between suspended particles and the porous media; these differences are brought about by differences in the chemistry of the aqueous phase.

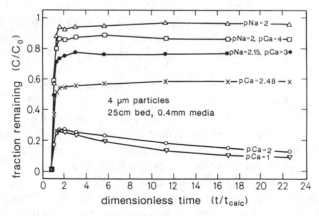

Figure 2. Transport and deposition of latex microspheres through laboratory filter columns, from Tobiason and O'Melia (1988). Particle concentration (dimensionless) plotted as a function of filtration time (dimensionless); $U = 120$ m day^{-1}, $L = 0.25$ m, $a_c = 0.02$ cm, $a_p = 2$ μm, $\rho_p = 1.05$ g cm^{-3}, $\varepsilon = 0.40$, $T \sim 25°C$, pH ~ 7, mean hydraulic residence time in column = 2.7 min.

These laboratory results are used here to provide a bridge between the field experiments described previously and theoretical results summarized subsequently. Discussion is directed at the following questions. How do the results of these laboratory experiments compare with present theories for the kinetics of particle–particle interactions and particle deposition in porous media? How do they compare with the aquifer experiments of Harvey et al. (1989)?

The kinetics of deposition in porous media have been described by the expression (Yao et al., 1971)

$$\frac{dC}{dL} = -\frac{3}{4}\alpha(p, c)\,\eta(p, c)\frac{(1-\varepsilon)}{a_c}C \tag{1}$$

where C is the number concentration of particles of a specific size and L is the length of travel of the particles in the porous media (e.g., the distance between injection and sampling wells in an aquifer experiment or the depth of a filter column). The travel distance is considered as the product of the average flow velocity and the travel time. The radius of the grains comprising the porous media is denoted as a_c; the grains are assumed spherical in this case. The porosity of the porous media is given as ε. The terms $\alpha(p, c)$ and $\eta(p, c)$ collectively describe a kinetic coefficient for the deposition process of suspended particles (p) on stationary collectors (c) or media. They are defined here as follows:

$$\eta(p, c) = \frac{\text{particle flux reaching the collector surface}}{\text{particle flux approaching the collector from upstream}} \tag{2}$$

$$\alpha(p, c) = \frac{\text{particle flux attaching to the collector surface}}{\text{particle flux reaching the collector surface}} \tag{3}$$

The product of these coefficients, $\alpha(p, c) \times \eta(p, c)$, is the ratio of the flux of particles actually adhering to a stationary collector to the flux of suspended particles approaching it from upstream. Each of these coefficients can have both theoretical and experimental or empirical values, denoted here by subscripts such as in $\alpha(p, c)_{exp}$ and $\eta(p, c)_{theor}$.

The coefficients $\alpha(p, c)$ and $\eta(p, c)$ describe chemical and physical effects on the kinetics of deposition. The transport of particles from the bulk of the flowing fluid to the surface of a collector or media grain by physical processes such as Brownian diffusion, fluid flow (direct interception), and gravity are incorporated into theoretical formulations for $\eta(p, c)$, together with corrections to account for hydrodynamic retardation or the lubrication effect as the two solids come into close proximity. Chemical effects are usually considered in evaluating $\alpha(p, c)$. These include interparticle forces arising from electrostatic interactions and steric effects originating from interactions between adsorbed layers of polymers and polyelectrolytes on the solid surfaces.

Rajagopalan and Tien (1976) have reported the results of a theoretical study of the effects of diffusion, fluid flow, gravity, hydrodynamic retardation, electrostatic effects, and London–van der Waals forces on particle deposition in porous media. The following correlating equation was developed to summarize the results of numerical calculations in the absence of electrostatic effects [α(p, c) = 1, sometimes called "favorable deposition"]:

$$\eta(\mathrm{p, c})_{\mathrm{theor}} = 4.0 \, A_s^{1/3} N_{\mathrm{Pe}}^{-2/3} + A_s N_{\mathrm{Lo}}^{1/8} N_{\mathrm{R}}^{15/8} + (3.38 \times 10^{-3}) \, A_s N_{\mathrm{G}}^{1.2} N_{\mathrm{R}}^{-0.4} \quad (4)$$

The dimensionless groups in Eq. 4 are defined in Table 1. Equation 4 is empirical in form, but it is an expression developed to fit numerical calculations based solely on theoretical hydrodynamic and physical models of the interactions between a spherical particle suspended in a flowing fluid and a spherical stationary collector in a packed bed, in the absence of any repulsive chemical interaction. This relationship is used subsequently to evaluate the kinetics of particle contacts with porous media in the aquifer experiments by Harvey et al. (1989) and the laboratory studies of Tobiason and O'Melia (1988).

The first group of terms on the right-hand-side of Eq. 4 describes particle transport to a collector surface by Brownian diffusion. N_{Pe} is the Peclet number, a ratio of particle transport by fluid advection to transport by molecular or viscous processes. The term A_s is introduced to account for the effects of neighboring collectors or media grains on the fluid flow around a collector of interest. The results here assume Happel's model (Happel, 1958) for flow around a sphere in a packed bed; A_s depends only on the porosity of the bed (Table 1). The derivation for diffusive transport is based on the early work of Levich (1962).

TABLE 1. Dimensionless Groups and Parameters in Eq. 4

Parameter	Definition
N_{R}	$a_{\mathrm{p}}/a_{\mathrm{c}}$
N_{G}	$[2(\rho_{\mathrm{p}} - \rho_{\mathrm{w}})g a_{\mathrm{p}}^2]/(9\mu U)$
N_{Lo}	$H/[9\pi a_{\mathrm{p}}^2 U]$
N_{Pe}	$(2U a_{\mathrm{c}})/D_\infty = (12\pi\mu a_{\mathrm{c}} a_{\mathrm{p}} U)/(kT)$
A_s	$[2(1 - p^5)]/(2 - 3p + 3p^5 - 2p^6)$
p	$(1 - \varepsilon)^{1/3}$
a_{p}	Particle radius
a_{c}	Collector radius
D_∞	Stokes–Einstein diffusion coefficient, $kT/(6\pi\mu a_{\mathrm{p}})$
ε	Bed porosity
g	Gravitational acceleration
H	Hamaker constant (e.g., for glass–water–latex)
k	Boltzmann's constant
T	Temperature
U	Superficial or approach velocity of flow
$\rho_{\mathrm{p}}, \rho_{\mathrm{w}}$	Density of particle, water
μ	Viscosity of water

In the second group of terms on the right-hand side of Eq. 4, which describes particle transport by fluid flow (termed *direct interception*), N_R is the interception number, the ratio of the sizes of the suspended particle and the stationary collector, and N_{Lo} reflects the London–van der Waals attractive force between the particle and the collector and the drag force on the suspended particle. As a moving particle approaches close to a stationary collector, the drag force on the moving particle increases and it becomes increasingly difficult to drain fluid from the region separating the two solids. Considering only hydrodynamic interactions, particle–collector contacts cannot occur. However, as the solids come into close proximity, the London–van der Waals attractive force also increases rapidly. The combined effects of increasing hydrodynamic drag preventing contacts and increasing London forces enhancing them is termed *hydrodynamic retardation*. The overall effect is slight for Brownian diffusion, but it can be significant for transport by interception and gravity (Spielman and FitzPatrick, 1973).

The third group of terms on the right-hand side describes particle transport to the collector by gravity forces acting on the suspended particle. Hydrodynamic retardation is included. Here, N_G is a gravity number, that is, the ratio of the Stokes' settling velocity of a suspended particle to the superficial or approach velocity of flow.

With theoretical values of particle transport $[\eta(p, c)_{theor}]$ available with Eq. 4, experimental values of the attachment coefficient or sticking probability $[\alpha(p, c)_{exp}]$ can be obtained from experimental measurements of particle removal in porous media with the following relationship, obtained by integrating Eq. 1 with boundary conditions that $C = C_0$ at $L = 0$ and $C = C_L$ at $L = L$:

$$\alpha(p, c)_{exp} = -\ln\left(\frac{C_L}{C_0}\right)\left(\frac{4a_c}{3(1-\varepsilon)\eta(p, c)_{theor} L}\right) \tag{5}$$

The results presented in Figure 2 from the laboratory experiments reported by Tobiason and O'Melia (1988) yield experimental sticking probabilities that are listed in Table 2.

Since physical parameters were held constant in these experiments, the theoretical single collector efficiency, $\eta(p, c)_{theor}$, is constant at 0.00256. The experimental attachment efficiency, $\alpha(p, c)_{exp}$, however, varies from 0.014 to 0.94, depending on the chemical composition of the solution. In the presence of a high concentration of Ca^{2+}, the attachment coefficient approaches 1. This means that, in the absence of a repulsive chemical interaction, the mass-transport rate as calculated with Eq. 4 successfully describes the performance of these laboratory columns. At low ionic strength ($p_{Na} = 3.0$), the sticking coefficient is reduced to a value of 0.014 by repulsive chemical interactions (presumably primarily electrostatic) between the suspended latex particles and the stationary glass collectors. Only 1.4% of the contacts produced by mass transport lead to attachment and deposition of the latex particles from the suspension.

TABLE 2. Particle Transport and Attachment in Laboratory Filters[a]

pNa[b]	pCa[b]	$C_L/C_0{}^b$	$\eta(p, c)_{theor}$	$\alpha(p, c)_{exp}$
3.0		0.98	0.00256	0.014
2.0		0.94	0.00256	0.043
2.0	4.0	0.88	0.00256	0.089
2.15	3.0	0.78	0.00256	0.17
	2.48	0.56	0.00256	0.40
	2.0	0.29	0.00256	0.86
	1.0	0.26	0.00256	0.94

[a] Parameters: pH = 6.7 to 7; $U = 120$ m day^{-1}; $a_c = 0.02$ cm; $a_p = 2$ μm; $T \sim 25°C$; $L = 25$ cm; $\varepsilon = 0.4$; $\rho_p = 1.05$ g cm^{-3}; H assumed as 10^{-20} J.
[b] Data from Tobiason and O'Melia (1988).

The results of the laboratory experiments (Fig. 2 and Table 2) are taken to confirm the ability of Eq. 4 to predict the physical or mass-transport aspects of the kinetics of particle deposition in porous media and to provide an estimate of the effects of chemical interactions in retarding deposition under inorganic solution conditions that are representative of the ionic strengths of freshwaters. These results can then be used with and compared to similar analyses of the field observations of Harvey et al. (1989).

Calculations of particle transport and attachment in the field experiments of Harvey et al. (1989) are presented in Table 3. Theoretical mass-transfer coefficients are calculated with Eq. 4 and experimental sticking probabilities are then calculated with Eq. 5. In making these calculations, a "superficial" or approach

TABLE 3. Particle Transport and Attachment in Groundwater Aquifer[a]

Particle Diameter (μm)[b]	Type[b]	$C_L/C_0{}^b$	$\eta(p, c)_{theor}$	$\alpha(p, c)_{exp}$
0.23	Carboxyl	0.0001	0.215	0.0033
0.53	Carboxyl	0.0004	0.123	0.0049
0.93	Carboxyl	0.0006	0.086	0.0067
1.35	Carboxyl	0.0012	0.069	0.0076
0.6	Uncharged	0.0005	0.114	0.0052
0.85	Polyacrolein	0.031	0.090	0.0030

[a] Parameters: $U = 0.125$ m day^{-1}; $a_c = 0.025$ cm; $T \sim 11°C$; $L = 6.9$ m; $\varepsilon = 0.38$; DOC = 1–2 g m^{-3}; ρ_p assumed as 1.05 g cm^{-3}; H assumed as 10^{-20} J.
[b] Data from Harvey et al. (1989).

velocity (U) of 0.125 m day^{-1} was used, based on the travel velocity of 0.33 m day^{-1} and the aquifer porosity of 0.38 reported by Harvey et al. (1989). Very high removals were observed in the field experiments, due to the high mass-transport rates [$\eta(\mathrm{p}, \mathrm{c})_{\mathrm{theor}}$] and the long travel distance. The dominant transport mechanism is convective Brownian diffusion, and the high values of $\eta(\mathrm{p}, \mathrm{c})_{\mathrm{theor}}$ result from the low flow velocity in the aquifer. The observed attachment probabilities, $\alpha(\mathrm{p}, \mathrm{c})_{\mathrm{exp}}$, are remarkably uniform, varying by a factor of only about 2.5, while the observed residual fractions (C_{L}/C_0) vary by a factor of 310. In fact, the particles with the best retention [0.23-μm carboxyl latex, C_{L}/C_0 = 0.0001, $\alpha(\mathrm{p}, \mathrm{c})_{\mathrm{exp}}$ = 0.0033] had essentially the same attachment probability as the particles with the poorest retention [0.85-μm polyacrolein latex, $\alpha(\mathrm{p}, \mathrm{c})_{\mathrm{exp}}$ = 0.0030]. This arises in part because of differences between the two particle sizes in mass transport by convective diffusion, but is due primarily to the first-order character of the overall kinetics of particle deposition (Eqs. 1 and 5). For extensive deposition (low C_{L}/C_0), the residual fraction is very sensitive to the product $\eta(\mathrm{p}, \mathrm{c})_{\mathrm{theor}} \times \alpha(\mathrm{p}, \mathrm{c})_{\mathrm{exp}}$. The difference between residual fractions of 0.0001 and 0.031 in Table 3 is accomplished by differences in the product $\eta(\mathrm{p}, \mathrm{c})_{\mathrm{theor}} \times \alpha(\mathrm{p}, \mathrm{c})_{\mathrm{exp}}$ of only (ln 0.0001)/(ln 0.031) or 2.65.

The attachment probabilities determined for the field experiments range from only 0.003 to 0.0076 despite the use of six different latex suspensions with three different types of surface characteristics. This suggests that the surface properties of the latex particles that pertain to attachment depend primarily on the solution chemistry of the aqueous suspending phase. The ionic strength of the ground-water is in the order of 10^{-3} M based on conductivity determinations (Harvey et al., 1989). The observed attachment probabilities in the groundwater aquifer range from about 20 to 55% of the value measured for 10^{-3} M NaCl in the laboratory experiments (0.014, Table 2). This reduction could be due to the presence of macromolecular dissolved organic substances originating in the treated wastewater discharge. It is probable that dissolved organic matter in the groundwater, including organic substances originating in the treated wastewater discharge, enhance colloidal stability and reduce $\alpha(\mathrm{p}, \mathrm{c})_{\mathrm{exp}}$. Similar effects discussed subsequently have been observed in lakes.

The experimental conditions of the laboratory and field experiments are compared in Table 4. They are remarkable primarily for their large differences. Flow velocities vary by about three orders of magnitude between the two sets of experiments. The dominant mass-transport mechanism is different, with diffusion dominating in the aquifer and direct interception determining transport of the non-Brownian particles in the laboratory experiments. The length of flow differed by well over an order of magnitude between the two systems. Finally, the extent of particle removal or deposition was much less in the short laboratory columns operated at high velocities than in the aquifer with a longer travel distance and a lower flow velocity. Despite all of these differences, the estimated values of $\alpha(\mathrm{p}, \mathrm{c})_{\mathrm{exp}}$ in the two systems agree very well. Stated another way, the results in Tables 2 and 3 indicate that theories of the kinetics of particle deposition in porous media that have been tested in the laboratory are able to

TABLE 4. Comparison of Experimental Conditions and Results
of Laboratory[a] and Field[b] Particle Deposition Experiments

Parameter	Laboratory[a]	Field[b]
Type of input	Continuous	Pulse
Superficial flow velocity	120 m day^{-1}	0.125 m day^{-1}
Length of flow	0.25 m	6.9 m
Temperature	25°C	11°C
Media porosity	0.40	0.38
Media diameter	0.04 cm	0.05 cm
Particle diameter	4 μm	0.23–1.35 μm
Fraction passing	0.26–0.98	0.0001–0.031
Dominant mass transport mechanism	Direct interception	Brownian diffusion

[a] Data from Tobiason and O'Melia (1988).
[b] Data from Harvey et al. (1989)

describe particle transport and deposition in the aquifer rather well. At present, $\alpha(p, c)_{exp}$ can be considered as a fitting parameter that, if known or determined experimentally, permits useful estimates of particle transport and deposition in field situations such at the aquifer considered here.

In the field experiment, an important characteristic of the concentration profiles presented in Figure 1 is the apparent delay or retardation in the passage of the 0.23 and 0.91-μm particles relative to the 1.35-μm particles and the chloride tracer. From the main part of the figure, however, it can be seen that the concentrations of the two smaller particles decrease slightly (0.91-μm size) or dramatically (0.23-μm size) with time for a period before they eventually rise, after about 27 days. This rise occurs after most of the large (1.35-μm) particles have been removed in the aquifer. The continuous decrease in the concentration of the 0.23-μm particles with time over the period from 18 to 27 days is characteristic of a process known in packed-bed filtration as "filter ripening." It can be observed, for example, in the laboratory experiments presented in Figure 2 for the runs in which chemical retardation of deposition is substantially reduced or eliminated by high concentrations of Ca^{2+}. This is because particles that are deposited from a flowing suspension at one time can then become filter media for subsequent particles that are traveling in the pores of the media. The removal of particles from suspension in a groundwater aquifer can be considered as an accumulation of new filter media in the aquifer.

Consideration of this ripening process provides an explanation for the apparent delay in the travel of the 0.23- and 0.91-μm particles relative to the 1.35-μm particles and the chloride tracer. The largest particles act as collectors or filter media for the smaller ones. Although not stated in the paper by Harvey et al. (1989), these large particles were present in substantially lower *number* concentrations than were the smaller ones. Data in Figure 1 indicate, for example, that the initial number concentration of the 0.23-μm particles at the injection well was over 50 times larger than that of the 1.35-μm particles. Removal of the larger particles aided in the removal of the smaller ones, but, after most of the larger ones were removed, the smaller particles were able to travel more easily through the aquifer. This description of a process of filter ripening in the aquifer provides an explanation for the continuous decrease in the concentration of the 0.23-μm particles over the time period from 18 to 27 days after injection and is consistent with a rise in the concentration of the smaller sizes after that time, since most of the larger particles had been retained in the aquifer.

In summary, the results of the field study, the laboratory experiments, and present theory for the kinetics of particle deposition in porous media are internally consistent. The field observations in Figure 1 are consistent with a perspective that views the groundwater aquifer as a packed-bed filter and uses theories for the kinetics of particle deposition in porous media to describe the process. The apparent retardation of the smaller particles is consistent with a view of the process that considers "filter ripening" to occur in the aquifer during the experiments. The principal factor affecting particle deposition in the system is the chemical composition of the aqueous solution. This apparently establishes the colloid chemical properties of the latex particles to a greater extent than their initial surface characteristics prior to introduction into the aquifer. Macromolecular dissolved organic substances, including those present in treated wastewater discharges, may be particularly important in this regard.

Having a model that has a good theoretical basis, that has been validated in laboratory experiments, and that is consistent with field observations, it is advisable to make some predictions about particle deposition in systems of interest. An example is presented in Figure 3, adapted from the work of Tobiason (1987). The travel distance in an aquifer required to deposit 99% of the particles from a suspension is termed L_{99} and is plotted as a function of the diameter of the suspended particles for two different values of $\alpha(p, c)$, specifically: 1.0 (favorable deposition) and 0.001 (deposition with significant chemical retardation of the particle–collector interaction, termed "unfavorable deposition"). Assumptions include $U = 0.1$ m day^{-1}, $T = 10°C$, $d_c = 0.05$ cm, $\varepsilon = 0.40$, $\rho_p = 1.05$ g cm^{-3}, and $H = 10^{-20}$ J. These results indicate the dependence of the kinetics of deposition on the size of the particles in suspension that has been predicted and observed in many systems. Small particles are transported primarily by convective Brownian diffusion, and large particles in this system are transported primarily by gravity forces. A suspended particle with a diameter of about 3 μm is most difficult to transport. Nevertheless, in the absence of chemical retardation, a travel distance of only about 5 cm is all that is needed to deposit 99% of such particles in a clean aquifer, that is, an aquifer that has not received and retained previous particles.

The results in Figure 3 indicate that particle passage or retention in groundwater aquifers can be more dependent on chemical factors than physical ones. The required travel distance for 99% deposition increases as $\alpha(p, c)$ decreases. For deposition that is sufficiently unfavorable so that $\alpha(p, c)$ equals 10^{-3}, L_{99} is

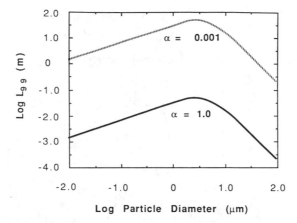

Figure 3. Simulated travel distance (L_{99}) required to accomplish 99% deposition of suspended particles in a groundwater aquifer as a function of the diameter of the suspended particles (μm) for favorable [α(p, c) = 1] and unfavorable [α(p, c) = 10^{-3}] chemical conditions. Assumptions include U = 0.1 m day^{-1}, a_c = 0.025 cm, ε = 0.40, T = 10°C, ρ_p = 1.05 g cm^{-3}, and H = 10^{-20} J.

about 50 m. Although not shown in Figure 3, the previously noted sensitivity of particle passage as indicated by C_L/C_0 to α(p, c) is repeated here. An increase in α(p, c) by a factor of only 2 is sufficient to increase the extent of deposition from 99 to 99.99%, a reduction in C_L/C_0 by a factor of 100. Good models for η(p, c)$_{theor}$ and good measurements of α(p, c)$_{exp}$ are necessary for reliable predictions of C_L/C_0. At present the reliability of models for η(p, c)$_{theor}$ exceeds the number and reliability of measurements of α(p, c)$_{exp}$.

3. PARTICLE AGGREGATION AND SEDIMENTATION IN LAKES

With hydraulic residence times ranging from months to years, lakes constitute efficient sedimentation basins for particles. Lacustrine sediments are sinks for nutrients that support biological growth and for pollutants such as those toxic metals and synthetic organic compounds that associate with particulate matter. Natural aggregation increases particle sizes and thus particle settling velocities, accelerates particle removal to bottom sediments, and decreases the concentrations of particles and particle-reactive pollutants in the water column of a lake.

Field studies to evaluate coagulation and sedimentation in lakes were conducted by Weilenmann et al. (1989). Some results of particle size and concentration measurements at several depths in the water column of Lake Zürich in summer are presented in Figure 4. Particle concentrations are expressed as the incremental change in particle volume concentration (ΔV) that is observed in an incremental logarithmic change in particle diameter ($\Delta \log d_p$). Plots such as Figure 4 have the useful characteristic that the area under a curve between any two particle sizes is related to the total volume concentration of

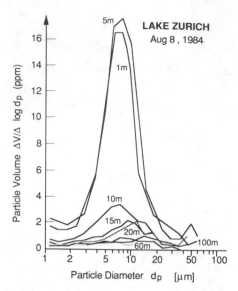

Figure 4. Observed particle volume concentration distributions in the water column of Lake Zürich. Samples obtained on August 8, 1984 at the depths indicated. From Weilenmann et al. (1989).

particles between those two particles. Aerosol size distribution data are commonly presented in this manner (Friedlander, 1977); processes in aquatic and sediment environments (Lerman, 1979) and oceanic particle size distributions (McCave, 1984) have also been represented in this form.

The particle volume distributions in this lake have a single size peak that increases somewhat with depth. Particle volume concentrations decrease substantially with depth from the epilimnion (the upper 5 m) through the thermocline and then into the hypolimnion. The total particle volume concentration in the epilimnion (as indicated by the area under the curves for depths of 1 or 5 m) is about $10 \, \text{cm}^3 \, \text{m}^{-3}$ (ppm); the distributions in the bottom waters indicate total volume concentrations of about $1 \, \text{cm}^3 \, \text{m}^{-3}$ in the hypolimnion.

Discussion of these results is directed at the following question. How do these field results compare with present theories for the kinetics of aggregation and sedimentation in aquatic systems? In answering this question, some laboratory determinations of attachment probabilities will be used in modelling simulations of the kinetics and effects of colloid chemical processes in lakes.

A model for the kinetics of aggregation and sedimentation in lakes has been presented elsewhere (O'Melia, 1980); a short summary is given here. The approach begins with a particle balance for the epilimnion of a lake:

$$\frac{dn_k}{dt} = \frac{1}{2} \sum_{i+j=k} \alpha(i, j)_S \, \lambda(i, j)_S \, n_i n_j - n_k \sum_{i=1}^{\infty} \alpha(i, k)_S \, \lambda(i, k)_S \, n_i$$

$$- \frac{v_k}{z_e} n_k \pm W + \frac{q_{in}}{z_e} n_{k, in} - \frac{q_{out}}{z_e} n_k \tag{6}$$

Here n_i, n_j, and n_k are the number concentrations of particles of sizes i, j, and k in the epilimnion and $n_{k, \text{in}}$ is the number concentration of k-size particles in river inflows. The term $\lambda(i, j)_S$ incorporates most of the effects of physical processes on the rate at which particles of size i and j come into close proximity. The subscript S is used to indicate that Smoluchowski's approach (1917) to the kinetics of particle transport has been adopted. Smoluchowski did not consider hydrodynamic retardation in his early analysis, and it has not been included here in $\lambda(i, j)_S$. A more rigorous approach is possible (Valiolis and List, 1984a, b). The term $\alpha(i, j)_S$ incorporates chemical factors that retard the kinetics of aggregation between particles of size i and j and also those aspects of the kinetics of particle transport that are not included in Smoluchowski's analysis. The Stokes' settling velocity of a particle of size k is denoted as v_k; the mean depth of the epilimnion is z_e; q_{in} and q_{out} refer to river flows into and out of the lake expressed as volume of water per unit of lake surface area and time (the sum of such inflows or outflows is also termed the "areal hydraulic loading" of the lake). The symbol W refers to all processes of production or destruction of particles in the epilimnion; it can include a variety of chemical and biological processes.

The term on the left-hand-side of Eq. 6 is the net rate of change in the number concentration of particles of size k in the epilimnion. The first two terms on the right-hand side are Smoluchowski's (1917) expressions for the kinetics of particle aggregation. The third term is the rate of loss of particles of size k from the epilimnion to the bottom waters by sedimentation. The fourth term W, includes all particle inputs to the lake save for those in riverine inflows, and also particle destruction processes in the epilimnion. The final two terms on the right-hand side describe rates of particle inputs and discharge by river flows.

The terms $\alpha(i, j)_S$ and $\lambda(i, j)_S$ collectively describe a kinetic coefficient for the coagulation or aggregation of suspended particles of sizes i and j. They have analogies with but are not identical to the terms $\alpha(p, c)$ and $\eta(p, c)$ used previously in describing the kinetics of particle deposition processes in porous media. Like $\eta(p, c)$, the term $\lambda(i, j)_S$ incorporates information about various processes of particle transport, although as used here hydrodynamic retardation is not considered. Unlike $\eta(p, c)$, $\lambda(i, j)_S$ is not a ratio of fluxes. It is a rate coefficient that includes most physical aspects the second-order coagulation reaction. Like $\alpha(p, c)$, the term $\alpha(i, j)_S$ incorporates chemical aspects of the interactions between two colliding solids; however, as used here, the effects of hydrodynamic retardation are subsumed in $\alpha(i, j)_S$. The term $\alpha(i, j)_S$ is a ratio defined here as follows:

$$\alpha(i, j)_S = \frac{\text{attachment rate between particles of sizes } i \text{ and } j}{\text{contact rate for particles of sizes } i \text{ and } j \text{ after Smoluchowski}} \quad (7)$$

Three particle transport processes that bring about interparticle contacts are considered here: Brownian diffusion (thermal effects), fluid shear (flow effects), and differential settling (gravity effects). Following Smoluchowski's approach, the appropriate individual transport coefficients for these three processes are as follows:

$$\lambda(i,j)_{\text{S, diff}} = \frac{2kT(a_i + a_j)^2}{3\mu(a_i a_j)} \tag{7a}$$

$$\lambda(i,j)_{\text{S, shear}} = \frac{4(a_i + a_j)^3}{3} G \tag{7b}$$

$$\lambda(i,j)_{\text{S, sed}} = \frac{2\pi(\rho_p - \rho_w)}{9\mu}(a_i + a_j)^3 |a_i - a_j| \tag{7c}$$

Here the radii of the two colliding particles are termed a_i and a_j; the mean velocity gradient inducing contacts by fluid shear is given as G. Other parameters are as defined in Table 1.

Theoretical values of the transport coefficient in Eq. 6 are obtained by summing the contributions of Brownian diffusion, fluid shear, and gravity as follows:

$$\lambda(i,j)_{\text{S, theor}} = \lambda(i,j)_{\text{S, diff}} + \lambda(i,j)_{\text{S, shear}} + \lambda(i,j)_{\text{S, sed}} \tag{8}$$

Measurements are performed to determine experimental values of the attachment coefficient, $\alpha(i,j)_{\text{S, exp}}$. These experiments involve aggregation studies using natural particles and samples of lake water in laboratory studies where interparticle contacts are produced predominantly by fluid shear. Details are given by Ali (1985) and Weilenmann (1986). Illustrative results for five lakes are given in Table 5, together with information about the concentrations of dissolved organic carbon (DOC) and Ca^{2+} in these lakes.

TABLE 5. Experimental Particle Stability Factors in Lakes

Lake	$\alpha(i,j)_{\text{S, exp}}$ (n)[a]	DOC (mg dm^{-3})	Ca^{2+} (M)
Zürich (Switzerland)	0.091[b] (5)	~1	~1.2×10^{-3}
Sempach (Switzerland)	0.011[b] (3)	~4	~1.2×10^{-3}
Luzern (Switzerland)	0.055[b] (1)	~1	~0.9×10^{-3}
Greifen (Switzerland)	0.047[b] (1)	~4	~2.0×10^{-3}
Loch Raven Reservoir (USA)	0.036[c] (10)	~2	~5.0×10^{-4}

[a] Numbers in parentheses denote number of different lake samples tested.
[b] Data from Weilenmann et al. (1989)
[c] Data from Ali et al. '1985).

Following the methodology described previously for particle deposition in porous media, simulations of the effects of particle aggregation and sediment-ation in lakes can be made using theoretical predictions of particle transport $[\lambda(i,j)_{S,\text{theor}}]$ and experimental measurements of particle attachment proba-bilities $[\alpha(i,j)_{S,\text{exp}}]$ in Eq. 6. Unlike the aquifer experiment with latex particles, however, particle inputs to Lake Zürich are not known; it is necessary to provide an estimate of particle production in the epilimnion. Based on considerations of primary production, $CaCO_3$ precipitation, and observed sediment fluxes collec-ted in sediment traps, a flux ranging from 3 to 12 cm^3 m^{-2} day^{-1} of particulate matter is produced in the epilimnion of the lake. The size range of the particle production is assumed to be from 0.3 to 30 μm and to have a power law particle size distribution (Friedlander, 1977; Lerman, 1979; McCave, 1984) as follows:

$$n(d_p) = \frac{dN}{d(d_p)} = Ad_p^{-\beta} \qquad (9)$$

Here $n(d_p)$ is a particle size distribution function, dN is the number concentration of particles in the size range from d_p to $[d_p$ to $d(d_p)]$, and A is a coefficient related to the total number concentration of particles. The exponent β describes the particle size distribution; a value of 4 is assumed here.

Results of some simulations of particle volume concentration distributions by M. Wiesner of Rice University are presented in Figure 5. A net production of particles in the epilimnion of 8.2 cm^3 m^{-2} day^{-1} is assumed, corresponding to a total input volume concentration of 95 cm^3 m^{-3} with an areal hydraulic loading of 10^{-4} cm s^{-1} for Lake Zürich. In the absence of coagulation and sediment-ation [i.e., if $\alpha(i,j)_{S,\text{exp}} = 0$ and $\rho_p = \rho_w$], the particle volume concentration distribution would assume a rectangular shape identical to the input and the total particle volume concentration in the lake (the rectangular area in Fig. 5) would be 95 cm^3 m^{-3}, considerably more than the 10 cm^3 m^{-3} observed in Figure 4. In the absence of coagulation and sedimentation, the shape of the simulated particle volume concentration distribution and the simulated total volume concentration of the epilimnion in Figure 5 do not resemble the observations presented in Figure 4.

In the presence of sedimentation but with no coagulation [i.e., if $\alpha(i,j)_{S,\text{exp}} = 0$ and $\rho_p = 1.05$ g cm^{-3}], larger particles ($d_p > 3$ μm) are removed from the epilimnion to the bottom waters by gravity but smaller sizes remain in the epilimnion to be removed by discharge downstream in the river. The total particle volume concentration in the epilimnion is about 57 cm^3 m^{-3}, again considerably more than the 10 cm^3 m^{-3} observed by Weilenmann et al. (1989). Here again, the shape of the simulated particle volume concentration distri-bution and the simulated total volume concentration of the epilimnion in Figure 5 do not resemble the observations presented in Figure 4.

Particle aggregation in Lake Zürich is considered here to have an attachment probability of 0.1 (Table 5). When both coagulation and sedimentation are occurring [i.e., if $\alpha(i,j)_{S,\text{exp}} = 0.1$ and $\rho_p = 1.05$ g cm^{-3}], smaller particles are

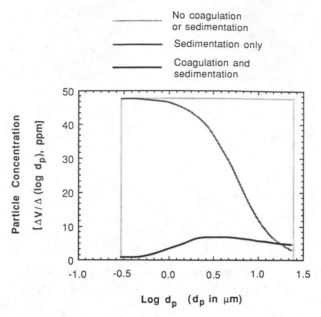

Figure 5. Simulated particle volume concentration distributions in the epilimnion of Lake Zürich, assuming a net particle production flux of $8.2 \text{ cm}^3 \text{ m}^{-2} \text{ day}^{-1}$ and an areal hydraulic loading for the lake of $10^{-4} \text{ cm s}^{-1}$. For no coagulation and sedimentation, $\alpha(i, j)_{\text{S, exp}} = 0$ and $\rho_p = \rho_w$; for sedimentation only, $\alpha(i, j)_{\text{S, exp}} = 0$ and $\rho_p = 1.05 \text{ g cm}^{-3}$; for coagulation and sedimentation, $\alpha(i, j)_{\text{S, exp}} = 0.1$ and $\rho_p = 1.05 \text{ g cm}^{-3}$. Other assumptions stated in the text. Simulations performed by M. Wiesner of Rice University, Houston, TX.

"removed" from their size class by aggregation to larger sizes and then removed from the epilimnion to the bottom waters by gravity. The simulated particle volume distribution has a single size peak of a few micrometers and the total particle volume concentration in the epilimnion is about $11 \text{ cm}^3 \text{ m}^{-3}$, similar to the observations of Weilenmann et al. (1989). When coagulation and sediment-ation are considered simultaneously, the shape of the simulated particle volume concentration distribution and the simulated total volume concentration in the epilimnion as presented in Figure 5 resemble the observations presented in Figure 4 rather well.

The model for aggregation and sedimentation in lakes (Eq. 6) has a conceptual basis and is consistent with some field observations. It is used here to make predictions about the kinetics and effects of coagulation and sedimentation in Lake Zürich. The responses of a lake to coagulation and sedimentation can be represented as mass flux distributions (O'Melia and Bowman, 1984). Simulations of mass fluxes by river flow, net production in the epilimnion, coagulation, and sedimentation for Lake Zürich are presented in Figure 6. The particle mass flux distributions, $\Delta J/(\Delta \log d_p)$, are plotted as functions of particle size (log scale) for mass fluxes by these processes in the epilimnion. A positive sign indicates a flux of mass into a given size class.

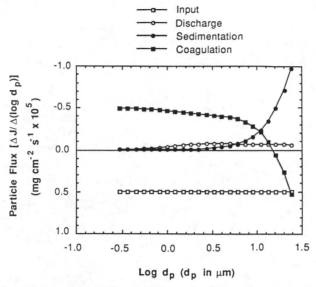

Figure 6. Simulated net mass fluxes with coagulation and sedimentation occuring in the epilimnion of Lake Zürich, assuming a net particle production flux of 8.2 cm^3 m^{-2} day^{-1}, an areal hydraulic loading of 10^{-4} cm s^{-1}, $\alpha(i, j)_{S, exp} = 0.1$, and $\rho_p = 1.05$ g cm^{-3}. Other assumptions stated in the text. Simulations performed by M. Wiesner of Rice University, Houston, TX.

These results show the following. First, most small particles leave their size class by coagulation; only small fractions are carried from the epilimnion by river discharge or settled to the bottom waters by gravity. Second, many large particles enter their size class by aggregation; for the largest particles the aggregation flux even exceeds the net production flux. Third, many more large particles are removed by sedimentation from the epilimnion than are introduced into it. With respect to removal of particulate material from the epilimnion, sedimentation is the dominant process with hydraulic discharge a distant second. Fourth, the shape of the mass flux curve for hydraulic discharge is proportional to the volume concentration distribution in the water. Similarly, the shape of the mass flux curve for sedimentation is proportional to the volume concentration distribution of materials recovered from sediment traps (Weilenmann et al., 1989), and differs substantially from the volume concentration distribution in the water column.

Additional results of model simulations of the effects of coagulation and sedimentation in Lake Zürich are presented in Figure 7, adapted from Weilenmann et al. (1989). Total particle volume concentrations in the epilimnion and hypolimnion are plotted as functions of the net particle production flux in the epilimnion. Observed particle volume concentrations are 5–10 cm^3 m^{-3} in the epilimnion and ~1 cm^3 m^{-3} in the hypolimnion. Model simulations indicate that these concentrations can result from a net particle production flux of 5–10 cm^3 m^{-2} day^{-1} and a particle or aggregate density ranging from 1.05 to

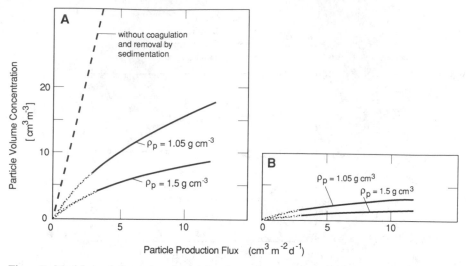

Figure 7. Model simulations of total particle volume concentrations in the summer as functions of net particle production flux in the epilimnion of Lake Zürich, adapted from Weilenmann et al. (1989). Predictions are made for the epilimnion (*A*) and the hypolimnion (*B*). Simulations are made for input particle size distributions ranging from 0.3 to 30 μm described by a power law (Eq. 9) with an exponent (β) of 4. Two particle densities (ρ_p) are considered, 1.05 and 1.5 g cm^{-3}; $\alpha(i,j)_{S,\,exp} = 0.1$. The dashed line represents conditions when coagulation and sedimentation do not occur [$\alpha(i,j)_{S,\,exp} = 0$ and $\rho_p = \rho_w$].

1.5 g cm^{-3}. Aggregates with a density of 1.5 g cm^{-3} are considered to result from coagulation of calcite particles with phytoplankton and agree with density estimates from sediment trap data. The dashed line in Figure 7 describes the total particle volume concentration that would occur in the epilimnion and in the lake outflow as a function of the net particle production, but in the absence of coagulation and sedimentation. It is considerably in excess of observations and model simulations that consider these processes.

In summary, the results of a field study (Weilenmann et al., 1989), coupled with laboratory determinations of $\alpha(i,j)_{S,\,exp}$ and present theory for the kinetics of particle aggregation in aqueous suspensions, are internally consistent. They indicate that coagulation can alter particle transport in lakes, directing small particles and particle-reactive substances to lake sediments rather than leaving them in suspension for eventual removal in lake discharges. The process is probably more complicated than particle deposition in groundwater aquifers because of uncertainties in the magnitude and size distribution of particle inputs and the occurrence of biological and chemical processes for particle destruction. Nevertheless, good predictions of particle concentrations in the water column and particle fluxes to sediments depend on the product $\lambda(i,j) \times \alpha(i,j)$. Estimates of interparticle contact rates can be improved by considering hydrodynamic retardation. The effects of diffusion to a sedimenting particle, introduced by Levich (1962), have been considered as an additional transport mechanism by

Jackson (1989) and could be added to Eqs. 7 and 8. At present, however, models for $\lambda(i,j)_{\text{theor}}$ are more reliable than models for $\alpha(i,j)_{\text{theor}}$. There are few measurements of $\alpha(i,j)_{\text{exp}}$ for natural aquatic systems; most are included in Table 5. Experimental evidence shows that $\alpha(i,j)_{\text{S,exp}}$ depends primarily on solution chemistry. Major divalent cations such as Ca^{2+} increase $\alpha(i,j)_{\text{S,exp}}$, and dissolved macromolecular organic substances decrease it. As noted previously for particle deposition in aquifers, the organic substances in wastewater discharges may be important in retarding the kinetics of particle aggregation in surface waters.

4. MODELING APPROACH AND ASSESSMENT

The quantitative description of the kinetics of particle–particle interactions and of processes such as particle aggregation and deposition in aquatic systems can be considered in five sequential stages:

1. Formulation of the reactions at the solid–solution interface among a solid, solutes, and the solvent (water).
2. Description of the structure of the interfacial region surrounding a single solid particle.
3. Consideration of the physical processes accomplishing particle transport in aquatic systems.
4. Quantitative assessment of the interaction energies or forces that exist when two solid particles and their associated interfacial regions are brought into close proximity.
5. Integration of the effects of these two-body interactions to the scale of the system of interest, such as a lake, groundwater aquifer, or water treatment system.

This sequence describes a perspective covering a range of distances that spans 13 orders of magnitude or more, beginning at bond lengths with dimensions of angstroms and expanding to aquatic systems having dimensions of kilometers or larger. It is a sequence in which all steps have received considerable attention, but it is also a sequence that has not been successfully traversed from beginning to end.

4.1. Surface Speciation

An essential beginning in a comprehensive and quantitative assessment of the kinetics of colloid chemical processes is the speciation at a surface that results from the reactions of solutes with surface sites. In some cases this formulation has been accomplished by a description of surface charge, but the diversity of solids and solutes in aquatic systems necessitates a broader assessment. An example is the adsorption of selenite on the surface of goethite (Hayes et al., 1987) in which a selenium atom is directly coordinated to two oxygen atoms that are, in turn, coordinated to two iron atoms. In aquatic systems, the principal origins of colloidal stability and instability involve some form of bonding at the surface. Several models for the surface speciation of oxides incorporating the adsorption of many metals and several ligands have been developed and used successfully (Westall and Hohl, 1980; Dzombak and Morel, 1987; Schindler and Stumm, 1987). They incorporate some views of the three-dimensional structure of the solid–water interface and have names such as the "diffuse layer", "constant capacitance", or "triple-layer model." Each of these models can be used to simulate surface speciation (including surface charge) as a function of pH, ionic strength, and metal, ligand, and sorbent concentrations. The results have been used successfully in mechanistic studies of the kinetics of surface reactions, including mineral weathering and redox processes. Surface speciation models can provide good assessment of surface speciation and charge, but they do not provide accurate views of the three-dimensional structure of the interfacial region.

4.2. Interfacial Structure

A frequent and important representation of the structure of the interfacial region is the Stern–Guoy–Chapman model illustrated in Figure 8. Most reactivity is accomplished by point charges; in the diffuse layer all reactivity is assigned in this manner. When two similar particles and their associated electrical double layers approach each other, their diffuse atmospheres overlap and a repulsive force is produced. The classic DLVO theory (Verwey and Overbeek, 1948) considers their repulsive electrostatic interaction and the London–van der Waals attractive force that is also essentially electrostatic, arising from dipole fields reflected among atoms. It is important to recognize, however, that DLVO theory has had only limited success in describing the kinetics of colloid chemical reactions. It does *not*, for example, successfully predict the Schulze–Hardy rule for the concentration of electrolyte required to bring about rapid or favorable particle–particle interactions (Lyklema, 1978), probably because of specific chemical interactions between multivalent solutes and surface sites. With regard to the kinetics of aggregation and deposition reactions, DLVO theory predicts dependencies of attachment probabilities [$\alpha(i, j)_{\text{theor}}$ and $\alpha(p, c)_{\text{theor}}$] on particle size and electrolyte valence that are not observed in experiments. It also leads to predictions of these attachment probabilities that are substantially less than observed values [i.e., $\alpha(i, j)_{\text{theor}} < \alpha(i, j)_{\text{exp}}$ and $\alpha(p, c)_{\text{theor}} \ll \alpha(p, c)_{\text{exp}}$].

There is extensive evidence from freshwaters, estuaries, and the oceans that the surface properties, colloidal stability, and the kinetics of aggregation reactions in natural waters are affected by naturally occurring organic substances dissolved in these waters. These effects of natural organic substances in establishing colloidal stability in aquatic systems are anticipated to occur in subsurface environments and to affect the kinetics of particle and pollutant passage and retention in subsurface systems. The kinetics, extent, and significance of

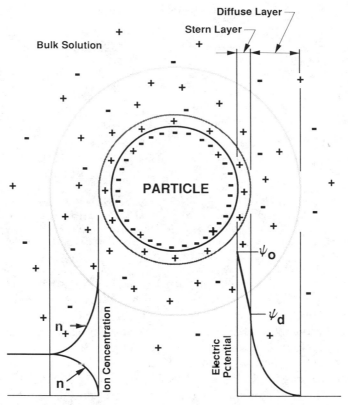

Figure 8. Schematic representation of an electrical double layer (EDL) surrounding a solid particle in aqueous solution, where ψ_0 and ψ_d are surface and diffuse layer potentials, respectively, and n_+ and n_- are cation and anion concentrations. [Adapted from Hirtzel and Rajagopalan (1985).]

aggregation and deposition reactions in aquatic environments depend on solution chemistry, and can therefore be expected to vary widely among environmental systems.

The enhanced stability of natural particles in natural waters containing natural organic matter is a consistent observation without a clear cause. Some speculation is presented here. Humic substances constitute the principal fraction of dissolved organic matter in most natural waters. These molecules can be considered as flexible polyelectrolytes with anionic functional groups (Cornel et al., 1986); they also have hydrophobic components. A schematic representation of the effects of pH and ionic strength on the configuration of polyelectrolytes such as humic substances is presented in Figure 9, adapted from the work of Yokoyama et al. (1989). In freshwaters such molecules assume extended shapes as a result of intramolecular electrostatic repulsive interactions. When adsorbed at interfaces at low ionic strength, they assume flat configurations (Lyklema and Fleer, 1987). Adsorption on inorganic surfaces such as metal oxides could result

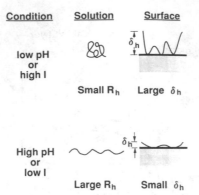

Figure 9. Schematic description of the effects of pH and ionic strength (I) on the conformations of an anionic polyelectrolyte in solution and on a surface, where R_h denotes the hydrodynamic radius of a polyelectrolyte in solution and δ_h denotes the hydrodynamic thickness of an adsorbed polyelectrolyte. [Adapted from Yokoyama et al. (1989).]

from the ligand exchange of functional groups on the humic substances (carboxylic, phenolic) with surface hydroxyl groups on the metal oxides, supplemented by a hydrophobic interaction involving nonpolar components of the humic molecules. The result would be an accumulation of negative charge on the surface of the oxide due to the adsorbed organic substances. At low ionic strength the Debye length (diffuse layer thickness) could exceed the thickness of the flat adsorbed organic layer and the particles would be stabilized electrostatically. The *origin* of the stabilization is in the chemical and hydrophobic interactions between the humic substances, the water, and the surface; the *effect* of these interactions is an electrostatic type of stabilization of the particles. In this perspective, calcium ions could interact specifically with functional groups on the adsorbed molecules, reducing the net charge on the particle–polyelectrolyte composite; a reduction in the range of diffuse layer interactions would also occur. An increase in attachment probability would result. For organic surfaces such as algal cells and detritus, the process could differ somewhat, but particle stability would again involve adsorption of anionic natural organic substances through specific chemical and hydrophobic interactions to produce negatively charged particle–polyelectrolyte composites that are stabilized primarily by their charge.

An evaluation of the effects of natural macromolecules with reactive functional groups on the structure, speciation, and electrical properties of solid–water interfaces and of the influence of major divalent ions on interfacial properties is a necessary and underdeveloped component of a quantitative conceptual formulation of the kinetics of particle–particle interactions in aquatic systems.

4.3. Transport Processes

The effects of Brownian diffusion, fluid flow, and gravity on the transport of particles in aquatic systems have been described with some success, at least under conditions of laminar flow, and the results permit assessment of particle transport in some natural systems with at least moderate accuracy. Some unresolved questions include the effects of initial deposition on the hydro-dynamic aspects of subsequent deposition in porous media, the effects of size heterogeneity of suspended particles and stationary collectors on aggregation or deposition, the breakup of aggregates in suspension and the dislodgement of attached deposits in porous media, and the effects of fluid flow between colliding particles on the chemical and electrochemical characteristics of the two interact-ing interfacial regions. These and other unresolved issues are important and interesting; nevertheless, present understanding of particle transport mechan-isms is a strong component in the overall process of characterizing the kinetics of colloid chemical processes in aquatic systems.

4.4. Interaction Forces

A schematic example for the forces acting on a particle approaching a large spherical stationary collector is shown in Figure 10. Forces due to fluid flow (F_{drag}), gravity (F_{gravity}), London–van der Waals interactions between the particles (F_{LVDW}), and chemical interactions (F_{chem}) between the two interfacial regions are illustrated. Other processes (e.g., Brownian diffusion) and other

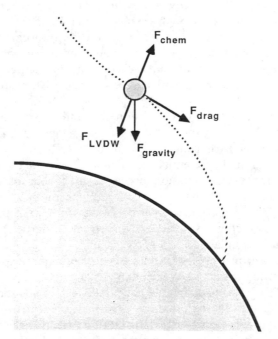

Figure 10. Forces acting on a suspended particle in the neighborhood of a stationary collector: F_{gravity} = gravity force F_{drag} = drag force; F_{LVDW} = London–van der Waals force, F_{chem} = chemical forces. [Adapted from Tobiason and O'Melia (1988).]

forces may be considered in such analyses. As used here, "chemical" interactions include all processes arising from the conformational, electrostatic, hydrophobic, structural, and specific chemical characteristics of the two interacting interfacial regions.

As illustrated by the results presented in Figure 2 and in Table 2 at high ionic strength and high Ca^{2+} for favorable particle–particle interactions (e.g., in the deposition of non-Brownian particles, $\Sigma F = F_{gravity} + F_{drag} + F_{LVDW}$; $F_{chem} = 0$), transport models based on physical and hydrodynamic characteristics of a system can predict the initial kinetics of aggregation and deposition processes in aquatic systems quantitatively. In the presence of repulsive chemical interactions, however, quantitative theoretical predictions of such kinetics are very inaccurate and even many qualitative predictions are not observed. The determination of F_{chem} in aquatic systems merits study and development; it is necessary for the quantitative prediction of the kinetics of colloid chemical processes in these systems.

4.5. Aquatic Systems

Two examples of the integration of two-body interactions to the scale of natural aquatic systems have been presented in this chapter. The interaction between a single suspended particle and a single stationary collector in a porous media (Fig. 10) has been expanded to provide a useful quantitative understanding of the passage of colloidal particles through a sandy aquifer. In this view, an aquifer is a filter that removes particles from suspension with transport by convective diffusion, interception, and gravity, even under chemical conditions in the solution that produce unfavorable particle–collector interactions. "Ripening" of the aquifer may even occur, leading to increased particle deposition over a period of time.

The interaction in a two-body collision in a dilute suspension has been expanded to provide a useful and quantitative understanding of the aggregation and sedimentation of particulate matter in a lake. In this view, Brownian diffusion, fluid shear, and differential sedimentation provide contact opportunities that can change sedimentation processes in a lake, particularly when solution conditions are such that the particles attach readily as they do in Lake Zürich [high $\alpha(i, j)_{exp}$]. Coagulation provides a conceptual framework that connects model predictions with field observations of particle concentrations and size distributions in lake waters and sediment traps, laboratory determinations of attachment probabilities, and measurements of the composition and fluxes of sedimenting materials (Weilenmann et al., 1989).

Integration of such two-body interactions to the scale of aquatic systems is not done without difficulties. At least three types of questions are encountered. What is α? What are the inputs to the system? In addition to colloid chemical reactions, what other processes may be important? As suggested subsequently in this chapter, laboratory experiments can provide estimates of $\alpha(i, j)_{exp}$ and $\alpha(p, c)_{exp}$ for use in simulations. Inputs of particles to aquatic environments are usually quite difficult to determine, much more so than is sometimes considered. In the aquifer study considered previously in this chapter, particle inputs were known when they were added to the system. In the lake study, particle inputs were assumed based on measurements in sediment traps and other information, but the inputs themselves were not measured. This is perhaps the most poorly defined factor in this study. There are several other processes that can affect particle concentrations and fluxes in aquatic systems in addition to colloid chemical ones. Such processes can be incorporated into large-scale modeling if their kinetics can be described. This subject will not be considered further here.

5. EXPERIMENTAL AND THEORETICAL DIRECTIONS

Two components in the overall process of determining the kinetics of colloid chemical processes in aquatic systems limit the value of quantitative and even some qualitative theoretical predictions of these kinetics. First, the structure and chemistry of an interfacial region in an aquatic environment containing inorganic ions together with organic macromolecules and polyelectrolytes are not known. Solid surfaces may be inorganic (oxides, carbonates, clays) or organic (algae, bacteria, detritus, biological and synthetic membranes). Solutes may be metals (H^+, Ca^{2+}, Fe^{3+}) or ligands (OH^-, SO_4^{2-}, NTA), organic (oxalate, amino acids) or inorganic (F^-, PO_4^{3-}), monomers (phenol) or polymers (polysac-charides), and hydrophobic (oils), hydrophilic (sugars), or amphiphilic (proteins, humic acids). Aquatic interfacial chemistry is diverse. Second, the forces (F_{chem}) or energies arising from the interactions of two interfacial regions in aquatic environments are not known accurately. This arises in part because the structure and chemistry of these regions is not known, as stated above. In addition, it should be recognized that these chemical forces are not sufficiently well understood to permit quantitative or qualitative theoretical description of the *kinetics* of aggregation or deposition in much simpler systems for which models of the interfacial region have been proposed.

Until such theories can be developed, laboratory experiments can be performed to determine chemical effects in aquatic colloid chemical processes for actual situations. This is suggested by the analysis presented in this chapter of the aquifer study by Harvey et al. (1989) and is illustrated for Lake Zürich by the study of Weilenmann et al. (1989). Since mass transport can be described with some success [e.g., $\eta(p, c)_{theor}$ and $\lambda(i, j)_{s, theor}$], this knowledge can be combined with laboratory determinations of attachment probabilities such as those illustrated in Table 2 for $\alpha(p, c)_{exp}$ and listed in Table 5 for $\alpha(i, j)_{s, exp}$ to describe the kinetics of deposition and aggregation (e.g., Eqs. 5 and 6) in aquatic systems.

These laboratory experiments must use controlled and characterized conditions of mass transport, such as the column studies by Tobiason (1987) and the stirred reactor studies by Ali (1985) and Weilenmann (1986). It is not necessary that the physics of the system of interest be reproduced experimentally, but it is necessary that the physics of the laboratory system be known. Since solution chemistry is a dominant and perhaps controlling factor for attachment probabilities in aquatic environments, it is necessary that water from the system to be studied be used. Insofar as possible, such experiments should also use natural particles and collectors.

Some research directions for studies of the chemistry of particle–particle interactions in aquatic systems are suggested here. First, the origins and effects of natural organic substances on interfacial structure and chemistry in aquatic systems should be established conceptually and quantitatively. Because of the diversity of organic substances in natural systems, this task has many components. It involves assessment of electrostatic, van der Waals, hydrophobic, and specific chemical interactions, together with consideration of structural rearrangements and conformational changes of adsorbing species. Second, the effects of macromolecules and polyelectrolytes on the speciation and distribution of inorganic species in the interfacial region, and the accompanying effects of inorganic species, particularly multivalent ones, on organic substances at interfaces should be characterized. Third, the dynamics of the interaction between the double layers surrounding two solids has received attention recently (e.g., Dukhin and Lyklema, 1987), and such efforts should be extended. Dynamic studies could involve a coupling of hydrodynamics with species concentrations in the region between particles, within which both separating distance and fluid velocity are altered as the particles approach each other. Finally, theories for the structure and chemistry of the interfacial region should be tested by their ability to quantitatively describe the *kinetics* of colloid chemical processes such as aggregation and deposition.

Acknowledgments

The hospitality of the Director, scientists, students, and staff of the EAWAG during my sabbatical there in 1988/89 is primarily responsible for this chapter. John Tobiason, Mark Weisner, Ulrich Weilenmann, Menachem Elimelech, and Christine Tiller contributed with their ideas and endeavors. The financial support of the EAWAG, the U.S. Environmental Protection Agency through grant number R812760, and the Lyonnaise des Eaux are gratefully acknowledged.

REFERENCES

Ali, W. (1985), "Chemical Aspects of Coagulation in Lakes," unpublished doctoral dissertation, The Johns Hopkins University, Baltimore, MD, 222 pp.

Ali, W., C. R. O'Melia, and J. K. Edzwald (1985), "Colloidal Stability of Particles in Lakes: Measurement and Significance," *Water Sci. Technol.* **17**, 701–712.

Cornel, P. K., R. S. Summers, and P. V. Roberts (1986), "Diffusion of Humic Acid in Aqueous Solution," *J. Colloid Interface Sci.* **110**, 149–164.

Dukhin, S. S., and J. Lyklema (1987), "Dynamics of Colloid Particle Interaction," *Langmuir* **3**, 94–98.

Dzombak, D. A., and F. M. M. Morel, (1987), "Adsorption of Inorganic Pollutants in Aquatic Systems," *J. Hydraulic Eng.* **113**, 430–475.

Friedlander, S. K. (1977), *Smoke, Dust and Haze*, Wiley-Interscience, New York.

Happel, J. (1958), "Viscous Flow in Multiparticle Systems: Slow Motion of Fluids Relative to Beds of Spherical Particles," *Am. Inst. Chem. Eng. J.* **4**, 197–201.

Harvey, R. W.., L. H. George, R. L. Smith, and D. R. Leblanc (1989), "Transport of Microspheres and Indigenous Bacteria through a Sandy Aquifer: Results of Natural and Forced-Gradient Tracer Experiments," *Environ. Sci. Technol.* **23**, 51–56.

Hayes. Kim F., L. Roe, G. E. Brown, Jr., K. O. Hodgson, J. O. Leckie, and G. A. Parks (1987), "In Situ X-Ray Absorption Study of Surface Complexes: Selenium Oxyanions on α-FeOOH," *Science* **238**, 783–786.

Hirtzel, C. S., and R. Rajagopalan, (1985), *Colloidal Phenomena*, Noyes, Park Ridge, NJ.

Jackson, G. A. (1989), "Simulation of Bacterial Attraction and Adhesion to Falling Particles in an Aquatic Environment," *Limnol. Oceanogr.* **34**, 514–530.

Lerman, A. (1979), *Geochemical Processes Water and Sediment Environments*, Wiley-Interscience, New York.

Levich, V. G. (1962), *Physiochemical Hydrodynamics*, Prentice-Hall, Englewood Cliffs, NJ.

Lyklema, Johannes (1978), "Surface Chemistry of Colloids in Connection with Stability," in K. J. Ives, Ed., *The Scientific Basis of Flocculation*, Sijthoff and Noordhoff, The Netherlands, 3–36.

Lyklema, J. and G. J. Fleer (1987), "Electrical Contributions to the Effect of Macromolecules on Colloid Stability," *Colloids Surfaces*, **25**, 357–368.

McCave, I. N. (1984), "Size Spectra and Aggregation of Suspended Particles in the Deep Ocean," *Deep-Sea Res.* **31**, 329–352.

O'Melia, C. R. (1980), "Aquasols: The Behavior of Small Particles in Aquatic Systems," *Environ. Sci. Technol.* **14**, 1052–1060.

O'Melia, C. R., and K. S. Bowman (1984). "Origins and Effects of Coagulation in Lakes," *Schweizerische Z. Hydrol.* **46**, 64–85.

Rajagopalan, R., and C. Tien (1976), "Trajectory Analysis of Deep-Bed Filtration with the Sphere-in-cell Porous Media Model," *Am. Inst. Chem. Eng. J.* **22**, 523–533.

Schindler, P. W., and W. Stumm (1987), "The Surface Chemistry of Oxides, Hydroxides, and Oxide Minerals," in W. Stumm, Ed., *Aquatic Surface Chemistry*, Wiley-Interscience, New York, pp. 83–110.

Smoluchowski, M. (1917), "Versuch einer mathematischen Theorie der Koagulationskinetik kolloidaler Lösungen," *Z. Phys. Chem.* **92**, 129–168.

Spielman, L. A., and J. A. FitzPartick (1973), "Theory of Particle Capture under London and Gravity Forces," *J. Colloid Interface Sci.* **42**, 607–623.

Tobiason, J. E. (1987), "Physiochemical Aspects of Particle Deposition in Porous Media," unpublished doctoral dissertation, The Johns Hopkins University, Baltimore, Maryland, 280 pp.

Tobiason, J. E., and C. O'Melia (1988), "Physicochemical Aspects of Particle Removal in Depth Filtration," *J. Am. Water Works Assoc.* **80**, (12), 54–64.

Valioulis, I. A., and E. J. List (1984a), "Numerical Simulation of a Sedimentation Basin. 1. Model Development," *Environ. Sci. Technol.* **18**, 242–247.

Valioulis, I. A., and E. J. List (1984b), "Numerical Simulation of a Sedimentation Basin. 2. Design Application," *Environ. Sci. Technol.* **18**, 248–253.

Verwey, E. J. W., and J. Th. G. Overbeek (1948), *Theory of the Stability of Lyophobic Colloids*, Elsevier, Amsterdam.

Weilenmann, U. (1986), "The Role of Coagulation for the Removal of Particles by Sedimentation in Lakes," Ph.D. thesis No. 8018, Eidgenössische Technischen Hochschule Zürich, Zurich, 163 pp.

Weilenmann, U., C. R. O'Melia, and W. Stumm (1989), "Particle Transport in Lakes: Models and Measurements," *Limnol. Oceanogr.* **34**, 1–18.

Westall, J., and H. Hohl (1980), "A Comparison of Electrostatic Models for the Oxide/Solution Interface," *Adv. Colloid Interface Sci.* **12**, 265–294.

Yao, K. M., M. T. Habibian, and C. R. O'Melia (1971), "Water and Wastewater Filtration: Concepts and Applications," *Environ. Sci. Technol.* **5** 1105–1112.

Yokoyama, A., K. R. Srinivasan, and H. S. Fogler (1989), "Stabilization Mechanism by Acidic Polysaccharides, Effects of Electrostatic Interactions on Stability and Peptization," *Langmuir* **5**, 534–538.

17

KINETICS OF CHEMICAL WEATHERING: A COMPARISON OF LABORATORY AND FIELD WEATHERING RATES

Jerald L. Schnoor

*Department of Civil and Environmental Engineering,
The University of Iowa, Iowa City, Iowa*

1. INTRODUCTION

"Chemical weathering" can be defined as the dissolution of minerals by the action of water and its solutes. It is an important feature of the global hydrogeochemical cycle of elements, whereby rocks and primary minerals become transformed to solutes and soils and, eventually, to sediments and sedimentary rocks. In this cycle, water occupies a central position serving as both a reactant and a transporting agent of suspended and dissolved material. The sea is the ultimate receptacle of weathered material, and the atmosphere provides a reservoir of weak acids (CO_2) and oxidants.

In addition to chemical weathering by dissolved carbon dioxide, strong acids, such as H_2SO_4 and HNO_3 deposited as " acid rain," may promote dissolution of rocks and minerals. The occurrence of acid deposition results from the anthropogenic disturbance of cycles that couple atmosphere, land, and water. Redox conditions in the atmosphere are disturbed by the oxidation of carbon, sulfur and nitrogen resulting from fossil-fuel combustion. Oxidation reactions exceed reduction reactions in the elemental cycles, and a net production of hydrogen ions in atmospheric precipitation is a necessary consequence. The disturbance is transferred to the terrestrial and aquatic environment where, either acidity is neutralized via acid–base and reduction reactions, or the disturbance is transported downstream.

Kinetics of chemical weathering are important to understand the rates of rock dissolution, sediment formation, and acid neutralization in the environment. In this chapter, there are three principal objectives: (1) to describe the theoretical kinetics of chemical weathering for pure minerals, (2) to discuss the relative advantages and disadvantages of various experimental apparatus for measuring those kinetics in soils, and (3) to make comparisons between laboratory and field measurements of weathering rates and solute transport. A case study of laboratory and field measurements at Bear Brook Watershed, east of Orono, Maine, at Lead Mountain, will be used to illustrate the principles discussed in this chapter.

2. KINETIC MODELS

Many models have been used to describe chemical weathering. These models fall into two general categories: diffusion-controlled dissolution and surface reaction-controlled dissolution. "Diffusion-controlled dissolution" refers to the formation of a thin layer that limits transport, either in the liquid film surrounding the mineral or as a solid-phase layer that is depleted in certain cations. Transport through a stagnant liquid film surrounding the mineral surface is too rapid to be the rate-limiting step, based on comparisons of mineral dissolution rates and transport of solutes by molecular diffusion through liquid film. However, migration of ions through a solid layer may occur for a limited period of time during incongruent dissolution. Transport of ions to the bulk solution through a cation-depleted solid layer eventually becomes as slow as the surface-controlling dissolution reaction. At this time, a pseudo-steady state develops in which the rate of transport through the solid layer of cations is equal to the surface-controlled dissolution of the central metal ion. Detachment of the surface coordinated complex [ligand(s) and the central metal ion] to the bulk solution is the rate-determining step in dissolution. Dissolution proceeds congruently after an initial incongruent period.

2.1.   Initial Incongruent Dissolution

Figure 1 is a schematic of buildup of a cation-depleted layer of thickness y from a hypothetical mineral with constituents A and B. For the simplest case, a $1:1$ stoichiometry is assumed for A and B in mineral AB. Initially incongruent dissolution of AB results in the rapid migration (diffusion) of constituent B from the core of the mineral grain through a layer that is depleted in B (Eq. 1). Reaction of constituent A is controlled by detachment of the surface-coordinated complex. Equation 2 results from the rate-determining step of detachment of a hydrolyzed metal ion (Furrer and Stumm, 1983, 1986). A surface-coordinated complex may involve reaction of any ligand at the surface with a metal ion (e.g., an organic ligand and central ion), but only one reaction will limit the overall dissolution. In the following equations, it is assumed that attack by hydrogen

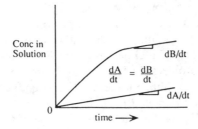

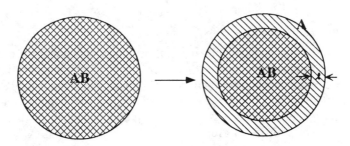

Figure 1. Schematic of initial incongruent dissolution. In the initial stages, a mineral grain may develop a cation depleted layer l, but eventually congruent dissolution is observed and the rate of dissolution of $A(dA/dt)$ must equal $B(dB/dt)$ for a 1:1 stoichiometry.

ions is the rate-determining step resulting in hydrolysis of the central metal ion:

$$\frac{dM_B}{dt} = -D_{BA} S \frac{dB_{AB}}{dr} \cong -D_{BA} S \frac{(B_{AB} - B_{surf})}{y} \tag{1}$$

$$\frac{dM_A}{dt} = -kS\{\blacksquare-H^+\}^m \tag{2}$$

where M_B = mass of B dissolving (mol)
     M_A = mass of A dissolving (mol)
     D_{BA} = molecular solid diffusion coefficient of B diffusing through A ($L^2 T^{-1}$) (where mol L, and T are hypothetical units of mass, length, and time, respectively)
     S = surface area (L^2)
     B_{AB} = concentration of ion B in solid AB (mol L^{-3})
     B_{surf} = concentration of ion B at the surface of solid AB (mol L^{-3})
     y = cation-depleted layer thickness (L)
     k = hydrolysis rate constant (T^{-1})
     t = time (T)
     m = valence of ion A
     r = radial distance (L)
    $\blacksquare-H^+$ = surface concentration of protons (mol L^{-2})

The concentration of constituent B becomes negligible at the surface of the mineral grain. Gradually, the rate of mass diffusion of B (Eq. 1) through an increasing depleted layer (y) becomes slower, and it is equal to the rate of surface-controlled dissolution of A (Eq. 2). Thus, a pseudosteady state is attained, and the depleted layer thickness stabilizes. For a 1:1 stoichiometry, the rates of reaction of solid layer diffusion (Eq. 1) and surface reaction (Eq. 2) become equal (Eq. 3). The thickness of the cation depleted layer is related to the concentration of B in the original mineral AB and the rate of detachment of the activated complex (Eq. 4).

$$D_{BA} S (B_{AB}/y) = kS \{\blacksquare-H^+\}^m \tag{3}$$

$$y = \frac{D_{BA} B_{AB}}{k\{\blacksquare-H^+\}^m} \tag{4}$$

At pseudo steady state, the thickness of the cation depleted layer is a function of pH. Thickness will decrease with increasing H^+ ion concentration in bulk solution. Congruent dissolution will result immediately from low pH solutions. In practical situations, the depleted layer thickness has been estimated to be 2–17 Å (Schott et al., 1979). For Al or Si dissolving from a feldspar grain at pH 4, typical parameters for Eq. 4 are

$$y = \frac{(10^{-22} \text{ cm}^2/\text{s})(0.03 \text{ mol/cm}^3)}{(10^{-12} \text{ mol/m}^2 - \text{s})} (10^4 \text{ cm}^2/\text{m}^2) \tag{5}$$

$$= 3 \times 10^{-8} \text{ cm}^2 = 3 \text{ Å}$$

An initial incongruent dissolution may last over a time scale of hours to days. Basic cations (Ca^{2+}, Mg^{2+}, Na^+, K^+) are depleted preferentially to Si and Al for most aluminosilicates. This period is followed by a long period of subsequent congruent dissolution.

If the mineral stoichiometry is generalized, $A_x B_z$, then pseudosteady state exists when

$$x \frac{dM_B}{dt} = z \frac{dM_A}{dt} \tag{6}$$

such that

$$y = \frac{x}{z} \frac{D_{BA} B_{AB}}{k\{\blacksquare-H^+\}^m} \tag{7}$$

Congruent surface-controlled dissolution follows after the initial incongruent period.

2.2.   Surface-Controlled Dissolution Kinetics

Chemical weathering requires the mass transport of solutes to a mineral surface to form a coordination complex. In a series of steps involving geometric and chemical principles at surface defects (kinks, edges, and screw dislocations), a detachment occurs and transport of the reactants into bulk solution follows. In surface-controlled dissolution reactions, detachment of an activated surface-coordinated complex is the rate-determining step. The general rate law may be expressed as (Wieland et al., 1988)

$$R = kX_a P_j S_{sites} \qquad (8)$$

where R is the proton or ligand-promoted dissolution rate (mol m^{-2} s^{-1}), k is the rate constant (s^{-1}), X_a denotes the mole fraction of dissolution active sites (dimensionless), P_j represents the probability of finding a specific site in the coordinative arrangement of the activated precursor complex, and S_{sites} is the total surface concentration of sites (mol m^{-2}). The rate expression in Eq. 8 is essentially a first-order reaction in the concentration of activated surface complex, C_j (mol m^{-2}):

$$C_j = X_a P_j S_{sites} \qquad (9)$$

Formulation of Eq. 9 is consistent with transition-state theory, where the rate of the reaction far from equilibrium depends solely on the activity of the activated transition-state complex (Wieland et al., 1988). Equations 8 and 9 are equivalent to Eq. 2 for proton attack, where C_j is equal to the surface concentration of activated complex.

The surfaces of oxides and aluminosilicate minerals in the presence of water are characterized by amphoteric surface hydroxyl groups. The surface OH$^-$ group has a complex-forming O-donor atom that coordinates with H$^+$ and metal ions (Schindler, 1982; Kummert and Stumm, 1980; Sigg and Stumm, 1980). The underlying aluminum or metal ion in the surface tetrahedra of the mineral and other cations are subject to coordination with OH$^-$ groups, which, in turn, weakens the bond to their structural oxygen atoms. Detachment of an activated complex removes the coordination complex and renews the surface for further hydrolysis, protonation, and dissolution. An example of hydrolysis, protonation, formation of the surface coordination complex, and detachment to solution is shown in Figure 2 for an aluminum oxide. Also, the underlying central metal ions can exchange their structural OH$^-$ ions for other ligands (e.g., organic anions) to form a strong surface coordination complex that can facilitate dissolution.

A very important result of recent laboratory studies has been the fractional order dependence of mineral dissolution on bulk phase hydrogen ion activity. If the dissolution reaction is controlled by hydrogen ion diffusion through a thin liquid film or residue layer, one would expect a first-order dependence on $\{H^+\}$. If the dissolution reaction is controlled by some other factor such as surface area alone, then the dependence on hydrogen ion activity should be zero-order. Rather, the dependence has been fractional order in a wide variety of studies, indicating a surface reaction controlled dissolution (Schott et al., 1981; Giovanoli, et al. 1989; Schnoor and Stumm, 1986; Busenberg and Plummer, 1982; Grandstaff, 1977; Furuichi et al., 1969).

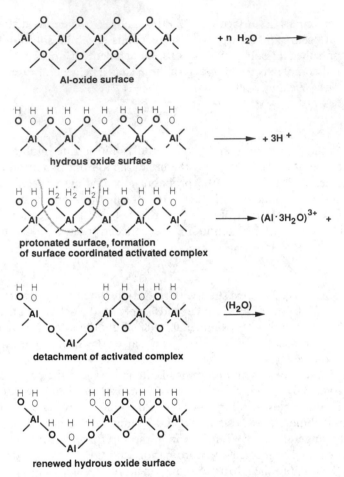

Figure 2. Schematic of an aluminum oxide mineral undergoing hydrolysis and protonation. Dissolution occurs when the surface-coordinated-activated complex $(Al \cdot 3H_2O)^{3+}$ detaches from the surface and goes into solution, thereby renewing the surface.

The dependence of mineral weathering on hydrogen ion activity in solution can be demonstrated using surface coordination chemistry as in

$$\blacksquare\text{-OH} + H^+ \xrightarrow[\text{fast}]{} \blacksquare\text{-OH}_2^+ \tag{10}$$

$$\blacksquare\text{-OH}_2^+ \xrightarrow[\text{slow}]{} Me(H_2O)_x^{z+} + \blacksquare\text{-} \tag{11}$$

where $\blacksquare$–OH = hydrous metal (Al, Si, Fe, etc.) oxide surface with a functional OH group
   $\blacksquare$–OH$_2^+$ = protonated surface group
   z = change on the central metal ion

The dissolution rate is proportional to the degree of protonation of the surface to the power of the central metal ion valence, with the following stoichiometry:

$$\text{Rate} \propto \{\blacksquare\text{-OH}_2^+\}^z \tag{12}$$

The degree of surface protonation becomes saturated by high concentrations of hydrogen ions in bulk solution, following a Langmuir-type adsorption isotherm. If the logarithm of the degree of surface protonation is plotted against pH of the bulk solution (Freundlich isotherm), the middle portion of the plot is of fractional order, $n < 1$:

$$\log\{\blacksquare\text{-OH}_2^+\} = n\log\{H^+\} + \log K_F \tag{13}$$

where n represents the slope of Freundlich isotherm and K_F is the intercept of Freundlich isotherm. By substituting Eq. 13 into 12, it is possible to express the rate of mineral dissolution in terms of the hydrogen ion concentration in bulk solution:

$$\text{Rate} \propto \{H^+\}^m \tag{14}$$

in which $m = n\,z$, fractional order dissolution rate. Experimentally, for pure δ-aluminum oxide dissolution, $m = 0.4$, $n = 0.13$, and $z = 3.1$, approximately the valence of Al^{3+} (Furrer and Stumm, 1983).

However, hydrogen ion attack on minerals (hydrolysis) is not the only means of mineral dissolution. Other ligands, such as organics in soil solution, are known to accelerate the dissolution of minerals. This process, too, can be viewed as a surface-coordination reaction with the following mechanism. For a hypothetical diprotic acid (H_2A) and a central metal ion (Me) of valence z, we obtain

$$\blacksquare\text{-OH} + H_2A \xrightarrow[\text{fast}]{} \blacksquare\text{-AH} + H_2O \tag{15}$$

$$\blacksquare\text{-AH} \xrightarrow[\text{slow}]{} \text{Me(AH)}^{(z-1)+} + \blacksquare\text{-} \tag{16}$$

The rate-determining step is detachment of the coordination complex at the surface, and the rate of dissolution is proportional to the degree of ligand binding on the mineral surface or the concentration of ligand in bulk solution to fractional power (p), the slope of the Freundich adsorption isotherm for [HA$^-$] on the mineral surface for a 1:1 metal-to-ligand complex:

$$\text{Rate} = k_1\{\blacksquare\text{-A}^-\} \propto [HA^-]^p \tag{17}$$

Two types of mineral dissolution (hydrolysis and ligand attack) give rise to mixed kinetics overall. These kinetics are (Furrer and Stumm, 1983)

$$\text{Rate} = k_1\{\blacksquare\text{-OH}_2^+\}^z + k_2\{\blacksquare\text{-A}^-\} \tag{18}$$

It is well known that organic complex-formers, such as simple organic acids (citric, oxalic, tartaric, salicyclic) formed by microorganisms in soils and humic or fulvic acids, "solubilize" mineral iron(III) and aluminum. These complex-formers not only increase the solubility of these minerals but also are able to form chelates on hydrous oxide surfaces and thus, in turn, catalyze the dissolution of oxides and aluminum silicates (Kummert and Stumm, 1980; Sigg and Stumm, 1980). The downward vertical displacement of Al and Fe, as it is observed in the podsolization of soils, can be accounted for by considering the effect of pH and of complex formers on both solubility equilibria and dissolution rates (Schnoor and Stumm, 1985; Schnoor and Stumm, 1986).

With highly siliceous minerals (pure quartz as the limiting case), the hydrogen ion effect on weathering rate is not as great because release of silica and depolymerization of silica tetrahedra are difficult bonds (Si–O) to break. Silica depolymerization increases with hydroxide ion concentration but not much with hydrogen ion. Other organic ligands may increase the rate of surface controlled dissolution in such cases.

3.   SEM/EDX OBSERVATIONS

Seminal studies of mineral dissolution in the laboratory have reported "parabolic" dissolution rates (linear functions of the square root of time), a result consistent with diffusional transport limited kinetics (Luce et al., 1972; Helgeson, 1971, 1972; Paces, 1973; Wollast, 1967; Busenberg and Clemency, 1976). Recent work has suggested that parabolic dissolution kinetics are an artifact of mineral grinding (Berner and Holdren, 1977, 1979; Berner et al., 1980; Holdren and Berner, 1979; Schott et al., 1981). Disrupted grain surfaces and ultrafine mineral particles dissolve initially at a high rate. Dissolution rates are actually linear in time.

Scanning electron microscopy and energy dispersive X-ray spectroscopy (SEM/EDX) have allowed observations of mineral surfaces during hydrolysis. While secondary mineral formation is sometimes observed, a coating of the primary mineral surface does not occur. Dissolution reactions occur preferentially along crystal defects, screw dislocations, and etch pits. These results have been reported for a wide range of minerals including microcline, feldspars, augite, diopside, hypersthene, and hornblende (Berner and Holdren, 1977, 1979; Berner et al., 1980; Holdren and Berner, 1979; Schott, et al., 1981; Giovanoli et al., 1989).

Figure 3 shows photomicrographs of sediments from Lake Cristallina in the Maggia Valley of Ticino, Switzerland, on the southern face of the Alps at 2300-m elevation. The area receives moderate acidic deposition (pH 4.6–4.9). Because of its crystalline rock geology (granitic gneiss bedrock) and flashy hydrograph, the lake is slightly acidic (pH 5.0–5.4). Plagioclase feldspars and biotite are the primary weathering minerals based on scanning electron microscopy and energy dispersive X-ray spectrometry. Figure 3a shows ferric hyroxide that is deposited in the lake sediment at certain locations and on some mineral grains, such as

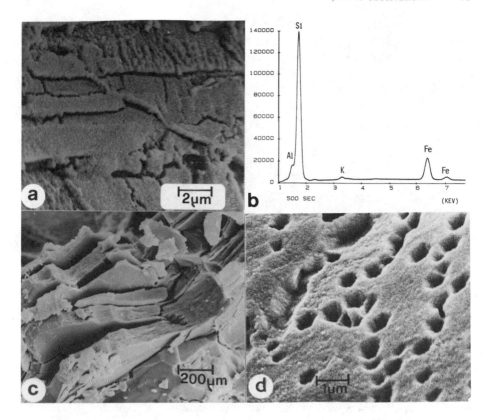

Figure 3. SEM/EDX images of (*a*) ferric hydroxide coating on a feldspar grain, (*b*) EDX spectra of ferric hydroxide on a quartz grain, (*c*) chemical weathering of biotite platelets, and (*d*) etch pits on a plagioclase mineral grain. All grains were taken from surface sediments of Lake Cristallina, Switzerland. Photographs are courtesy of Professor Rudolf Giovanoli, Laboratory of Electron Microscopy, University of Bern.

feldspars and quartz. Amorphous $Fe(OH)_3$ is the oxidation product and precipitate on biotite weathering found in acid lakes with pH<6, where oxidation of ferrous iron is slow enough to allow reduced iron to find its way to the lake for subsequent precipitation. The chemical equation for biotite weathering in the presence of strong acid is

$$KMg_x Fe_{(3-x)}AlSi_3O_{10}(OH)_{2(s)} + (1+2x)H^+ + (9-2x)H_2O \rightarrow$$
$$K^+ + xMg^{2+} + 3H_4SiO_4 + Al(OH)_3 + (3-x)Fe(OH)_2$$

where x = mole fraction of magnesium. On reaction with dissolved oxygen in water, ferrous hydroxide becomes oxidized to ferric hydroxide.

$$(3-x)Fe(OH)_{2(aq)} + [(3-x)/2]O_{2(aq)} \leftrightarrow (3-x)Fe(OH)_{3(s)}$$

Figure 3*b* is the energy-dispersive X-ray of a quartz grain covered with Fe(OH)$_3$ precipitate. Figure 3*c* shows the weathering of biotite sheets, and Figure 3*d* shows etch pits on plagioclase grains indicating surface reaction control. (If diffusion was controlling the dissolution reaction, minerals should be rounded.) Dissolution of plagioclase feldspar in the presence of strong acids may be written

$$Na_{(1-x)}Ca_xAl_{(1+x)}Si_{(3-x)}O_{(s)} + (1+x)H^+ + (7-x)H_2O \rightarrow$$
$$(1-x)Na^+ + xCa^{2+} + (3-x)H_4SiO_4 + (1+x)Al(OH)_3$$

where $x =$ mole fraction of calcium.

4.   EXPERIMENTAL APPARATUS

Various reactor types have been used to determine the kinetics and rates of chemical weathering, including batch pH-stats, flow-through columns, fluidized-bed reactors, and recirculating columns.

Figures 4–7 are schematics of various automated reactors that have been used in this research to determine the kinetics of chemical weathering. Each reactor has advantages and disadvantages. Results from batch studies at constant pH are readily amenable to determine the H$^+$ ion dependence of weathering rates. Solute concentrations are not constant during the course of the experiment with a pH-stat, and solutes may influence the results, especially if the concentration increases to the point of secondary mineral formation. Flow-through columns provide a realistic flow regime and solids concentration, but it is difficult to obtain kinetic formulations because of the increasing solute concentrations over the length of the reactor. Fluidized-bed reactors with recycle allow steady-state concentrations of solutes to be achieved, but operating conditions are sometimes limited by the need to keep mineral grains suspended in the reactor. Recirculating-column reactors offer many of the advantages of the fluidized-bed and column designs, and they may be operated with rapid recycle and a small aspect ratio (length : diameter) in order to ensure constant pH and solute concentrations within the reactor.

Figure 4 shows a jacketed water bath with two 2-L batch reaction vessels that are continuously monitored for pH, temperature, and conductivity. Discrete samples can be taken for ion analyses, and the volume is replaced by deionized water or fresh titrant. A feedback controller is used with a micrometering pump to maintain the set-point pH and to keep track of titrant volume. Data capture is achieved by an IBM-PC-XT computer, and plots of H$^+$ ion consumption rate versus time are immediately generated and stored. Advantages of continuous data logging include the ability to conduct long-term unsupervised experiments and the ability to investigate pulses/transients (Etringer, 1989).

Figure 5 illustrates an upflow soil column (18 × 1-in. inner diameter) with screw cap fittings and watertight O-ring seals. Soil ($\sim$250 g) is supported by

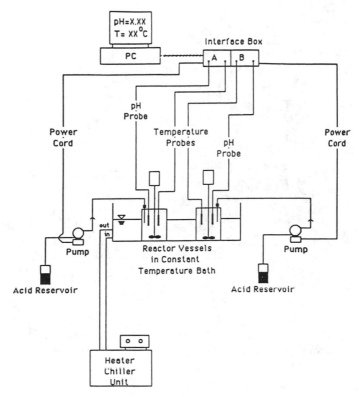

Figure 4. Schematic diagram of the pH-stat system. Special features include the suspended paddle stirring to reduce abrasion, continuous data capture suitable for transient studies, and feedback control from the pH probe to the acid reservoir pump through the interface box and PC computer.

aquarium glass filter fiber, and a Whatman 45-μm microfiber filter is installed at the outflow to retain fine soil particles. A peristaltic pump (0.5 mL min^{-1}) with silicone tubing was adequate for influent feed. Continuous data logging of both the influent and effluent was performed with a Leading Edge D-PC with a 30-megabyte hard disk and Intel 8087 math coprocessor for Hercules graphics emulation. A Metrabyte DASH-16 data acquisition board allows 50,000 samples per second to be taken with 16 single-ended or eight differential analog sensors (Wallace, 1989).

Figure 6 is a schematic of a fluidized-bed reactor first proposed for mineral dissolution by Chou and Wollast (1985) and further developed by Mast and Drever (1987). The principle is to achieve a steady-state solute concentration in the reactor (unlike the batch pH-stat, where solute concentrations gradually build up). Mineral grains must be suspended by upflow feeding in order to fluidize the bed. This limits operation somewhat because the size and density of the mineral grains must be balanced by the desired flow-rate : recycle ratio.

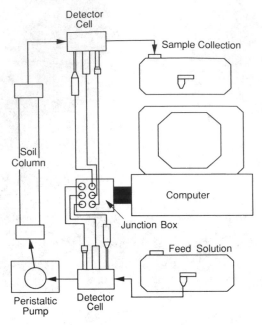

Figure 5. Experimental apparatus for the soil column experiments. Special features include continuous data capture on the influent and effluent from the column for transient studies.

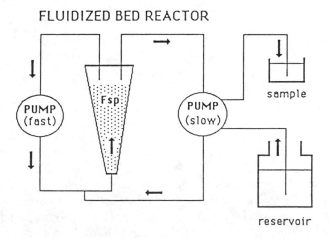

Figure 6. Schematic diagram of the fluidized-bed reactor (from Mast and Drever, 1987). Special features include fast pump recycle to suspend all particles in the reactor and the reservoir for transient event studies or to maintain steady state solute concentrations.

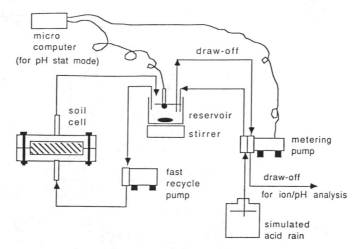

Figure 7. Schematic diagram of the recirculating column reactor. Reactor has advantages of both the soil column and recirculating fluidized-bed reactor.

Recycle is necessary to achieve the flowrate to suspend the bed and to allow solute concentrations to build to a steady state. A "blowdown" of solutes is provided via sample collection. Back reactions should be prevented and experiments are designed to be "far from equilibrium". Chou and Wollast (1985) found that dissolved aluminum inhibited further dissolution rates of albite even at concentrations below the solubility limit for gibbsite. With the fluidized-bed apparatus (Fig. 6), they could control the aluminum concentrations to a low level at steady state by withdrawing sample at a high-rate.

Figure 7 is a schematic of a recirculating column reactor that combines many of the advantages of the soil column and fluidized bed. It can be operated at realistic flow rates and solids concentrations, and steady-state solute concentrations can be achieved as in the fluidized bed. Unlike once-through soil columns or thin-layer experiments, sufficient weathering rates of even crystalline, slow-to-dissolve minerals can be obtained using rapid recycle.

5. SITE DESCRIPTION

Most experiments on chemical weathering have been performed on pure minerals under carefully controlled laboratory conditions. However, under natural conditions in the field, mixtures of minerals of different abundances are reacting simultaneously, together with amorphous material and organic material subject to ion exchange in soils and sediments. It is instructive to conduct experiments on natural soils in the laboratory and then to compare these results with small plot experiments in the field and the prototype watershed. The site of Bear Brook Watershed at Lead Mountain, Maine (latitude 44°51′75″, longitude 68°6′25″) is shown in Figure 8. It contains two almost

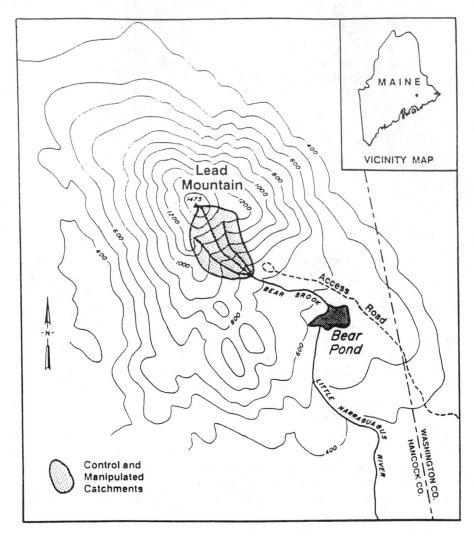

Figure 8. Location of the Bear Brook catchments. Dotted area shows the twin stream catchments, West Bear and East Bear Brook.

identical streams, draining approximately 37 acres each, with a southern exposure from the top of Lead Mountain, 1475 ft above mean sea level. It is a part of U.S. Environmetal Protection Agency's (EPA's) Watershed Manipulation Project, a long-term experimental manipulation in which the western catchment will be artificially acidified with applications of ammonium sulfate, and the eastern catchment will remain as a reference. Bedrock at the site consists of metamorphosed quartz sandstones with granitic sills and dikes, overlain by glacial till 4–5 m thick at the lower elevations. Exposed bedrock is visible at the

TABLE 1. Elemental Analyses of Soil Samples from Bear Brook Watershed, Maine, by X-Ray Fluorescence and SEM/EDX Microprobe Surface Analyses (% by wt)

| Element | Horizon | | | | | B Surface [SEM/EDX] |
	B_1 <62 μm	B_2 125–250 μm	C_1 <62 μm	C_2 62–125 μm	C_3 125–250 μm	63–125 μm
Na as Na$_2$O	1.7	1.5	2.2	2.0	2.0	1.00
Mg as MgO	0.7	0.6	1.0	0.8	0.8	0.05
Al as Al$_2$O$_3$	13	11	11	10	10	12.63
Si as SiO$_2$	76	68	73	78	68	71.34
P as P$_2$O$_5$	0.09	0.07	0.1	0.07	0.06	
K as K$_2$O	1.8	2.1	1.9	2.2	2.4	3.02
Ca as CaO	0.9	0.8	1.3	1.2	1.1	2.15
Ti as TiO$_2$	0.9	0.06	0.7	0.6	0.5	
Mn as MnO	0.03	0.03	0.04	0.04	0.03	
Fe as Fe$_2$O$_3$	4.1	3.4	2.9	2.4	2.3	9.80
Lost at 1050°C	15	10	3.2	2.2	2.2	

top of Lead Mountain. Mineralogy by X-ray diffraction of the soils (B- and C-horizons) consists of quartz as the major mineral, followed by feldspar as a significant mineral [$X(Si, Al)_4O_8$, $X = Na$, K, Ca, Ba], and minor minerals muscovite $KAl_3Si_3O_{10}(OH)_2$, chlorite $(Mg, Fe, Al)_6(Si, Al)_4O_{10}(OH)_8$, and hornblende $X_2Y_5(Si, Al)_8O_{22}(OH)_2$, $X = Na,K,Ca,Mn,Fe,Y = Mg,Al,Ti,Mn,Fe$. The soils are quite deficient in calcium (0.8–1.3% by weight as oxide equivalents) as seen in Table 1. In the fine size fractions of the upper B-horizon, there is quite a lot of iron oxides (4.1% by weight) and organic carbon (15% as measured by loss on ignition at 1050°C). There is little chemical difference among the various size fractions otherwise. The B-horizon soils have a little more Fe and Al than do C-horizon soils, as expected for acid, podzolic soils.

6.   LABORATORY RESULTS

Soils were homogenized from the B-horizon at Bear Brook Watershed by repeated quartering and reaggregating according to standard soil procedures. They were subsequently size-fractionated, rinsed in deionized water (similar to the procedure of Holdren and Berner (1979)) and air-dried. The size fraction 63–125 μm was used in the experiments because it was of high surface area and contained the weatherable minerals in the till. Mineralogy was approximately 60% quartz, 25% plagioclase (albite), 5% K–feldspar, and 10% mica (Drever and Swoboda-Colberg, in press). The most weatherable minerals might be considered plagioclase, biotite, and hornblende, together with amorphous material and clays (chlorite) from the B-horizon. In batch pH-stat experiments, it took 100–200 h for a steady slope to be achieved on plots of H^+ ion consumption and ion release versus time (Etringer, 1989). The pH was controlled with dilute titrant solutions of sulfuric acid. During the first 100 h, soils came to ion-exchange equilibrium with the bulk solution. Following exchange, linear disso-lution rates were observed for all ions indicative of chemical weathering. Actually, the pH-stat was operated in a semibatch mode because it was necessary to draw off samples periodically (~ 2% of the volume per day) and replace the sample volume with fresh titrant.

Figure 9 shows that total dissolved aluminum was released very rapidly at pH 2.7 by ion exchange and dissolution of amorphous aluminum hydroxide present in the soil. After approximately 100 h, aluminum release was constant at 0.04 μmol g^{-1} soil per hour. This was about three times greater than the release of total dissolved silica, which likely serves as a tracer for aluminosilicate weathering in the pH range above 2.7. Calcium was released very rapidly by exchange reactions, but it showed a constant rate of weathering after 100 h in a 1 : 10 ratio of Ca : Si. Magnesium ions were not released by ion exchange initially because of a lower selectivity coefficient, but chemical weathering of mafic minerals and chlorite was thought to be responsible for the subsequent high rate of Mg^{2+} release in a ratio of 3 : 1 for Mg : Ca. This was also observed in soil column studies later on.

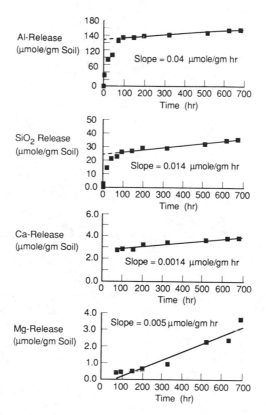

Figure 9. Element release rates from the batch pH-stat at pH 2.7, 10 g of bulk B-horizon soil, 15°C, and 1 L of H_2SO_4 solution.

To test the interpretation of results that ion exchange predominates during the first 100 h and chemical weathering predominates thereafter, an experiment was run for 360 h at pH 3.5, whereupon it was stopped, the solution was centrifuged, and the liquid was decanted off. Fresh titrant solution was added and the experiment was restarted. The linear slope of the chemical weathering rate and H^+ ion consumption did not change. Because the surfaces of the solids were previously exposed to the sulfuric acid solution, it did not respond with ion exchange as initially. The experiment indicated that chemical weathering is responsible for the linear release rates of ions after 100–200 h, and that the experiment was sufficiently far from equilibrium that back reactions did not inhibit dissolution rates substantially.

Figure 10 shows that in the initial period, there was a large amount of $Al(OH)_3$ dissolution, and the ratio of dissolved Al in solution was large relative to dissolved Si. Once natural gibbsite ($K_{SO} = 10^{+8.77}$) solubility was reached, which occurred only at pH 4 after 200 h, the ratio of the *rate* of release of Al : Si became more nearly equal to the ratio of aluminosilicate minerals in the soil. The ratio of

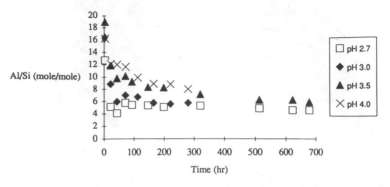

Figure 10. Aluminum : silica ratio at various pH values, H_2SO_4 solution at 15°C, with 10-g bulk B-horizon soil.

Al : Si in elemental analysis of the soil (Table 1) is roughly 1 : 1, assuming that 60% of the mineral mass is nonreactive quartz. If the experiments had been continued beyond the point of $Al(OH)_3$ saturation, probably Si and Al would have been released in a stoichiometric ratio, but the Al^{3+} ions would reprecipitate as gibbsite [or amorphous $Al(OH)_3$] and silica concentrations would increase until kaolinite saturation prevailed. In the long-term experiment at pH 3.5 (Fig. 10), the solution approached kaolinite and natural gibbsite saturation after 500 h. Formation of secondary products likely buffered the dissolved Al and Si concentrations after that time at 6 : 1 Al : Si. The ratio 6 : 1 was so large because, in a batch reactor, aluminum released from the beginning of the experiment was not flushed away but remained throughout. Results suggest that there is a large amount of amorphous Al(hydr)oxides present in the B-horizon soils at Bear Brook, and silica release may not be a conservative estimate of aluminosilicate weathering under such conditions. Silica can readsorb to amorphous Al(hydr)oxides. The experiment after 500 h at pH 3.5 points out some of the limitations of the pH-stat. The rate of removal of samples (semibatch mode) began to determine the calculated weathering rate when secondary products buffered Al and Si concentrations in solution. On the other hand, the pH-stat allowed observation of reaction progress over the entire range of ion exchange, chemical weathering, and secondary mineral formation.

Figure 11 is a plot of H^+ ion consumption versus pH for batch pH-stat experiments at pH 2.7, 3.0, 3.5, and 4.0 (Etringer, 1989). The pH was controlled with dilute sulfuric acid. Rates were determined after 150 h in the linear portion of the weathering experiment. Fractional order dissolution rates with respect to bulk hydrogen ion activity in solution were demonstrated for these B-horizon soils. Fractional order weathering with hydrogen ion activity is consistent with the proposed surface controlled dissolution reaction for chemical weathering.

Soil column results are shown in Figures 12–14. The same soil was used as for the pH-stat experiments. The column was run in a saturated mode with step-function changes in pH of the influent (pH 4, 3.3, 3.0, 2.3, and 2.1 sulfuric acid)

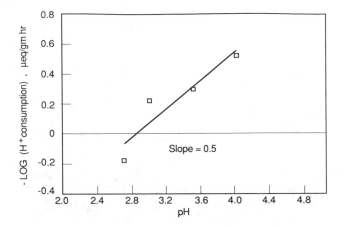

Figure 11. Log (acid consumption rate) versus pH bulk B-horizon soil at 15°C.

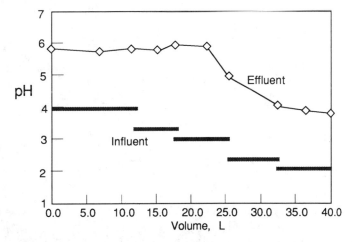

Figure 12. Soil column experiment pH results. Difference in pH between the influent and effluent indicates neutralization reactions taking place in the soil column.

after steady state was established at each setting. Flow rate was 0.79 L day^{-1} with a hydraulic residence time of 0.15 day^{-1}. In Figure 12, the effluent pH tracked the influent pH quite well, but the effluent was ~1.5–2.0 pH units greater, indicating neutralization reactions occurring in the column via ion exchange, chemical weathering, and sulfate sorption. At low pH, sulfate sorption was important suggesting pH-dependent sorption of sulfate on protonated surfaces of amorphous iron and aluminum oxides or precipitation of aluminum sulfate compounds such as jurbanite.

With pH 3.0 influent at 35 days (Fig. 13), aluminum release was very large, suggesting the breakdown of amorphous alumina and clays. Silica release also

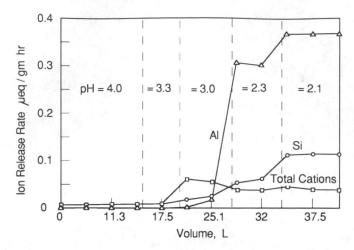

Figure 13. Soil column experiment ion production rates (Si in micromoles per gram per hour). Base cations and dissolved silica were released at influent pH 4, 3.3, and 3.0; Al release dominated at pH 2.3; and silica was dissolved at influent pH 2.1.

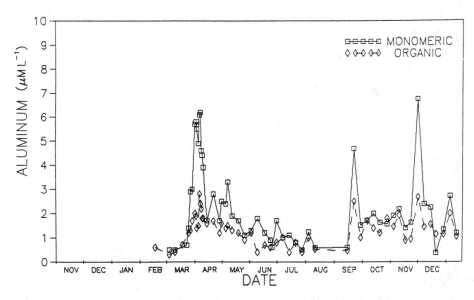

Figure 14. East Bear Brook Aluminum. Aluminum concentrations in the field are relatively low compared to laboratory reactors and small plot studies at more acidic pH values. Peak concentrations are due to snowmelt and runoff events. Data courtesy of Charles Driscoll, Syracuse University and Steve Norton, University of Maine.

increased dramatically and, at pH 2.1 influent, the molar ratio of Al:Si was approximately 1:1 as in kaolinite. Base cation release (Ca^{2+}, Mg^{2+}, Na^+, K^+) did not increase after pH 3 influent except for release of Mg^{2+} at the most acidic loading (pH 2.1). It is reasoned that Mg-rich chlorite minerals were decomposing at low pH. When the influent pH to the soil column was less than 3.0, the weathering reactions were not very relevant to field conditions.

At an influent pH of 4.0, the soil column effluent consisted mostly of dissolved silica and base cations, on a molar basis roughly in stoichiometric ratio to a plagioclase feldspar with $x = 0.75$. This is richer in calcium than elemental analyses would suggest (Table 1), and it indicates that some calcium is coming from the exchange complex or by preferential weathering of Ca–plagioclase (Clayton, 1986). Otherwise, ion ratios and concentrations were very indicative of the mineralogy and of the field prototype when the soil column was fed with pH 4 influent. The soil column had a distinct advantage over the pH-stat in portraying reactions that occur in the natural system because solutes were flushed out of the reactor and weathering stoichiometries were more reflective of the minerals presumed to be weathering. Silica was a good tracer of aluminosilicate weathering in the flow-through soil column experiments above pH 3. One disadvantage of the soil column is the difficulty of estimating kinetic formulations (as in Fig. 11 for the pH-stat reactor) because the concentrations within the column are varying and difficult to sample.

7. FIELD RESULTS

Laboratory experiments are meant to provide insight for the natural system, in this case, Bear Brook Watershed, Maine. It is interesting to consider how closely laboratory measurements of chemical weathering rates using soils from Bear Brook Watershed compare to actual field results. Figures 14–16 show the export of dissolved aluminum (both inorganic monomeric and organic Al), dissolved silica, and base cations for Bear Brook. The average pH of the twin streams is approximately pH 5.8, with snowmelt events occurring in March as low as pH 5.0. Gran alkalinity titrations average around 20 μequiv L^{-1} with minimum values of ~ 0.0 during snowmelt events. The two streams are very similar in size, discharge rate, and chemistry.

During snowmelt and storm events in 1987 at Bear Brook (Fig. 14), there is a breakthrough of monomeric aluminum that could be toxic to fish, ~ 4 μmol L^{-1} (Driscoll and Schecher, 1988). However, the aluminum concentrations observed in the field are much smaller than those measured in the laboratory reactors, except for soil column results with pH 4 influent and pH 5.8 effluent, which were quite similar to field conditions. The release of aluminum during hydrologic events is due to increased solubility of amorphous $Al(OH)_3$, much like the initial phases of the batch pH-stat experiments. At the relatively high pH of Bear Brooks during most of the year, aluminum solubility is relatively low and Al:Si ratios are on the order of 1:50 (Figure 15). This is characteristic of the soil column experiment with pH 4 influent and laboratory experiments at high pH (4–10) as well (Wollast, 1967). Figure 16 shows base cation concentrations in the stream which were similar to those in the batch pH-stat and soil column experiments. Sodium ions were large in concentration due to sea salt inputs at

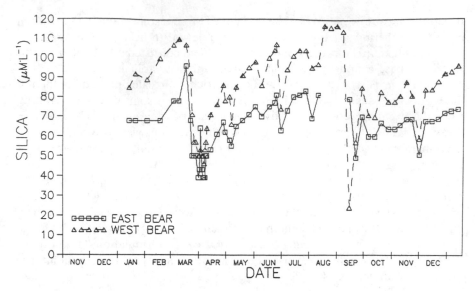

Figure 15. Bear Brooks Watershed Dissolved Silica. Silica concentrations were slightly higher in West Bear Brook. Highest concentrations occurred during base flow periods when groundwater was contributing most of the water to the streams. Dissolved silica decreased during snowmelt and runoff events due to dilution by "new" water.

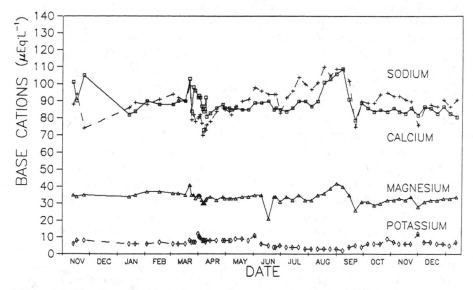

Figure 16. West Bear Brook Cations. Sodium and calcium are the predominant cations but most of the sodium comes from sea salt. Cation concentrations are relatively constant with time, compared to aluminum and silica in the Brook.

Bear Brook, and Na$^+$ due to chemical weathering was only ~ 20 μequiv L^{-1} if chloride concentration was subtracted. Concentrations of the base cations were similar to the laboratory experiments, but the ratio of Ca:Mg is larger in the field, suggesting that mafic minerals break down excessively in the laboratory below pH 4.0. Experiments below pH 3.5 were not very representative of field conditions.

Table 2 is a comparison between weathering rates determined in the laboratory on pure minerals and estimated weathering rates from catchment studies including this one. Estimates of weathering rates in the field depend on assumptions of the effective water depth in the terrestrial environment, surface area of the minerals, and relative abundances of the weatherable minerals. Even with such uncertainty, the field-estimated weathering rates are within two orders of magnitude of the laboratory rates. Generally, field-estimated weathering rates are one to two orders of magnitude smaller than the laboratory measurements, because of a number of difficulties discussed in the literature (Paces, 1983; Velbel, 1985; Drever and Swoboda-Colberg, 1989). Drever and Swoboda-Colberg (1989) have stated that the most likely reason for the discrepancy between laboratory and field measurements of chemical weathering rates is due to (1) unsaturated flow through soil macropores, which limits the amount of weatherable minerals exposed to water, and (2) dissolved aluminum concentrations, which inhibit aluminosilicate dissolution rates in soil macropores.

In order to examine the first hypothesis, field and laboratory studies were compiled for similar aluminosilicate minerals with reference to their weathering rates and flow rates or discharge measurements. Table 3 shows the results for eight examples, five of which were with soils from Bear Brook Watershed (BBW) in experiments of different scale. Site 1 refers to silica export measured at the discharge from East Bear Brook. Site 2 was a small (1.4×1.4-m) weathering plot experiment at BBW with HCl applications of pH 2, 2.5, and 3.0 (Drever and Swoboda-Colberg, 1989). Sites 3–5 were all laboratory experiments at pH 3–4 on Bear Brook size-fractionated soils. Site 6 was Coweeta Watershed 27 in the Southern Blue Ridge mountains of North Carolina (Velbel, 1985). Coweeta soils were composed of three main weatherable minerals: plagioclase, garnet, and biotite. Site 7 was Filson Creek in northern Minnesota, a large watershed of 25 km^2 with waters of pH 6 and plagioclase and olivine as the predominant weatherable minerals in the till (Siegel and Pfannkuch, 1984). Site 8 was Lake Cristallina in the Swiss Alps with plagioclase, biotite, and epidote as predominant weathering minerals (Giovanoli et al., 1988). These examples ranged over 10 orders of magnitude in dissolved silica export and flow rate.

When silica export is plotted against flow rate on a log–log scale (Fig. 17), most points lie near the line with a slope of 1.0, corresponding to approximately 60 μM dissolved silica. A strong correlation exists in part because flow rate is a multiplier on both the ordinate and the abscissa. Even though the sites and experiments differ over 10 orders of magnitude in the amount of water flowing, they are all within about one order of magnitude in dissolved silica concentration (25–300 μM), except for the fluidized-bed experiment. Thus, one conclusion is

TABLE 2. Comparison of Laboratory and Field Weathering Rates

Mineral	Laboratory Weathering Rate (mol Si m^{-2} s^{-1})	Field-Estimate Weathering Rate (mol Si m^{-2} s^{-1})	Field-Measured Cation Export (equiv ha^{-1} yr^{-1})	Cation(s)	Notes	Reference
Plagioclase (oligoclase)	5×10^{-12a}	3×10^{-14}	210	Na$^+$	Trnavka River Basin (CZ)	Paces (1983)
Plagioclase (oligoclase)	5×10^{-12a}	8.9×10^{-13}	350	Na$^+$, Ca^{2+}	Coweeta Watershed, NC (USA)	Velbel (1985)
Almandine		3.8×10^{-12}	300	Mg^{2+}, Ca^{2+} K$^+$, Mg^{2+}		
Biotite		1.2×10^{-13}	150			
Plagioclase (bytownite)	5×10^{-12a}	5×10^{-15}	330	Ca^{2+}, Na$^+$	Filson Creek, MN (USA)	Siegel and Pfannkuch (1984)
Olivine	7×10^{-12b}	1×10^{-13}	310	Mg^{2+}	Cristallina, Switzerland	Giovanoli et al. (1989)
Plagioclase epidote, biotite		6×10^{-14}	200	Ca^{2+}, Na$^+$, K$^+$		
Plagioclase, biotite	6×10^{-12}	9×10^{-15}	960	Ca^{2+}, Na$^+$, Mg^{2+}	Bear Brooks Watershed, Maine (USA)	This study

Assumption: 50 cm of saturated regolith, 0.5 m^2 g^{-1} surface area of mineral grains measured, 40% of mineral grains active in weathering

[a] Rate determined in the laboratory by Busenberg and Clemency (1976) and Mast and Drever (1987) at pH 4.
[b] Rate determined in the laboratory on beach sand by Grandstaff (1986) at pH 4.5.

TABLE 3. Laboratory and Field Studies of Silicon Export and Release Rates (Si RR) and Flow Rates

Site	Si Conc. (μM)	Si Export (mol s^{-1})	Flow Rate (L s^{-1})	Si RR (mol m^{-2} s^{-1})	Flow Rate/ Mass Ratio (L day^{-1} g^{-1})	Reference
1. Bear Brook	70	2.9×10^{-4}	4.2	9×10^{-15}	2×10^{-6}	This chapter
2. Small plot	300	2.9×10^{-8}	9.7×10^{-5}	3×10^{-13}	1×10^{-5}	Drever and Swoboda-Colberg (in press)
3. Soil column	30	2.8×10^{-10}	9.2×10^{-6}	6×10^{-12}	3×10^{-3}	This chapter
4. pH-stat	100	1.2×10^{-11}	1.2×10^{-7}	6×10^{-11}	1×10^{-3}	This chapter
5. Fluidized bed	3	5.4×10^{-12}	1.7×10^{-6}	3×10^{-12}	4×10^{-2}	Drever and Swoboda-Colberg (in press)
6. Coweeta	42	1.4×10^{-3}	33	7×10^{-13}	2×10^{-5}	Velbel (1985)
7. Filson Creek	167	4.2×10^{-2}	250	5×10^{-15}	1×10^{-6}	Siegel and Pfannkuch (1984)
8. Cristallina	25	1.6×10^{-4}	6.5	6×10^{-14}	2×10^{-5}	Giovanoli et al. (1989)

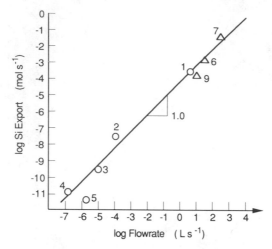

Figure 17. Dissolved silica export versus flow rate (log–log). Circles represent BBW soils, and triangles are other field sites. Numbers refer to site numbers listed in Table 3.

that experiments can be scaled to field conditions by ensuring that the dissolved silica concentration is of the same order of magnitude as the prototype ecosystem. Weathering export rates are correlated to flow rate in log–log plots in many natural systems (Schnoor et al., 1984). This implies that a surface coordination reaction could not be controlling the rate of weathering in natural systems because there would be no increase in ion export rates with flow rate. Indeed, in laboratory experiments 4 and 5 of Table 3, there was no change in weathering rate with changing flow rate or stirring rate. The laboratory experimental points lie near the origin of Figure 14, and they may indicate a break in the linear relationship in systems with very high flow rate : solids ratios.

The silica release rate and flow-rate : mass ratio of Figure 18 depend on estimations of the wetted surface area of reacting minerals and the mass of wetted soil. As discussed by Paces (1983), there is much uncertainty in estimating wetted surface areas of reacting minerals. In Figure 18 and Table 3, it was assumed that 40% of the surface area of soil was weatherable minerals; the soil surface area was $0.5 \text{ m}^2 \text{ g}^{-1}$ (except at Coweeta, where Velbel's (1985) procedure was used), the bulk density of the soil was 1.5 kg L^{-1}, and the equivalent saturated water depth was 0.5 m (except at Coweeta and Cristallina, where it was taken to be 0.1 m). All of these parameters were measured at Bear Brook Watershed and applied to the other sites.

If the ratio of flow rate to mass of wetted soil ($\text{L day}^{-1} \text{ g}^{-1}$) is plotted on the abscissa, and silica release rate [silica export per square meter of wetted mineral ($\text{mol m}^{-2} \text{ s}^{-1}$)] is plotted on the ordinate, an asymptotic relationship for silica release rate is determined (Fig. 18). Silica release rates reach a maximum of 10^{-11} to 10^{-12} at a flow-rate : mass ratio greater than $10^{-3.4} \text{ L day}^{-1} \text{ g}^{-1}$. This corresponds to the flow regime and weathering rates most frequently reported in

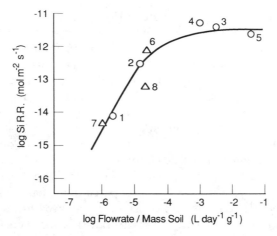

Figure 18. Dissolved silica release rate (weathering rate) versus flow-rate:mass ratio. Circles represent BBW soils and triangles are other field sites. Numbers refer to site numbers listed in Table 3.

laboratory studies of weathering of pure minerals. The flow-rate:mass ratio of $10^{-3.4}$ is a flushing rate of approximately 1.0 volumes of water per volume of soil per day, and it represents a limit above which surface reactions control mineral dissolution rates and below which hydrology controls (limits) the rate of weathering. Hydrologic control may exist because of unsaturated macropore flow through soils, which results in insufficient flow to wet all the available minerals and to carry away the dissolved solutes. Uncertainty in weathering rates measured in the laboratory is one order of magnitude, but limitations of hydrology can result in two orders of magnitude lower estimates of weathering rates.

8. SUMMARY

Aluminosilicate minerals undergo surface-controlled dissolution reactions that are fractional order with respect to bulk H^+ ion concentration in solution. Kinetics of chemical weathering can be explained by surface coordination chemistry, where the detachment of an activated complex is the rate-determining step in dissolution. Dissolution reactions may occur incongruently at first, but eventually the build-up of a cation-depleted solid layer will provide for stoichiometric mineral weathering. Protons and organic ligands may accelerate the rate of mineral weathering.

Results of a field study at Bear Brook Watershed, Maine, were compared with laboratory kinetic experiments using size-fractionated soils from the same location. It is a forested, glaciated region with thin podzolic soils and granitic gneiss bedrock. Fractional order dependence ($m = 0.5$) of weathering rates on H^+

ion activity were demonstrated for batch pH-stat experiments of B-horizon soils in the pH range 2.7–4.0. The soils tested were rich in aluminum, and B-horizon samples yielded high concentrations of dissolved aluminum (0.2–1.0 ,mmol L^{-1}) during the first 50 h of reaction. This was attributed to ion exchange and a rapid dissolution of amorphous aluminum hydroxide. As a result of the initial rapid reaction, H^+ ion consumption and release rates of other solutes (Si, Al, Ca, Mg, Na) were asymptotic with time. Release rates of ions became constant after 200 h, and weathering rates were determined in the range 200–400 hours. Flow-through column experiments were also run on the same soils. Soil column results provided the best simulation of field conditions with pH 4 influent (sulfuric acid), and the effluent resulted in a pH of 5.8, similar to Bear Brook. The advantage of the soil column reactor was its realism compared to the natural setting, and the advantage of the pH-stat was that constant pH allowed for estimation of the kinetic dependence on hydrogen ion activity. Both reactors were fully automated with feedback control and a computer for data capture and acquisition. Advantages of recirculating reactors (the fluidized bed and soil cell) were to maintain constant solute concentrations far from geochemical equilibrium with secondary phases.

A comparison of laboratory weathering rates and estimates from the field indicated that laboratory rates were one to two orders of magnitude greater than field estimates of chemical weathering. Based on dissolved silica as a conservative tracer of weathering, laboratory rates were on the order of 10^{-12} to 10^{-11} mol m^{-2} s^{-1}, while field weathering rates were 10^{-14} to 10^{-12} mol m^{-2} s^{-1}. The discrepancy is likely due to the difficulty of estimating a suitable wetted surface area of weatherable (reacting) minerals in the field, and the possibility of hydrologic control, due to macropore flow through soils.

Acknowledgments

The author thanks several students and colleagues at The University of Iowa, without whom this chapter would have been impossible, including Professor Richard Valentine, Dr. Kent Johnson, Mark Etringer, Scott Wallace, and Shyam Asolekar. Our cooperators on the Watershed Manipulation Project at Bear Brooks Watershed have supplied much of the field data: in particular, Professor Steve Norton, Dr. Lindsey Rustad, and Dr. Steve Kahl, University of Maine; Professor Charles Driscoll and Dr. Drew McAvoy, Syracuse University; and Professor James I. Drever and Dr. Norbert Swoboda-Colberg, University of Wyoming. The Environmental Protection Agency (EPA) Environmental Research Laboratory at Corvallis, Oregon, was helpful in project coordination and quality assurance and quality control, especially Drs. Jim Wigington (Project Officer), Timothy Strickland (Project Leader), Dan McKenzie, Larry Hughes, and Heather Erickson. Indispensable support and encouragement for this research was also given by Professor Werner Stumm during a sabbatical leave in 1988 at the Swiss Federal Institute of Technology from The University of Iowa. Professor Rudolf Giovanoli, University of Bern, took the photomicrographs in Figure 2 and contributed ideas on chemical weathering. This research was supported by a cooperative agreement with the U.S. EPA, Corvallis Environmental Research Laboratory. It has not been subjected to the Agency's "peer and policy" review, and no official endorsement should be inferred.

REFERENCES

Berner, R. A., and G. R. Holdren, Jr., (1977), "Mechanism of Feldspar Weathering: Some Observational Evidence," *Geology*, **5**, 369–372.

Berner, R. A., and G. R. Holdren, Jr. (1979), "Mechanism of Feldspar Weathering—II. Observations of Feldspars from Soils," *Geochim. Cosmochim. Acta*, **43**, 1173–1186.

Berner, R. A., E. L. Sjoberg, M. A. Velbel, and M. D. Krom, (1980), "Dissolution of Pyroxenes and Amphiboles During Weathering," *Science* **207**, 1205–1206.

Bevin, K., and P. Germann (1982) "Macropores and Water Flow in Soils," *Water Resources Res.* **18**, 1131–1325.

Busenberg, E., and C. V. Clemency (1976), "The Dissolution Kinetics of Feldspars at 25°C and 1 Atm CO_2 Partial Pressure," *Geochim. Cosmochim. Acta* **41**, 41–49.

Busenberg, E., and L. N. Plummer (1982), "The Kinetics of Dissolution of Dolomite in CO_2–H_2O Systems at 1.5 to 65°C and 0 to 1 atm pCO_2," *Am. J. Sci.* **282**, 45–78.

Chou, L., and R. Wollast (1985), "Steady State Kinetics and Dissolution Mechanisms of Albite," *Am. J. Sci.* **285**, 963–993.

Clayton, J. L. (1986), in J. I. Drever, Ed., *Rates in Chemical Weathering of Rocks and Minerals*, Academic Press, Orlando, FL, pp. 453–467.

Drever, J. I., and N. Swoboda-Colberg (in press), "Mineral Weathering Rates in Acid Sensitive Catchments: Extrapolation of Laboratory Experiments to the Field".

Driscoll, C. T., and W. D. Schecher (1988) in H. Sigel and A. Sigel, Eds., *Metal Ions in Biological Systems*, Vol. 24, Dekker, New York.

Etringer, M. A. (1989), M.S. thesis, University of Iowa, Iowa City

Furrer, G., and W. Stumm (1983), "The Role of Surface Coordination in the Dissolution of δ-Al_2O_3 in Dilute Acids," *Chimia* **37**, 338–341.

Furrer, G., and W. Stumm (1986), "The Coordination Chemistry of Weathering: I. Dissolution Kinetics of δ-Al_2O_3 and BeO," *Geochim. Cosmochim. Acta* **50**, 1847–1860.

Furuichi, R., N. Sato, and G. Okamoto (1969), "Reactivity of Hydrous Ferric Oxide Containing Metallic Cations," *Chimia* **23**, 455.

Giovanoli, R., J. L. Schnoor, L. Sigg, W. Stumm, and J. Zobrist. (1989), "Chemical Weathering of Crystaline Rocks in the Catchment Area of Acidic Ticino Lakes, Switzerland," *Clays Clay Min.* **36**, 521–529.

Grandstaff, D. E. (1977), "Some Kinetics of Bronzite Orthopyroxene Dissolution," *Geochim. Cosmochim. Acta* **41**, 1097–1103.

Helgeson, H. C. (1971), "Kinetics of Mass Transfer Among Silicates and Aqueous Solutions", *Geochim. Cosmochim. Acta* **35**, 421–429.

Helgeson, H. C. (1972), "Kinetics of Mass Transfer Among Silicates and Aqueous Solutions: Correction and Clarification," *Geochim. Cosmochim. Acta* **36**, 1067–1070.

Holdren, G. R., Jr., and R. A. Berner (1979), "Mechanism of Feldspar Weathering—I. Experimental Studies," *Geochim. Cosmochim. Acta* **43**, 1161–1171.

Kummert, R., and W. Stumm (1980), "The Surface Complexation of Organic Acids on Hydrous δ-Al_2O_3," *J. Colloid Interface Sci.* **75**, 373.

Luce, R. W., R. W. Bartlet, and G. A. Parks (1972), "Dissolution Kinetics of Magnesium Silicates", *Geochim. Cosmochim. Acta* **36**, 35–50.

Mast, M. A., and J. I. Drever (1987), "The Effect of Oxalate on the Dissolution Rates of Oligoclase and Tremolite," *Geochim. Cosmochim. Acta* **51**, 2559–2568.

Paces, T. (1973) "Steady-State Kinetics and Equilibrium Between Groundwater and Granitic Rock," *Geochim. Cosmochim. Acta* **37**, 2641–2663.

Paces, T. (1983), "Rate Constants of Dissolution Derived from the Measurements of Mass Balance in Hydrological Catchments," *Geochim. Cosmochim. Acta* **47**, 1855–1863.

Schindler, P. W. (1982), in M. A. Anderson, and A. Rubin, Eds., *Adsorption of Inorganics at Oxide Surfaces*, Ann Arbor Science Publishers, Ann Arbor, MI, pp. 1–49.

Schnoor, J. L., and W. Stumm (1985), in W. Stumm Ed., *Chemical Processes in Lakes*, Wiley, New York, pp. 311–338.

Schnoor, J. L., and W. Stumm, (1986), "The Role of Chemical Weathering in the Neutralization of Acidic Deposition," *Schweiz. Z. Hydrol.* **48**, 171–193.

Schnoor, J. L., W. D. Palmer, Jr. and G. E. Glass (1984), in J. L. Schnoor, Ed., *Modeling of Total Acid Precipitation Impacts*, Butterworth, Boston.

Schott, J., R. A. Berner, and E. L. Sjöberg (1979), "Mechanism of Pyroxene and Amphibole Weathering—I. Experimental Studies," *Geochim. Cosmochim. Acta* **43**, 1161–1171.

Schott, J., R. A. Berner, and E. L. Sjöberg (1981), "Mechanism of Pyroxene and Amphibole Weathering—I. Experimental Studies of Iron-Free Minerals," *Geochim. Cosmochim. Acta* **45**, 2123–2135.

Siegel, D. I., and H. O. Pfannkuch (1984), "Silicate Dissolution Influence on Filson Creek Chemistry, Northeastern Minnesota," *Geol. Soc. Am. Bull.* **95**, 1446–1453.

Sigg, L., and W. Stumm (1980), "The Interaction of Anions and Weak Acids with the Hydrous Geothite (α-FeOOH) Surface," *Colloids Surf.* **2**, 101.

Velbel, M. A., (1985), Geochemical Mass Balances and Weathering Rates in Forested Watersheds of the Southern Blue Ridge," *Am. J Sci.*, **285**, 904–930.

Wallace, S. D. (1989), M.S. thesis, University of Iowa, Iowa City, 131 pp.

Wieland, E., B. Wehrli, and W. Stumm (1988), "The Coordination Chemistry of Weathering: III. A Generalization on the Dissolution Rates of Minerals", *Geochim. Cosmochim. Acta* **52**, 1969–1981.

Wollast, R. (1967), "Kinetics of the Alteration of K-Feldspar in Buffered Solutions at Low Temperatures," *Geochim. Cosmochim. Acta* **31**, 635–648.

18

TRANSPORT AND KINETICS IN SURFICIAL PROCESSES

Abraham Lerman

Department of Geological Sciences, Northwestern University, Evanston, Illinois

1. INTRODUCTION

Transport of materials in the surficial environment involves water and air, mineral and biogenic solids, and aqueous and gaseous species produced in biogeochemical reactions taking place between the waters, solids, and atmosphere. Chemical evolution of the surface environment of the earth has been viewed as an ongoing acid–base titration reaction of the earth's crust, with some of the acids originally released by degassing of the early planet's interior, such as the release of HCl, and subsequently with other acids formed by inorganic and biogeochemical oxidation reactions, such as CO_2, HNO_3, and H_2SO_4 (Sillén, 1961, p. 551; Henriksen, 1980, in Drever, 1988; Stumm and Morgan, 1981; Holland, 1984). Interwoven with the global acid–base titration of a system consisting of the earth's crust, soils, waters, and atmospheric precipitation are the oxidation and reduction reactions that involve many of the chemical elements, of which quantitatively he most important in the surficial environment are the following four groups:

$$Fe^{2+} \rightleftharpoons Fe^{3+}$$

in sediments, soils, waters, and intracellular solutions of the higher organisms;

$$C^{4-}, C^0 \rightleftharpoons C^{4+}$$
$$S^{2-}, S^0 \rightleftharpoons S^{6+}$$

and

$$N^{3-}, N^0 \rightleftharpoons N^{5+}$$

in most of the processes involving the land and aquatic biota, sediments, waters, and the atmosphere.

The oxidation states of these four elements represent the more common nominal valences in their naturally occurring chemical species, without regard to the intermediate oxidation states of sulfur and nitrogen that occur in the atmospheric and biological cycles of these elements. Many of the natural redox reactions—those involving ozone, UV radiation, metal catalysis, enzymes (Hoigné, Chapter 2; Hoffman, Chapter 3; Price and Morel, Chapter 8; Sulzberger, Chapter 14; all in this volume) and such trace metals as Mn, V, and U—complement the point made in the preceding paragraph about the quantitative importance of the redox cycles of Fe, S, C, and N in the inorganic and living worlds.

Acid–mineral reactions, and oxidation and reduction reactions that variably contribute to mineral dissolution or precipitation (Hering and Morel, Chapter 5; Schnoor, Chapter 17; Schott, Chapter 12; Stumm and Wieland, Chapter 13; Wehrli, Chapter 11; all in this volume) all result in net transport of materials on a macroscopic scale within the system crustal rocks–hydrosphere–atmosphere.

In the continental–aquatic environment, the total material transport flux is a net sum of fluxes driven by chemical and physical forces:

$$F_T = F_{chem} + F_{phys}$$

where the subscripts chem and phys designate processes that are primarily chemical or physical in nature.

Each of the fluxes is an algebraic sum of other fluxes that determine the magnitudes of the net transport flows. Thus the flux driven by major physical forces is the flux of erosion or mechanical denudation of the land surface, and it may be viewed as a net sum of two terms, a water-transported and wind-transported material components:

$$F_{phys} = F_{water} + F_{wind}$$

The chemical flux is a net balance of dissolution of crustal minerals, input of solutes from atmospheric precipitation, changes in the pools of cations and anions stored in the standing crop of vegetation and soils, and formation of new minerals in the soil profiles. This relationship becomes

$$F_{chem} = F_{dissolution} + F_{volatilization} - F_{deposition} \pm F_{biota} \pm F_{soils}$$

Each individual flux in the two preceding equations is a function of combinations of environmental factors. These factors, and the coupling between the chemical and macroscopic transport processes in the surficial environment, will be discussed in the subsequent sections of this chapter. The discussion will focus on the coupling between the chemical, hydrological, and physical processes and will deal with the global average fluxes of water, solid and dissolved materials that are products of the three interacting classes of forces and processes.

2. COUPLING OF TRANSPORT FLUXES

A conventional conceptual model of transport views a flux of material (F) as being driven by a force X through a medium of some characteristic material conductance L:

$$F = LX \tag{1}$$

Numerous fluxes obey the empirical relationship in Eq. 1, such as the flux of electric charges in a conductor (Ohm's law), molecular diffusion (Fick's law), conduction of heat (Fourier's law), or flow of water in an aquifer (Darcy's law). For two or more simultaneous transport fluxes, the driving forces of one may affect the other, resulting in mutually dependent fluxes in a *coupled process*:

$$F_1 = L_{11}X_1 + L_{12}X_2 + \cdots$$
$$F_2 = L_{21}X_1 + L_{22}X_2 + \cdots$$

If there is no coupling between the two fluxes, then the cross-coefficients $L_{12} = L_{21} = 0$, and each flux depends only on one driving force, as in Eq. 1. The extensive theory behind the coupling of transport fluxes is usually given in texts on irreversible and statistical thermodynamics. Discussions of the general theory of transport based on interatomic forces, and its generalizations to fundamental chemical and physical processes, are also given in more specialized publications dealing with the physical properties of crystals (Dhagavantam, 1966), general theory of molecular transport (Montroll, 1967), and heat and mass transfer in porous media (Liukov, 1966). The latter source in particular discusses the rules behind the permissible and not permissible coupling between different driving forces, known as the *Curie principle*—permissible coupling, if the driving forces are tensors of the same rank or differ in rank by an even number, and not permissible coupling, when the driving forces are tensors differing by one rank. The theoretically permissible or not permissible coupling bears on material fluxes due to such processes as molecular diffusion, thermal diffusion, Darcian flow, and chemical reactions in systems where transport and reaction occur simultaneously. In principle, the coupling of the fluxes in the surficial aquatic environment requires understanding of the effects of various external forces on reaction kinetics and macroscopic transport. A list of such driving forces includes mechanical stresses acting on sections of the solid crust on physically very different scales, temperature gradients, the gradients of chemical, electrical or gravitational potentials, and perhaps the magnetic fields.

It is, however, simpler to consider for the totality of chemical kinetic and transport processes that operate on different physical and time scales in the surficial environment three main classes of driving forces—physical, hydro-logical, and chemical—and responses to these forces in various physical, hydrological, and biogeochemical processes. This is illustrated schematically in Table 1, which gives a matrix of 3×3 couplings between physical, hydrological, and chemical forces and their environmental effects as far as these concern transport.

TABLE 1. Coupling among Macroscopic Transport Processes and Physical, Hydrological, and Chemical Driving Forces

Processes	Driving Forces		
	Physical	Hydrological	Chemical
Physical	Crustal uplift and subsidence Solid-state deformation	Erosion by water and wind	Rock disaggregation by chemical and biological attack Solid–solution–gas reactions Dissolution, precipitation, cementation, corrosion
Hydrological	Gravitational water flows Water and atmosphere circulation Sedimentation, suspensions Thermal water and gas flows	Evaporation, condensation, precipitation	Chemical, density-driven currents Chemical stratification of waters Evaporative mixing Chemical potential-driven flows, diffusion
Chemical	Temperature, pressure, and mechanical stress effects on reactions	Transport of reaction products Media for heterogeneous phase reactions	Homogeneous phase reactions, solutions and gases

It may be noted in the matrix of Table 1, that the diagonal boxes represent internal couplings between the physical (P→P), hydrological (H→H), and chemical (C→C) forces and processes that occur in essentially homogeneous environments. Cross-couplings between physical forces and hydrological processes—P→H—are the results of physical forces acting on surface waters, the atmosphere and oceans, and these produce circulation in the hydrosphere and atmosphere that reflects itself in water flow, wind, oceanic and lake circulation, and turbulence. A reverse coupling—H→P—represents the forces of flowing water and wind on the land surface that result in erosion or mechanical denudation.

The major mechanical forces that affect chemical processes—coupling P→C—are mechanical stresses or pressures that deform crystal lattices, affect the solid density and chemical potentials, and cause disaggregation or aggregation of solid particles on a macroscopic scale. The reverse coupling of the chemical effects on solids—C→P—includes a very broad category of chemical reactions in a solid state, reactions of mineral or biogenic solids with waters and atmospheric gases, and corrosion of metals.

The internal chemical coupling C→C includes homogeneous reactions in a gaseous or aqueous phase, and transport processes driven by chemical forces, such as molecular or ionic diffusion, and those heterogeneous processes related to it that do not involve large-scale transport: for example, certain types of surface reactions or interactions between particles in colloidal suspensions (O'Melia, Chapter 16, this volume) that are driven primarily by electrochemical forces fall in the class of C→C interactions.

The coupling between the hydrological forces and chemical processes —H→C—is responsible for transport of solutes, and this coupling is also to some extent reflected in the water-flow-rate dependence of some mineral dissolution reactions. The reverse coupling between chemical forces and hydrological processes—C→H—is seen in such phenomena as chemical density stratification in lakes, evaporative mixing caused by solute concentration increase in a surface water layer, and chemical density-driven water currents.

The interactions of the chemical kinetic and macroscopic transport processes depend to a variable extent on each of the nine couples given in Table 1. For example, transport of solutes by water runoff is a result of atmospheric precipitation, land exposure to water, chemical reactivity of solids in an aqueous solution, and flow of water over the continental surface. The net result of this process is controlled by coupling among the physical, hydrological and chemical entities in Table 1, as shown below:

P→P and H→H	→	P→H and C→P	→	H→P and H→C
land uplift and atmospheric precipitation		water flow and solution–mineral reactions		transport of solids and solutes

3. BACKGROUND RATES OF CHEMICAL KINETIC AND TRANSPORT PROCESSES

The individual processes listed within each of the coupling boxes in the P–H–C matrix of Table 1 differ in their magnitudes greatly one from another. A measure of a magnitude of a process is either a mass flux (F, in centimeter–gram–second units)

$$F \quad g\,cm^{-2}\,s^{-1} \quad or \quad mol\,cm^{-2}\,s^{-1}$$

or a rate of linear change in a unit of time, equal to the ratio of the flux to density or concentration of material

$$\text{Linear rate} = \frac{F}{\rho}\,cm\,s^{-1} \tag{2}$$

where ρ is in units of either grams per cubic centimeter or moles per cubic centimeter,

Linear rates of change for the more important chemical, hydrological, and physical processes are plotted in Figure 1. The main types of coupling among the driving forces and processes are also indicated in the figure. These linear rates are characteristic of macroscopic processes and net effects of chemical reactions: not shown in the figure, as being outside the scope of this chapter, are the linear rates for chemical processes on atomic scales, such as, for example, molecular velocities in gases or soluions where thermal velocities of an order of $(RT/M)^{1/2}$ over angstrom-long distances can be much higher that the net velocities of such processes as recession of a surface of a dissolving crystal, metal–surface oxidation, or corrosional dissolution.

Water flow over the continental surface and underground is a transport agent of major importance, and its flow velocities are generally fast in comparison to the rates of other coupled processes shown in Figure 1. The low end of the water-flow rates represents deeper groundwaters where linear flow rates may be of a magnitude 10^{-1} to 10^0 m y^{-1}, and the mean flow rates through deep lakes of long water-residence times. For such large lakes, the linear rate of water renewal is the ratio (mean depth)/(water renewal time) at this can be in the vicinity of 10^0–10^1 m yr^{-1}, as discussed further in Section 4.1.

The range of fast linear flow rates also includes the flows of air in the subsurface, usually mixed with water vapor and other gases, where such flows may be driven by vertical temperature gradients.

In mineral dissolution reactions, net effects of dissolution can be translated into linear rates of recession of crystal surfaces in a direction perpendicular them. For dissolution of silicate minerals in water near 25°C under laboratory conditions, the linear dissolution rates are generally $\leq 10^{-3}$ mm yr^{-1} or ≤ 1 mm ka^{-1}. For the carbonate minerals calcite and dolomite, the linear dissolution rates are higher, $10^{0\pm1}$ mm yr^{-1}, and for long period of time this translates into very high dissolution rates of more than 1 meter in 1000 years.

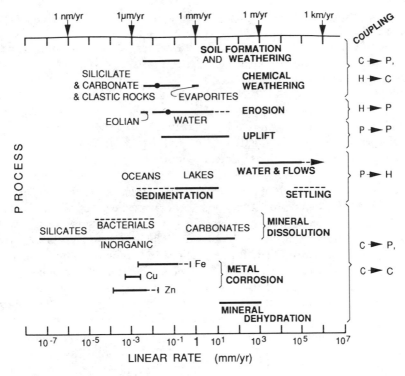

Figure 1. Linear rates (in millimeters per year) of chemical kinetic and macroscopic transport processes in surficial aquatic and sedimentary environments. Individual processes are coupled to the driving forces, identified as three main groups of chemical (C), hydrological (H), and physical (P) driving forces. Data sources: mineral dissolution rates, Tables 4 and 5, and Berthelin (1988); mineral dehydration, compilation in Bodek and Lerman (1988); metal corrosion, Coburn (1968), Costa (1982), and Haynie and Upham (1970); uplift, Lajoie (1986), Stallard (1988); physical erosion, Table 3; chemical weathering, soil formation and chemical denudation, Table 6.

Mineral dehydration is a reaction involving loss of H_2O from the crystal structure and a concomitant formation of a new anhydrous phase:

$$ML \cdot nH_2O(s) = ML(s) + nH_2O(v)$$

Dehydration reactions, such as the gypsum to anhydrite transition, occur in saline brines, where the thermodynamic activity of water changes with the solute concentration and temperature, and they may also occur in subaerial weathering under conditions of low humidity. The experimentally determined rates of such reactions, as shown in Figure 1, indicate that the growth of anhydrous crystalline phases at the expense of the hydrated minerals is generally fast. The higher rates may be due, however, to the often practised experimental procedure of measuring mineral dehydration rates in partial vacuum, and applicability of such rates to natural conditions at atmospheric pressure is uncertain. Corrosion rates of metals exposed to atmospheric precipitation and dry fallout for variable periods of time under different environmental conditions show that zinc (Zn) and copper (Cu) corrode, forming new solid phases in the process, slower than does iron (Fe). Corrosion rates of latter extend into the 0.1–0.5-mm yr^{-1} range in some industrial and coastal oceanic areas.

The products of erosion or mechanical denudation of the land surface are solid particles of a wide range of sizes that are ultimately deposited as clastic (or detrital) sediments on land, in riverbeds, lakes, estuaries, and on the ocean floor. Biogenic sediments include mainly skeletons of planktonic organisms consisting of calcium carbonate minerals (calcite, aragonite, magnesian calcites) and silica, characteristic of diatoms. The differences between the rates of mechanical denudation and deposition of clastic sediments in different sections of the world are accounted for by combinations of several factors: clastic sediments eroded from a large drainage area may be deposited in a relatively small area, resulting in high net sediment accumulation rates; conversely, dispersal of eroded materials from a relatively small area, such as a mountain range, over much greater areas would result in a smaller sedimentation rate; clastic sediment piles are usually less dense than crustal rocks, and the vertical rate of deposition of bulk sediments, composed of solids and pore space, may therefore be higher. In biologically productive lakes and oceanic sections, land-derived materials mix with the biogenic and organic materials forming in the water column and on the bottom, possibly resulting in relatively higher rates of sediment accumulation.

A fraction of the eroded mass made of small-size particles is transported by winds over subglobal distances, to be ultimately deposited in the open ocean and on other continents. Rates of sediment accumulation in the open oceans are much lower than for sediments in continental bodies of water, where in lakes the rates are typically of an order of $10^{0 \pm 1}$ mm yr^{-1} for terrigenous sediments and are higher for deposition of organic-matter-rich muds in areas of strong primary productivity.

Gravitational settling of particles is a process distinct from accumulation of sediments, as discussed in the preceding paragraph. Rates of particle settling in water are generally fast, except for uncoagulated clay-size and colloidal materials (of linear dimensions < 2 μm), approaching settling velocities of meters per day or 10^2 to 10^3 m yr^{-1}.

Chemical denudation of the land surface is a net result of mineral dissolution, as will be discussed in more detail in Section 6. The rates of chemical and mechanical weathering on land range over values of a comparable magnitude; erosion is as a whole somewhat faster than chemical denudation. The global means of the two processes indicate that physical denudation is about five times faster than chemical denudation. This difference reflects the mass ratio of about 5 of the physical:chemical weathering fluxes, characteristic of the present-day conditions on the earth's surface. The process of denudation of the continental surface is driven mainly by a hydrological-physical or H→P coupling (Table 1), where the role of the chemicophysical or chemicophysicohydrological coupling —C→P or C→P→H—is subordinate in determining the magnitude of material transport fluxes, although chemical and biological activity is likely to be important in the initial stages of disaggregation and crack development in exposed rocks.

Soils, consisting of residual minerals from the bedrock, newly formed minerals, and organic materials, form and erode at rates comparable to the rates of chemical weathering of bedrocks. The rates of soil development have been estimated from thicknesses of soil layers developed within historical or longer periods of time, identifiable by certain chronological markers (Brunsden, 1979; Slaymaker, 1988), and by a number of other physical and chemical techniques (Colman and Dethier, 1986). The rates of material loss by chemical weathering, however, are often based on shorter-term observations of months to a few years, conducted in experimental facilities or based on chemical composition of stream and soil waters [e.g., see Schnoor, Chapter 17, this volume, with additional references cited therein; Wright, (1988), with references cited therein].

4. WATER TRANSPORT FLUXES

4.1. Global Discharge and Linear Flow Rates

The annual water discharge from the continents to the oceans, $Q = 3.74 \times 10^4$ km yr^{-1} (Table 2), translates into a value of global mean runoff (q) of

$$q = \frac{Q}{A} \approx 0.37 \text{ m yr}^{-1} \tag{3}$$

This value represents a mean difference between atmospheric precipitation and evapotranspiration, and it also assumes that there is no increase in volume of groundwater from surface recharge.

A fraction of the total runoff from the continents to oceans passes through freshwater lakes. From the data on global volume, surface and drainage basin area, and mean water residence time for freshwater lakes, a mean linear rate of flow through lakes is

$$\text{Lake flow} = \frac{V_L}{\tau_L (A_L + A_{DB})} \approx 7.6 \text{ m yr}^{-1}$$

The preceding estimate is sensitive to the relatively short residence time of water in lakes, $\tau_L = 5.5$ years, inclusion of the lake-basin drainage area in computation, taken as about 2.5 times the total lake surface area (see footnotes to Table 2 regarding this factor), and the mean depth represented by the ratio

$$\bar{z} = \frac{V_L}{A_L} = \frac{1.25 \times 10^5}{8.55 \times 10^5} = 146 \text{ m}$$

TABLE 2. Water Discharge and Residence Times on Land

Parameter	Value	References
Global water discharge to oceans (Q_T)	3.74 $\times 10^4$ km^3 yr^{-1}	Baumgartner and Reichel (1975), Meybeck (1984)
Continental drainage area for discharge to oceans (A_T)	99.913 $\times 10^6$ km^2	
Continental nonglaciated	133.113 $\times 10^6$ km^2	
Median residence time for discharge to oceans (τ)	(0.3)–3 yr	Lerman (in press)
Freshwater lakes		Herdendorf (1982) and additional references cited in Lerman and Hull (1987)
Volume (V_L)	1.25 $\times 10^5$ km^3	
Surface area (A_L)	8.55 $\times 10^5$ km^2	
Mean depth (V_L/A_L)	146 m	
Drainage basin area (A_{DB})[a]	$\approx 2.5 \times A_L$ km^2	
Mean volume[b]	41 km^3	
Mean depth	21 m	
Water residence times		
In freshwater lakes	≈ 5.5 yr	Lerman (in press)
In rivers without major lakes in their basins	(0.65)–6.5 yr	

[a] Area ratios A_{DB}/A_L for small lakes are usually much bigger (Linthurst et al., 1986), but most of the water volume resides in the large lakes.

[b] Mean volume and depth based on a sample of 98 lakes of volume $V > 1$ km^3 (Lerman and Hull, 1987).

A somewhat different estimate is obtained from the mean depth of a sample of freshwater lakes, of volumes $V > 1$ km^3 (Lerman and Hull, 1987): logarithmic means for these lakes are about 41 km^3 for the volume and 21 m for the mean depth. Thus the linear rate of water flow through lakes is

$$\text{Lake flow} = \frac{\bar{z}}{\tau_L} \approx \frac{21}{5.5} = 3.8 \text{ m yr}^{-1}$$

The preceding discussion brackets the linear rates of water flow from continents to oceans within an order of magnitude at $10^{0 \pm 0.5}$ m yr^{-1}, with the global mean runoff at the lower end of the range.

4.2. Surface and Groundwater Flows

Surface water runoff from land to the oceans, by latitudinal zones 5° wide as shown in Figure 2, has a median value of about 0.3 m yr^{-1}, in good agreement

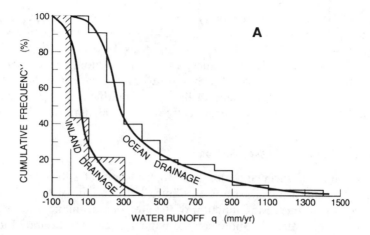

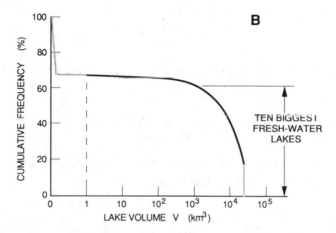

Figure 2. (*A*) Water runoff (*q*, mm yr^{-1}) to the oceans from 5°-wide latitudinal zones, 80°N–80°S. Complementary cumulative frequency distribution: frequency (in per cent) of runoff greater than the runoff value on the abscissa. [Data source: Baumgartner and Reichel (1975).] (*B*) Cumulative frequency distribution of freshwater lake volumes. Lakes of *V* > 1 km^3 account for 67%, and 10 largest lakes for 61% of the global volume. From data compiled in Lerman and Hull (1987).

with the simple mean discussed in the preceding section. The median of the cumulative distribution of zonal water flows indicates a 50% probability dividing line for zonal discharges greater or smaller than 0.3 m yr^{-1}. This number may be compared with the much faster flows in large rivers (a magnitude of flow that impresses an observer), measured in centimeters to tens of centimeters per second, or 10^2 to 10^3 km yr^{-1}.

Very slow linear flow velocities characteristic of deep groundwater aquifers extend from meters per year and up. Such slow velocities are generally derived from model calculations based on the known physical characteristics of the porous rocks (hydraulic permeability, temperature, hydrostatic pressure gradients, and distribution of the probable flow paths). At another extreme are the fast velocities of water runoff from dense or nonporous rocks exposed on the surface (Reddy, 1988). For example, the times of contact are short between rain water and smooth surfaces consisting of such natural or man-made materials as metals, metal alloys, and structures made of limestone, marble, sandstone, crystalline rocks, and concrete containing chemically reactive calcium–silicate phases.

4.3.   Water Flow and Dissolved Load

The importance of surficial water flow to transport of solids and dissolved materials is schematically illustrated in Figure 3; the volume of flow or discharge (Q) is generally an increasing function of the area of a drainage basin when the discharge per unit area (q) is about constant under given climatic and lithological conditions:

$$Q = \sum_{i=1}^{i=n} Q_i = \sum_{i=1}^{i=n} q_i A_i$$

A load carried by running water is a mass flux made of contributions from the smaller areas:

$$W = \sum_{i=1}^{i=n} C_i q_i A_i \bigg/ \sum_{i=1}^{i=n} A_i$$

Concentration (C_i) in flow from any areal section is a function of input from source rocks and soils, and water discharge:

$$C_i = \frac{R_i A_i}{Q_i} = \frac{R_i}{q_i} \tag{4}$$

where R_i is a rate of material input to runoff (mass $cm^{-2} s^{-1}$) per unit of land surface area A_i.

The relationships between river discharge and concentrations of dissolved materials fall into three main classes, which are shown in Figure 3 and discussed below with reference to Eq. 4.

Concentrations of dissolved materials may be approximately constant and independent of the specific discharge q. This relationship holds for rivers draining more or less homogeneous rock-type terrains or the same climatic zones. A constant concentration in river flow in this case is not a universal indication of an equilibrium-controlled dissolution of source minerals, and the chemical kinetic reasons for this phenomenon will be addressed in Section 6.2. Concentration ranges of SiO_2(aq) and Ca^{2+} for a wide range of water discharges in cold and tropical regions, and discharges from crystalline and sedimentary rocks are shown in Figure 3a.

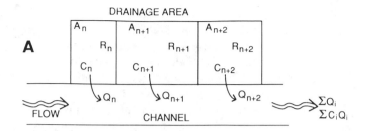

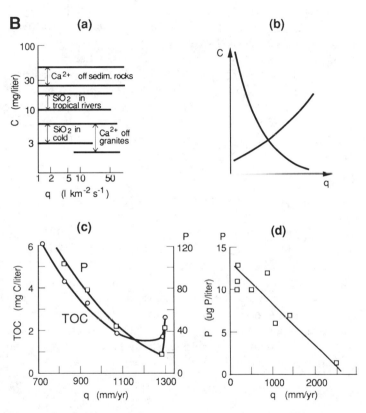

Figure 3. (*A*) Increase in water volume flow (discharge *Q*) and dissolved load of streams (*CQ*) with an increasing surface area of the drainage basin; A_i are areas of the basin sections, R_i are mass input rates of dissolved materials, and C_i are concentrations in discharge from the individual sections. (*B*) Relationships between concentrations of dissolved species (*C*) and specific water discharge (*q*). (*a*) Concentrations vary within fairly narrow bands for wide range of river discharge values. $SiO_2(aq)$ in rivers from cold and tropical regions. Calcium in rivers draining granitic and gneissic, and sedimentary rock terrains. [From data in Meybeck (in press).] (*b*) Concentrations increase or decrease with an increasing discharge (Davis and Zobrist, 1978). (*c*) Total organic carbon (TOC, mg C L^{-1}) and dissolved inorganic phosphorus (P, μg P/L^{-1}) in five rivers of the Rhine and Rhone drainage basins, Switzerland (Davis and Zobrist, 1978). (*d*) Dissolved inorganic phosphorus (P, μg P/L^{-1}) in eight rivers from temperate and tropical climatic zones (from Meybeck, 1982, Table 1).

When concentration in flow increases with an increasing discharge, this indicates that input from the drainage terrain increases faster than discharge at high volume flow. As the pore space of soils is only intermittently filled or saturated with water, at times of high discharge greater dissolution may be expected of such relatively soluble species as nutrient elements in fertilizers. A decrease of concentration in flow is a case of inputs located only in some parts of the drainage basin upstream (for example, a change in terrain lithology) or of dilution of point sources, such as of localized industrial or municipal discharges. Combinations of the increasing and decreasing concentrations as functions of discharge have been reported for rivers of seasonally variable discharge (Davis and Zobrist, 1978). These relationships between solute concentrations and discharge are shown schematically in Figure 3b.

Decreasing concentrations with an increasing river discharge q are shown in Figures 3c and 3d for total organic carbon (TOC) and dissolved phosphate in two different sets of rivers and drainage basins. Phosphate concentrations are generally higher in the rivers discharging from Switzerland to the Rhine (North Sea) and Rhone (Mediterranean) drainage basins. Phosphate values are lower in a sample of eight rivers and drainage basins from temperate and tropical zones on different continents, as shown in Figure 3d.

5. EROSION AND TRANSPORT OF SOLIDS

Solids eroded from the continental surface account for about 80% of the material mass transport (Table 3). Most of the solid material flux, about 95% of it, is carried by water discharge and the remaining 5% are transported by winds. The total suspended material load of rivers and the dust load of the atmosphere include products of the bedrock erosion as well as materials eroded from clastic and biogenic sediments exposed on the continents. Equating the sum of the riverine and atmospheric transport fluxes of solids with the rate of continental erosion or mechanical denudation implies that a part of the eroded mass that is stored in clastic sediments on land remains constant (Fig. 4):

(Surface erosion rate) = (transport flux) ± (storage or release in surface sediments)

Physical disaggregation of rocks transfers minerals to the soil layer developing on them, and it frees solid particles to be carried away by water and wind. Initial disaggregation and loosening of the crystalline aggregates in rocks are likely to be results of chemical and bacterial activity on the rock surface and in minute fractures.

The process of physical erosion of elevated terrains and subsequent transport and deposition of the eroded material downstream or downwind from the source are shown schematically in Figure 4. While erosion of the land relief and transport of the eroded materials by rivers are essentially continuous processes driven by the uplift of the continental surface, considerable uncertainties exist in

TABLE 3. Erosional Fluxes

Parameter	Value[a]	References and Comments
River Transport		
Mean sediment flux	$150 \, \mathrm{g \, m^{-2} \, yr^{-1}}$	Milliman and Meade (1983), Milliman (in press)
Mean sediment flux	$175 \, \mathrm{g \, m^{-2} \, yr^{-1}}$	Meybeck (in press)
Linear erosion rate	$60–70 \, \mathrm{mm \, ka^{-1}}$	From preceding two global estimates, $\rho = 2.5 \, \mathrm{g \, cm^{-3}}$
Eolian Transport		
Mass total of eolian dust from land	$0.7 \times 10^9 \, \mathrm{ton \, yr^{-1}}$	Rea et al. (in press), Prospero (1981)
Mean mass flux	$5–7 \, \mathrm{g \, m^{-2} \, yr^{-1}}$	From preceding estimate and two values of land area in Table 2
Linear erosion rate	$2–3 \, \mathrm{mm \, ka^{-1}}$	Same, $\rho = 2.5 \, \mathrm{g \, cm^{-3}}$

[a] $\mathrm{ka} = 1000 \, \mathrm{yr.}$

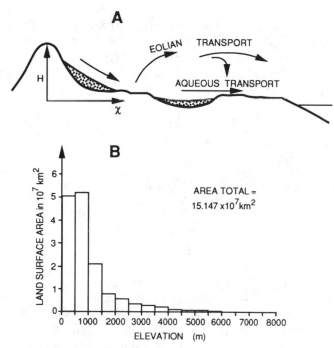

Figure 4. (*A*) Erosional fluxes and transport of erosional products by eolian and aqueous flows (*H*—land elevation, *x*—horizontal distance; see also Table 3). (*B*) Elevation of continental land areas in 500-m intervals (Gates and Nelson, 1975).

the estimates of time of the storage of eroded materials on land (Walling and Webb, 1983; Mead, 1988). Some of the transported materials are deposited in lakes, in the floodplanes of rivers, in inland areas, and in estuaries. The time scale for storage in such repositories may range from seasonal to a few decades for river valleys, and time scales are considerably longer for storage in lakes and estuaries, and areas of internal drainage only.

The existence of positive relief features and mountains is an almost self-explanatory indication that the rates of uplift of some continental sections have been much higher than the rates of denudation (Fig. 2). The rates of erosion that about balance or are slightly below the rates of uplift characterize the stable shield areas of South America, 10–20 mm ka^{-1}, and a tectonically active island Formosa (Taiwan), where uplift rates are as high as 10^3–10^4 mm ka^{-1} (Stallard, 1988).

A mean rate of continental denudation by erosion is a mean rate of a decrease of land elevation above some reference level:

$$-\frac{dH}{dt} = \frac{W_{rs} + W_{es}}{\rho(1-\phi)} \tag{5}$$

$$\left[\frac{dH}{dt}\right]_{total} = \left[\frac{dH}{dt}\right]_{rivers} + \left[\frac{dH}{dt}\right]_{wind} \tag{6}$$

where H is mean elevation of land (cm), t is time (years), W_{rs} is flux of suspended materials in rivers (g cm^{-2} yr^{-1}), W_{es} is the eolian flux of eroded solids, and ρ is mean density of eroded materials (g cm^{-3}), and ϕ is porosity of the rock. Erosion of mountain terrains has been modeled as a first-order kinetic process (Pinet and Surian, 1988; Flemings and Jordan, in press),

$$-\frac{dH}{dt} = kH \tag{7}$$

where $k = 4 \times 10^{-8}$ to 4×10^{-7} yr^{-1}. Dispersal of eroded materials and their deposition as clastic sediments on the continental surface has been treated analogously to a process of diffusion:

$$\frac{dH}{dt} = D_s \frac{\partial^2 H}{\partial x^2} \tag{8}$$

where $D_s \approx 0.1 - 1$ cm^2 s^{-1} for mountains and $D_s \approx 10$ cm^2 s^{-1} for sedimentary basins. The higher value of the dispersal coefficient D_s for the sedimentary basins of the geologic past possibly reflects dispersal and deposition of sediment over areas greater than the source terrain. The preceding rate constants and dispersal rates for diminution of mountain relief translate into linear rates of erosion that are comparable to the global mean rates and some regional rates mentioned earlier in this section.

Mean elevation of the continents is 840 m (Lagrula, 1965; Sverdrup et al., 1942; Gates and Nelson, 1975). A modest mountain relief of 2000–2500 m occupies about 5% of the unglaciated continental area (Fig. 4B), and its linear rate of erosion is, from Eq. 7

$$-\frac{dH}{dT} = \left(\frac{4 \times 10^{-8}}{4 \times 10^{-7}} \, yr^{-1}\right) \times 2 \times 10^3 \, m \approx \frac{80}{800} \, mm \, ka^{-1}$$

From the dispersal model, for the same elevation $H = 2000$ m and transport over distances of 10^2–10^3 km, the rate of erosion is

$$-\frac{dH}{dt} \approx \frac{D_s}{\Delta x}\left(-\frac{\Delta H}{\Delta x}\right) = 10 \div \left(\frac{10^7}{10^8}\right) \times \left(\frac{2 \times 10^{-2}}{2 \times 10^{-3}}\right) = \frac{60}{600} \, mm \, ka^{-1}$$

The share of eolian dust transport in continental erosion is submerged in the greater range of uncertainties associated with the transport rates, as discussed in the preceding paragraphs.

Physical denudation rates of 80–800 mm ka^{-1} would cause erosion of the land mass to the present-day mean elevation of continents, 840 m, in a period of 10^6 10^7 years if there were no crustal uplift compensating for, or exceeding, the rates of erosion.

6. MINERAL DISSOLUTION AND CHEMICAL WEATHERING

6.1. Mineral Dissolution Rates

Dissolution rates of common rock-forming silicate and carbonate minerals, plotted against the solution pH, fall into two distinct bands, as shown in Figure 5; the dissolution rates of calcite and dolomite are distinctly higher than those of the silicate minerals, among which particularly noteworthy are the low dissolution rates of the common components of surficial rocks, quartz (SiO_2), and Na–feldspar albite ($NaAlSi_3O_8$). However, the differences between the dissolution rates of individual minerals are very large, up to a factor of 10^5 (Table 4).

The pH dependence of the dissolution rates of silicates is a subject of intensive theoretical interest, based on transition-state and surface-reaction rate theories (e.g., Schott and Petit, 1987; Wollast and Chou, 1988; Stumm and Wieland, this volume). The features of the pH dependence of the silicate dissolution rates that are relevant to this section are the reported dependence of the rate in the acidic solution range (pH ≤ 5.5) on a power of the hydrogen ion concentration, $R \propto [H^+]^{0.5}$ to $[H^+]^{1.0}$, and its dependence in the alkalilne range (pH ≥ 7.5) on $R \propto [H^+]^{-0.3}$.

For the carbonates, calcite and dolomite, the dissolution rates increase with an increasing [H$^+$] ion concentration. A change in the dissolution rate by a factor

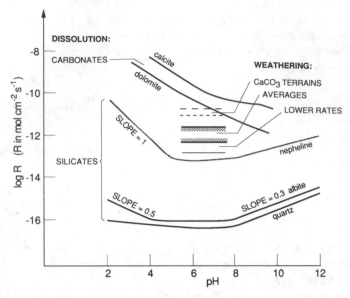

Figure 5. Dissolution rates (R, mol cm^{-2} s^{-1}) of silicate and carbonate minerals in H$_2$O near 25°C under laboratory conditions as a function of the solution pH. Data sources listed in Tables 4 and 5. Global averages for chemical denudation of the continents, and soil denudation and formation rates are shown for near-neutral pH region.

of 10^2–10^3 over a range of the pH values common to continental waters, $4 \leq pH \leq 9$, is indicated by the data in Table 5.

A field for global mean chemical weathering rates, and an overlapping field for the rates of soil weathering or soil formation are also shown in Figure 5. The ranges of dissolution rates for carbonate and silicate minerals are difficult to reconcile with the chemical weathering rates of the crustal rocks and soils that are based on concentrations in river water and river discharge. The increases in the dissolution rates by a factor of 3–10, in response to an increase in the solution [H$^+$] concentration by a factor of 10, or by 1 pH unit, would be far outweighed by the uncertainties in the relative contributions of the individual minerals and their effective dissolution rates in the weathering zone, as will be discussed in the next section. For calcite and dolomite, the experimentally determined dissolution rates are too fast to account for the calcium concentrations in rivers. In a natural setting near pH $= 7$, effective calcite dissolution rates several orders of magnitude lower than the reported rate near 10^{-10} mol cm^{-2} s^{-1} would be needed to account for the calcium ion concentrations in surface waters.

For a possible temperature dependence of the chemical weathering rates, the following should be noted: in rivers occurring from the tundra to the humid tropics, a temperature range of about 20–25°C, concentrations of SiO$_2$(aq) increase by a factor of 5, from about 3 to 15 mg L^{-1} (Meybeck, in press). An comparable increase by a factor of 5 characterizes the rates of net carbon fixation

TABLE 4. Dissolution Rates of Silicate Minerals

Mineral	Formula	Dissolution Rate R^a (mol SiO_2 cm^{-2} s^{-1}) log R	R dependence on $[H^+]^n$		Reference
			n for Acidic Range	n for Alkaline Range	
Albite	$NaAlSi_3O_8$	-15.8 to -14.3	0.5	-0.3	Chou and Wollast (1985)
Anorthite	$CaAl_2Si_2O_8$	-15.5 to -13	1.0	-0.3	Brady and Walther (in press), with additional references cited therein
Chrysotile	$Mg_3Si_2O_5(OH)_4$	-16 to -15.5		-0.3	Bales and Morgan (1985)
Diopside	$CaMgSi_2O_6$	-14.5 to -12	0.5		Schott et al. (1981)
Forsterite	Mg_2SiO_4	-13.8 to -12	0.5	-0.3	Blum and Lasaga (1988)
Quartz	SiO_2	-16.3 to -14.5	~ 0	-0.3	Brady and Walther (in press)
Nepheline	$NaAlSiO_4$	-12.5 to -10.5	1.0	-0.3	Tole et al. (1986)

[a] Lower values of R apply to the near-neutral pH range, from about 5.5 to 7.5. Rates increase as the pH becomes lower and higher than near neutral (see Fig. 5).

TABLE 5. Rates of Dissolution of Carbonate Minerals and Limestones

	Mass Rate R (mol cm^{-2}s^{-1}) log R	Linear Rate R/ρ (mm ka^{-1})	References and Comments
Calcite	-10 to -8.2	1×10^3–7×10^4	Plummer et al. (1978); pH $= 7$ to 4
	-10.5 to -8.2	4×10^2–7×10^4	Chou et al. (in press); pH $= 9$ to 4
Dolomite	-11.5 to -9.5		Same
Limestones		10–120	Jennings (1983); rate may include mechanical erosion, rate increases with runoff
	-13 to -12	1–20	Chemical weathering of limestone terrains; from Meybeck (in press)
Limestones, marbles		50–100	Tombstones (Yorkshire) [in Slaymaker (1988)]
		~ 5–12	Estimated from Reddy's (1988) data on dissolution of slabs by rain

by land plants, from the tundra to the tropics (Lieth, 1975), which may correlate with a general level of bacterial productivity in soils that increases mineral dissolution rates (Post et al., 1982; Eckhardt, 1985; Berthelin, 1988). Coincidentally, the dissolution rates of silicate minerals, characterized by activation energies (Schott and Petit, 1987) in the range of

$$\Delta E \approx 55 \pm 15 \text{ kJ mol}^{-1}$$

indicate for a temperature change from 5°C to 25°C an increase in dissolution rates by a factor of 3–8:

$$\frac{R_{25}}{R_5} = 5.5 \pm 2.5$$

6.2.   Constraints on Background Weathering Fluxes

It was pointed out in the preceding section that a lack of agreement between the mineral dissolution rates, as reported from laboratory experiments, and the chemical weathering rates, as derived from the composition and volumes of surficial runoff, suggest a number of bracketing values that can be placed on the parameters in the models for chemical weathering and transport.

A net rate of chemical weathering or chemical denudation may be defined as a product of specific discharge (q) and concentration of solutes (C) originating from mineral dissolution in rocks and soils:

$$W = Cq \ \text{mol cm}^{-2} \text{s}^{-1} \tag{9}$$

which is equivalent to a linear rate (W_h):

$$W_h = \frac{C \times q \times n \times \text{gfw}}{\rho(1 - \phi)} \ \text{cm s}^{-1} \tag{10}$$

where n is a stoichiometric factor for conversion of molar concentration in solution to a mineral formula, gfw is gram formula weight of the mineral (g mol^{-1}), ρ is mineral density (g cm^{-3}), and ϕ is porosity of the weathering material. Values of W and W_h are given in Table 6.

A net rate of weathering may also be defined in terms of mineral dissolution rates (R), when dissolution occurs within a porous layer of some thickness, and the dissolution products are transported by water flow in the pore space and further out to the surficial runoff (Fig. 6). In this formulation, the net weathering rate W is

$$W = \frac{RS}{A} \ \text{mol cm}^{-2} \text{s}^{-1} \tag{11}$$

where S (cm^2) is the solid surface area within a porous layer of thickness h and porosity ϕ, and A (cm^2) is a horizontal or map-surface area.

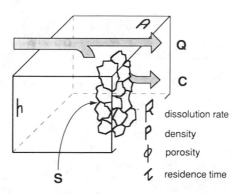

R   dissolution rate
ρ   density
φ   porosity
τ   residence time

Figure 6. Dissolution and transport in the weathering zone (W is net weathering rate): C is concentration in discharge, Q is volume discharge, A is land surface area, h is thickness of reactive zone, ϕ is its porosity, and τ is water residence time within the reactive zone space, where S is the reactive surface area of minerals. For a weathering rate consisting of contributions from individual minerals, R_i is mineral dissolution rate, S_{gi} is mineral surface area per gram, and ρ_i is mineral density. Depth of the reactive zone and volume of water in pore space ϕ may be functions of the water discharge volume—$h(q)$, $v(q)$.

TABLE 6. Chemical Weathering Rates (W) for Different Rock and Soil Types, from Different Areas

	Mass Flux (W)	Linear Rate (W_h)	References and Comments
Global mean	$33-40 \, \mathrm{g \, m^{-2} \, yr^{-1}}$	$14 \pm 1 \, \mathrm{mm \, ka^{-1}}$	Meybeck (1988); mass rate converted to linear rate for bulk density $\rho \approx 2.5 - 2.7 \, \mathrm{g \, cm^{-3}}$
Shields and sediments		$5-30 \, \mathrm{mm \, ka^{-1}}$	Stallard (1988)
Silicate rocks		$80 \, \mathrm{mm \, ka^{-1}}$	Stallard (1988); Andean drainage
Carbonates and shales		$200 \, \mathrm{mm \, ka^{-1}}$	Stallard (1988); Andean drainage
Evaporites		$700 \, \mathrm{mm \, ka^{-1}}$	Stallard (1988); Andean drainage
Volcanics	$18 \, \mathrm{g \, m^{-2} \, yr^{-1}}$	$7 \, \mathrm{mm \, ka^{-1}}$	Drever (1988); Andesitic terrain in Absaroka Mts., Wyoming
Small stream watersheds and catchment areas	$0.024-0.20 \, \mathrm{equiv \, m^{-2} \, yr^{-1}}$	$4-30 \, \mathrm{mm \, ka^{-1}}$	Fölster (1985); data compiled for areas in USA, Norway, Sweden, England, and Germany
Silicate minerals (plagioclase, almandine, biotite)		$37-160 \, \mathrm{mm \, ka^{-1}}$	Velbel (1986)
Various soils			
Sandy–silty soils	$0.05-0.27 \, \mathrm{equiv \, m^{-2} \, yr^{-1}}$	$7-40 \, \mathrm{mm \, ka^{-1}}$	Cronan (1985)
Loess soils	$0.04-0.10 \, \mathrm{equiv \, m^{-2} \, yr^{-1}}$	$6-15 \, \mathrm{mm \, ka^{-1}}$	Fölster (1985)
Soils in lake drainage basins	$0.06 \, \mathrm{equiv \, m^{-2} \, yr^{-1}}$	$9 \, \mathrm{mm \, ka^{-1}}$	Wright (1988); literature data for Adirondack Mts., New York and
Deglaciated terrain	$0.93 \, \mathrm{equiv \, m^{-2} \, yr^{-1}}$	$140 \, \mathrm{mm \, ka^{-1}}$	Cascade Mts., Oregon

Adapted from Lerman (1988), with additions.

Concentrations of solutes in discharge are approximately constant over wide ranges of river runoff values, as was mentioned in Section 4.3. In terms of Eq. 9, à constant C and an increasing q result in increasing values of the net chemical weathering rate W (Dethier, 1986; Berner and Berner, 1987; Meybeck, in press). From Eqs. 9 and 11, the weathering rates are equal:

$$Cq = \frac{RS}{A} \text{ mol cm}^{-2}\text{s}^{-1} \tag{12}$$

$$CQ = RS \text{ mol cm}^{-2}\text{s}^{-1} \tag{13}$$

If C is constant while q or Q is increasing, then either the dissolution rate R or the effective solid surface area S or both must also increase. It seems reasonable that at a higher discharge and at higher flow velocities water may penetrate deeper below the land surface, thereby coming in contact with a greater surface area of reactive minerals and resulting in an increase in S.

6.3. Dissolution and Transport Flow Model

Equations 12 and 13 define a balanced dissolution process, the products of which are removed by discharge through the porous surface layer and transported out to the river channels and beyond. The simple, if not too oversimplified, nature of this model is in the treatment of the balance

(Input from dissolution) = (removal by flow)

without consideration of other terms in chemical weathering, such as precipit- ation of secondary minerals and/or changes in the reservoirs of ions held by the exchangeable minerals in soils, organic matter, and vegetation. This simple model, however, allows us to define lower or upper limits for reasonable values of the parameters to make it consistent with the reported mean concentrations in river waters. Concentration of SiO_2 (aq) in the range from 3 to 15 mg L^{-1} was reported for rivers in the cold, temperate, and tropical regions, as mentioned previously (Meybeck, in press). Thus a guideline value of $10 \pm 5 \times 10^{-6}$ g SiO_2 cm^{-3} or about 2×10^{-7} mol cm^{-3} will be used for estimation of the bracketing values of other parameters.

Equation 13 can be rearranged to include implicitly measurable or easily identifiable parameters in the form of

$$C = \frac{\tau RS_g \rho (1 - \phi)}{\phi} \tag{14}$$

which follows from the two relationships discussed below (Fig. 6).

Discharge Q is the volume of a porous rock layer of a unit area, divided by the water residence time within it:

$$Q = \frac{\phi h A}{\tau} \text{ cm}^3 \text{ s}^{-1} \tag{15}$$

The reactive surface area S within a layer of thickness h is a product of the solid surface area per gram, S_g, solid density, and its volume fraction:

$$S = S_g \rho h A (1 - \phi) \text{ cm}^2 \tag{16}$$

Equation 14 is a result of substitution of Eqs. 15 and 16 in Eq. 13. Thus concentration in discharge is a function of the mineral dislution rate (R), water residence time in the reactive zone (τ), specific surface area of the reactive solids (S_g), their density (ρ), and porosity (ϕ) of the reactive zone. The following estimates of the individual parameters may be used:

$\tau \approx 0.3$–3 yr, as estimated from zonal latitudinal runoff data for runoff through a porous layer of assumed net thickness 10 to 100 cm (Lerman, in press).
$S_g = 1$–10 m^2 g^{-1} for common types of sedimentary materials, lower S_g values being characteristic of the coarser-grain sediments and larger values applying to fine-grained materials, as determined usually by gas adsorption method. Thus $S_g = 10^4 - 10^5$ cm^2 g^{-1}.
$R = 10^{-13}$–10^{-16} mol cm^{-2} s^{-1} are taken for silica dissolution rates from silicate minerals, as summarized in Figure 5 and Table 4.
$\rho \approx 3$ g cm^{-3} and $\phi \approx 50\%$, making the ratio $(1 - \phi)/\phi = 1$.

Using the above ranges of values in Eq. 14, we obtain for the dissolved silica concentration C:

$$C = \left(\frac{10^7}{10^8}\right) \times \left(\frac{10^{-16}}{10^{-13}}\right) \times \left(\frac{10^4}{10^5}\right) \times 3 \times 1 \gg 10^{-7} \text{ mol cm}^{-3} \tag{17}$$

The preceding result is a few orders of magnitude higher than the reported silica concentrations in rivers. A term that may conceivably be much smaller than the range of estimates used above is the effective or reactive surface area of minerals in the weathered zone. Its lower value may reflect such factors as low abundance of dissolving minerals, occurrence of coatings interfering with the dissolution reactions, or chemical reactivity of particle aggregates that is not representative of the gas-adsorption surface areas of sedimentary materials. For spherical particles of a mean radius r and density ρ, the specific surface area is

$$S_g = \frac{3}{r\rho} \approx \frac{1}{r} \text{ cm}^2 \text{ g}^{-1} \tag{18}$$

Alternatively,

$$r \approx \frac{1}{S_g} \approx \frac{10^{-5}}{10^{-4}} \, cm = \frac{0.1}{1} \, \mu m$$

as follows from the range of S_g values used earlier. Weathered crystalline rocks and residual soils forming on them, as far as data are available, indicate that mean particle sizes (determined by granulometric analyses) are much larger than the micrometer range:

$$r = 0.01 \quad \text{to} \quad 0.1 \, cm$$

is a range determined for a set of weathered granite and residual soils (Lerman, 1979). Smaller mean particle sizes, 0.001–0.01 cm, were reported for tephra deposits from a recent volcanic eruption (King, 1986), and 0.002–0.006 cm for different types of loess (Catt, 1988).

Concentration can now be redefined in terms of the mean particle size r by using S_g from Eq. 18 in Eq. 14:

$$C = \frac{3\tau R}{r} \frac{(1-\phi)}{\phi} \tag{19}$$

For larger particles and, consequently, smaller values of the specific surface areas, Eq. 19, produces

$$C = 3 \times \left(\frac{10^7}{10^8}\right) \times \left(\frac{10^{-16}}{10^{-13}}\right) \div \left(\frac{10^{-2}}{10^{-1}}\right)$$

$$= \frac{3 \times 10^{-7}}{3 \times 10^{-4}} \, mol \, cm^{-3} \tag{20}$$

Thus the lower end of the estimated concentration range agrees within a order of magnitude with the observed silica concentrations in a range of world rivers, $10^{-7.0\pm0.3} \, mol \, cm^{-3}$. Comparing the results in Eqs. 18 and 20, only the larger particle sizes or smaller specific surface areas were used in the latter. The mineral dissolution rates and water residence times in the reactive zone are the same in both estimates. There is no evidence for the mineral dissolution rates to be lower by a factor of 100–1000 than the values used in the preceding computation (Fig. 5, Table 4). For the mean residence time of water discharge on the continents, taken as bracketed by the values 0.3 and 3 years, much shorter residence times, on the order of 1–10 days, would also make concentration C in Eq. 17 compatible with concentrations of dissolved silica in river water.

The species SiO_2(aq) may be taken as a measure of dissolution and weathering of silicate minerals in the surficial zone of the continental crust, but the same does not apply to other dissolved constituents of continental runoff. All the major cations (Ca^{2+}, Mg^{2+}, Na^+, and K^+) originate from different rock types and in different proportions among the plutonic and metamorphic crystalline rocks, volcanic rocks, carbonates, clastic sediments, and evaporites (Garrels and Mackenzie, 1971; Wollast and Mackenzie, 1983; Wollast and Chou, 1988; Meybeck, 1984, in press). For calcite and limestones, the very wide range of dissolution and weathering rates (Table 5) suggests that physical erosion of limestones may be a significant component of limestone terrain denudation, in addition to its dissolution.

7. SUMMARY

The broad classes of the physical (P), hydrological (H), and chemical (C) driving forces in the surficial environment control through coupling between them the progress of chemical reactions and material transport fluxes of the reaction products. One major effect of the chemical driving forces is the coupling CP that represents mineral dissolution. Macroscopic transport of reaction products is effected by the coupling HC between the hydrological and chemical driving forces, as in the mass flux of dissolved materials in water flow. Additional processes involving chemical kinetics and macroscopic transport are grouped in the interactive couples among the P, H, and C driving forces and processes listed in Table 1.

The magnitudes of chemical kinetic and macroscopic transport processes, evaluated as their linear rates [linear rate = (mass flux)/(concentration or density) = F/ρ], indicate that great differences exist between the mineral dissolution rates, as reported from laboratory measurements, and the rates derived from river-water composition and volume flow. These differences point to an important role of the physical structure of the weathering zone and water residence time within it that control mineral dissolution fluxes and transport of the reaction products. An additional factor responsible for the faster rates of chemical weathering could be bacterial, activity which may be expected to vary from lower levels in the cold regions to the higher levels in the tropics, in parallel with the rates of net primary productivity.

A global mean for the rate of net chemical denudation of the continental surface is about 14 mm 1000 yr^{-1} or 14 μm yr^{-1}. In comparison to the corrosion rates of metals exposed to a range of environmental conditions, the global continental surface is less resistant to "corrosion" than zinc and copper, but it is considerably more resistant than iron exposed to coastal oceanic and industrial-area atmospheric conditions.

A linear dependence of the chemical weathering rates on water runoff may be accounted for by deeper percolation of water at high discharge volumes, resulting in a greater contact area with reactive mineral surfaces.

The present-day flux of solids eroded from exposed rocks, sediments, and soils carries four to five times more mass than the flux of solutes in rivers, as reported by a number of investigators over a period of years. Physical erosion is favored over chemical weathering by a combination of such factors as existence of a high relief on continents where about 25% of the nonglaciated surface occurs at elevations higher than 1000 m, and a relatively large fraction of the continental surface covered by siliceous sediments and rocks, as compared to only about 16% of the nonglaciated area consisting of carbonate outcrops. The prevalence of the physically over chemically controlled mode of global material transport is a phenomenon of geologically long duration on scales of 10^6 to 10^7 yr.

REFERENCES

Bales, R. C., and J. J. Morgan (1985), "Dissolution Kinetics of Chrysotile at pH 7 to 10," *Geochim. Cosmochim. Acta* **49**, 2281–2288.

Baumgartner, A., and E. Reichel (1975), *The World Water Balance*, Elsevier, Amsterdam, 179 pp.

Berner, E. K., and R. A. Berner (1987), *The Global Water Cycle: Geochemistry and Environment*, Prentice-Hall, Englewood Cliffs, NJ.

Berthelin, J. (1988), "Microbial Weathering Processes in Natural Environments," in A. Lerman and M. Meybeck, Eds., *Physical and Chemical Weathering in Geochemical Cycles*, Reidel, Dordrecht, Netherlands, pp. 33–60.

Bhagavantam, S. (1966), *Crystal Symmetry and Physical Properties*, Academic Press, New York, 230 pp.

Blum, A., and A. Lasaga (1988), "Role of Surface Speciation in Low-Temperature Dissolution of Minerals," *Nature* **331**, 431–433.

Bodek, I., and A. Lerman (1988), "Kinetics of Selected Processes," in I. Bodek, W. J. Lyman, W. F. Reehl, and D. H. Rosenblatt,, Eds., *Environmental Inorganic Chemistry*, Pergamon press, New York, Chapter 3.

Brady, P. V., and J. V. Walther (in press), "Controls of Silicate Dissolution Rates in Neutral and Basic pH Solutions at 25°C," *Geochim. Cosmochim. Acta.*

Brunsden, D. (1979), "Weathering," in C. Embleton and J. Thornes, Eds., *Process in Geomorphology*, Arnold, London, pp. 73–129.

Catt, J. A. (1988), "Loess—Its Formation, Transport and Economic Significance," in A. Lerman and M. Meybeck, Eds., *Physical and Chemical Weathering in Geochemical Cycles*, Kluwer Academic Publishers, Dordrecht, Netherlands, pp. 113–142.

Chou, L., and R. Wollast (1985), "Steady-State Kinetics and Dissolution Mechanisms of Albite," *Am. J. Sci.* **285**, 963–993.

Chou, L., R. M. Garrels, and R. Wollast (in press), "Comparative Study of the Kinetics and Mechanisms of Dissolution of Carbonate Minerals," *Chem. Geol.*

Coburn, S. K. (1968), "Corrosiveness of Various Atmospheric Test Sites as Measured by Specimens of Steel and Zinc," in *Metal Corrosion in the Atmosphere*, Vol. 435, ASTM STP, Philadelphia, p. 360.

Colman, S. M., and D. P. Dethier, Eds. (1986), *Rates of Chemical Weathering of Rocks and Minerals*, Academic Press, Orlando, FL.

Costa, L. P. (1982), "Atmospheric Corrosion of Copper Alloys Exposed for 15 to 20 Years," in S. W. Dean and E. C. Rhea, Eds., *Atmospheric Corrosion of Metals*, Vol. 767, ASTM STP, Philadelphia, pp. 106–115.

Cronan, C. S. (1985), "Chemical Weathering and Solution Chemistry in Acid Forest Soils: Differential Influence of Soil Type, Biotic Processes, and H$^+$," in J. I. Drever, Ed., *The Chemistry of Weathering*, Reidel, Dordrecht, Netherlands, pp. 175–196.

Davis, J. S., and J. Zobrist (1978), "The Interrelationships among Chemical Parameters in Rivers—Analysing the Effect of Natural and Anthropogenic Sources," *Progr. Water Technol.* **10**, 65–78.

Dethier, D. P. (1986), "Weathering Rates and the Chemical Flux from Catchments in the Pacific Northwest, U.S.A.," in S. M. Colman and D. P. Dethier, Eds., *Rates of Chemical Weathering of Rocks and Minerals*, Academic Press, New York, pp. 503–530.

Drever, J. I. (1989), *The Geochemistry of Natural Waters*, Prentice-Hall, Englewood Cliffs, NJ.

Eckhardt, F. E. W. (1985), "Solubilization, Transport, and Deposition of Mineral Cations by Microorganisms—Efficient Rock Weathering Agents," in J. I. Drever, Ed., *The Chemistry of Weathering*, Reidel, Dordrecht, Netherlands, pp. 161–174.

Flemings, P. B., and T. E. Jordan (1989), "A Synthetic Stratigraphic Model of Foreland Basin Development," *J. Geophys. Res.* **94**, 3851–3866.

Fölster, H. (1985), "Proton Consumption Rates in Holocene and Present-Day Weathering of Acid Forest Soils," in J. I. Drever, Ed., *The Chemistry of Weathering*, Reidel, Dordrecht, Netherlands, pp. 197–210.

Garrels, R. M., and F. T. Mackenzie (1971), *Evolution of Sedimentary Rocks*, Norton, New York.

Gates, W. L., and A. B. Nelson (1975), *A New (Revised) Tabulation of the Scripps Topography on a 1° Global Grid. Part I: Terrain Heights*, R-1276-1-ARPA, Rand Corp., Santa Monica, CA.

Haynie, F. H., and J. B. Upham (1970), "Effect of Atmospheric Sulfur Dioxide on the Corrosion of Zinc," *Mat. Prot. Perf.* **9**, 35–40.

Henriksen, A. (1980), "Acidification of Fresh Waters—A Large Scale Titration," in D. Drablos and A. Tollan, Eds., *Ecological Impact of Acid Precipitation*, SNSF Project, Oslo-As, pp. 68–74.

Herdendorf, C. E. (1982), "Large Lakes of the World," *J. Great Lakes Res.* **8** (3), 379–412.

Holland, H. D. (1984), *The Chemical Evolution of the Atmosphere and Oceans*, Princeton University Press, Princeton, NJ.

Jennings, J. N. (1983), "Karst Landforms," *Am. Scientist* **71**, 578–585.

King, R. H. (1986), "Weathering of Holocene Airfall Tephras in the Southern Canadian Rockies," in S. M. Colman and D. P. Dethier, Eds., *Rates of Chemical Weathering of Rocks and Minerals*, Academic Press, New York, pp. 239–262.

Lagrula, J. (1966), "Hypsographic Curve," in R. W. Fairbridge, Eds., *The Encyclopedia of Oceanography, Encyclopedia of Earth Sciences Series*, Vol. 1, Reinhold, New York, pp. 365–366.

Lajoie, K. R. (1986), "Coastal Tectonics," in *Active Tectonics*, Studies in Geophysica, National Academy of Sciences Press, Washington. DC, pp. 95–124.

Lerman, A. (1979), *Geochemical Processes—Water and Sediment Environment*, Wiley, New York.

Lerman, A. (1988), "Weathering Rates and Major Transport Processes," in A. Lerman and M. Meybeck, Eds., *Physical and Chemical Weathering in Geochemical Cycles*, Kluwer Academic Publishers, Dordrecht, Netherlands, pp. 1–10.

Lerman, A. (in press), "Surficial Weathering Fluxes and Their Geochemical Controls," *NRC Studies in Geophysics*, National Academy of Sciences, Washington, DC.

Lerman, A., and A. B. Hull (1987), "Background Aspects of Lake Restoration: Water Balance, Heavy Metal Content, Phosphorus Homeostasis," *Schweiz Z. Hydrol.* **49**, 148–169.

Lerman, A., and W. Stumm (1989), "CO_2 Storage and Alkalinity Trends in Lakes," *Water Res.* **23** (2), 139–146.

Lieth, H. (1975), "Modeling the Primary Productivity of the World," in H. Lieth and R. H. Whittaker, Eds., *Primary Productivity of the Biosphere*, Springer, New York, pp. 237–263.

Linthurst, R. A., D. H. Landers, J. M. Eilers, D. F. Brakke, W. S. Overton, E. P. Meier, and R. E. Crowe (1986), "Characteristics of Lakes in the Eastern United States," *Population Descriptions and Physico-Chemical Relationships*, Vol. 1, U.S. Environmental Protection Agency, Washington, DC, 136 pp.

Liukov, A. V. (1966), *Heat and Mass Transfer in Capillary–Porous Bodies*, Pergamon Press, New York, 523 pp.

Mead, R. H. (1988), "Movement and Storage of Sediment in River Systems," in A. Lerman and M. Meybeck, Eds., *Physical and Chemical Weathering in Geochemical Cycles*, Kluwer Academic Publishers, Dordrecht, Netherlands, pp. 165–180.

Meybeck, M. (1982), "Carbon, Nitrogen and Phosphorus Transport by World Rivers," *Am. J. Sci.* **282**, 401–450.

Meybeck, M. (1984), "Les Fleuves et le Cycle Géochimique des éléments," Thèse d'État (No. 84–35), École Normal Supérieure, Laboratoire de Géologie, Université Pierre et Marie Curie, Paris, France.

Meybeck, M. (in press), "Origins and Variability of Present Day Riverborne Material," *NRC Studies in Geophysics*, National Academy of Sciences, Washington, DC.

Milliman, J. D. (in press), "Discharge of Fluvial Sediment to the Oceans Global, Temporal and Athropogenic Implications," *NRC Studies in Geophysics*, National Academy of Sciences, Washington, DC.

Milliman, J. D., and R. H. Meade (1983), "World-wide Delivery of River Sediment to the Oceans," *J. Geol.* **91**, 1–21.

Montroll, E. W. (1967), "Principles of Statistical Mechanics and Kinetic Theory of Gasses," in E. U. Condon and H. Odishaw, Eds., *Handbook of Physics*, McGraw-Hill, New York, pp. 5-31–33.

Pinet, P., and M. Sourian (1988), "Continental Erosion and Large Scale Relief," *Tectonics* **7** (3), 563–582.

Plummer, L. N., T. M. L. Wigley, and D. L. Parkhurst (1978), "The Kinetics of Calcite Dissolution in CO_2–Water Systems at 5° to 60°C and 0.0 to 1.0 atm CO_2," *Am. J. Sci.* **278**, 179–216.

Post, W. M., W. R. Emmanuel, P. J. Zinke, and A. G. Stangenberger (1982), "Soil Carbon Pools and World Life Zones," *Nature* **298**, 156–159.

Prospero, J. M. (1981), "Eolian Transport to the World Ocean," in C. Emilianai, Ed., *The Sea*, Vol. 7, *The Oceanic Lithosphere*, Wiley, New York, pp. 801–874.

Rea, D. K., S. A. Hovan, and T. R. Janecek (in press), "Late Quaternary Flux of Eolian Dust to the Pelagic Ocean," *NRC Studies in Geophysics*, National Academy of Sciences, Washington, DC.

Reddy, M. M. (1988), "Acid Rain Damage to Carbonate Stone: A Quantitative Assessment Based on the Aqueous Geochemistry of Rainfall Runoff from Stone," *Earth Surface Processes Landforms* **13**, 335–354.

Schnoor, J. L., and W. Stumm (1986), "The Role of Chemical Weathering in the Neutralization of Acidic Deposition," *Schweiz Z. Hydrol.* **48**, 171–195.

Schott, J., R. A. Berner, and E. L. Sjöberg (1981), "Mechanism of Pyroxene and Amphibole Weathering.—I. Experimental Studies of Iron-Free Minerals," *Geochim. Cosmochim. Acta* **45**, 2123–2135.

Schott, J., and J.-C. Petit (1987), "New Evidence for the Mechanism of Dissolution of Silicate Minerals," in W. Stumm, Ed., *Aquatic Surface Chemistry*, Wiley, New York, pp. 293–315.

Sillén, L. G. (1961), "The Physical Chemistry of Sea Water," in M. Sears, Ed., *Oceanography*, American Association for the Advancement of Science, No. 6, Washington, DC, pp. 549–581.

Slaymaker, O. (1988), "Slope Erosion and Mass Movement in Relation to Weathering and Geochemical Cycles," in A. Lerman and M. Meybeck, Eds., *Physical and Chemical Weathering in Geochemical Cycles*, Kluwer Academic Publishers, Dordrecht, Netherlands, pp. 83–112.

Stallard, R. F. (1988), "Weathering and Erosion in the Humid Tropics," in A. Lerman and M. Meybeck, Eds., *Physical and Chemical Weathering in Geochemical Cycles*, Kluwer Academic Publishers, Dordrecht, Netherlands, pp. 225–246.

Stumm, W., and J. J. Morgan (1981), *Aquatic Chemistry*, Wiley, New York.

Sverdrup, H. U., M. W. Johnson, and R. H. Fleming (1942), *The Oceans*, Prentice Hall, New York, 1087 pp.

Tole, M. P., A. C. Lasaga, C. Pantano, and W. B. White (1986), "Factors Controlling the Kinetics of Nepheline Dissolution," *Geochim. Cosmochim. Acta* **50**, 379–392.

Velbel, M. A. (1986), "The Mathematical Basis for Determining Rates of Geochemical and Geomorphic Processes in Small Forested Watersheds by Mass Balance: Examples and Implications," in S. M. Colman and D. P. Dethier, Eds., *Rates of Chemical Weathering of Rocks and Minerals*, Academic Press, New York, pp. 439–449.

Walling, D. E., and B. W. Webb (1983), "The Dissolved Loads of Rivers: A Global Overview," *Dissolved Loads of Rivers and Surface Water Quantity/Quality Relationships*, IAHS, 3–20.

Wollast, R., and L. Chou (1988), "Rate Control of Weathering of Silicate Minerals at Room Temperature and Pressure," in A. Lerman and M. Meybeck, Eds., *Physical and Chemical Weathering in Geochemical Cycles*, Kluwer Academic Publishers, Dordrecht, Netherlands, pp. 11–32.

Wollast, R., and F. T. Mackenzie (1983), "The Global Cycle of Silica," in S. R. Aston, Ed., *Silicon Geochemistry and Biogeochemistry*, Academic Press, Orlando, FL, pp. 39–76.

Wright, R. F. (1988), "Influence of Acid Rain on Weathering Rates," in A. Lerman and M. Meybeck, Eds., *Physical and Chemical Weathering in Geochemical Cycles*, Kluwer Academic Publishers, Dordrecht, Netherlands, pp. 181–196.

INDEX

(*continued from front*)